T0310158

POLYSACCHARIDE BUILDING BLOCKS

POLYSACCHARIDE BUILDING BLOCKS

A Sustainable Approach to the Development of Renewable Biomaterials

Edited by

YOUSSEF HABIBI
LUCIAN A. LUCIA

A JOHN WILEY & SONS, INC., PUBLICATION

For general information on our other products and services or for technical support,
please contact our Customer Care Department within the United States at (800) 762-2974,
outside the United States at (317) 572-3993 or fax (317) 572-4002.

Wiley also publishes its books in a variety of electronic formats. Some content that appears in print
may not be available in electronic formats. For more information about Wiley products,
visit our web site at www.wiley.com.

Library of Congress Cataloging-in-Publication Data:

Polysaccharide Building Blocks: A Sustainable Approach to the Development of Renewable Biomaterials
/ edited by Youssef Habibi, Lucian A. Lucia.
      pages cm
   Includes bibliographical references and index.
   ISBN 978-0-470-87419-6
1.   Polysaccharides.   I.   Habibi, Youssef, editor of compilation.   II.   Lucia,
Lucian A., editor of compilation.
   QP702.P6P6395 2012
   572′.566–dc23

                                                      2011046734
Printed in the United States of America

ISBN: 9780470874196

10 9 8 7 6 5 4 3 2 1

CONTENTS

FOREWORD

The first polysaccharide was discovered in edible fungi while assessing their nutritional value in 1811, and was later named chitin. The bicentennial has been celebrated in a review article (*Carbohydrate Polymers*, 2012, **87**: 995–1012) that enables us to appreciate the immense spiritual resources of the western countries; in the context of the American and French revolutions, they elaborated new scientific interests, research methodologies, and communication means. The botanical studies made in Europe on thousand of unknown plants imported from Australia, New Zealand, Canada, and other countries explored at that time, stimulated the characterization of cellulose, lignin, pectin, and starch. Research was often aimed at alleviating food shortages, not to say famine that the European populations had to face. One century later, those polysaccharides together with alginates, xanthans, and others were to play major roles as technological commodities and food ingredients.

The unique structures of polysaccharides combined with appealing properties such as atoxicity, hydrophilicity, biocompatibility, multichirality, and multifunctionality imparted by hydroxyl, amino, acetamido, carboxyl, and sulfate groups conferred additional importance to polysaccharides as valuable and renewable resources, chemically amenable to elaborated specialties.

In the last quarter of the twentieth century, prominent research topics among others have been the following: technology (textiles, personal care items, drug delivery); food technology (quality of foods and drinks, functional foods, dietary supplements); biochemistry (hemostatics, blood anticoagulants, wound dressing, bone regeneration, glycosaminoglycans); enzymatic modification and inhibition of the biosynthesis (synthases, hydrolases, insecticides); environmental protection (industrial waste reclamation); combination with synthetic polymers and inorganics (grafting, polyelectrolyte complexes, spontaneous association, composites).

As a consequence of the great steps forward made in the elucidation of the enzyme structures and in the understanding of molecular recognition, in those years some research groups brought forward the concept that poly- and oligosaccharides had to be seen as components of supramolecular structures, and that associations with proteins (glycoproteins/proteoglycans) are necessary *in vivo* for molecular recognition, association, adhesion, bioactivity, and more. For example, the ordered and most elaborated structures of aggrecan and other compounds present in the extracellular matrix, elegantly confirm that the single components are to be seen as natural building blocks. Moreover, the complex structure of the carbohydrate moieties of glycoproteins contains biochemical messages. This way of thinking provides inspiration today for targeted drug delivery, tissue engineering and imaging, transfection, and other complex biotechnological manipulations that are prolonging our life expectation and contribute to our welfare. Polysaccharides (alone or in reciprocal combination) in the form of hydrogels, aerogels, and membranes, for instance, exert control over differentiation of stem cells, proliferation and phenotype preservation, and provide most satisfactory results in tissue engineering.

At present advanced technologies permit to isolate and produce large amounts of nanosized crystalline building blocks, particularly those of cellulose and chitin. The nanofibrils obtained via mechanochemical disassembly of said polysaccharides can be manufactured with good yield but minimum environmental impact and energy expenditure. Extremely long nanofibers can be manufactured by electrospinning: in both cases the materials obtained have enormous and unprecedented specific surface area, suitable for enhanced performances.

Notwithstanding the technological advances made, the daily life problems faced two centuries ago still afflict the majority of the exceedingly large contemporary world's population. The early research subjects in the areas of personal care, food production and preservation, innovative agriculture, plant protection, fishing activities and related issues are still pivotal for envisaging sustainable green solutions with the aid of polysaccharides.

RICCARDO A. A. MUZZARELLI
Emeritus Professor of Enzymology
University of Ancona, Italy

PREFACE

This book is a succinct and in-depth account of the progress and evolution of specific scientific concepts that fall under the umbrella of polysaccharide-based renewable biomaterials. Our aspiration is that the book will provide a panoramic snapshot of highly important developments in the field of science and engineering of polysaccharides. These are materials that to a certain extent have not received the attention they merit, especially because they currently occupy an important place in the emerging biomaterials and bioenergy disciplines.

A fundamental overview and treatment of important recent advances in cellulosic chemistry will comprise the basis for the first three chapters. These chapters will provide an in-depth account of the importance of cellulose, its wonderfully adaptable chemistry for providing a myriad of by-products, and ultimately explore two of these by-products, namely aerocellulose and nanocellulose.

The panoply of polysaccharide materials for research applications is not just limited to cellulosics but also includes chitin. The following five chapters are among the most thorough and written by several of the most involved researchers in the arena of chitin/chitosan. Chitin may be one of the most abundant polysaccharides in the biosphere today, but only recently have we begun to realize its potential as a valuable biomaterial especially in structural and functional nanocomposite applications, as well as in biomedical applications including bioadhesives/hemostatics. We have two entire chapters devoted to exploring the role of chitin and chitosan as unique building blocks for a number of highly valuable transformations. We then provide a much more in-depth investigation of a few of the many important functions that are mentioned, including the ability of chitosan and its derivatives to provide a potential water remediation role. This represents an altogether novel and highly economic and beneficial function that has very little precedent in the annals of renewable materials bioremediation efforts. In the same vein, we then explore the role of these polysaccharides within the field of nutrition science. Coma overviews the basic constructs that are involved in food preservation and how chitin/chitosan products can elegantly address almost all of the needs in this unique and highly valuable area. Finally, Yamazaki and Hudson will examine one of chitosan's most important and singularly attractive functions, namely that of bioadhesion and hemostasis.

We move on to some unique, very high value functions of polysaccharides in the chapters 9 and 10. In Chapter 9, Rußler and Rosenau explore electrical conductivity aspects of polysaccharides with respect to the necessary modifications that can be pursued to impart electro-active character to this abundant class of materials. In Chapter 10, Shuttleworth et al. explore the emerging area of porous materials with attention to the ability of polysaccharides to deliver new high value functionality in the area of absorption, storage, and delivery.

Chapters 11 and 12 examine starch, which is one of the most scientifically explored and manipulated polysaccharides in the research community. These chapters examine the fundamental properties of starch-based material, why they are so attractive for research applications, their utility now and in the future, and also provide a compilation of highly useful starch-based bionanocomposites, their chemical and physical properties, and the techniques and approaches for their processing.

Finally, the last three chapters explore the kingdom of the heteropolysaccharides (hemicelluloses), a species of polysaccharides that are amorphous in their structural dispersity. They are highly available and abundant biomaterials as are cellulose, chitin, and starch, but again they suffer from a relative paucity of advanced applications within the biomaterials community. The chapters will examine xylans, one of the most abundant hemicelluloses in the biosphere, which are well represented within angiosperms. Xylans and other related and nonxylan-based hemicellulosics are treated not only in terms of their structural aspects, properties, uses as viable starting material resources but also as highly functional precursors for advanced nanotechnological and other applications.

Our sincere hope is that the work represented herein serves as a useful and functional platform for practitioners in the art and for researchers who intend to explore the extraordinary "plasticity" and adaptability of polysaccharides. Henry Ford, the great U.S. Industrialist, once said, "An idealist is a person who helps other people to be prosperous." The editors of this book and all those associated with it from the publishers to the authors may be classified as idealists whose sincere hope and trust is that the collection of knowledge contained herein will serve to make our readers prosperous. We believe that the polysaccharide materials that nature offers us provide us with that possibility—it is our sacred duty, all of us, therefore, to explore and exploit these remarkable materials for the betterment of humankind.

YOUSSEF HABIBI
LUCIAN A. LUCIA

CONTRIBUTORS

Luc Avérous, LIPHT-ECPM, Université de Strasbourg, Strasbourg Cedex 2, France

James H. Clark, Green Chemistry Centre of Excellence, Department of Chemistry, University of York, Heslington, York, UK

Véronique Coma, University of Bordeaux, CNRS, LCPO, UMR 5629, F-33600 Pessac, France

Anna Ebringerova, Institute of Chemistry, Center for Glycomics, Slovak Academy of Sciences, Bratislava, Slovakia

Ram B. Gupta, Chemical Engineering Department, Auburn University, Auburn, AL, USA

Youssef Habibi, Department of Forest Biomaterials, North Carolina State University, Raleigh, NC, USA

Emmerich Haimer, Departments of Material Sciences and Process Engineering, and Chemistry, University of Natural Resources and Applied Life Sciences, Vienna, Austria

Natanya Hansen, Department of Chemical and Biochemical Engineering, Technical University of Denmark, DK-2800 Kgs. Lyngby, Denmark

Thomas Heinze, Center of Excellence for Polysaccharide Research, Institute of Organic Chemistry and Macromolecular Chemistry, Friedrich Schiller University of Jena, Jena, Germany; Finnish Distinguished Professor at Åbo Akademi/ University, Åbo, Finland

Samuel M. Hudson, Fiber and Polymer Science Program, College of Textiles, Centennial Campus, North Carolina State University, Raleigh, NC, USA

Falk Liebner, Department of Chemistry, University of Natural Resources and Applied Life Sciences, Vienna, Austria

José F. Louvier-Hernández, Chemical Engineering Department, Instituto Tecnológico de Celaya, Celaya, Guanajuato, México

Lucian A. Lucia, Department of Forest Biomaterials, North Carolina State University, Raleigh, NC, USA

Avtar Matharu, Green Chemistry Centre of Excellence, Department of Chemistry, University of York, Heslington, York, UK

Aji P. Mathew, Department of Applied Physics and Mechanical Engineering, Division of Wood and Bionanocomposites, Luleå University of Technology, Luleå, Sweden

Kristiina Oksman, Department of Applied Physics and Mechanical Engineering, Division of Wood and Bionanocomposites, Luleå University of Technology, Luleå, Sweden

Katrin Petzold-Welcke, Center of Excellence for Polysaccharide Research, Institute of Organic Chemistry and Macromolecular Chemistry, Friedrich Schiller University of Jena, Jena, Germany

David Plackett, Department of Chemical and Biochemical Engineering, Technical University of Denmark, DK-2800 Kgs. Lyngby, Denmark

Visakh P. M., Department of Applied Physics and Mechanical Engineering, Division of Wood and Bionanocomposites, Luleå University of Technology, Luleå, Sweden; Centre for Nanoscience and Nanotechnology, Mahatma Gandhi University, Kottayam, Kerala, India

Antje Potthast, Department of Chemistry, University of Natural Resources and Applied Life Sciences, Vienna, Austria

Mohammed Rhazi, Laboratory of Natural Macromolecules, E. N. S., Marrakech, Morocco

Thomas Rosenau, Department of Chemistry, University of Natural Resources and Applied Life Sciences, Vienna, Austria

Axel Rußler, Department of Chemistry, University of Natural Resources and Life Sciences, Vienna, Austria

Peter S. Shuttleworth, Green Chemistry Centre of Excellence, Department of Chemistry, University of York, Heslington, York, UK

Sabu Thomas, Centre for Nanoscience and Nanotechnology, Mahatma Gandhi University, Kottayam, Kerala, India

Abdelouhad Tolaimate, Laboratory of Natural Macromolecules, E. N. S., Marrakech, Morocco

Mai Yamazaki, Fiber and Polymer Science Program, College of Textiles, Centennial Campus, North Carolina State University, Raleigh, NC, USA

1

RECENT ADVANCES IN CELLULOSE CHEMISTRY

THOMAS HEINZE AND KATRIN PETZOLD-WELCKE

1.1 INTRODUCTION

The chemical modification of polysaccharides is still underestimated regarding the structure and hence property design of materials based on renewable resources. At present, the cellulose derivatives commercially produced in large scale are limited to some ester with C_2–C_4 carboxylic acids, including mixed esters and phthalic acid half-esters as well as ethers with methyl-, hydroxyalkyl-, and carboxymethyl functions. In general, organic chemistry of cellulose opens a wide variety of products by esterification and etherification. In addition, novel products may be obtained by nucleophilic displacement reactions, unconventional chemistry like "click reactions," introduction of dendrons in the cellulose structure, and regiocontrolled reactions within the repeating units and along the polymer chains. The aim of this chapter is to highlight selected recent advances in chemical modification of cellulose for the synthesis of new products with promising properties as well as alternative synthesis paths in particular under homogeneous conditions, that is, starting with dissolved polymer considering own research results adequately.

1.2 TECHNICAL IMPORTANT CELLULOSICS

The application of the glucane cellulose as a precursor for chemical modifications was exploited extensively even before its polymeric nature was determined and well

Polysaccharide Building Blocks: A Sustainable Approach to the Development of Renewable Biomaterials, First Edition. Edited by Youssef Habibi and Lucian A. Lucia.
© 2012 John Wiley & Sons, Inc. Published 2012 by John Wiley & Sons, Inc.

understood. Cellulose nitrate (commonly misnomered nitrocellulose) of higher nitrogen content was one of the most important explosives. Its partially nitrated ester was among the first polymeric materials used as a "plastic" well known under the trade name of Celluloid. Today, cellulose nitrate is the only inorganic cellulose ester of commercial interest (Balser et al., 1986). Further cellulose products like methyl-, ethyl-, or hydroxyalkyl ethers or cellulose acetate, and, in addition, products with combinations of various functional groups, for example, ethylhydroxyethyl and hydroxypropylmethyl cellulose, cellulose acetopropionates, and acetobutyrates are still important, many decades after their discovery. Ionic cellulose derivatives are also known since a long time. Carboxymethyl cellulose, up to now the most important ionic cellulose ether, was first prepared in 1918 and produced commercially in the early 1920s in Germany (Brandt, 1986). Various cellulose derivatives are produced in large quantities for diversified applications. Their properties are primarily determined by the type of functional group. Moreover, they are influenced significantly by adjusting the degree of functionalization and the degree of polymerization (DP) of the polymer backbone (Table 1.1).

1.3 NUCLEOPHILIC DISPLACEMENT REACTIONS (S_N)

It is well known from the chemistry of low molecular alcohols that hydroxyl functions are converted to a good leaving group for nucleophilic displacement reactions by the formation of the corresponding sulfonic acid esters (Heinze et al., 2006a). Moreover, cellulose derivatives obtained by S_N reactions are suitable starting materials for the preparation of novel products by unconventional chemistry like "click reactions." Even selectively dendronized celluloses could be prepared.

1.3.1 Cellulose Sulfonates

Typical structures of sulfonic acid esters used in polysaccharide chemistry are shown in Figure 1.1. The synthesis of sulfonic acid esters is realized heterogeneously by reaction of cellulose with sulfonic acid chlorides in aqueous alkaline media (NaOH, Schotten–Baumann reaction), or is most efficiently completely homogeneous in a solvent like N,N-dimethylacetamide (DMA)/LiCl. The main drawback of heterogeneous procedures is a variety of side reactions, including undesired nucleophilic displacement reactions caused especially by long reaction times and high temperatures required. In contrast, the homogeneous process using cellulose dissolved in DMA/LiCl yields well soluble sulfonic acid esters (McCormick and Callais, 1987).

The p-toluenesulfonic (tosyl) and the methanesulfonic (mesyl) acid esters of cellulose are the most widely used sulfonic acid esters, due to their availability and hydrolytic stability (Heinze et al., 2006a). The homogeneous reaction of cellulose in DMA/LiCl with p-toluenesulfonyl chloride permits the preparation of cellulose tosylate with defined degree of substitution (DS) easily controlled by the molar ratio reagent to anhydroglucose unit (AGU) with almost no side reactions (McCormick and Callais, 1986, 1987; Rahn et al., 1996; Siegmund and Klemm, 2002).

TABLE 1.1 Examples of Important Cellulose Esters and Ethers Commercially Produced

Product	Worldwide Production (t/a)	Functional Group	Degree of Functionalization	Examples of Solubility
Cellulose acetate	900,000	$-C(O)CH_3$	0.6–0.9	Water
			1.2–1.8	2-Methoxyethanol
			2.2–2.7	Acetone
			2.8–3.0	Chloroform
Cellulose nitrate	200,000	$-NO_2$	1.8–2.0	Ethanol
			2.0–2.3	Methanol, acetone, methyl ethyl ketone
			2.2–2.8	Acetone
Cellulose xanthate	3,200,000	$-C(S)SNa$	0.5–0.6	Aqueous NaOH
Carboxymethyl cellulose	300,000	$-CH_2COONa$	0.5–2.9	Water
Methyl cellulose	150,000	$-CH_3$	0.4–0.6	4% Aqueous NaOH
			1.3–2.6	Cold water
			2.5–3.0	Organic solvents
Ethyl cellulose	4,000	$-CH_2CH_3$	0.5–0.7	4% Aqueous NaOH
			0.8–1.7	Cold water
			2.3–2.6	Organic solvents
Hydroxyethyl cellulose	50,000	CH_2CH_2OH	0.1–0.5	4% Aqueous NaOH
			0.6–1.5	Water

Adapted from Heinze and Liebert (2001).

3

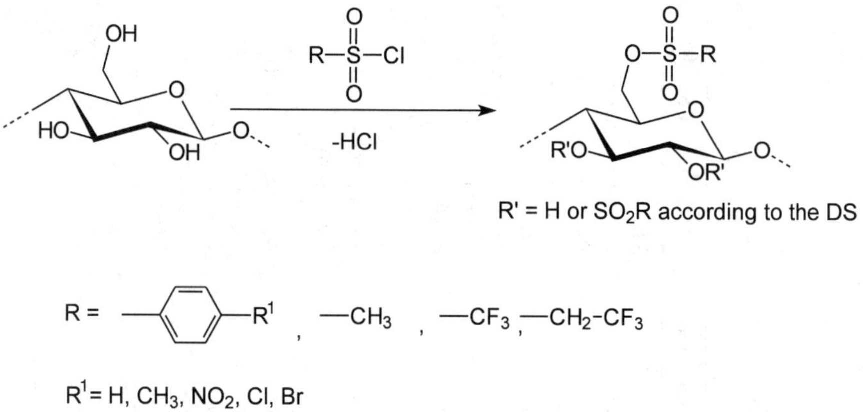

R' = H or SO$_2$R according to the DS

R = —⟨benzene⟩—R^1 ,   —CH$_3$,   —CF$_3$,—CH$_2$-CF$_3$

R^1= H, CH$_3$, NO$_2$, Cl, Br

FIGURE 1.1 Typical sulfonic acid esters of cellulose.

The structure of the product may depend on both the reaction conditions and the workup procedure used (McCormick et al., 1990). The tosyl chloride may react with DMA in a Vilsmeier–Haak-type reaction forming the O-(p-toluenesulfonyl)-N,N-dimethylacetiminium salt, which attacks the OH groups of the cellulose depending on the reaction conditions used. For a higher efficiency of tosylation of cellulose, stronger bases such as triethylamine (pK$_a$ 10.65) or 4-(dimethyl-amino)-pyridine (pK$_a$ 9.70) are necessary, which react with the O-(p-toluenesul-fonyl)-N,N-dimethylacetiminium salt building a quaternary ammonium salt and hence lead to the formation of tosyl cellulose without undesired side reactions (Figure 1.2) (McCormick et al., 1990). On the contrary, the use of a weak organic base like pyridine (pK$_a$ 5.25) or N,N-dimethylaniline (pK$_a$ 5.15) for the reaction with cellulose yields a reactive N,N-dimethylacetiminium salt, which may form chlorodeoxy celluloses at high temperatures or cellulose acetate after aqueous workup (Heinze et al., 2006a).

Various cellulose materials with degree of polymerization in the range of 280–1020 were transformed to the corresponding tosyl esters (Rahn et al., 1996). DS values in the range of 0.4–2.3 with negligible incorporation of chlorodeoxy groups were obtained at reaction temperatures of 8–10°C for 5–24 h (Table 1.2).

Cellulose tosylates are soluble in various organic solvents; beginning at DS of 0.4, solubility in aprotic dipolar solvents like DMA, N,N-dimethylformamide (DMF), and dimethylsulfoxide (DMSO) occurs. The cellulose tosylates become soluble in acetone and dioxane at a DS value of 1.4 and solubility in chloroform and methylene chloride appears at DS of 1.8. Position 6 reacts faster compared to the secondary OH groups at positions 2 and 3, which can be characterized by means of FTIR and NMR spectroscopy of cellulose tosylate (Rahn et al., 1996).

1.3.2 S$_N$ Reactions with Cellulose Sulfonates

Cellulose sulfonates are studied for a broad variety of S$_N$ reactions, as dis-cussed in various review papers (Belyakova et al., 1971; Hon, 1996; Siegmund

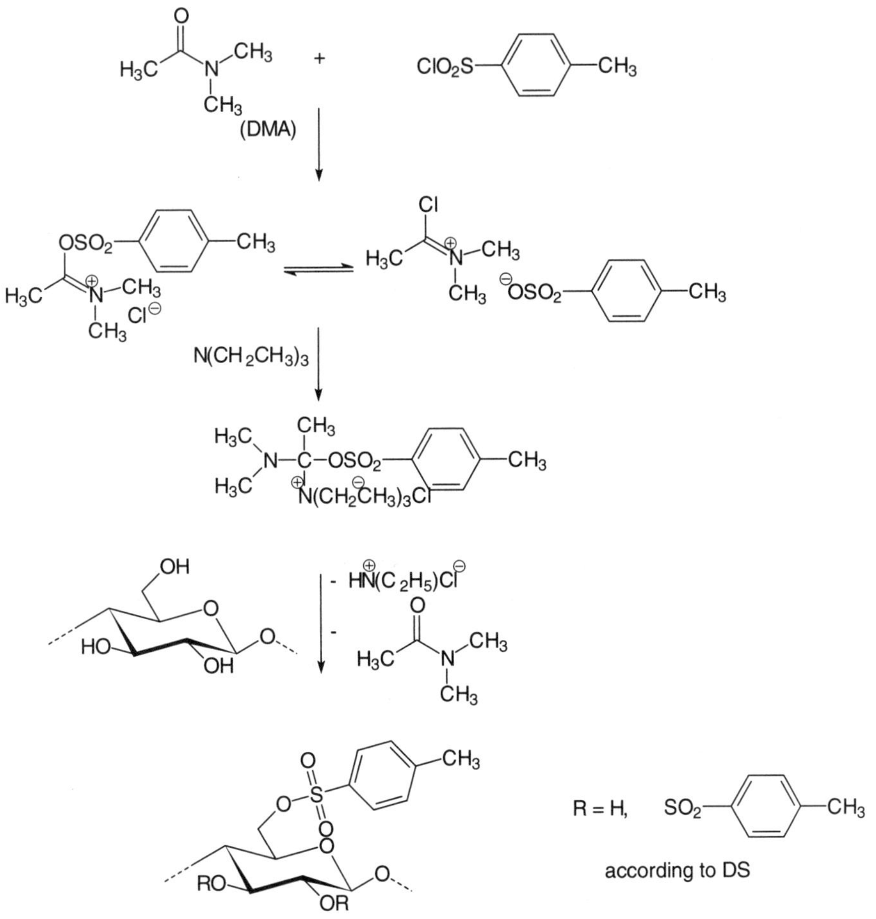

FIGURE 1.2 Mechanism of the reaction of cellulose with *p*-toluenesulfonyl chloride in DMA/LiCl in the presence of triethylamine. Adapted from McCormick et al. (1990).

and Klemm, 2002). Usually the S$_N$ reaction occurs selectively at the primary sulfonates. The mechanism (S$_N$1 versus S$_N$2) of nucleophilic substitution reaction of cellulose derivatives is still a subject of discussion. A remarkable finding is that a treatment of partially substituted cellulose tosylates (DS 1.2–1.5) with strong nucleophiles like azide or fluoride ions leads to a substitution of both primary and secondary tosylates (Siegmund and Klemm, 2002; Koschella and Heinze, 2003).

Water-soluble 6-deoxy-6-*S*-thiosulfato celluloses (Table 1.3) form S–S bridges by oxidation with H$_2$O$_2$—in analogy to nonpolymeric compounds of this type (Milligan and Swan, 1962)—leading to waterborne coatings (Klemm, 1998).

Water-soluble 6-deoxy-6-thiomethyl-2,3-carboxymethyl cellulose forms self-assembled monolayers at a gold surface (Wenz et al., 2005). The insoluble products yielded by S$_N$ reactions of cellulose tosylate with iminoacetic acid have high water

TABLE 1.2   Results and Conditions of the Reaction of Cellulose with *p*-Toluenesulfonyl Chloride (TosCl) in DMA/LiCl Applying Triethylamine as Base (2 mol/mol TosCl) for 24 h at 8°C

Cellulose	Degree of Polymerization	Molar Ratio TosCl/AGU[a]	Cellulose Tosylate		
			DS[b]	S (%)	Cl (%)
Microcrystalline	280	1.8	1.36	11.69	0.47
Spruce sulfite pulp	650	1.8	1.34	11.68	0.44
Cotton linters	850	0.6	0.38	5.51	0.35
		1.2	0.89	9.50	0.50
		2.1	1.74	12.90	0.40
		3.0	2.04	13.74	0.50
Beech sulfite pulp	1020	1.8	1.52	12.25	0.43

Adapted from Rahn et al. (1996).
[a] AGU anhydroglucose unit.
[b] Degree of substitution, calculated on the basis of sulfur content.

retention values of up to 11,000% (Heinze, 1998). A number of aminodeoxy celluloses are accessible. The nucleophilic displacement reaction with various amines results in water-soluble 6-deoxy-6-trialkylammonium cellulose (Koschella and Heinze, 2001). The initial chirality of the cellulose has no significant influence on its reactivity with

TABLE 1.3   Examples of Products Yielded by Nucleophilic Displacement Reactions of Cellulose Tosylate

Reagent	Product	References
Na₂S₂O₃	6-Deoxy-6-*S*-thiosulfato cellulose	Klemm (1998)
NaSCH₃ (subsequent carboxymethylation)	6-Deoxy-6-thiomethyl-2,3-di-carboxymethyl cellulose	Wenz et al. (2005)
NaSO₃	Sodium deoxysulfate-*co*-tosylate cellulose	Arai and Aoki (1994); Arai and Yoda (1998)
Iminodiacetic acid	6-Deoxy-6-iminodiacetic acid cellulose sodium salt	Heinze (1998)
Triethylamine	6-Deoxy-6-triethylammonium cellulose	Koschella and Heinze (2001)
N,N-Dimethyl-1,3-diaminopropane	6-Deoxy-6-(*N,N*-dimethyl-3-aminopropyl)ammonium cellulose	Koschella and Heinze (2001)
2,4,6-Tris(*N,N*-dimethylaminomethyl) phenol	6-Deoxy-6-(2,6-di(*N,N*-dimethylaminomethyl)phenol-4-methyl-*N,N*-dimethylamino cellulose	Koschella and Heinze (2001)
R(+)-, *S*(−)-, and racemic 1-phenylethylamine	6-Deoxy-6-(1-phenylethyl) amino cellulose	Heinze et al. (2001)
Aminomethane	6-Deoxy-6-methylamino cellulose	Knaus et al. (2003)

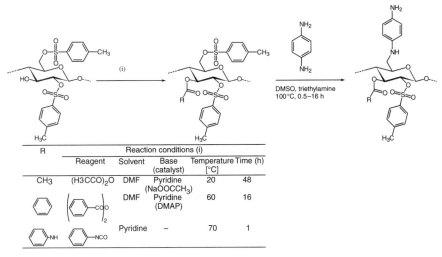

R	Reaction conditions (i)				
	Reagent	Solvent	Base (catalyst)	Temperature [°C]	Time (h)
CH$_3$	(H$_3$CCO)$_2$O	DMF	Pyridine (NaOOCCH$_3$)	20	48
⬡	⬡-C-O with 2	DMF	Pyridine (DMAP)	60	16
⬡-NH	⬡-NCO	Pyridine	–	70	1

FIGURE 1.3 Reaction path for the synthesis of 6-deoxy-6-amino cellulose ester derivatives by subsequent acylation and nucleophilic displacement with phenylenediamine of tosyl cellulose. Adapted from Tiller et al. (2000).

the two enantiomeric amines by the S$_N$ reaction of cellulose tosylate with $R(+)$-, $S(-)$- and racemic 1-phenylethylamine (Heinze et al., 2001). Methylamino celluloses are suitable as hydrophilic polymer matrices for immobilization of ligands for extracorporeal blood purification, for example, quaternary ammonium groups (Knaus et al., 2003). Conversion of cellulose tosylate with diamines or oligoamines yields polymers of the type P-CH$_2$-NH-(X)-NH$_2$ (P = cellulose, (X) = alkylene, aryl, aralkylene, or oligoamine) at position 6 and solubilizing groups at positions 2 and 3, which form transparent films that may be applied for the immobilization of enzymes like glucose oxidase (GOD), peroxidase, and lactate oxidase (Figure 1.3). The products are useful as biosensors (Tiller et al., 1999, 2000; Berlin et al., 2000, 2003; Becher et al., 2004).

Water-soluble and film-forming amino cellulose tosylates from alkylene-diamines can be used as enzyme support matrices with Cu^{2+} chelating properties (Jung and Berlin, 2005). The synthesis of 6-deoxy-6-amino cellulose via azido derivative is described in detail (Figure 1.4). The reaction conditions for a complete functionalization at position 6 are optimized, as well as various subsequent reactions of the product are studied (e.g., N-carboxymethylation, N-sulfonation) (Liu and Baumann, 2002; Heinze et al., 2006b).

1.3.2.1 Huisgen Reaction: "Click Chemistry" with Cellulose   Recently, Sharpless introduced click chemistry, that is, a modular approach that uses only the most practical and reliable transformation, which are experimentally simple, needing no protection from oxygen, requiring only stoichiometric amounts of starting materials, and generating no by-products (Kolb et al., 2001). The 1,3-dipolar cycloaddition of an azide moiety and a triple bond (Huisgen reaction) is the most popular click reaction to date (Rostovtsev et al., 2002; Lewis et al., 2002). Sharpless describes the

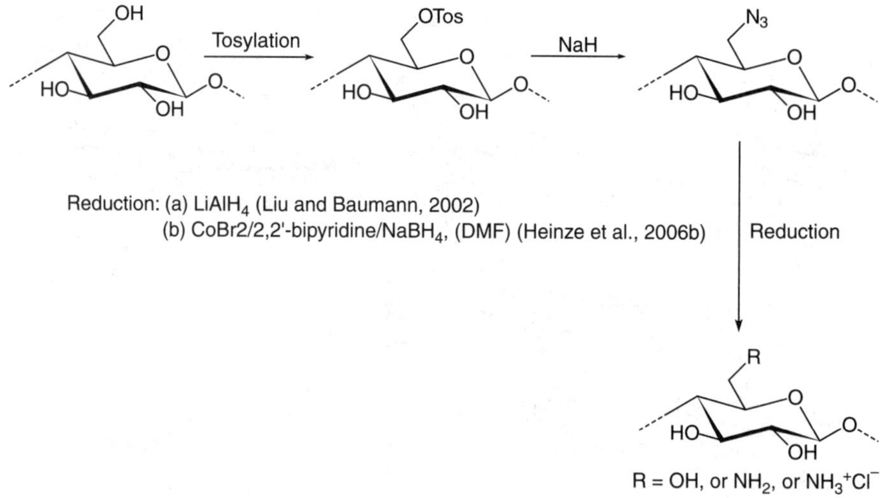

Reduction: (a) LiAlH$_4$ (Liu and Baumann, 2002)
          (b) CoBr2/2,2'-bipyridine/NaBH$_4$, (DMF) (Heinze et al., 2006b)   Reduction

R = OH, or NH$_2$, or NH$_3^+$Cl$^-$

FIGURE 1.4 Scheme of the synthesis of 6-deoxy-6-amino cellulose via cellulose tosylate and reduction of 6-deoxy-6-azido cellulose.

Huisgen reaction as "the cream of the crop" of click chemistry. The path of tosylation, S_N with sodium azide and subsequent copper-catalyzed Huisgen reaction, has significantly broaden the structural diversity of polysaccharide derivatives because the method yields products that are not accessible via etherification and esterification, the most commonly applied reactions (Liebert et al., 2006). The preparation of 6-deoxy-6-azido cellulose and subsequent copper-catalyzed Huisgen reaction of 1,4-disubstituted 1,2,3-triazols formed as linker lead to novel cellulose derivatives with methylcarboxylate, 2-aniline, and 3-thiophene moieties (Figure 1.5). No side reactions occur, the synthesis leads to pure and well-soluble derivatives with conversion efficiency of the azido moiety of 75–98% depending on the reaction temperature and the molar ratio (Table 1.4).

As can be concluded from the ^{13}C NMR spectra exemplified for the spectrum of 6-deoxy-6-methylcarboxytriazolo celluloses (DS 0.81) acquired in DMSO, no structure impurities are present (Figure 1.6). The signals at 48.5 ppm represent the methyl ester, and at 160.6 ppm the signals of the carbonyl group appear. The C-atoms of the triazole moieties give signals at 138.6 and 129.9 ppm, and peaks in the range of 51.6–110 ppm are related to the carbons of the repeating unit. A weak signal at about 60 ppm reveals the existence of remaining OH groups at position 6.

The 1,3-dipolar cycloaddition reaction of 6-azido-6-deoxycellulose with acetylene-dicarboxylic acid dimethyl ester and subsequent saponification with aqueous NaOH yield bifunctional cellulose-based polyelectrolytes (Figure 1.7) (Koschella et al., 2010).

Up to 62% of the azide moieties are converted. Starting with a 6-azido-6-deoxy cellulose with a DS of 0.84, the reaction is completed within 4 h using 2 mol of acetylenecarboxylic acid dimethyl ester per mole modified AGU to get 6-deoxy-6-(1-triazolo-4,5-disodiumcarboxylate) cellulose with a DS up to 0.52. The products form water-insoluble complexes with multivalent metal ions and organic

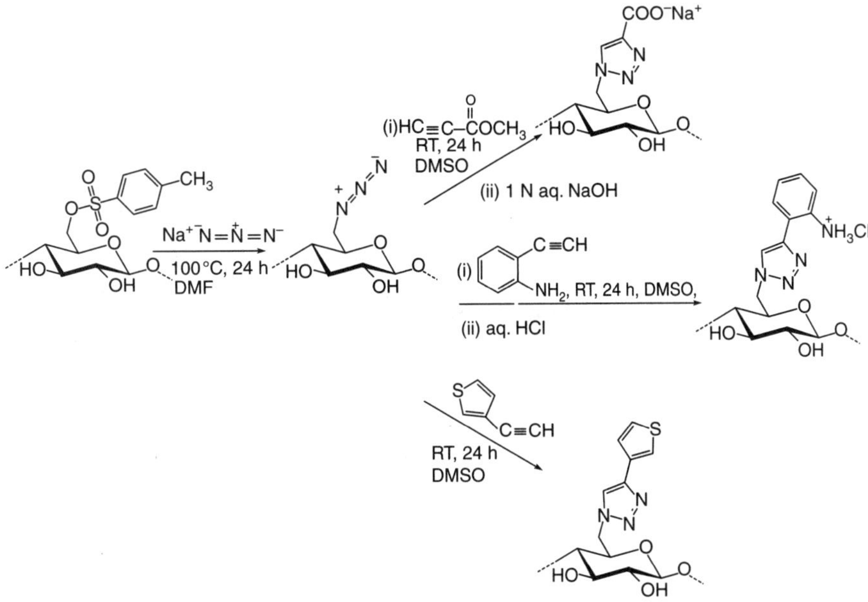

FIGURE 1.5 Reaction path for the preparation of 6-deoxy-6-azido cellulose and subsequent copper-catalyzed Huisgen reaction of 1,4-disubstituted 1,2,3-triazols used as linker for the modification of cellulose with methylcarboxylate, 2-aniline, and 3-thiophene moieties.

polycations that may possess different shapes; metal salts like calcium(II) chloride or aluminum(III) chloride yield a bagel-like shape. The polyelectrolyte complex with poly(diallyldimethylammonium) chloride is very smooth and unstable "tubes" are formed (Figure 1.8).

A promising approach for the synthesis of unconventional cellulose products is the introduction of dendrons in the cellulose backbone, which are easily accessible through the convergent synthesis of dendrimers (Vögtle et al., 2007). Apart from

TABLE 1.4 Conditions of the Copper-Catalyzed Huisgen Reaction of 6-Deoxy-6-Azido Cellulose (Azido Cellulose) and Degree of Substitution (DS) of the Products

Azido Cellulose	Reagent			Product	
DS	Type	Molar Ratio[a]	Temperature (°C)	DS	Conversion efficiency (%)
0.88	Methyl propiolate	1	25	0.86	98
0.88	2-Ethynylaniline	1	25	0.67	76
0.99	2-Ethynylaniline	3	25	0.80	81
0.99	3-Ethynylthiophene	3	25	0.91	92
0.99	3-Ethynylthiophene	3	70	0.93	94

Adapted from Liebert et al. (2006).
[a] Mole reagent per mole repeating unit of 6-deoxy-6-azido cellulose, reaction time 24 h.

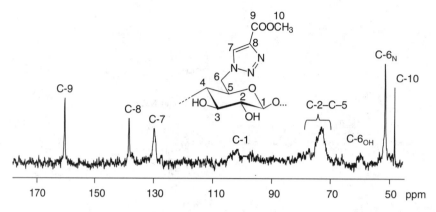

FIGURE 1.6 ¹³C NMR spectrum of methylcarboxytriazolo celluloses (DS 0.81) in DMSO-d_6. Reproduced with permission from Wiley–VCH, Liebert et al. (2006).

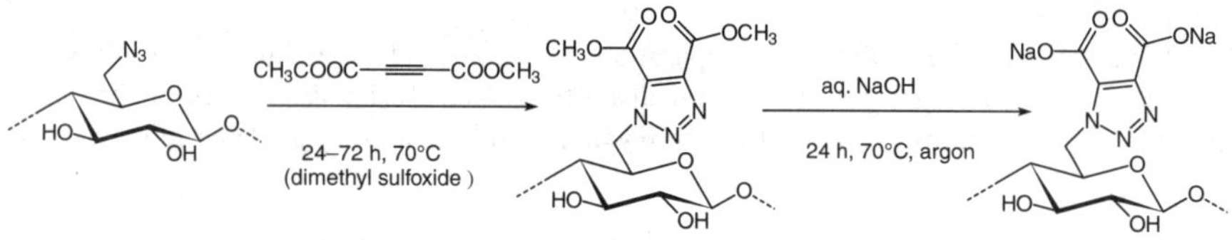

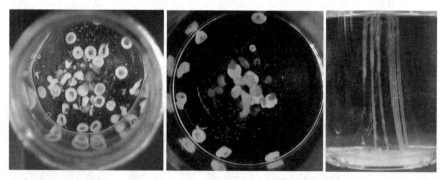

FIGURE 1.7 1,3-Dipolar cycloaddition of 6-azido-6-deoxycellulose with acetylenecarboxylic acid dimethyl ester.

FIGURE 1.8 Ionotropic gels of 6-deoxy-6(1-triazolo-4,5-disodiumcarboxylate) cellulose (DS 0.51) with aqueous calcium chloride (5 w/v), aqueous aluminum(III) chloride (5% w/v), and poly(diallyldimethylammonium) chloride (1% w/v). Reproduced with permission from Elsevier, Koschella et al. (2010).

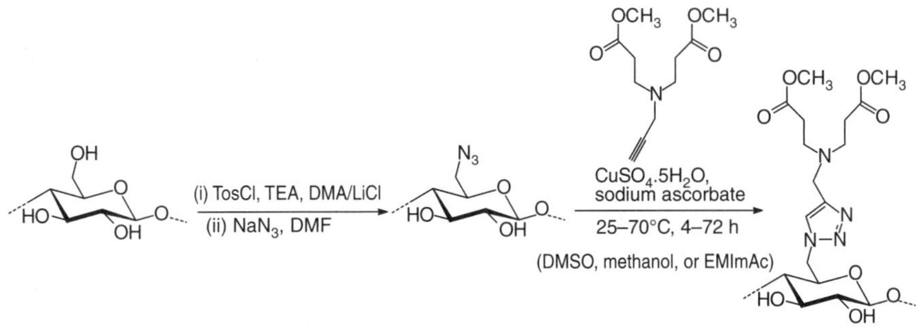

FIGURE 1.9 Reaction path for the conversion of cellulose with propargyl-PAMAM dendron of first generation via tosylation, nucleophilic displacement by azide, and conversion with the dendron.

the first described amino triester-based dendrons (Behera's amine) with an isocyanate moiety (Hassan et al., 2004, 2005), carboxylic acid-containing dendrons (Heinze et al., 2007; Pohl et al. 2008a) are explored, which are allowed to react with cellulose or cellulose derivatives like ethyl cellulose (Khan et al., 2007), hydroxypropyl cellulose (Oestmark et al., 2007), or carboxymethyl cellulose (CMC) (Zhang and Daly, 2005, 2006; Zhang et al., 2006a). Regioselective introduction of dendrons in cellulose is achieved by the reaction of 6-deoxy-6-azido cellulose with propargyl-polyamidoamine (PAMAM) dendron homogeneously in DMSO or heterogeneously in methanol in the presence of CuSO$_4$·5H$_2$O/sodium ascorbate (Figure 1.9, Table 1.5)

TABLE 1.5 Degree of Substitution (DS) of Dendritic PAMAM-Triazolo Cellulose Derivatives of First (1), Second (2), and Third (3) Generations Synthesized Homogeneously in Dimethylsulfoxide (DMSO) or 1-Ethyl-3-Methylimidazolium Acetate (EMImAc) as well as Heterogeneously in Methanol by Reacting 6-Deoxy-6-Azido Cellulose (DS 0.75) with Propargyl Polyamidoamine Dendrons of First, Second, and Third Generations via Copper-Catalyzed (CuSO$_2$·5H$_2$O/Sodium Ascorbate) Huisgen Reaction

Conditions				Product	
Molar Ratio	Solvent	Temperature (°C)	Time (h)	Generation	DS
1:1	DMSO	25	48	First	0.68
1:3	DMSO	25	24	First	0.65
1:3	DMSO	60	24	First	0.69
1:3	DMSO	25	24	Second	0.56
1:3	DMSO	25	72	Third	0.31
1:1	EMImAc	25	24	First	0.52
1:2	EMImAc	25	24	First	0.55
1:3	EMImAc	25	24	Second	0.48
1:1	EMImAc	25	72	Third	0.28
1:1	Methanol	25	72	First	0.63
1:3	Methanol	25	24	First	0.63

(Pohl et al., 2008b). Even ionic liquids (ILs) like 1-ethyl-3-methylimidazolium acetate (EMImAc) could successfully be applied as reaction medium due to the solubility of 6-deoxy-6-azido cellulose (Figure 1.9, Table 1.5) (Heinze et al., 2008a; Schöbitz et al., 2009).

Under homogeneous conditions, 6-deoxy-6-azido cellulose reacts with propargyl-PAMAM dendrons of first to third generation. The structure character-ization of the dendritic PAMAM-triazolo celluloses succeeded by FTIR and NMR spectroscopy, including two-dimensional techniques. The HSQC-DEPT NMR spectrum of second-generation PAMAM-triazolo celluloses (DS 0.59) allows the complete assignment of the signals of the protons of the substituent in ^{1}H NMR spectra (Figure 1.10).

In Figure 1.11, a comparison of ^{13}C NMR spectra of first-, second-, and third-generation PAMAM-triazolo celluloses synthesized in EMImAc demonstrate the possibility to assign the signals of the dendrons and the AGU. However, the intensity of the peaks of the carbon atoms of the repeating unit decreases due to the large number of branches and corresponding carbon atoms.

Water-soluble deoxy-azido cellulose derivatives could be obtained by hetero-geneous carboxymethylation applying 2-propanol/aqueous NaOH as medium. Starting from the cellulose derivatives with different DS values of the azide moiety (0.58–1.01), various DS values of the carboxymethyl functions (1.01–1.35) could be realized (Pohl, 2009a). The carboxymethyl deoxy-azido cellulose provides a convenient starting material for the selective dendronization of cellulose via Huisgen reaction yielding water-soluble carboxymethyl 6-deoxy-(1-N-(1,2,3-tria-zolo)-4-PAMAM) cellulose derivatives of first (DS 0.51) (Figure 1.12), second (DS 0.44), and third generation (DS 0.39).

The conformation and the flexibility of the dissolved polymer are estimated qualitatively using conformation zoning and quantitatively using the combined global method. Sedimentation conformation zoning shows a semiflexible coil conformation and the global method yields persistence length in the range of 2.8–4.0 nm with no evidence of any change in flexibility after dendronization (Table 1.6).

6-Deoxy-(1-N-(1,2,3-triazolo)-4-PAMAM) cellulose of the 2.5th generation (DS 0.25) is a promising starting polymer for biofunctional surfaces (Pohl et al., 2009b), either by embedding the dendronized cellulose in cellulose acetate (DS 2.5) matrix or by modifying the deoxy-azido cellulose film heterogeneously with the dendron (Figure 1.13). The heterogeneously functionalized cellulose solid support provides the higher amount of amino groups (determined by acid orange 7). The enzyme immobilization on the dendronized cellulose films after activation with glutardialde-hyde is demonstrated using glucose oxidase (GOD) as a model enzyme. The specific enzyme activity of immobilized GOD (28.73 mU/cm^{2}) and the coupling efficiency (2.2%) are rather small compared to the blend of dendronized cellulose and cellulose acetate (135.16 mU/cm^{2}, 27.2%). Nevertheless, the heterogeneous approach of dendronization with propargyl-polyamidoamine dendron of 2.5th generation affords an interesting possibility for biofunctionalized surfaces and thus protein attachment.

Chemoselective synthesis of dendronized cellulose may be realized with regio-selectively functionalized propargyl cellulose at position 6 (Pohl and Heinze, 2008)

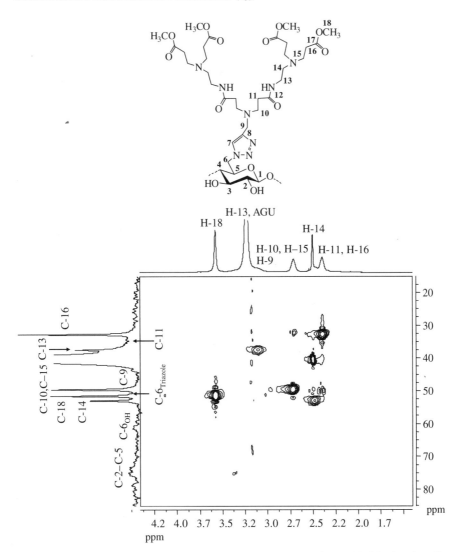

FIGURE 1.10 HSQC-DEPT NMR spectrum of second-generation PAMAM-triazolo cellu-loses (DS 0.59). Adapted from Pohl et al. (2008b).

or at position 3 (Fenn et al., 2009) (see Section 1.5.3). By nucleophilic displacement reaction of 6-O-tosyl cellulose (DS 0.58) with propargyl amine, 6-deoxy-6-amino-propargyl cellulose is formed that provides an excellent starting material for the dendronization of cellulose via the copper-catalyzed Huisgen reaction yielding 6-deoxy-6-amino-(4-methyl-(1,2,3-triazolo)-1-propyl-polyamido amine) cellulose derivatives of first (DS 0.33) and second (DS 0.25) generation (Figure 1.14).

3-Mono-O-propargyl cellulose could be synthesized by treatment of 2,6-di-O-thexyldimethylsilyl (TDS) cellulose with propargyl bromide in the presence of

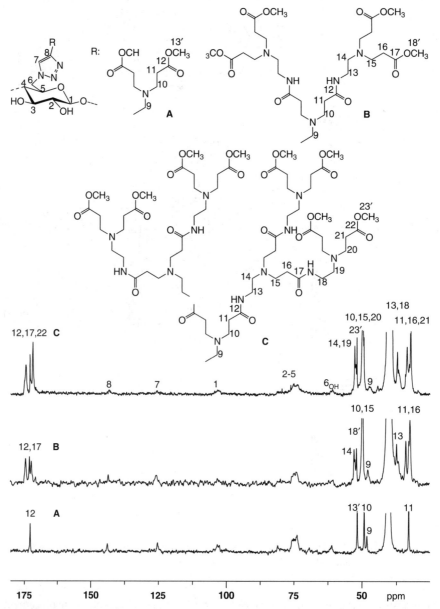

FIGURE 1.11 ^{13}C NMR spectra of **A** first (DS 0.60), **B** second (DS 0.48), and **C** third (DS 0.28) generation PAMAM-triazolo celluloses in DMSO-d_6 at 60°C.). Reproduced with permission from John Wiley & Sons, Inc., Heinze et al. (2008a).

sodium hydride and the subsequent complete removal of the silicon-containing group of the 3-mono-O-propargyl-2,6-di-O-thexyldimethylsilyl cellulose with tetrabutyl-ammonium fluoride trihydrate (see Section 5.3). Copper-catalyzed Huisgen reaction with azido-propyl-polyamidoamine dendron of first and second generation leads to

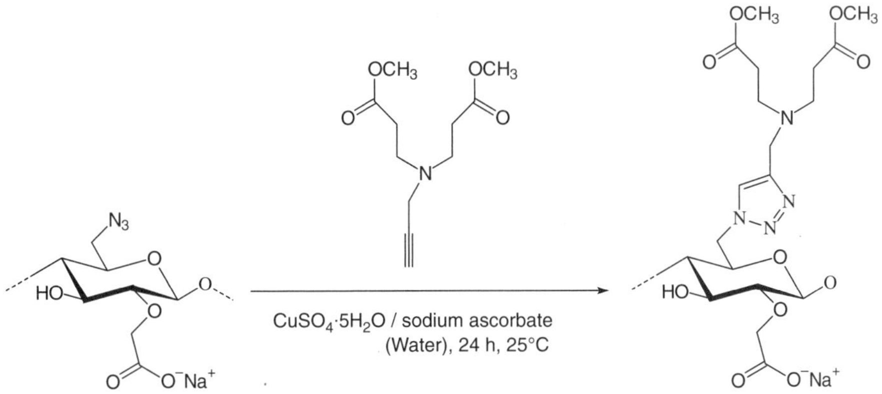

FIGURE 1.12 Homogeneous conversion of carboxymethyl 6-deoxy-6-azidocellulose (DS_{Azide} 0.81, DS_{CM} 1.25) with first generation of propargyl-polyamidoamine dendron via the copper-catalyzed Huisgen reaction.

regioselectively functionalized 3-O-(4-methyl-1-N-propyl-polyamidoamine-(1,2,3-triazole)) cellulose (Figure 1.15) (Fenn et al., 2009).

1.4 SULFATION OF CELLULOSE

Polysaccharide sulfuric acid half-esters are often referred to as polysaccharide sulfates (PSS), constituting a complex class of compounds occurring in living organisms. They possess a variety of biological functions, for example, inhibition of blood coagulation, or may be present as component of connective tissues (Fransson, 1985). These polysaccharides are usually composed of different sugars, including

TABLE 1.6 Conformational Parameters for Carboxymethyl Deoxy-Azido Cellulose and Dendronized Carboxymethyl 6-Deoxy-(1-N-(1,2,3-Triazolo)-4-PAMAM) Celluloses

Sample	DS_{Azide}	DS_{CM}	DS_{Dend}/ Generation	f/f_0	L_P (nm)	M_L (g/(mol nm))	Conformation Zone
CM-deoxy-azido cellulose	0.81	1.25	–	3.6	3.0	310	C
CM-6-(1,2,3-TA) (PAMAM)C	0.30	1.25	0.51/First	2.9	2.8	470	C
CM-6-(1,2,3-TA) (PAMAM)C	0.37	1.25	0.44/Second	4.3	3.7	450	C
CM-6-(1,2,3-TA) (PAMAM)C	0.42	1.25	0.39/Third	5.0	4.0	420	C

Zone C = semiflexible conformation.
CM: carboxymethyl; CM-6-(1,2,3-TA)(PAMAM)C: carboxymethyl 6-deoxy-(1-N-(1,2,3-triazolo)-4-PAMAM) cellulose.
Adapted from Pohl et al. (2009a).

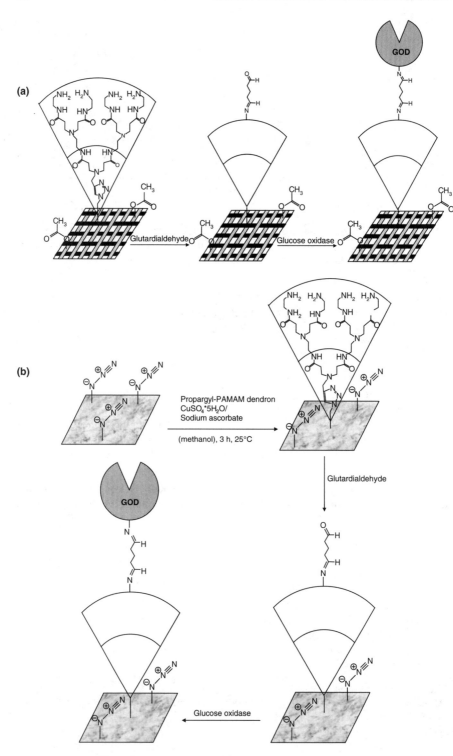

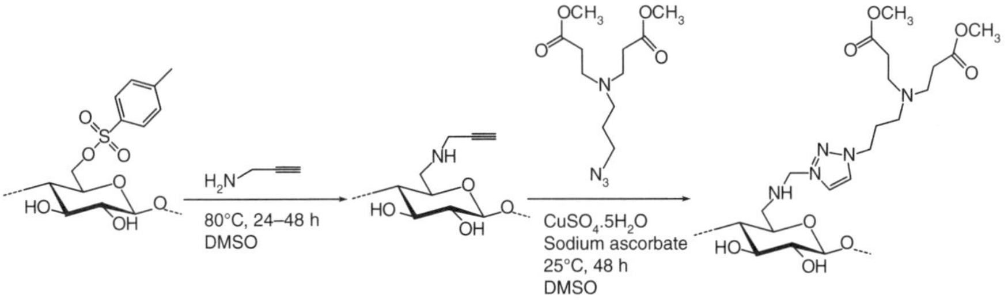

FIGURE 1.14 Reaction path for the synthesis of 6-deoxy-6-amino-(4-methyl-(1,2,3-triazolo)-1-propyl-polyamido amine) cellulose derivatives of first generation (DS 0.33) via 6-deoxy-6-aminopropargyl cellulose. Adapted from Pohl and Heinze (2008).

aminodeoxy and carboxy groups containing derivatives, for example, β-D-glucuronic acid, α-L-iduronic acid, and N-acetyl-β-D-galactosamine (Nakano et al., 2002). Heparan sulfate, chondroitin-6-sulfate, and dermatan sulfate are among the most important naturally occurring PSS (Figure 1.16).

Promising biological properties are not only observed for naturally occurring PSS but also for semisynthetic ones that can be received by introduction of sulfate groups into the polymer backbone of polysaccharides such as cellulose, dextran, pullulan, or chitosan. They have a number of advantages over their natural occurring counterparts. The isolation of the naturally occurring PSS often requires high cost due to intensive enrichment, extraction, and purification procedures, while homopolysaccharides suitable for sulfation are often easily available by biotechnological processes (dextran, pullulan) or even by industrial scale production (cellulose, xylan, and chitosan). On one hand, natural PSS constitute of very complex structures making it difficult to elucidate structure–property correlations, while on the other hand, ease of chemical modification of polysaccharides in combination with modern structure characterization methods offers a broad structural diversity of semisynthetic PSS with well-defined chemical structures. These products can mimic the structure and biological activity of naturally occurring PSS and are intensively studied regarding their applications in various fields especially in biotechnology and medicine.

Many procedures for the preparation of cellulose sulfates (CS) have been developed (Figure 1.17). The properties of CS, like water solubility, superstructure formation, and biological activity, strongly depend on the DS, on the molecular weight, and on the distribution of substituents within the repeating unit and along the polymer chain, which are ascertained by the course of reaction. The influence of the pattern of substitution on the properties is especially distinct at very low DS values.

FIGURE 1.13 Scheme preparation of biofunctional surfaces. (a) Blend of 6-deoxy-6-(1,2,3-triazolo)-4-polyamidoamine cellulose (DS 0.25) and cellulose acetate (DS 2.50). (b) Heterogeneous functionalization of deoxy-azido cellulose film with propargyl-polyamidoamine dendron of 2.5th generation via copper-catalyzed Huisgen reaction and for both subsequent surface activation with glutardialdehyde for covalent immobilization of glucose oxidase. Adapted from Pohl et al. (2009b).

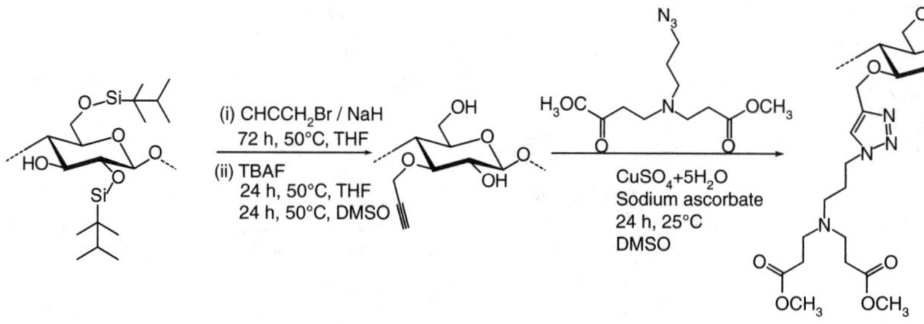

FIGURE 1.15 Reaction scheme for the synthesis of 3-*O*-(4-methyl-1-*N*-propyl-polyamidoa-mine-(1,2,3-triazole)) cellulose of first generation via 3-*O*-propargyl cellulose. Adapted from Fenn et al. (2009).

Complete heterogeneous sulfation of cellulose is carried out with mixtures of H_2SO_4 and propanol (Yao, 2000; Lukanoff and Dautzenberg, 1994). The course of the reaction is largely committed by the equilibrium formation of propylsulfuric acid. Cooling to about $-10°C$ is necessary in order to limit acid-catalyzed chain cleavage. The CSs yielded are not uniformly substituted and contain large amounts of water-insoluble parts without previous activation of the cellulose. An increase of the DS due to increasing reaction time, temperature, and amounts of H_2SO_4 is in general not only combined with a decrease of insoluble parts but also with considerable polymer degradation and hence lower solution viscosities of the aqueous solutions of the products (Figure 1.18) (Lukanoff and Dautzenberg, 1994).

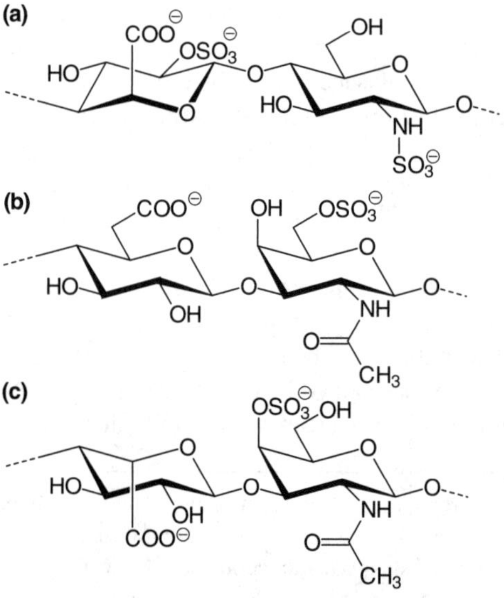

FIGURE 1.16 Typical repeating units of heparan sulfate (a), chondroitin-6-sulfate (b), and dermatan sulfate (c).

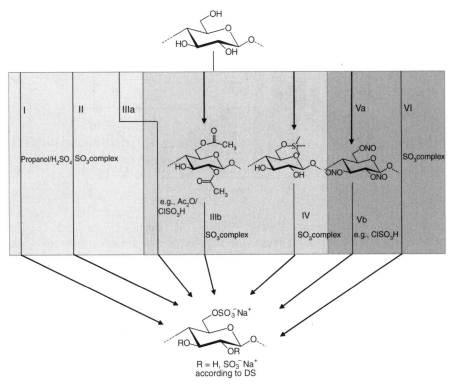

FIGURE 1.17 Overview of different approaches for the preparation of cellulose sulfate under heterogeneous (light gray), quasi-homogeneous (medium gray), and homogeneous (dark gray) conditions. I: heterogeneous sulfation with propanol/H$_2$SO$_4$; II: sulfation in DMF or pyridine under heterogeneous starting conditions; IIIa: parallel acetylation and sulfation of cellulose; IIIb: sulfation of cellulose acetate; IV: sulfation of trimethylsilyl cellulose; Va: dissolution of cellulose in N$_2$O$_4$/DMF; Vb: sulfation of cellulose trinitrite in N$_2$O$_4$/DMF solution; and VI: direct sulfation in ionic liquids.

Sulfation of cellulose suspended in DMF with a SO$_3$ complex starts under heterogeneous conditions and leads to the dissolution of the CS formed at a certain DS. This method is suitable for the preparation of CS with high DS > 1.5 only. Lower substituted derivatives are sulfated in the swollen amorphous parts of the cellulose, while the crystalline parts remain unfunctionalized. Thus, water insolubility and nonuniform sulfation among the cellulose chains, that is, different amounts of water-soluble parts (high DS) and water-insoluble parts (low DS) are formed (Schweiger, 1972).

Sulfation of cellulose derivatives, in particular cellulose acetate, cellulose nitrite, and trimethylsilyl cellulose (TMSC), and subsequent cleavage of the initial functional group are a valuable quasi-homogeneous route for the CS preparation. The major drawbacks are the requirement of large amounts of chemicals and the additional effort necessary for both the reaction and purification processes.

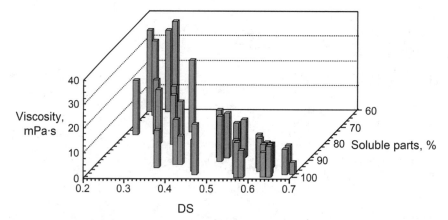

FIGURE 1.18 Correlation of DS value, viscosity, and amount of water-soluble part of cellulose sulfates obtained by heterogeneous sulfation of cellulose using propanol/H_2SO_4 mixtures. Adapted from Lukanoff and Dautzenberg (1994), and with permission from Nova Science Publishers, Heinze et al. (2010a).

In this context, N_2O_4/DMF was intensively studied as derivatizing cellulose solvent for the preparation of CS, although it is very hazardous. The intermediately formed cellulose nitrite is attacked by various reagents (SO_3, $ClSO_3H$, SO_2Cl_2, and H_2NSO_3H), resulting in CSs via transesterification with DS values ranging from 0.3 to 1.6 after cleavage of the residual nitrite moieties during the workup procedure under protic conditions (Schweiger, 1974; Wagenknecht et al., 1993). The regioselectivity of the transesterification reaction can be controlled by reaction conditions used (Table 1.7). The polymer degradation is rather low-yielding products that form high-viscous solutions. Besides their promising properties, the application of CS prepared in N_2O_4/DMF, especially for biomedical application, is limited due to the high toxicity of the solvent and the by-products formed (nitrosamines).

TABLE 1.7 Regioselectivity of Sulfation of Cellulose Nitrite with Different Reagents (2 mol/mol AGU) Depending on the Reaction Conditions

Reaction Conditions			Cellulose Sulfate			
			Partial DS			
Reagent	Time (h)	Temperature (°C)	2	3	6	Total DS
$NOSO_4H$	4	20	0.04	0	0.31	0.35
NH_2SO_3H	3	20	0.10	0	0.30	0.40
SO_2Cl_2	2	20	0.30	0	0.70	1.00
SO_3	3	20	0.26	0	0.66	0.92
SO_3	1.5	−20	0.45	0	0.10	0.55

DS values were determined by means of NMR spectroscopy.
Adapted from Wagenknecht et al. (1993).

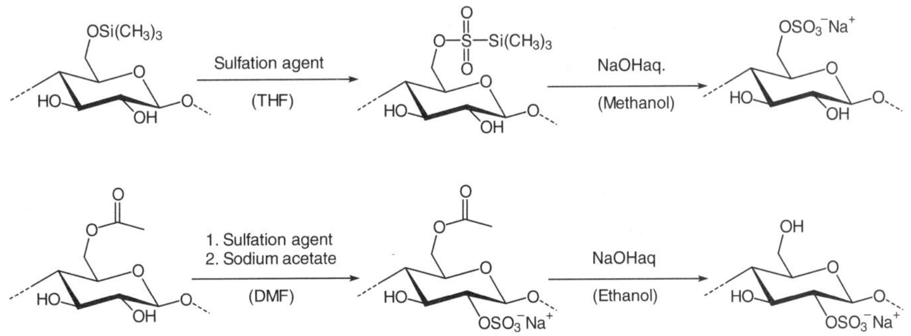

FIGURE 1.19 Synthesis of cellulose sulfate via trimethylsilyl cellulose (upper scheme) and starting from cellulose acetate (lower scheme).

In order to avoid the toxic N_2O_4/DMF solvent, TMS cellulose was used, which is soluble in various organic solvents, for example, DMF and tetrahydrofuran (THF), that readily reacts with SO_3–pyridine or SO_3–DMF complex (Wagenknecht et al., 1992). The synthesis of TMS cellulose is quite easy and can be achieved by homogeneous reaction of cellulose in DMA/LiCl and ionic liquids with hexamethyldisilazane or heterogeneously in DMF/NH$_3$ with trimethylchlorosilane (DS $<$ 1.5) (Köhler et al., 2008; Mormann and Demeter, 1999; Heinze, 1998).

Similar to the sulfation of cellulose nitrite, the TMS group acts as leaving group. The first step consists of an insertion of SO_3 into the Si–O bond of the silyl ether (Figure 1.19). The instable intermediate formed is usually not isolated. Subsequent workup with aqueous NaOH results in a cleavage of the TMS group under formation of CS (Richter and Klemm, 2003).

It is described that due to the course of reaction, the $DS_{Sulfate}$ is limited by the DS_{TMS} of the starting TMS cellulose and can be adjusted in the range of 0.2–2.5. Typical reaction conditions and DS values are summarized in Table 1.8. The sulfation reaction is fast and takes about 3 h with negligible depolymerization. Thus, products of high molar mass are accessible if TMS cellulose of high DP was applied as starting material. For instance, the specific viscosity of a CS with DS 0.60 is 4900 (1% in H_2O) (Wagenknecht et al., 1992).

The sulfation could be carried out in a one-pot reaction, that is, without isolation and redissolution of the TMS cellulose (Wagenknecht et al., 1992). After silylation of cellulose in DMF/NH$_3$, the excess of NH$_3$ is removed under vacuum followed by separation of the NH$_4$Cl formed. The sulfating agent, for example, SO_3 or ClSO$_3$H, dissolved in DMF is added and the CS is formed.

Carboxylic acid esters of cellulose, in particular commercially available cellulose acetate with DS 2.5 as well as cellulose formiate both dissolved in DMF, are useful intermediates for the preparation of CS (Philipp et al., 1990). In case of cellulose formiate, the $DS_{Sulfate}$ can be higher than the amount of remaining OH groups partly because of the displacement of formiate moieties by sulfate groups. In contrast, no transesterification appears during sulfation of cellulose acetate. The acetyl groups act as protecting group and the sulfation with SO_3–pyridine, SO_3–DMF complex, or

TABLE 1.8   Sulfation of Cellulose via TMS Cellulose

Reaction Conditions				Product
TMS Cellulose		Sulfating Agent		
DS_{TMS}	Solvent	Type	mol/mol AGU	$DS_{Sulfate}$
1.55	DMF	SO_3	1.0	0.70
1.55	DMF	SO_3	2.0	1.30
1.55	DMF	SO_3	6.0	1.55
1.55	DMF	$ClSO_3H$	1.0	0.60
1.55	DMF	$ClSO_3H$	2.0	1.00
1.55	DMF	$ClSO_3H$	3.0	1.55
2.40	THF	SO_3	1.0	0.71
2.40	THF	SO_3	1.7	0.90
2.40	THF	SO_3	3.3	1.84
2.40	THF	SO_3	9.0	2.40

Adapted from Wagenknecht et al. (1992).

acetylsulfuric acid proceeds exclusively at the remaining hydroxyl functions (Figure 1.19). The cellulose acetosulfate formed is neutralized with sodium acetate and subsequently treated with NaOH in ethanol to cleave the acetate moieties in an inert atmosphere within 16 h at room temperature (RT).

Regioselective deacetylation of cellulose acetate at position 2 is achieved by treating the starting polymer (DS 2.5) with amines of low basicity, such as hexamethylene diamine, together with certain amounts of water at 80°C (Wagenknecht, 1996). Thus, CS with preferred sulfation at position 2 could be isolated (Table 1.9).

Acetosulfation, that is, competitive esterification of cellulose suspended in DMF, DMA, or N-methylpyrrolidone (NMP) with a mixture of acetic anhydride

TABLE 1.9   Partial DS Values of Cellulose Sulfate obtained by Conversion Position 2 and 3 of partly in 1 Deacetylated Cellulose Acetate with NH_2SO_3H in DMF (80°C, 2 h)

Cellulose Acetate, DS					Cellulose Sulfate, DS			
Overall	Position 2	Position 3	Position 6	mol/mol AGU	Overall	Position 2	Position 3	Position 6
2.65^a	0.80	0.85	1.0	1	0.26	0.17	0.08	0.0
1.86^a	0.45	0.45	0.90	2	0.92	0.55	0.20	0.17
1.48^a	0.20	0.20	0.85	3	1.15	0.74	0.13	0.28
2.40^b	0.85	0.80	0.75	1	0.52	0.17	0.15	0.20

Adapted from Wagenknecht (1996).
[a] Regioselectively deacetylated cellulose acetate.
[b] Statistical cellulose acetate.

and SO_3 or $ClSO_3H$, is another route toward CS with distinct sulfation at position 6 (Hettrich et al., 2008; Zhang et al., 2009, 2010a, 2010b). The synthesis involves the formation of a mixed cellulose acetosulfate combined with dissolution of the polysaccharide derivative in the dipolar aprotic solvent. Acetylating agent up to 6–11 mol and sulfating agent up to 3 mol are needed to yield CS with $DS_{sulfate}$ up to about 2. As a result of this quasi-homogeneous reaction, water solubility of CS is achieved at rather low DS > 0.3. In addition, high solution viscosities can be observed when celluloses with high DP such as cotton linters are used.

Homogeneous sulfation of dissolved cellulose can also overcome the problem of irregular substituent distribution. Although widely used for the esterification of cellulose with carboxylic acids, DMA/LiCl is not the solvent of choice for sulfation, because insoluble products of low DS are obtained due to gel formation by addition of the sulfating agent (Klemm et al., 1998a). Several other cellulose solvents including N-methylmorpholine-N-oxide (NMNO) have also been investigated for the homogeneous sulfation of cellulose, but showed coagulation of the reaction medium yielding badly soluble CS (Wagenknecht et al., 1985).

Promising solvents for the sulfation of cellulose are ionic liquids. This group of salt-like compounds with melting points below 100°C turned out to be excellent media for shaping and functionalization of cellulose (Swatloski et al., 2002; El Seoud et al., 2007). Cellulose dissolved in 1-butyl-3-methylimidazolium chloride (BMIMCl)/cosolvent mixtures can be easily transformed into CS by using SO_3–pyridine and SO_3–DMF complex or $ClSO_3H$ (Gericke et al., 2009a). Highly substituted CSs are described for sulfation in BMIMCl at 30°C (Wang et al., 2009), but it has to be noted that cellulose/IL solutions slowly turned solid upon cooling to room temperature depending on cellulose and moisture content. Furthermore, they have rather high solution viscosities, which make it very difficult to ensure sufficient miscibility and to guarantee even accessibility of the sulfating agent to the cellulose backbone. Consequently, the synthesis of CS with a uniform distribution of sulfate groups along the polymer chains demanded a dipolar aprotic cosolvent that drastically reduces the solution viscosity (Gericke et al., 2009a). The reactivity of the sulfating agent is not influenced by the cosolvent. At a molar ratio of 2 mol SO_3–DMF complex per mole AGU, the sulfation of microcrystalline cellulose in BMIMCl and BMIMCl/DMF mixtures yields CS with comparable DS values of about 0.9. While the CS synthesized without cosolvent shows water insolubility, the other one readily dissolves in water (Gericke et al., 2009a). On one hand, the increase of temperature results in a considerable decrease of the viscosity of cellulose/IL solutions, which improves solution miscibility (Gericke et al., 2009b), while on the other hand, high temperatures favor the acid-catalyzed chain degradation leading to rather low solution viscosities of aqueous solutions of the resulting CSs of about 2 mPa·s (1%).

The homogeneous sulfation in IL allows tuning of CS properties by simply adjusting the amount of sulfating agent and choosing different types of cellulose (Table 1.10). The reaction proceeds with almost no polymer degradation if conducted at room temperature (Gericke et al., 2009a). This makes the procedure very valuable for the preparation of water-soluble CS over a wide DS range. Especially CS with low DS could be prepared efficiently in IL/cosolvent mixture that is of interest for the bioencapsulation (see pp. 25).

TABLE 1.10   DS Values and Water Solubility of Cellulose Sulfates Obtained by Sulfation of Spruce Sulfite Pulp Dissolved in BMIMCl/DMF at Different Conditions

Reaction conditions				Product	
Sulfating Agent					
Type	mol/mol AGU	Time (h)	Temperature (°C)	DS	Water solubility
$SO_3–Py^a$	0.7	2	25	0.14	No
$SO_3–Py$	0.8	2	25	0.25	Yes
$SO_3–Py$	0.9	2	25	0.48	Yes
$SO_3–Py$	1.1	2	25	0.58	Yes
$SO_3–Py$	1.3	1	80^b	0.52	Yes
$SO_3–Py$	1.4	2	25	0.81	Yes
$SO_3–DMF$	1.0	2	25	0.34	Yes
$SO_3–DMF$	1.5	2	25	0.78	Yes
$ClSO_3H$	1.0	3	25	0.49	Yes

Adapted from Gericke et al. (2009a).
[a] SO_3–pyridine complex.
[b] Without cosolvent.

Sulfation of cellulose in BMIMCl/DMF yields a preferred 6-sulfated product. A typical ^{13}C NMR spectrum of CSs with DS 0.48 prepared in BMIMCl/DMF and the assignment of the peaks are shown in Figure 1.20. The signal at 67.3 ppm corresponds to sulfation at position 6. Peaks in the region of 82 ppm that would correspond to sulfated position 2 or 3 are missing in the spectrum and no splitting of the C-1 signal can be observed, which would indicate sulfation at position 2.

Disadvantages of IL are their costs and the high viscosities. These drawbacks are compensated by the ease of recycling due to their negligible vapor pressure. Reusability of IL for sulfation has already been reported for BMIMCl leading to

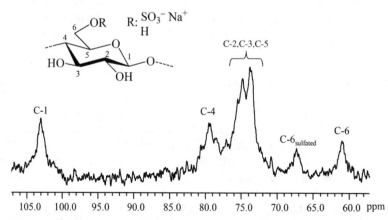

FIGURE 1.20 ^{13}C NMR spectrum (in D_2O) of cellulose sulfates (DS = 0.48) obtained by sulfation in BMIMCl/DMF with $SO_3–DMF$. Reproduced with permission from Wiley–VCH, Gericke et al. (2009a).

similar DS values compared to fresh IL (Gericke et al., 2009a). Furthermore, the use of cosolvents and the development of low viscous task-specific IL, bearing additional functional groups, can lead to further improvement of the homogeneous sulfation of cellulose. It should be noted that IL can act as "noninnocent" solvents that participate in the reaction. For instance, sulfation of cellulose in the room-temperature IL 1-ethyl-3-methylimidazolium acetate yields cellulose acetate instead of CS (Liebert et al., 2009). Similar side reactions were previously observed for acylation, tritylation, and tosylation of cellulose in EMIMAc (Köhler et al., 2007).

Thus, acetosulfation and homogeneous sulfation in IL are suitable pathways for the preparation of well-soluble, high-molecular weight CS under lab-scale conditions. Commercial application of acetosulfation, however, is limited due to the large amounts of acetylating agent necessary and the inefficient workup.

Besides the bioactivity, sulfates of polysaccharides were investigated toward their ability to form polyelectrolyte complexes (PEC) with various synthetic (Li and Yao, 2009; Renken and Hunkeler, 2007a, 2007b; Zhang et al., 2005, 2006b) and natural polycations (Xie et al., 2009). The process is based on the sequential deposition of interactive polymers from their solutions by electrostatic, van der Waals, and hydrogen bonding, as well as charge transfer interactions (Decher, 1996). These interactions can be applied to create layer-by-layer (LbL) assemblies of functional material surfaces with defined biodegradability or bioactivity (Heinze et al., 2006c).

Microencapsulation of biological material within PEC capsules based on CS has been studied for numerous applications, especially in biotechnology and medicine (Dautzenberg et al., 1999). CSs for bioencapsulation are preferably synthesized via homogeneous sulfation in IL because water-soluble products with high solution viscosities and rather low DS values of 0.3–0.7 are required (Gericke et al., 2009a). CSs of higher DS lead to complexes of low stability (Dautzenberg et al., 1993). Capsular PECs can be achieved with diameters in millimeter till 100 µm scale by dropping a polyanion solution into a polycation precipitation bath, such as poly (diallyldimethylammonium chloride) (polyDADMAC) (Figure 1.21).

Immobilization of enzymes allows simple recovery and improves their mechanical stability drastically during agitation, which makes this technique very attractive for large-scale biotechnological processes where high durability is required (Hanefeld et al., 2009). After encapsulation within PEC capsules, the velocity of substrate conversion is determined by diffusion through the PEC membrane leading to a decrease of relative enzyme activity. For GOD-containing capsules, they retain 14% of their initial activity after encapsulation within CS-polyDADMAC (Gericke et al., 2009a). Higher capsule stability without further loss of GOD activity is maintained using a more complex four-component system composed of CS, sodium alginate (SA), $CaCl_2$, and the cation poly(methylene-co-guanidine) (PMCG) (Vikartovská et al., 2007). The capsule preparation includes the formation of calcium alginate gel and subsequently polyanion–polycation complexation. Relative activity of GOD-CS-SA-PMCG capsules is 13%. Another elegant two-component approach for enhancing PEC properties applies water-insoluble CS with very low DS values of about 0.15, dissolved in ionic liquid, for the formation of CS–polyDADMAC capsules with increased mechanical stability (Figure 1.22a) (Gericke et al., 2009c). The increase of mechanical

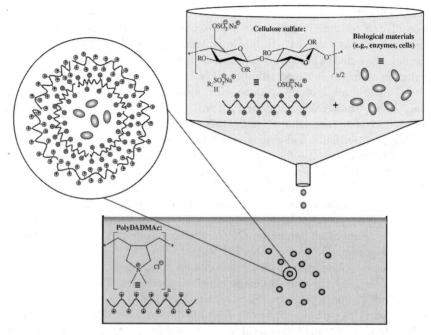

FIGURE 1.21 Scheme of the microencapsulation of biological material in polyelectrolyte complex capsules based on cellulose sulfate and poly(diallyldimethylammonium chloride) (polyDADMAC). Reproduced with permission from Wiley–VCH, Gericke et al. (2009a).

stability can be attributed to reestablished hydrogen bonds of the low substituted CS in addition to the electrostatic interaction of polyanion–polycation (Figure 1.22b). Despite the fairly harsh conditions, enzyme entrapment is also realized with IL-based CS/polyDADMAC capsules. The membrane properties, determining matter transfer between the inner core of the PEC capsule and the outer medium, are comparable to common capsules from water-soluble CS, because the same relative activity of 14% of the initial activity is found. Thus, sulfation, *in situ* PEC formation, and encapsulation in a one-pot procedure may be established. After completed reaction, GOD is suspended in the reaction mixture and GOD–PEC capsules are prepared by dropping the mixture directly into aqueous polyDADMAC, omitting time- and energy-consuming isolation and purification steps.

   The high biocompatibility and lack of cytotoxicity of CS-polyDADMAC capsules make them ideal candidates for the encapsulation of cells (Pelegrin et al., 1998; Wang et al., 1997). PEC layer is inert to metabolic breakdown, can survive for several months *in vivo*, and prevents recognition and attack of the protected cells by the immune system (Pelegrin et al., 1998). Thus, xenotransplantation of encapsulated cells may become a powerful therapeutic tool for the treatment of various diseases, including cancer and diabetes. Encapsulation of insulin producing porcine islet cells demonstrates that the CS-polyDADMAC PEC membrane is permeable for vital nutrients as well as oxygen, allowing glucose-dependent cell growth

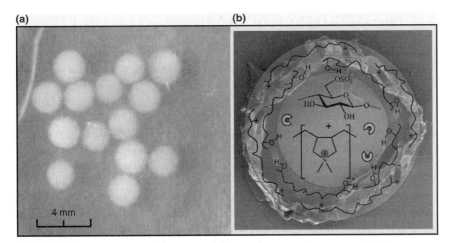

FIGURE 1.22 (a) Polyelectrolyte capsules prepared from water-insoluble cellulose sulfate (CS). (b) Scheme of the *in situ* polyelectrolyte complex formation and enzyme encapsulation using water-insoluble CS and ionic liquids. Reproduced with permission from Nova Science Publishers, Heinze et al. (2010a).

(Schaffelner et al., 2005). Moreover, PEC immobilization is used to protect cells during cryopreservation (Stiegler et al., 2006).

PEC capsules with trigger-controlled release have been studied applying cellulose-producing cells (Fluri et al., 2008). Transgenic mammalian cells, which exhibit doxycycline (DOX)-controlled $(1 \rightarrow 4)$-β-glucanase expression, are encapsulated within CS-based PEC capsules. The removal of the inducer molecule DOX suppressing cellulase accumulation enabled time-dependent capsule rupture and discharge of therapeutic proteins (Figure 1.23).

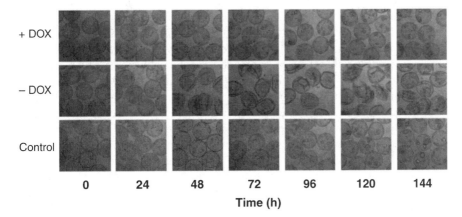

FIGURE 1.23 Encapsulated $(1 \rightarrow 4)$-β-glucanase secreting mammalian cells cultivated in the presence (+DOX) and absence (−DOX) of doxycycline. Control cell line produces no cellulose. Reproduced with permission from Nova Science Publishers, Heinze et al. (2010a).

1.5 REGIOSELECTIVELY FUNCTIONALIZED CELLULOSE ETHER

Etherification is a very important branch of commercial cellulose functionalization. Cellulose ethers are prepared in technical scale by reaction of alkali cellulose with alkylating reagents, for example, epoxides, alkyl-, and carboxymethyl halides (Brandt, 1986; Klemm et al., 1998b). In heterogeneous synthesis, the accessibility of the hydroxyl groups is determined by hydrogen bond-breaking activation and by interaction with the reaction media (Klemm et al., 1998b, 2005). The reaction of cellulose with reagents of low steric demand leads to a random distribution of ether functions within AGU and along the polymer chain provided a sufficient activation is carried out. It is well known that not only DS but also the pattern of substitution may influence the properties of cellulose ethers (Heinze, 2004). To gain detailed information about the influence of the structures on properties of cellulose derivatives, not only a comprehensive structure characterization but also cellulose ethers with a defined distribution of the functional groups (i.e., regioselective functionalization pattern) are indispensable for the establishment of the structure–property relationships. "Regioselectivity" in cellulose chemistry means an exclusive or significant preferential reaction at one or two of the three positions 2, 3, and 6 of AGU as well as along the polymer chain (Figure 1.24).

Up to now, the most important approach to control the functionalization within the repeating unit is the application of protecting groups (Figure 1.25). Other methods comprising, for example, selective cleavage of primary substituents by chemical or enzymatic treatment play a minor role (Deus et al., 1991; Wagenknecht, 1996; Altaner et al., 2003). Examples are the deacetylation of cellulose acetate under aqueous acidic or alkaline conditions or in the presence of amines (see Section 1.4). In addition, activating groups like the tosyl moiety may also be disposed for selective reactions (see Section 1.3.1).

The most widely used protecting group for the primary OH group is the triphenylmethyl (trityl) moiety (Figure 1.26). Heterogeneous introduction of the trityl groups starts with an activated polymer obtained either by deacetylation of cellulose acetate (Harkness and Gray, 1990; Kondo and Gray, 1991) or by mercerization of cellulose (Kern et al., 2000) followed by a conversion in anhydrous pyridine. More efficient tritylation of cellulose yielding polymers with DS values of 1.0 takes place in DMA/LiCl (preferred solvent) and DMSO/SO_2/diethylamine (Kasuya and Sawatari, 2000; Hagiwara et al., 1981). Methoxy-substituted triphenylmethyl compounds are more effective protecting groups for the primary hydroxyl group of cellulose (Camacho et al., 1996). The conversion of cellulose dissolved in DMA/LiCl with 4-monomethoxytriphenylmethyl chloride is 10 times faster than the reaction with unsubstituted trityl chloride. Complete functionalization of the primary hydroxyl groups occurs within 4 h and 70°C. Even after long reaction times, excess of the reagent, and elevated temperatures, alkylation at the positions 2 and 3 is less than 11%, which is in the same range as for the unsubstituted trityl function. Moreover, the detritylation proceeds 20 times faster (Heinze, 2004).

Trialkyl- (with at least one bulky alkyl moiety) and triarylsilyl groups are known to protect the primary groups of cellulose. Pawlowski describes the synthesis of

(a)

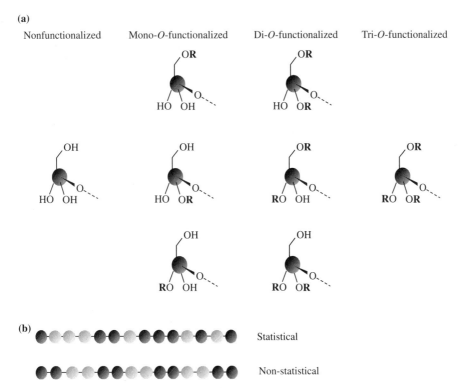

FIGURE 1.24 Distribution of the functional groups by regiocontrolled synthesis of cellulose derivatives (a) within the anhydroglucose units and (b) along the polymer chain. With permission from Elsevier, Heinze and Petzold (2008).

tert-butyldimethylsilyl cellulose with a DS of 0.68 in DMA/LiCl with functionalization at position 6 (Pawlowski et al., 1988). Among this type of derivatives, 6-*O*-thexyldimethylsilyl cellulose is most suitable (Figure 1.26) (Koschella and Klemm, 1997; Petzold et al., 2003). The synthesis starts with a heterogeneous phase reaction in ammonia-saturated polar aprotic liquids at −15°C by conversion of the cellulose with TDS chloride leading to a specific state of dispersion after evaporation of the ammonia at about 40°C, which does not permit any further reaction of the secondary hydroxyl groups, even with a large reagent excess, increased temperature, or long reaction time (Petzold et al., 2003). The degree of

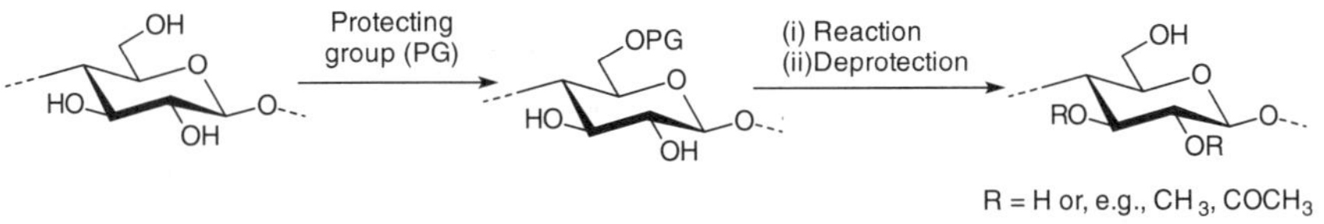

R = H or, e.g., CH₃, COCH₃

$R = H$ or, e.g., CH_3, $COCH_3$

FIGURE 1.25 Scheme of protecting group technique as the main tool for regioselective functionalization of cellulose exemplified for the 2,3-di-*O*-cellulose derivatives.

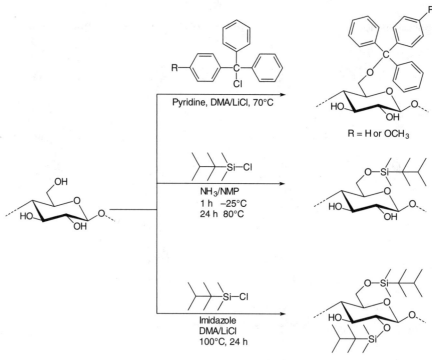

FIGURE 1.26 Protecting group techniques: tritylation with trityl chloride or derivatives in *N,N*-dimethyl acetamide (DMA)/LiCl, silylation with thexyldimethylchlorosilane in *N*-methyl-2-pyrrolidone (NMP)/ammonia for the regioselective blocking of the primary OH group, and silylation in DMA/LiCl to protect the 6 and 2 positions simultaneously.

TDS groups introduced by homogeneous silylation in DMA/LiCl to a total DS value of 0.99 is determined to be 85% at position 6 only (GC/MS analysis). However, the homogeneous reaction in DMA/LiCl may yield 2,6-di-*O*-TDS cellulose (Koschella and Klemm, 1997; Heinze, 2004; Fenn et al., 2007).

The structural uniformity and regioselectivity of the silylated cellulose products are characterized by means of one- and two-dimensional NMR techniques after subsequent acetylation of the remaining hydroxyl groups (Figure 1.27) (Hagiwara et al., 1981; Petzold et al., 2003) or after permethylation of the residual OH groups and chain degradation by means of HPLC and GC-MS (Mischnik et al., 1995; Camacho et al., 1996; Koschella and Klemm, 1997; Kern et al., 2000).

Recently, *tert*-butyldimethylsilyl cellulose with a degree of substitution of up to 2 could be obtained by homogeneous conversion of the biopolymer with *tert*-butyl-dimethylchlorosilane in DMA/LiCl in the presence of imidazole. The cellulose derivatives are characterized in detail by means of two-dimensional NMR spectroscopic techniques, including subsequent derivatization of the original polymer by consecutive methylation–desilylation–acetylation (Figure 1.28). The very well-resolved NMR

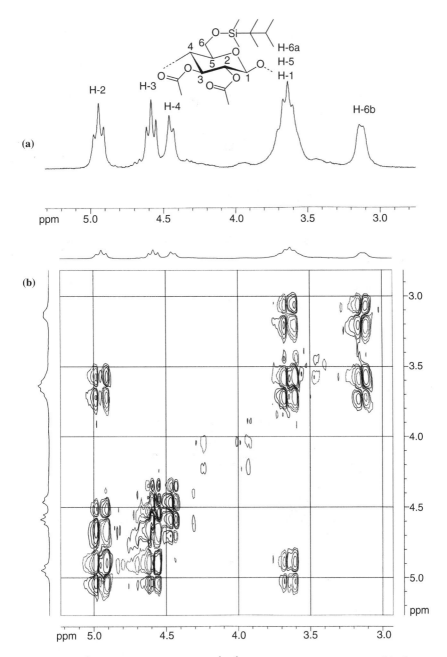

FIGURE 1.27 ^{1}H NMR spectrum (a) and ^{1}H/^{1}H-correlated NMR spectrum (b) of perace-tylated 6-O-thexyldimethylsilyl celluloses in CDCl$_3$: assigned cross-peaks of the anhydroglucose unit (2,3-O-Ac-6-O-TDS) (Ac = acetyl). Reproduced with permission from Wiley–VCH, Petzold et al. (2003).

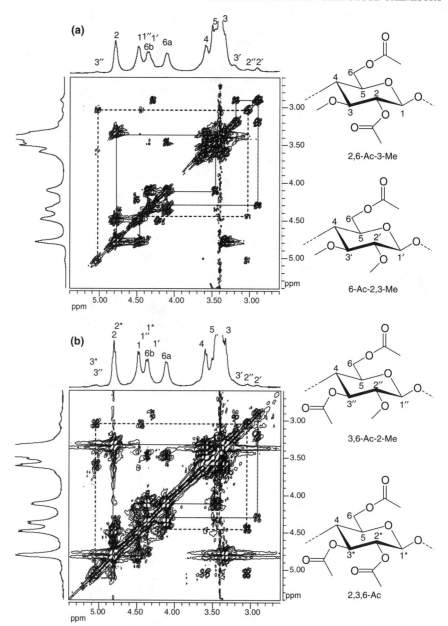

FIGURE 1.28 ^{1}H,^{1}H COSY NMR spectra of two acetyl methyl celluloses in CDCl$_3$. Silylation conditions for (a) DS$_{Si}$ 1.98: molar ratio 1:3.5:4.2 (cellulose:TBS chloride:imidazole), 24 h, 20°C; (b) DS$_{Si}$ 2.11: molar ratio 1:4.0:4.8 (cellulose:TBS chloride:imidazole), 24 h, 100°C; assigned cross-peaks: — cross-peaks of the unit 2,6-di-O-acetyl-3-mono-O-methyl (2,6-Ac-3-Me); ···· cross-peaks of the unit 6-mono-O-acetyl-2,3-di-O-methyl (6-Ac-2,3-Me), positions marked with $'$, (dashed lines) cross-peaks of the unit 3,6-di-O-acetyl-2-mono-O-methyl (3,6-Ac-2-Me), positions marked with $''$, — — cross-peaks of the unit 2,3,6-tri-O-acetyl (2,3,6-Ac), positions marked with *. Adapted from Heinze et al. (2008b).

spectra indicate that depending on the reaction temperature, 2,6-di-*O*-*tert*-butyldimethylsilyl moieties are the main repeating units. 3,6-Di-*O*- and 6-mono-*O*-functionalized repeating units are identified in very small amounts if the reaction is carried out at room temperature. In addition, 2,3,6-tri-*O*-silylated functions appear if reaction is carried out at temperature of 100°C (Heinze et al., 2008b).

1.5.1 2,3-Di-*O*-Ethers of Cellulose

The 6-mono-*O*-trityl cellulose or the more efficient 6-*O*-mono-*O*-(4-mono-methoxy) trityl derivative and the 6-mono-*O*-TDS cellulose are used to synthesize regioselectively functionalized cellulose ethers at positions 2 and 3 after the exclusive cleavage of the protecting groups (Table 1.11). The deprotection is carried out most efficiently with HCl in a suitable solvent (e.g., THF) in case of trityl derivatives. Tetrabutylammonium fluoride in THF is most successful for the cleavage of the silyl groups in TDS-protected derivatives.

The alkylation of the 6-*O*-trityl cellulose is carried out in DMSO with solid NaOH as base and the corresponding alkyl halides at 70°C within several hours. Interestingly, a small amount of water in the mixture (about 1 mL per 60 mL DMSO) increases the conversion up to a nearly complete functionalization of the secondary hydroxyl groups (Kondo and Gray, 1991). Ionic 2,3-*O*-carboxymethyl cellulose is obtained with sodium monochloroacetate as etherifying reagent in the presence of solid NaOH in DMSO and detritylation with gaseous HCl in dichloromethane for 45 min at 0°C or alternatively with aqueous hydrochloric acid in an ethanol slurry (Heinze et al., 1994). 2,3-*O*-CMCs up to a DS of 1.91 were accessible while solubility in water appears at a DS of 0.3 (Liu et al., 1997). 2,3-Di-*O*-hydroxyethyl (HEC) and 2,3-di-*O*-hydroxypropyl celluloses (HPC) are synthesized by heterogeneous etherification of 6-*O*-(4-monomethoxytrityl) cellulose (MMTC) with alkylene oxide in a 2-propanol/10% NaOH–water mixture providing anionic and nonionic detergent is used in addition due to the very hydrophobic character of MMTC (Yue and Cowie, 2002; Schaller and Heinze, 2005). The polymers become water soluble, starting with a MS 0.25 (HEC) and 0.50 (HPC), while a conventional HPC is water soluble with MS > 4. 2,3-*O*-HEC samples without oxyethylene side chains are synthesized up to DS values of 0.87 and compared with 2,3-*O*-hydroxyethyl cellulose with side chains (MS 0.83) using 6-*O*-trityl cellulose, which is allowed to react on one hand with the protected etherifying agent 2-(2-bromoethoxy)tetrahydropyran and with 2-bromoethanol on the other (Petzold-Welcke et al., 2010b). One- and two-dimensional NMR spectroscopy is efficiently used for the characterization of the substitution pattern of the repeating units of the 2,3-*O*-hydroxyethyl celluloses after perpropionylation of the remaining OH groups. In addition, differences between the oxyethylene chain-containing HEC and the HEC without side chains are clearly evaluated by peak assignment of the carbon and proton signals of the substituents using the cross-peaks in the two-dimensional NMR spectra (Figure 1.29). The formation of oxyethylene side chains influences the properties of the HEC like the solubility of the products.

Cellulose ethers can exhibit the phenomenon of thermoreversible gelation that strongly depends on the functionalization pattern as demonstrated for selectively

TABLE 1.11 Examples of Regioselectively Functionalized 2,3-Di-*O*-Cellulose Ethers

	R	DS	Protecting group	References
2,3-Di-*O*-methyl cellulose	—CH$_3$	Up to 1.77	Trt	Kondo and Gray (1991); Petzold-Welcke et al. (2010a)
		2.0	Trt	Kern et al. (2000); Kondo (1997)
		2.0	TDS	Koschella and Klemm (1997)
2,3-Di-*O*-ethyl cellulose	—CH$_2$—CH$_2$	Up to 1.93	Trt	Kondo and Gray (1991)
2,3-Di-*O*-CM cellulose	—CH$_2$—C(O)ONa	0.054–1.07	Trt MMTr	Liu et al. (1997)
		Up to 1.91		Heinze et al. (1994)
2,3-Di-*O*-hydroxyethyl cellulose	—(CH$_2$—CH$_2$—O)$_n$—OH	Up to MS 2.0	MMTrt	Schaller and Heinze (2005); Petzold-Welcke et al. (2010b)
2,3-Di-*O*-hydroxypropyl cellulose	—CH$_2$—CH(OH)—CH$_2$—CH$_3$	Up to MS 2.0	MMTrt	(Schaller and Heinze, 2005)
2,3-Di-*O*-allyl cellulose	—CH$_2$CH=CH$_2$	2.0	Trt	Kondo (1993, 1994)
2,3-Di-*O*-benzyl cellulose	—CH$_2$—C$_6$H$_5$	2.0	Trt	Kondo (1993, 1994)
2,3-Di-*O*-octadecyl cellulose	—(CH$_2$)$_{17}$CH$_3$	2.0	Trt	Kasai et al. (2005)

Trt: trityl, TDS: thexyldimethylsilyl, MMTrt: 4-monomethoxytrityl.

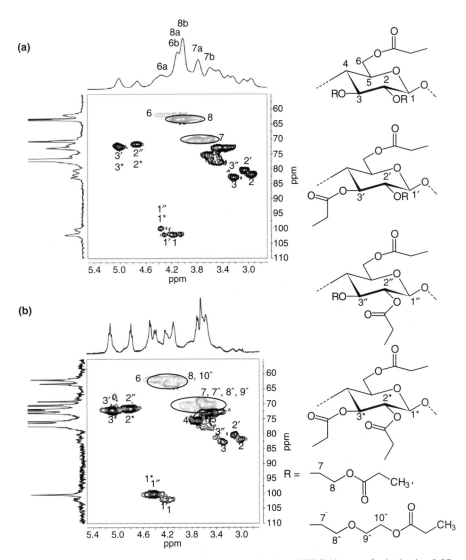

FIGURE 1.29 HSQC-DEPT spectra of (a) perpropionylated HEC (degree of substitution 0.87, synthesized with 2-(2-bromoethoxy)tetrahydropyran) and (b) perpropionylated HEC (molecular degree of substitution 0.83, synthesized with 2-bromoethanol) in CDCl3; negative signals scaled in gray. Reproduced with permission from Wiley–VCH, Petzold-Welcke et al. (2010b).

methylated cellulose (Hirrien et al., 1998). A series of 2,3-O-methyl celluloses (2,3-O-MC) are prepared starting from 6-O-trityl- and 6-O-monomethoxytrityl celluloses. Although there are differences in the total DS, it turned out that the thermal events are in strong correlation with the polymer composition (functionalization pattern) (Kern et al., 2000). In a polymer containing tri-O-methylated glucose units in combination with monomethylated ones, a distinct thermal behavior is found, that is, the methyl cellulose shows thermoreversible gelation. In contrast, a MC mainly functionalized at position 2 and 3 shows no thermal gelation. It becomes obvious that the 2,3-di-O-methyl glucose units do not significantly affect intermolecular interactions that are necessary for the gelation. The methylation of the 6-O-TDS cellulose with methyl iodide and NaH in THF leads to a very structurally uniform 2,3-O-MC, which is insoluble in water but dissolves in NMP and methanol/chloroform in contrast to a MC prepared via 6-O-trityl cellulose (Koschella and Klemm, 1997).

1.5.2 6-Mono-O-Ethers of Cellulose

Up to now, the only path to 6-O-cellulose ethers is a time-consuming synthesis, which comprises two different protecting groups (Kondo, 1993). The procedure includes the conversion of 6-O-trityl cellulose with allyl chloride in the presence of NaOH, which results in a complete functionalization at positions 2 and 3, subsequent detritylation and isomerization of the allyl groups to 2,3-O-(1-propenyl) substituents with potassium *tert*-butoxide, and alkylation at position 6 followed by the cleavage of the 1-propenyl groups at positions 2 and 3 with HCl in methanol. The investigation of blends of 6-O-MC with poly(ethyleneoxide) (PEO) and poly(vinyl alcohol) (PVA) shows that hydrogen bond engaged at position 6 of cellulose should be more favorable than that at positions 2 and 3 (Kondo et al., 1994; Shin and Kondo, 1998). Gelation of the cellulose derivatives in THF solutions does not take place for 6-O-MC and 6-O-benzyl cellulose and the trisubstituted derivatives, whereas the 2,3-O-ethers form gels (Itagaki et al., 1997). While the gylcosidic bond between two adjacent substituted units of 6-O-MC can be cleaved with *Trichoderma viride* cellulase to give oligomers with a degree of polymerization of about 8, 2,3-O-MC is not degraded (Nojiri and Kondo, 1996). Long-chain 6-O-alkyl ethers of cellulose could be transferred to ultrathin films by Langmuir–Blottget technique (Kasai et al., 2005).

1.5.3 3-Mono-O-Ethers of Cellulose

The conversion of 2,6-di-O-TDS cellulose with an excess of alkyl halides in THF in the presence of NaH affords the fully etherified polymers that can be desilylated with fluoride ions yielding 3-mono-O-functionalized cellulose ethers (Table 1.12) (Koschella et al., 2001, 2006; Petzold et al., 2004; Heinze and Koschella, 2008; Fenn and Heinze, 2009; Fenn et al., 2009; Schumann et al., 2009; Heinze et al., 2010b).

3-Mono-O-methyl cellulose swells in polar media like DMSO indicating a strong network of hydrogen bonds. It becomes soluble by addition of LiCl that destroys the interactions. Increasing the length of the alkyl chains leads to 3-mono-O-alkyl celluloses soluble in water (C_2), in aprotic dipolar solvents (up to C_5 alkyl chains), or in nonpolar solvents like THF for C_5–C_{12} alkyl chains. Light scattering investigations of 3-O-n-pentyl-, 3-O-iso-pentyl-, and 3-O-dodecyl in THF disclose a different aggregation behavior. Although 3-O-dodecyl cellulose forms molecularly dispersed solutions (at concentration less than 2 mg/L), the C_5 ethers show aggregation numbers of 6.5 (*iso*-pentyl) and 83 (*n*-pentyl) (Petzold et al., 2004).

TABLE 1.12 Solubility of 3-O-Mono-Alkyl Celluloses

$\underline{R}$	Solubility				Reference
	Ethanol	DMSO	DMA	H_2O	
$-CH_3$	−	−	−	−	Koschella et al. (2001)
$-CH_2-CH_3$	−	+	+	+	Koschella et al. (2006)
$-CH_2-CH=CH_2$	−	+	+	−	Koschella et al. (2001)
$-CH_2-CH_2-OH$	−	+	+	+	Fenn and Heinze (2009)
$-CH_2-CH_2-O-CH_3$	−	+	+	+	Heinze and Koschella (2008)
$-CH_2-CH_2-CH_2-OH$	−	+	+	+	Schumann et al. (2009)
$-CH_2-CH_2-CH_3$	+	+	+	+[a]	Heinze et al. (2010b)
$-CH_2-C\equiv CH$	−	+	−	−	Fenn et al. (2009)
$-(CH_2)_4-CH_3$	+	+	+	−	Petzold et al. (2004)
$-(CH_2)_2-CH-(CH_3)_2$	+	+	+	−	Petzold et al. (2004)
$-(CH_2)_{11}-CH_3$	−	−	−	−	Petzold et al. (2004)
$-(CH_2-CH_2-O)_n-CH_3$; $n=3, 7, 16$	−	−	−	−	Bar-Nir and Kadla (2009)

DMA: N,N-dimethyl acetamide; DMSO: dimethyl sulfoxide; soluble (+), insoluble (−).
[a] Soluble below 15°C, insoluble at room temperature.

Ethyl cellulose of different functionalization pattern possesses different temperatures for thermoreversible gelation. While ethyl cellulose with DS 0.7–1.7 already becomes water insoluble at about 30°C (Dönges, 1990), ethyl cellulose that is synthesized via induced phase separation (conversion of cellulose dissolved in DMA/LiCl with solid NaOH and ethyl iodide) possesses a distinct higher cloud point temperature of 56°C. A block-like distribution of substituents along the polymer chain is assumed as described for CMC and MC (Heinze, 1998; Liebert and Heinze, 1998). A similar value (63°C) is determined for the structurally uniform 3-mono-O-ethyl cellulose independent of DP (Sun et al., 2009). 3-Mono-O-methoxyethyl and 3-mono-O-hydroxyethyl cellulose show no thermoreversible gelation between 15°C and 95°C (Figure 1.30) (Sun et al., 2009; Fenn and Heinze, 2009).

Interestingly, 3-O-propyl cellulose shows water solubility below room temperature depending on the DS. 3-O-propyl cellulose with a DS of 1.02 has a cloud point at 15.2°C, whereas the sample with a DS of 0.71 possesses a slightly higher cloud point temperature of 23.5°C. Randomly distributed propyl celluloses with comparable DS values of 0.64 and 0.93 are insoluble in water (Heinze et al., 2010b).

One- and two-dimensional NMR spectroscopy demonstrates the uniform structure of the 3-O-alkyl ethers after peracetylation of the remaining OH groups, as shown in Figure 1.31, for 3-O-methoxyethyl-2,6-di-O-acetyl cellulose as a typical example (Heinze and Koschella, 2008).

The synthesis of 3-mono-O-hydroxyethyl cellulose is realized by conversion of the 2,6-TDS cellulose with 2-(2-bromoethoxy)tetrahydropyran via a completely functionalized derivative. The complete removal of the protecting group first involves

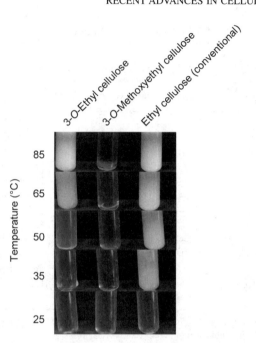

FIGURE 1.30 Photographs of cloud point of 3-*O*-ethyl cellulose and conventional ethyl cellulose compared with 3-*O*-methyoxyethyl cellulose. Note the slight banding for 3-*O*-ethyl cellulose and ethyl cellulose held at 85°C. Reproduced with permission from Wiley–VCH, Sun et al. (2009).

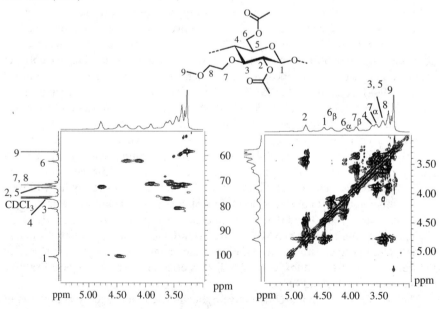

FIGURE 1.31 HMQC (left) and COSY (right) NMR spectra of peracetylated 3-*O*-methoxy-ethyl celluloses (CDCl$_3$). Reproduced with permission from Elsevier, Heinze and Koschella (2008).

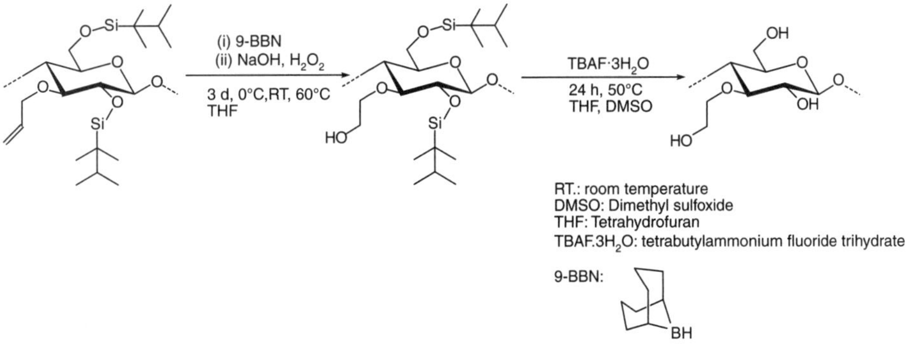

FIGURE 1.32 Reaction scheme for the preparation of 3-mono-O-($3'$-hydoxypropyl) cellulose. Adapted from Schumann et al. (2009).

the split off of the TDS function with tetrabutylammonium fluoride trihydrate and subsequently of the tetrahydropyran moieties with hydrochloric acid (Fenn and Heinze, 2009).

The conversion of the double bond of 3-mono-O-allyl-2,6-di-O-TDS cellulose with 9-borabicycl(3.3.1)nonane and subsequent alkaline oxidation lead to the $3'$-hydroxypropyl group. The treatment with tetrabutylammonium fluoride yields regioselectively functionalized 3-mono-O-($3'$-hydroxypropyl) cellulose (Figure 1.32) (Schumann et al., 2009).

FTIR spectroscopy of 3-mono-O-methyl cellulose in combination with curve fitting and deconvolution shows that the resulting two main bands indicate the existence of another intramolecular hydrogen bond between OH-2 and OH-6 instead of intramolecular hydrogen bonds between OH-3 and OH-5 (Figure 1.33) (Kondo et al., 2008). The large deconvoluted band at $3340\,cm^{-1}$ refers to strong interchain hydrogen bonds involving the hydroxyl groups at C-6. The crystallinity of 54% calculated from the WAXD also supports the dependency of the usually observed crystallization in cellulose of the hydroxyl groups at position 6 to engage in interchain hydrogen bonding.

1.5.4 2,6-Di-O-Ethers of Cellulose

The synthesis of 2,6-di-O-methyl cellulose involves the complete allylation of 2,6-dithexyldimethylsilyl cellulose followed by the desilylation with tetrabutyl ammonium fluoride to get pure 3-mono-O-allyl cellulose. The product is methylated using methyl iodide in the presence of sodium hydride to get 3-mono-O-allyl-2,6-di-O-methyl cellulose (Figure 1.34) (Kamitakahara et al., 2008).

Deallylation proceeds with palladium(II) chloride to give 2,6-di-O-methyl cellulose, which is not isolated after the deprotection step. It is acetylated with acetic anhydride, N,N-dimethylaminopyridine, and pyridine leading to 3-mono-O-acetyl-2,6-di-O-methyl cellulose. ^{1}H and ^{13}C NMR spectra of 3-mono-O-acetyl-2,6-di-O-methyl celluloses indicate complete desilylation at positions 2 and 6 and acetylation

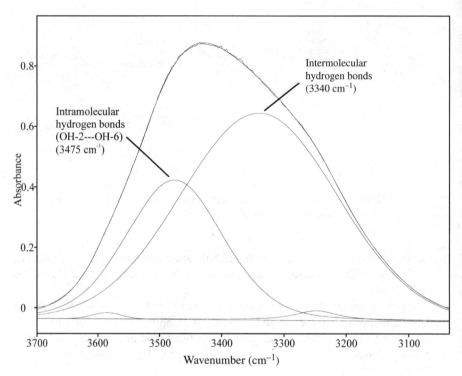

FIGURE 1.33 Curve fitting and peak assignments for the OH stretching region in 3-mono-*O*-methyl cellulose. Adapted from Kondo et al. (2008).

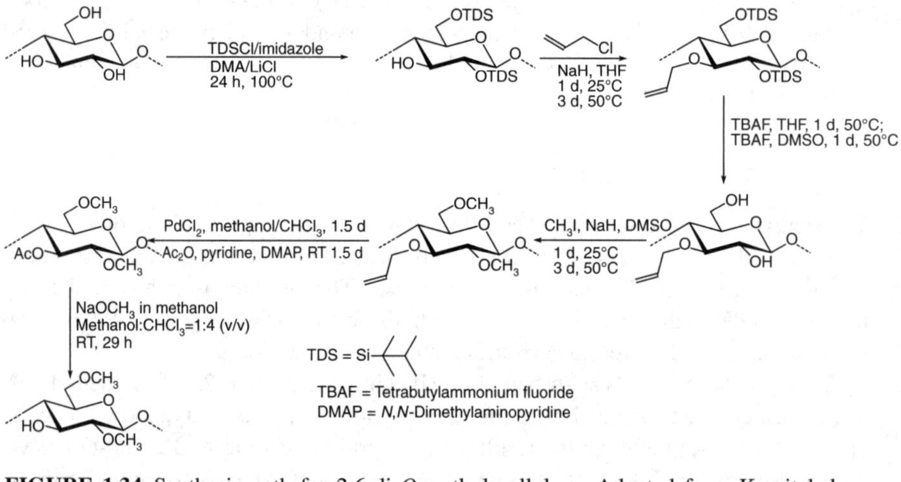

FIGURE 1.34 Synthesis path for 2,6-di-*O*-methyl cellulose. Adapted from Kamitakahara et al. (2008).

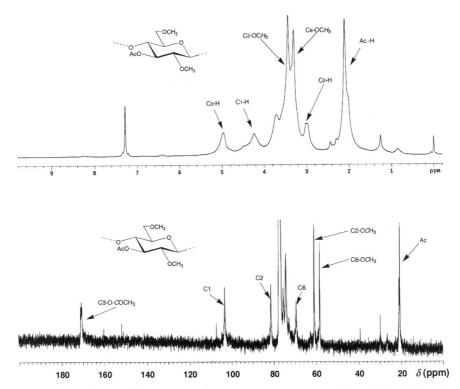

FIGURE 1.35 ^{1}H and ^{13}C NMR spectra of 3-mono-O-acetyl-2,6-di-O-methyl celluloses in CDCl$_3$. Adapted from Kamitakahara et al. (2008).

at position 3 by the low filed shift of the signal at 4.96 ppm (Figure 1.35). After deacetylation, 2,6-di-O-methyl cellulose is obtained, which is insoluble in all solvents tested.

1.6 SUMMARY

This chapter highlights unconventionally functionalized cellulose derivatives obtained by advanced synthesis paths like nucleophilic displacement (S$_N$) reactions, homogeneous sulfation in ionic liquids, and regioselective functionalization. A short overview about "classical" S$_N$ reactions is also given. Moreover, the review is focused on novel cellulose products obtained by click chemistry (copper-catalyzed Huisgen reaction), starting from deoxyazido cellulose yielding derivatives with methylcarboxylate-, 2-aniline-, 3-thiophene moieties, for example, and new selectively dendronized cellulose-based materials. Structure characterization and selected applications are also briefly reviewed. Biofunctionalized surfaces based on dendronized cellulose are prepared. The well-defined cellulose sulfates synthesized in ionic liquids may mimic the structure and biological activity of naturally occurring polysaccharide sulfates and are applied for microencapsulation of biological material in polyelectrolyte complex capsules. Regioselective derivatization of protected cellulosics leading to 3-*O*-, 2,3-*O*-, 6-*O*-, and 2,6-*O*-functionalized products is of recent interest because the products possess remarkable differences in properties compared to common cellulose derivatives and are hence important products for the establishment of "real" structure–property relationships. Moreover, these regioselectively functionalized cellulose derivatives are useful compounds to calibrate analytical techniques and for other research and application issues.

1.7 CONCLUSION AND FUTURE PERSPECTIVE

The examples discussed illustrate the enormous structural diversity and application potential of cellulose derivatives. Starting from cellulose sulfonates especially cellulose tosylate, SN reactions undoubtedly will lead to a variety of sophisticated products. Moreover, Huisgen reaction with cellulose is already successfully realized and new dendronized cellulose derivatives will appear. For instance, biofunctionalized surfaces based on dendronized cellulose are prepared either by embedding of dendritic 6-deoxy-6-(1,2,3-triazolo)-4-polyamidoamine cellulose (degree of substitution 0.25), obtained by homogeneous conversion of 6-deoxy-6-azido cellulose with propargyl-PAMAM dendron via the copper-catalyzed Huisgen reaction in a cellulose acetate (DS 2.50) matrix, or by the heterogeneous functionalization of deoxyazido cellulose film with the dendron. Cellulose sulfates with well-defined chemical structures can mimic the structure and biological activity of naturally occurring polysaccharide sulfates. Cellulose sulfates are applied for microencapsulation of biological material in polyelectrolyte complex capsules. Using ionic liquids as solvent, one-pot procedure for sulfation, *in situ* polyelectrolyte complex formation, and encapsulation are established. In the field of cellulose ethers, the regioselective introduction of the functional groups is still a synthetic challenge. These products will give new insights into the interaction of cellulose derivatives with each other, with other polymers and surfaces, and hence improve common applications and introduce cellulose-based materials in new application fields.

ACKNOWLEDGMENTS

This work was supported in part by the Finnish Funding Agency for Technology and Innovation (Tekes) and Åbo Akademi within the FiDiPro program.

REFERENCES

Altaner, C., B. Saake, and J. Puls (2003). Specificity of an *Aspergillus niger* esterase deacetylating cellulose acetate. *Cellulose* **10**:85–95.

Arai, K., and F. Aoki (1994). Preparation and identification of sodium deoxycellulosesulfonate. *Sen'i Gakkaishi* **50**:510–514.

Arai, K., and N. Yoda (1998). Preparation of water-soluble sodium deoxycellulosesulfonate from homogeneously prepared tosylcellulose. *Cellulose* **5**:51–58.

Balser, K., L. Hoppe, T. Eicher, M. Wandel, and H.-J. Astheimer (1986). Cellulose esters. In: W. Gerhartz, Y. S. Yamamoto, F. T. Campbell, R. Pfefferkorn, and J. F. Rounsaville, editors, *Ullmanns's Encyclopedia of Industrial Chemistry*, 5th ed., Vol. A5 Weinheim: VCH, p. 419.

Bar-Nir, B. B.-A., and J. F. Kadla (2009). Synthesis and structural characterization of 3-*O*-ethylene glycol functionalized cellulose derivatives. *Carbohydr. Polym.* **76**:60–67.

Becher, J., H. Liebegott, P. Berlin, and D. Klemm (2004). Novel xylylene diaminocellulose derivatives for enzyme immobilization. *Cellulose* **11**:119–126.

Belyakova, M. K., L. S. Gal'braikh, and Z. A. Rogovin (1971). Synthesis of cellulose stereoisomers by nucleophilic substitution reactions. *Cell. Chem. Technol.* **5**:405–415.

Berlin, P., D. Klemm, J. Tiller, and R. Rieseler (2000). A novel soluble aminocellulose derivative type: its transparent film-forming properties and its efficient coupling with enzyme proteins for biosensors. *Macromol. Chem. Phys.* **201**:2070–2082.

Berlin, P., D. Klemm, A. Jung, H. Liebegott, R. Rieseler, and J. Tiller (2003). Film-forming aminocellulose derivatives as enzyme-compatible support matrices for biosensor developments. *Cellulose* **10**:343–367.

Brandt, L. (1986). Cellulose ethers. In: W. Gerhartz, Y. S. Yamamoto, F. T. Campbell, R. Pfefferkorn, and J. F. Rounsaville, editors, *Ullmann's Encyclopedia of Industrial Chemistry*, 5th ed., Vol. A5 Weinheim: VCH, p. 461.

Camacho, J. A. G., U. W. Erler, D. O. Klemm (1996). 4-Methoxy substituted trityl groups in 6-*O* protection of cellulose: homogeneous synthesis, characterization, detritylation. *Macromol. Chem. Phys.* **197**:953–964.

Dautzenberg, H., B. Lukanoff, U. Eckert, B. Tiersch, and U. Schuldt (1993). Immobilisation of biological matter by polyelectrolyte complex formation. *Ber. Bunsenges. Phys. Chem.* **100**:1045–1053.

Dautzenberg, H., U. Schuldt, G. Grasnick, P. Karle, P. Müller, M. Löhr, M. Pelegrin, M. Piechaczyk, K. V. Rombs, W. H. Gunzburg, B. Salmons, and R. M. Saller (1999). Development of cellulose sulfate-based polyelectrolyte complex microcapsules for medical applications. *Ann. NY Acad. Sci.* **875**:46–63.

Decher, G. (1996). Layered nanoarchitectures via directed assembly of anionic and cationic molecules. In: J. P. Sauvage and M. W. Hosseini, editors, *Comprehensive Supramolecular Chemistry*, Vol. 9, Oxford: Pergamon Press, pp. 507–528.

Deus, C., H. Friebolin, and E. Siefert (1991). Partially acetylated cellulose: synthesis and determination of the substituent distribution via proton NMR spectroscopy. *Makromol. Chem.*, **192**:75–83.

Dönges, R. (1990). Nonionic cellulose ethers. *Br. Polym. J.* **23**:315–326.

El Seoud, O., A. Koschella, L. C. Fidale, S. Dorn, and T. Heinze (2007). Applications of ionic liquids in carbohydrate chemistry: a window of opportunities. *Biomacromolecules* **8**:2629–2647.

Fenn, D., and T. Heinze (2009). Novel 3-mono-*O*-hydroxyethyl cellulose: synthesis and structure characterization. *Cellulose* **16**:853–861.

Fenn, D., A. Pfeifer, and T. Heinze (2007). Studies on the synthesis of 2,6-di-*O*-thexyl-dimethylsilyl cellulose. *Cell. Chem. Technol.* **41**:87–91.

Fenn, D., M. Pohl, T. Heinze (2009). Novel 3-*O*-propargyl cellulose as a precursor for regioselective functionalization of cellulose. *React. Funct. Polym.* **69**:347–352.

Fluri, D. A., C. Kemmer, M. Daoud-El Baba, and M. Fussenger (2008). A novel system for trigger-controlled drug release from polymer capsules. *J. Control Release* **131**:211–219.

Fransson, L.-A. (1985). Mammalian glycosaminoglycans. In: G. O Aspinall,editor, *The Polysaccharides*, New York: Academic Press, pp. 337–415.

Gericke, M., T. Liebert, and T. Heinze (2009a). Interaction of ionic liquids with polysaccharides-8: synthesis of cellulose sulfates suitable for polyelectrolyte complex formation. *Macromol. Biosci.* **9**:343–353.

Gericke, M., K. Schlufter, T. Liebert, T. Heinze, and T. Budtova (2009b). Rheological properties of cellulose/ionic liquid solutions: from dilute to concentrated states. *Biomacromolecules* **10**:1188–1194.

Gericke, M., T. Liebert, and T. Heinze (2009c). Polyelectrolyte synthesis and *in situ* complex formation in ionic liquids. *J. Am. Chem. Soc.* **131**:13220–13221.

Hagiwara, I., N. Shiraishi, and T. Yokota (1981). Homogeneous tritylation of cellulose in a sulphur dioxide-diethylamine-dimethylsulfoxide medium. *J. Wood Chem. Technol.* **1**:93–109.

Hanefeld, U., L. Gardossi, and E. Magner (2009). Understanding enzyme immobilization. *Chem. Soc. Rev.* **38**:453–468.

Harkness, B. R., and D. G. Gray (1990). Preparation and chiroptical properties of tritylated cellulose derivatives. *Macromolecules* **23**:1452–1457.

Hassan, M. L., C. N. Moorefield, and G. R. Newkome (2004). Regioselective dendritic functionalization of cellulose. *Macromol. Rapid Commun.* **25**:1999–2002.

Hassan, M. L., C. N. Moorefield, K. K. Kotta, and G. R. Newkome (2005). Regioselective combinatorial-type synthesis, characterization, and physical properties of dendronized cellulose. *Polymer* **46**:8947–8955.

Heinze, T. (1998). New ionic polymers by cellulose functionalization. *Macromol. Chem. Phys.* **199**:2341–2364.

Heinze, T. (2004). Chemical functionalization of cellulose. In: S. Dumitriu,editor, *Polysaccharides: Structural Diversity and Functional Versatility*, 2nd ed., New York: Marcel Dekker, pp. 551–590.

Heinze, T., and T. Liebert (2001). Unconventional methods in cellulose functionalization. *Prog. Polym. Sci.* **26**:1689–1762.

Heinze, T., and A. Koschella (2008). Water-soluble 3-*O*-methoxyethyl cellulose: synthesis and characterization. *Carbohydr. Res.* **343**:668–673.

Heinze, T., and K. Petzold (2008). Cellulose chemistry: novel products and synthesis paths. In: M. N. Belgacem and A. Gandini,editors, *Monomers, Oligomers, Polymers and Composites from Renewable Resources*, Oxford: Elsevier, pp. 343–368.

Heinze, T., K. Röttig, and I. Nehls (1994). Synthesis of 2,3-*O*-carboxymethylcellulose. *Macromol. Rapid Commun.* **15**:311–317.

Heinze, T., A. Koschella, L. Magdaleno-Maiza, and A. S. Ulrich (2001). Nucleophilic displacement reactions on tosyl cellulose by chiral amines. *Polym. Bull.* **46**:7–13.

Heinze, T., T. Liebert, and A. Koschella (2006a). *Esterification of Polysaccharides*, Heidelberg: Springer, p. 117.

Heinze, T., A. Koschella, M. Brackhagen, J. Engelhardt, and K. Nachtkamp (2006b). Studies on non-natural deoxyammonium cellulose. *Macromol. Symp.* **244**:74–82.

Heinze, T., T. Liebert, B. Heublein, S. Hornig (2006c). Functional polymers based on dextran. *Adv. Polym. Sci.* **205**:199–291.

Heinze, T., M. Pohl, J. Schaller, and F. Meister (2007). Novel bulky esters of cellulose. *Macromol. Biosci.* **7**:1225–1231.

Heinze, T., M. Schöbitz, M. Pohl, and F. Meister (2008a). Interactions of ionic liquids with polysaccharides: IV. Dendronization of 6-azido-6-deoxy cellulose. *J. Polym. Sci. A* **46**:3853–3859.

Heinze, T., A. Pfeifer, and K. Petzold (2008b). Regioselective reaction of cellulose with *tert*-butyldimethylsilyl chloride in *N,N*-dimethyl acetamide/LiCl. *BioResources* **3**:79–90.

Heinze, T., S. Daus, M. Gericke, and T. Liebert (2010a). Semi-synthetic sulfated polysaccharides: promising materials for biomedical applications and supramolecular architecture. In: A. Tiwari,(ediror), *Polysaccharides: Development, Properties and Applications,* Hauppauge: Nova Science Publishers, pp. 213–259.

Heinze, T., A. Pfeifer, V. Sarbova, and A. Koschella (2010b). 3-*O*-Propyl cellulose: cellulose ether with exceptionally low flocculation temperature. *Polym. Bull.* **66**(9): 1219–1229.

Hettrich, K., W. Wagenknecht, B. Volkert, and S. Fischer (2008). New possibilities of the acetosulfation of cellulose. *Macromol. Symp.* **262**:162–169.

Hirrien, M., C. Chevillard, J. Desbrieres, M. A. V. Axelos, and M. Rinaudo (1998). Thermal gelation of methyl celluloses: new evidence for understanding the gelation mechanism. *Polymer* **39**:6251–6259.

Hon, D. N.-S. (1996). Chemical modification of cellulose. In: D. N.-S. Hon, *Chemical Modifications of Ligno-Cellulosic Materials,* New York: Marcel Dekker, pp. 97–127.

Itagaki, H., M. Tokai, and T. Kondo (1997). Physical gelation process for cellulose whose hydroxyl groups are regioselectively substituted by fluorescent groups. *Polymer* **38**:4201–4205.

Jung, A., and P. Berlin (2005). New water-soluble and film-forming aminocellulose tosylates as enzyme support matrices with Cu^{2+}-chelating properties. *Cellulose* **12**:67–84.

Kamitakahara, H., A. Koschella, Y. Mikawa, F. Nakatsubo, T. Heinze, and D. Klemm (2008). Syntheses and comparison of 2,6-di-*O*-methyl celluloses from natural and synthetic celluloses. *Macromol. Biosci.* **8**:690–700.

Kasai, W., S. Kuga, J. Magoshi, and T. Kondo (2005). Compression behaviour of Langmuir–Blodgett monolayers of regioselectively substituted cellulose ethers with long alkyl side chains. *Langmuir* **21**:2323–2329.

Kasuya, N., and A. Sawatari (2000). A simple and facile method for triphenylmethylation of cellulose in a solution of lithium chloride in *N,N*-dimethylacetamide. *Sen'i Gakkaishi* **56**:249–253.

Kern, H., S. W. Choi, G. Wenz, J. Heinrich, L. Ehrhardt, P. Mischnik, P. Garidel, and A. Blume (2000). Synthesis, control of substitution pattern and phase transitions of 2,3-di-*O*-methylcellulose. *Carbohydr. Res.* **326**:67–79.

Khan, F. Z., M. Shiotsuki, Y. Nishio, T. Masuda (2007). Synthesis and properties of amidoimide dendrons and dendronized cellulose derivatives. *Macromolecules* **40**:9293–9303.

Klemm, D. (1998). Regiocontrol in cellulose chemistry: principles and examples of etherification and esterification. In: T. J. Heinze and W. G. Glasser,editors, *Cellulose Derivatives: Modification, Characterization, and Nanostructures,* ACS Symposium Series, Vol. 688, American Chemical Society, pp. 19–37.

Klemm, D., B. Philipp, T. Heinze, U. Heinze, and W. Wagenknecht (1998a). Cellulose sulfates. In: *Comprehensive Cellulose Chemistry*, 1st ed., Vol. 2, Weinheim: Wiley–VCH, pp. 115–133.

Klemm, D., B. Philipp, T. Heinze, U. Heinze, and W. Wagenknecht (1998b). *Comprehensive Cellulose Chemistry*, 1st ed., Vols 1–2, Weinheim: Wiley–VCH.

Klemm, D., B. Heublein, H.-P. Fink, and A. Bohn (2005). Cellulose: fascinating biopolymer and sustainable raw material. *Angew. Chem., Int. Ed.* **44**:3358–3393.

Knaus, S., U. Mais, and W. H. Binder (2003). Synthesis, characterization and properties of methylaminocellulose. *Cellulose* **10**:139–150.

Köhler, S., T. Liebert, M. Schöbitz, J. Schaller, F. Meister, W. Günther, and T. Heinze (2007). Interactions of ionic liquids with polysaccharides-1: unexpected acetylation of cellulose with 1-ethyl-3-methylimidazolium acetate. *Macromol. Rapid Commun.* **28**:2311–2317.

Köhler, S., T. Liebert, and T. Heinze (2008). Interactions of ionic liquids with polysaccharides-6: Pure cellulose nanoparticles from trimethylsilyl cellulose prepared in ionic liquids. *J. Polym. Sci. A* **46**:4070–4080.

Kolb, H. C., M. G. Finn, and K. B. Sharpless (2001). Click chemistry: diverse chemical function from a few good reactions. *Angew. Chem., Int. Ed.* **113**:2004–2021.

Kondo, T. (1993). Preparation of 6-O-alkylcelluloses. *Carbohydr. Res.* **238**:231–240.

Kondo, T. (1994). Hydrogen bonds in regioselectively substituted cellulose derivatives. *J. Polym. Sci. B* **32**:1229–1236.

Kondo, T. (1997). The relationship between intramolecular hydrogen bonds and certain physical properties of regioselectively substituted cellulose derivatives. *J. Polym. Sci. B* **35**:717–723.

Kondo, T., and D. G. Gray (1991). The preparation of O-methyl- and O-ethyl-cellulose having controlled distribution of substituents. *Carbohydr. Res.* **220**:173–183.

Kondo, T., C. Sawatari, R. S. Manley, and D. G. Gray (1994). Characterization of hydrogen-bonding in cellulose synthetic polymer blend systems with regioselectively substituted methylcellulose. *Macromolecules* **27**:210–215.

Kondo, T., A. Koschella, B. Heublein, D. Klemm, and T. Heinze (2008). Hydrogen bond formation in regioselectively functionalized 3-mono-O-methyl cellulose. *Carbohydr. Res.* **343**:2600–2604.

Koschella, A., and T. Heinze (2001). Novel regioselectively 6-functionalized cationic cellulose polyelectrolytes prepared via cellulose sulfonates. *Macromol. Biosci.* **1**:178–184.

Koschella, A., and T. Heinze (2003). Unconventional cellulose products by fluorination of tosyl cellulose. *Macromol. Symp.* **197**:243–254.

Koschella, A., and D. Klemm (1997). Silylation of cellulose regiocontrolled by bulky reagents and dispersity in the reaction media. *Macromol. Symp.* **120**:115–125.

Koschella, A., T. Heinze, and D. Klemm (2001). First synthesis of 3-O-functionalized cellulose ethers via 2,6-di-O-protected silyl cellulose. *Macromol. Biosci.* **1**:49–54.

Koschella, A., D. Fenn, and T. Heinze (2006). Water soluble 3-mono-O-ethyl cellulose: synthesis and characterization. *Polym. Bull.* **57**:33–41.

Koschella, A., M. Richter, T. Heinze (2010). Novel cellulose-based polyelectrolytes synthesized via the click reaction. *Carbohydr. Res.* **345**:1028–1033.

Lewis, W. G., L. G. Green, F. Grynszpan, Z. Radic, P. R. Carlier, P. Taylor, M. G. Finn, and K. B. Sharpless (2002). Click chemistry *in situ*: acetylcholinesterase as a reaction vessel for the selective assembly of a femtomolar inhibitor from an array of building blocks. *Angew. Chem., Int. Ed.* **41**:1053–1057.

Li, M.-M., and S.-J. Yao (2009). Preparation of polyelectrolyte complex membranes based on sodium cellulose sulfate and poly(dimethyldiallylammonium chloride) and its permeability properties. *J. Appl. Polym. Sci.* **112**:402–409.

Liebert, T., and T. Heinze (1998). Induced phase separation: a new synthesis concept in cellulose chemistry. In: T. J. Heinze and W. G. Glasser,editors, *Cellulose Derivatives: Modification, Characterisation, and Nanostructures*, ACS Symposium Series, Vol. 688, American Chemical Society, pp. 61–72.

Liebert, T., C. Hänsch, and T. Heinze (2006). Click chemistry with polysaccharides. *Macromol. Rapid Commun.* **27**:208–213.

Liebert, T., J. Wotschadlo, M. Gericke, S. Köhler, P. Laudeley, and T. Heinze (2009). Modification of cellulose in ionic liquids towards biomedical applications. *ACS Symp. Ser.* **1017**:115–132.

Liu, C., and H. Baumann (2002). Exclusive and complete introduction of amino groups and their *N*-sulfo and *N*-carboxymethyl groups into the 6-position of cellulose without the use of protecting groups. *Carbohydr. Res.* **337**:1297–1307.

Liu, H.-Q., L.-N. Zhang, A. Takaragi, and T. Miyamoto (1997). Water solubility of regioselectively 2,3-*O*-substituted carboxymethylcellulose. *Macromol. Rapid Commun.* **18**:921–925.

Lukanoff, B., and H. Dautzenberg (1994). Sodium cellulose sulfate as a component in the creation of microcapsules through forming polyelectrolyte complexes: 1. Heterogeneous sulfation of cellulose by the use of sulfuric-acid propanol as reaction medium and sulfating agent. *Das Papier* **6**:287–298.

McCormick, C. L., and P. A. Callais (1986). Derivatization of cellulose in lithium chloride and *N,N*-dimethylacetamide solutions. *Polymer Prepr.* **27**:9192.

McCormick, C. L., and P. A. Callais (1987). Derivatization of cellulose in lithium-chloride and *N,N*-dimethylacetamide solutions. *Polymer* **28**:2317–2323.

McCormick, C. L., T. R. Dawsey, and J. K. Newman (1990). Competitive formation of cellulose *p*-toluenesulfonate and chlorodeoxycellulose during homogeneous reaction of *p*-toluenesulfonyl chloride with cellulose in *N,N*-dimethylacetamide-lithium chloride. *Carbohydr. Res.* **208**:183–191.

Milligan, B., and J. M. Swan (1962). Bunte salts ($RSSO_3Na$). *Rev. Pure Appl. Chem.* **12**:72–94.

Mischnik, P., M. Lange, M. Gohdes, A. Stein, and K. Petzold (1995). Trialkylsilyl derivatives of cyclomaltoheptaose, cellulose, and amylose: rearrangement during methylation analysis. *Carbohydr. Res.* **277**:179–187.

Mormann, M., and J. Demeter (1999). Silylation of cellulose with hexamethyldisilazane in liquid ammonia: first examples of completely trimethylsilylated cellulose. *Macromolecules* **32**:1706–1710.

Nakano, T., W. T. Dixon, and L. Ozimek (2002). Proteoglycans (glucosaminoglycans/mucopolysaccharides). In: S. De Baets, E. J. Vandamme, and A. Steinbüchel,editors, *Biopolymers Polysaccharides II*, Vol. 6, Weinheim: Wiley–VCH, pp. 575–604.

Nojiri, M., and T. Kondo (1996). Application of regioselectively substituted methylcelluloses to characterize the reaction mechanism of cellulase. *Macromolecules* **29**:2392–2395.

Oestmark, E., J. Lindqvist, D. Nystroem, E. Malmstroem (2007). Dendronized hydroxypropyl cellulose: synthesis and characterization of biobased nanoobjects. *Biomacromolecules* **8**:3815–3822.

Pawlowski, W. P., R. D. Gilbert, R. E. Fornes, and S. T. Purrington (1988). The liquid-crystalline properties of selected cellulose derivatives. *J. Polym. Sci. B* **26**:1101–1110.

Pelegrin, M., M. Marin, D. Noel, M. Del Rio, R. Saller, J. Stange, S. Mitzner, W. H. Gunzburg, and M. Piechaczyk (1998). Systemic long-term delivery of antibodies in immunocompetent animals using cellulose sulfate capsules containing antibody-producing cells. *Gene Ther.* **5**:828–834.

Petzold, K., A. Koschella, D. Klemm, and B. Heublein (2003). Silylation of cellulose and starch: selectivity, structure analysis, and subsequent reactions. *Cellulose* **10**:251–269.

Petzold, K., D. Klemm, B. Heublein, W. Burchard, and G. Savin (2004). Investigations on structure of regioselectively functionalized celluloses in solution exemplified by using 3-*O*-alkyl ethers and light scattering. *Cellulose* **11**:177–193.

Petzold-Welcke, K., M. Kötteritzsch, and T. Heinze (2010a). 2,3-*O*-Methyl cellulose: studies on synthesis and structure characterization. *Cellulose* **17**:449–457.

Petzold-Welcke, K., M. Kötteritzsch, D. Fenn, A. Koschella, and T. Heinze (2010b). Study on synthesis and NMR characterization of 2,3-*O*-hydroxyethyl cellulose depending on synthesis conditions. *Macromol. Symp.* **294**:133–140.

Philipp, B., W. Wagenknecht, I. Nehls, J. Ludwig, M. Schnabelrauch, H. R. Kim, and D. Klemm (1990). Comparison of cellulose formate and cellulose acetate under homogeneous reaction conditions. *Cellulose Chem. Technol.* **24**:667–678.

Pohl, M., and T. Heinze (2008). Novel biopolymer structures synthesized by dendronization of 6-deoxy-6-aminopropargyl cellulose. *Macromol. Rapid Commun.* **29**:1739–1745.

Pohl, M., J. Schaller, F. Meister, and T. Heinze (2008a). Novel bulky esters of biopolymers: dendritic cellulose. *Macromol. Symp.* **262**:119–128.

Pohl, M., J. Schaller, F. Meister, and T. Heinze (2008b). Selectively dendronized cellulose: synthesis and characterization. *Macromol. Rapid Commun.* **29**:142–148.

Pohl, M., G. A. Morris, S. E. Harding, and T. Heinze (2009a). Studies on the molecular flexibility of novel dendronized carboxymethyl cellulose derivatives. *Eur. Polym. J.* **45**:1098–1110.

Pohl, M., N. Michaelis, F. Meister, and T. Heinze (2009b). Biofunctional surfaces on dendronized cellulose. *Biomacromolecules* **10**:382–389.

Rahn, K., M. Diamantoglou, D. Klemm, H. Berghmans, and T. Heinze (1996). Homogeneous synthesis of cellulose *p*-toluenesulfonates in *N,N*-dimethylacetamide/LiCl solvent system. *Angew. Makromol. Chem.* **238**:143–163.

Renken, A., and D. Hunkeler (2007a). Polyvinylamine-based capsules: a mechanistic study of the formation using alginate and cellulose sulphate. *J. Microencapsul.* **24**:323–336.

Renken, A., and D. Hunkeler (2007b). Polymethylene-*co*-guadinine based capsules: a mechanistic study of the formation using alginate and cellulose sulphate. *J. Microencapsul.* **24**:20–39.

Richter, A., and D. Klemm (2003). Regioselective sulfation of trimethylsilyl cellulose using different SO_3-complexes. *Cellulose* **10**:133–138.

Rostovtsev, V. V., L. G. Green, V. V. Fokin, and K. B. Sharpless (2002). A stepwise Huisgen cycloaddition process: copper(I)-catalyzed regioselective "ligation" of azides and terminal alkynes. *Angew. Chem., Int. Ed.* **114**:2708–2711.

Schaffelner, S., V. Stadlbauer, P. Stiegler, O. Hauser, G. Halwachs, C. Lackner, F. Iberer, and K. H. Tscheliessnigg (2005). Porcine islet cells microencapsulated in sodium cellulose sulphate. *Transplant. P.* **37**:248–252.

Schaller, J., and T. Heinze (2005). Studies on the synthesis of 2,3-*O*-hydroxyalkyl ethers of cellulose. *Macromol. Biosci.* **5**:58–63.

Schöbitz, M., F. Meister, and T. Heinze (2009). Unconventional reactivity of cellulose dissolved in ionic liquids. *Macromol. Symp.* **280**:102–111.

Schumann, K., A. Pfeifer, and T. Heinze (2009). Novel cellulose ethers: synthesis and structure characterization of 3-mono-*O*-(3′-hydroxypropyl) cellulose. *Macromol. Symp.* **280**:86–94.

Schweiger, R. G. (1972). The conformation of amylose in alkaline salt solution. *Carbohydr. Res.* **21**:219–228.

Schweiger, R. G. (1974). Anhydrous solvent systems for cellulose processing. *TAPPI J.* **57**:86–90.

Shin, J.-H., and T. Kondo (1998). Cellulosic blends with poly(acrylonitrile): characterization of hydrogen bonds using regioselectively methylated cellulose derivatives. *Polymer* **39**:6899–6904.

Siegmund, G., and D. Klemm (2002). Cellulose sulfonates: preparation, properties, subsequent reactions. *Polym. News* **27**:84–90.

Stiegler, P., V. Stadlbauer, S. Schaffelner, G. Halwachs, C. Lackner, O Hauser, F. Iberer, and K. H. Tscheliessnigg (2006). Cryopreservation of insulin-producing cells microencapsulated in sodium cellulose sulfate. *Transplant P.* **38**:3026–3030.

Sun, S., T. J. Foster, W. MacNaughtan, J. R. Mitchell, D. Fenn, A. Koschella, and T. Heinze (2009). Self-association of cellulose ethers with random and regioselective distribution of substitution. *J. Polym. Sci. B* **47**:1743–1752.

Swatloski, R. P., S. K. Spear, J. D. Holbrey, and R. D. Rogers (2002). Dissolution of cellulose with ionic liquids. *J. Am. Chem. Soc.* **124**:4974–4975.

Tiller, J., P. Berlin, and D. Klemm (1999). Soluble and film-forming cellulose derivatives with redox-chromogenic and enzyme immobilizing 1,4-phenylenediamine groups. *Macromol. Chem. Phys.* **200**:1–9.

Tiller, J., P. Berlin, and D. Klemm (2000). Novel matrices for biosensor application by structural design of redox-chromogenic aminocellulose esters. *J. Appl. Polym. Sci.* **75**:904–915.

Vikartovská, A., M. Bučko, D. Mislovičcová, V. Potoprstý, I. Lacík, and P. Gemeiner (2007). Improvement of the stability of glucose oxidase via encapsulation in sodium alginate-cellulose sulfate-poly(methylene-*co*-guanidine) capsules. *Enzyme Microb. Technol.* **41**:748–755.

Vögtle F., G. Richardt, and N. Werner (2007). *Dendritische Moleküle, Konzepte, Synthesen, Eigenschaften und Anwendungen*, Wiesbaden: Teubner Studienbücher Chemie.

Wagenknecht, W. (1996). Regioselectively substituted cellulose derivatives by modification of commercial cellulose acetates. *Das Papier* **12**:712–720.

Wagenknecht, W., B. Phillipp, and M. Keck (1985). Zur Acylierung von Cellulose nach Auflösung in *O*-basischen Lösemittelsysteme. *Acta Polym.* **36**:697–698.

Wagenknecht, W., I. Nehls, A. Stein, D. Klemm, and B. Philipp (1992). Synthesis and substituent distribution of sodium cellulose sulfates via trimethylsilyl cellulose as intermediate. *Acta Polym.* **43**:266–269.

Wagenknecht, W., I. Nehls, and B. Philipp (1993). Studies on the regioselectivity of cellulose sulfation in a nitrogen oxide (N_2O_4)-*N,N*-dimethylformamide-cellulose system. *Carbohydr. Res.* **240**:245–252.

Wang, T., I. Lacik, M. Brissova, A.V. Anilkumar, A. Prokop, D. Hunkeler, R. Green, K. Shahrokhi, and A. C. Powers (1997). An encapsulation system for the immunoisolation of pancreatic islets. *Nat. Biotechnol.* **15**:358–362.

Wang, Z.-M., L. Li, K.-J. Xiao, and J.-Y. Wu (2009). Homogeneous sulfation of bagasse cellulose in an ionic liquid and anticoagulation activity. *Bioresour. Technol.* **100**:1687–1690.

Wenz, G., P. Liepold, and N. Bordeanu (2005). Synthesis and SAM formation of water soluble functional carboxymethylcelluloses: thiosulfates and thioethers. *Cellulose* **12**, 85–96.

Xie, Y.-L., M.-J. Ming, and S.-J. Yao (2009). Preparation and characterization of biocompatible microcapsules of sodium cellulose sulfate/chitosan by means of layer-by-layer self-assembly. *Langmiur* **25**:8999–9005.

Yao, S. (2000). An improved process for the preparation of sodium cellulose sulfate, *Chem. Eng. J.* **78**:199–204.

Yue, Z., and J. M. G. Cowie (2002). Preparation and chiroptical properties of a regioselectively substituted cellulose ether with PEO side chains. *Macromolecules* **35**:6572–6577.

Zhang, C., and W. H. Daly (2005). Synthesis and properties of second generation dendronized cellulose. *Polymer Prepr.* **46**:707–708.

Zhang C., and W. H. Daly (2006). Control of degree of substitution of Newkome's first generation dendron on cellulose backbone and evaluation of antibacterial properties of corresponding quaternized derivatives. *Polymer Prepr.* **47**:35–36.

Zhang, J., S.-J. Yao, and Y.-X. Guan (2005). Preparation of macroporous sodium cellulose sulphate/poly(dimethyldiallylammonium chloride) (PDMDAAC) capsules and their characteristics. *J. Memb. Sci.* **255**:89–98.

Zhang, C., L. M. Price, and W. H. Daly (2006a). Synthesis and characterization of a trifunctional aminoamide cellulose derivative. *Biomacromolecules* **7**:139–145.

Zhang, J., Y.-X. Guan, Z. Ji, and S.-J. Yao (2006b). Effects on membrane properties of NaCS-PDMDAAC capsules by adding inorganic salts. *J. Memb. Sci.* **277**:270–276.

Zhang, K., D. Peschel, E. Brendler, T. Groth, and S. Fischer (2009). Synthesis and bioactivity of cellulose derivatives. *Macromol. Symp.* **280**:28–35.

Zhang, K., D. Peschel, T. Klinger, K. Gebauer, T. Groth, and S. Fischer (2010a). Synthesis of carboxyl cellulose sulfate with various contents of regioselectively introduced sulfate and carboxyl groups. *Carbohydr. Polym.* **82**:92–99.

Zhang, K., E. Brendler, and S. Fischer (2010b). FT Raman investigation of sodium cellulose sulfate. *Cellulose* **17**:427–435.

2

CELLULOSIC AEROGELS

FALK LIEBNER, EMMERICH HAIMER, ANTJE POTTHAST,
AND THOMAS ROSENAU

2.1 FASCINATING AEROGELS: AN INTRODUCTION

Lightweight native materials have always been fascinating to human beings. In our early childhood, we used to take delight in catching the white seed of poplar trees and blow it far away carrying our most private wishes. Wondering, we blow a swarm of dandelion clock parachutes across the meadow, we track the small down feather of a chicken swaying down in a gentle summer breeze, and are enchanted by a large snow crystal landing on the back of our small hands and melting slowly on the arm skin.

Solid aerogels and foams are two types of highly porous materials that are lightweight similar to those of the examples mentioned above. These two classes of materials are distinguished by their pore types: While foams are usually built up by stable spherical bubbles, aerogels consist of an interconnected open porous network structure. As the pores of both types of materials are filled with a gas—mostly air—and their pore volumes can reach up to 99%, the density of foams and aerogels is usually very low. Typical examples of ultralightweight foams and aerogels are expanded polystyrene and silica aerogels with densities down to 16 and 5 mg/cm^3, respectively.

Today, the term aerogel is commonly mainly associated with those photogenic, bluish, shimmering, lightweight materials that are displayed if the search string "silica aerogel" is entered in an Internet search engine. Silica aerogels are the most prominent representatives as they have the longest history and are probably the most comprehensively studied subclass of aerogels. Their utilization by the

Polysaccharide Building Blocks: A Sustainable Approach to the Development of Renewable Biomaterials,
First Edition. Edited by Youssef Habibi and Lucian A. Lucia.
© 2012 John Wiley & Sons, Inc. Published 2012 by John Wiley & Sons, Inc.

National American Space Agency during several space missions and 15 entries in the *Guinness Book of Records* for their intriguing properties further contributed to their publicity.

However, in the shadow of silica aerogel's publicity, the synthetic inorganic and organic aerogels have caught up with their famous predecessors and have attracted considerable interest in material science since the early 1990s. Aerogels from natural resources such as wood, lignin, cellulose, starch, and so on are regarded as the third "young" generation of aerogels and research in this field have literally undergone a boom after the turn of the millennium.

2.1.1 Silica Aerogels: The Famous Representatives of a New Class of Materials

According to a legend, it was a bet between Steven Samuel Kistler from the College of the Pacific in Stockton, California and one of his colleagues Charles H. Learned in the early 1930s that led to the preparation of the very first aerogels. Steven Kistler who contended that it was possible to replace the liquid inside a jelly jar without causing any shrinkage won the bet. In the aftermath he discovered that highly porous aerogels from silicates, as well as from a multitude of organic materials such as nitrocellulose, rubber, or gelatine, can be obtained when the solvent of the corresponding lyogels is replaced by a supercritical fluid such as methanol (Hunt and Ayers, 2000). In contrast to the silica xerogels, for example, that had been only known as a scientific curiosity since the 1640s, but had found wide use as a desiccant and adsorbant for gases since World War I (Feldmann and Desrochers, 2003), silica aerogels feature a considerably higher porosity along with a very low density due to the application of the supercritical drying technique. Kistler's experimental contributions to the field of aerogel synthesis and characterization are nowadays considered as pioneer work leading to a new area of material research that has been rapidly developing especially during the past two decades.

Although the novel aerogels and in particular silica aerogels—obtained by acidic condensation of sodium silicate in aqueous solution, subsequent solvent exchange to methanol, and supercritical drying of the methanolic lyogels—had quite interesting properties, the number of industrial applications was rather small for many years. An interesting overview about the history of silica aerogels can be found in Hunt and Ayers (2000). Monsanto—which had a license agreement with Kistler—operated a large-scale production unit in Everett, MA between 1942 and 1970. Silica aerogels that had been sold under the trade names Santocel™, Santocel-C™, Santocel-54™, and Santocel-Z™ were mainly used as additives for paints and varnishes or as thermal insulation materials. However, "significant and unusual applications for Santocel, outside the flatting and insulation fields, were developed for civilian and military use. Among these were the US Department of Agriculture's approval of Santocel as a thickening agent for screwworm salves for sheep, and its use as a thickening agent in the jelly of the fiery Napalm bomb" (Hunt and Ayers, 2000). After 28 years, in about 1970, the Santocel production unit in Everett was abandoned, most likely due to the high production costs and insufficient competitiveness. Further drawbacks included the laborious solvent exchange to get rid of the sodium salts and the considerable energy requirements for the drying step due to the comparatively high critical point of methanol ($T_{crit} = 239°C$, $p_{crit} = 8.1$ MPa) that was used almost exclusively for half a century after Kistler's invention. Furthermore, the risky drying of the alcogels at high temperatures and pressure and a serious accident that destroyed the first pilot plant for the production of silica aerogel monoliths from tetramethoxysilane (TMOS) in Sjobo (Sweden, Lund group) in 1984 (Fricke, 1993) prevented the aerogels from a broader utilization during that time period.

Renewed interest in aerogels was seen only in the mid-1980s. It was prompted by two technological innovations that decisively fructified not only the aerogel research but also the material science as a whole: sol-gel chemistry based on metal alkoxides and supercritical carbon dioxide ($scCO_2$) drying.

It was Teichner in the late 1960s who developed and applied sol-gel chemistry to silica aerogel synthesis. Instead of using water-soluble alkali silicates as precursor compounds, silica sol-gel chemistry is nowadays based on alkoxysilanes such as TMOS or tetraethoxysilane (TEOS) that easily hydrolyze in the presence of water to metastable silanols that subsequently condensate to two- and three-dimensional oligomers and polymers. The great advantages of the new method were that (a) the alcogels no longer contained those large amounts of salts that required laborious rinsing before the final drying step, and (b) the time-consuming solvent replacement could be eliminated as gel formation was from now on accomplished in methanol or ethanol. Today, sol-gel chemistry is probably the most important technology for the synthesis of both aerogels and nanoscale particles, the latter being increasingly used in ceramics and highly effective surface coatings of plant components permanently exposed to high-temperature corrosion.

Supercritical fluid extraction (SFE) is a technique that has long been known but did not find wide use in technical applications. However, in 1985, the article "Supercritical Fluids: Still Seeking Acceptance" appeared in *Chemical Engineering* (Basta, 1985) when the first large plants for extracting coffee, hops, and spices were built in Germany, France, and the United Kingdom. In the following years, SFE techniques were rapidly further advanced and in addition to the above-mentioned applications, $scCO_2$ was also increasingly used, for example, for extracting cholesterol from egg yolks, milk, or meat. Henceforth, the available technical equipment for $scCO_2$ extraction along with the advances in the field of sol-gel technology provided the fertile ground for a rapid expansion of aerogel research in the 1990s, which is reflected in the large number of 50,000 publications published during this decade.

2.1.1.1 What Made Silica Aerogels So Famous?

The intriguing properties of silica aerogels are certainly one of the key factors that rendered them attractive to many applications and that contributed to the above-mentioned advances in aerogel research. A closer look at some of the properties of silica aerogels may give an idea of the potential of this class of materials.

The term silica aerogel refers to their chemical composition that can be described by the formula $[SiO(OH)_x(OR)_y]_z$, R being H or short-chain organic residues and x, y, z

depending on processing parameters. Chemically they are quite inert, nonflammable, and nontoxic.

Microscopically, aerogels are built of small particles usually 1–10 nm in diameter that aggregate to chains and subsequently to three-dimensional network structures. They exhibit a dendritic structure with large, readily accessible, mostly cylindrical open pores having diameters of 2–50 nm. The interchain distance is typically on the order of 10–100 nm. Despite the high porosity, silica aerogels feature a good mechanical strength and dimensional stability, which are also due to the extremely low thermal expansion coefficient of $\alpha(\mathrm{lin}) = 2 \times 10^{-6}\,\mathrm{K}^{-1}$ in the range of 20–80°C.

Silica aerogels scatter shortwave light more than longer wave light that makes them appear in bluish color against a dark background. They are optically transparent with a refractive index of 1.007–1.24, a typical value being 1.02. These properties gave them names like "blue smoke" or "frozen smoke" (Figure 2.1). Their comparatively high optical transmission of about 90% (632.8 nm) was one of the reasons of the Swedish Airglass company for building a large-scale production unit as the material was considered to find use, for example, as front windows in airplanes.

Silica aerogels furthermore exhibit a very large specific surface area of 600–1600 m²/g due to their high and well spread porosity. A large surface area in this order of magnitude would be beneficial to applications where aerogels are considered as catalyst supports, filter materials, or gas-adsorbing matrices.

The low heat transmission coefficient that is usually in the range of 0.016–0.03 W/(m²K) and the high melting point (about 1200°C) render silica aerogels excellent materials for thermal insulation of sensitive devices. They were used, for example, by the NASA during their Mars Pathfinder expedition (from December 1996 to September 1997) for protecting the electronics box on the Sojourner Mars rover from the extremely low Mars surface temperature of up to −120°C. Based on this success, Aspen Aerogels, Inc. (Northborough, MA)—one of the leading aerogel

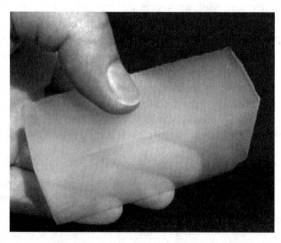

FIGURE 2.1 Shortwave light scattering and extremely low densities gave silica aerogels the byname "frozen smoke" (Picture: http://stardust.jpl.nasa.gov/photo/aerogel.html, last accessed 9/2010).

producers founded in 2001—intends to protect the astronauts of the manned expedition to Mars scheduled for 2018 by thin layers of silica aerogels.

The application of silica aerogels in the construction sector is still in its infancy and is limited to special applications such as light bands or saucer domes, even though the material is up to five times more effective than the other insulation materials (2010b). In certain constructions where the available space for insulation materials is very small, highly effective insulation materials are required. For this purpose, some of the aerogel companies such as the above-mentioned Aspen Aerogels, Inc. or the Cabot Corporation (Boston, MA) with its subsidiary Cabot Nanogel Ltd. (Frankfurt, Germany) offer thin blankets that serve as replacements for traditional fiberglass-, foam-, or cellulose-based insulation materials. In general, Aspen Aerogels products are claimed "... to enable customers to conserve energy in a variety of industries including building and construction, chemicals, transportation, and oil and gas." Quite recently, Aspen BASF Venture Capital GmbH, Ludwigshafen, has led a $21.5 million (about 15.7 million euros) round of investment in Aspen Aerogels, Inc. (2010a).

Another "famous" application of silica aerogels is based on their high shock absorbing potential that allows trapping hypervelocity particles within a very short stopping distance inside the aerogels without significantly converting kinetic into thermal energy that would prevent the particles from elucidating their original composition. Spherical soda glass beads (diameter 75–375 μm), for example, shot onto the surfaces of a silica aerogel (92.5 ± 0.5 mg/cm^3) at a velocity of 5.1 ± 0.2 km/s were found to get gradually slowed down to stop after a few millimeters without damaging them or altering their shape and chemical composition (Burchell et al., 1999). This effect was used during the NASA's Stardust mission (1999–2006) for trapping and distinguishing cosmic and interstellar dust by their different stopping distance inside the aerogel.

2.1.2 The Next Generation: Inorganic and Organic Aerogels

Sol-gel technology is not limited to silica chemistry and can be virtually applied to all chemical compounds that can be dissolved and have the tendency to form colloidal suspensions (sol) upon changing certain conditions (temperature, pH, solvent, etc.) or as a result of a chemical reaction (e.g., hydrolysis and derivatization). Aggregation of such colloids gives polymer-like semisolid network structures (gel) that usually form voids and can be further modified such as by cross-linking, derivatization, or conversion into composite materials.

Inorganic aerogels are mainly obtained from metals or metal oxides. They find use in the preparation of ceramics or ultrafine powders, the latter being used, for example, in high-temperature corrosion preventing applications where highly resistant, plasma-sputtered alloys (e.g., TiAl50) are additionally protected by covering them with an extremely dense layer of nanoscale SiO_2/B_2O_3 particles (DECHEMA, 2007).

Recently, alumina aerogels doped with chromium were demonstrated to produce fluorescent pulses that are proportional in intensity to heating. Therefore, this type of aerogels would be another excellent tool for distinguishing cosmic from interstellar dust as the two types of particles (0.1–10 μm) travel at different speed (< 10 versus

> 60 km/s), have different kinetic energy, and would thus generate fluorescent pulses of different intensity (Domínguez et al., 2003).

Chalcogels are sulfur- or selenium-based aerogels that are obviously the most recent addition to the family of inorganic aerogels. Co(Ni)–Mo(W)–S networks, for example, that feature extremely large surface areas and porosity can be obtained by coordinative reactions of $(MoS_4)^{2-}$ and $(WS_4)^{2-}$ with Co^{2+} and Ni^{2+} salts in nonaqueous solvents. In addition to their use in gas separation (CO_2 is much stronger bound than H_2) or hydrodesulfurization catalysis (Bag et al., 2009), sulfide-based aerogels were shown to reduce significantly the concentration of different heavy metal ions in river water (load drop up to 99.9%) that is due to the low solubility product of many heavy metal sulfides (Bag et al., 2007).

The immense progress in the field of organic polymer chemistry in the last few decades has also affected the aerogel research. Thus, organic aerogels have been obtained from a multitude of organic precursors, such as resorcinol/formaldehyde condensates (Pekala, 1989), mixtures of phenolic and *m*-cresolic resoles with methylolated melamines (Zhang et al., 2004), phenol/furfural resins (Pekala et al., 1995), or polyurethanes (Biesmans et al., 1998). Organic aerogels can further be converted into carbon aerogels by subsequent pyrolysis at 600–2000°C (Reynolds et al., 1995).

The beneficial properties of all three classes of aerogels—silica, organic, and carbon aerogels—include high nanoscale porosity, large specific surface, extremely low density, sound propagation, thermal expansion, heat transmission, large infrared optical adsorption coefficient, and good electrical conductivity. Thus, the new materials are very attractive for many applications (Akimov, 2002; Farmer et al., 1996; Fricke and Emmerling, 1992; Pierre and Pajonk, 2002). Aerogels are already used for thermal isolation (Hrubesh and Pekala, 1994; Nilsson et al., 1994) and sound insulation, as carriers for catalysts, as filters for removing toxic compounds in exhaust gas and effluent (Bag et al., 2007), and for electrochemical applications in high-performance batteries (Cherepy et al., 2006) and supercapacitors (Hwang and Hyun, 2004).

2.2 CELLULOSIC AEROGELS

2.2.1 Novel Functional Materials from a Renewable Source

The remark "we are slowly but surely running out of coal, oil and gas" is sometimes rebutted with statements like "scientists have been saying this since decades and decades, and we still have fossil fuels, actually more than we can use up. . .". It is true that many reports on shortages of fossil fuels and the resulting future scenarios were sometimes lacking scientific accuracy and were playing with emotions of people and trying to evoke anxiety about the future. But it is a hard and an undeniable fact—even if it is to occur in the far future—that fossil resources *will* be used up. When this point is eventually reached, the only source for any substance produced today by the petrochemicals-based industries will be renewable resources. While these resources are still being used today mainly for energy production (refer to catchwords as biofuels, biogas, and bioethanol), the utilization of renewables in the long run will shift from energy to materials/chemicals—the simple reason for this being that when the above-mentioned situation of exhausted fossil resources will have been reached, there would be several proper alternatives to energetic utilization of renewables, wind, water, solar, or nuclear energy, but no alternatives to chemical/material utilization. Taking into account this long-term perspective, renewables will become the base of the chemical industries (a place that has been largely—if not exclusively—occupied by fossil resources) for bulk chemical, special chemical, and material production. Therefore, many existing production lines and synthesis schemes will have to be adjusted so as to make them renewables based. The additional advantages of renewables with regard to environmental issues, CO_2 balance, and climate change are already evident today and are generally acknowledged, although the scarcity of fossil fuels has not yet been a decisive factor.

All the manifold approaches toward this future scenario are commonly described by the term "biorefinery", which almost inevitably became a rather blurred and ill-defined term. There are two general points that apply to all biorefinery scenarios: (a) Any sustainable solution for fractionation of biomass components must use up not only the cellulose but also the hemicellulose fraction and especially the lignin fraction, which is present in amounts comparable to that of cellulose; (b) The operating efficiency of all new biomass processing units and thus the sustainability of the current activities in the biorefinery sector largely depend on the profitability of all product lines and the value of individual products. Thus, many attempts have been made in the past years for the development of innovative, biopolymer-based materials. Stimulated by the enormous interest, the organic aerogels have attracted attention since the early 1990s and new applications for these materials as well as the initiation of a systematic research on aerogels from renewable sources were only a matter of time.

2.2.2 Recent Development of Cellulosic Aerogel Research

The first reports about the preparation of aerogels from native cellulose and cellulose derivatives were those of Jin et al. (2004), Tan et al. (2001), and Fischer et al. (2006). Kuga's group at the University of Tokyo obtained highly porous, nanofibrillary aerogels by regenerating nonderivatized cellulose from a solution in aqueous calcium thiocyanate and subsequent solvent-exchange drying (Jin et al., 2004). Embedded in the sixth framework program of the European Union, a joint project with the acronym Aerocell was initiated in Europe the same year, headed by Lenzing AG, with 10 partners from 5 European countries (Austria, Germany, France, England, and Italy). The objectives of this project comprised of various aspects of aerogel preparation, application, and commercialization (Innerlohinger et al., 2006a). Besides cellulose carbamates and acetates, the latter partially cross-linked with various amounts of a polyfunctional isocyanate (MDI), native cellulose of a DP ranging from 170–6000 was processed to shaped aerogels via direct dissolution in aqueous NaOH or N-methylmorpholine-N-oxide monohydrate (NMMO·H_2O), subsequent regeneration, and drying with supercritical carbon dioxide (Innerlohinger et al., 2006a, 2006b; Pinnow et al., 2008). A similar project initially aiming solely at the preparation of shaped aerogels from nonderivatized pulps was started around the same time in

Rosenau's group at the University of Natural Resources and Life Sciences, Vienna, Austria. It was based on the results of a comprehensive study of side reactions in the system NMMO·H$_2$O/cellulose/water/stabilizer, which had been conducted in the scope of a 7 years research cluster funded by the Austrian Christian Doppler Research Society between 1998 and 2005 (Liebner et al., 2007; Rosenau et al., 2001, 2005a).

Today, about 10 years after the preparation of the first cellulose aerogels, many groups not only across Europe but also overseas and in particular in Japan have established research in this particular field. Apart from underivatized cellulose I (I_α: bacterial cellulose (BC), I_β: plant cellulose) or cellulose II (regenerated cellulose), current approaches in preparing cellulosic aerogels include the utilization of cellulose derivatives, hemicelluloses, and lignocelluloses and their subsequent functionalization, cross-linking, or conversion into composite materials.

2.2.3 Aerogels from Regenerated Cellulose

Similar to the first attempts, aerogels from nonderivatized cellulose are predominantly prepared by dissolving cellulose in an appropriate cellulose solvent, shaping, and subsequent regeneration of the cellulose II polymorph by means of a cellulose antisolvent. The obtained lyogels are finally converted into an aerogel aiming at a far-reaching preservation of the highly porous network structure, as the latter is a prerequisite to many aerogel applications.

The number of cellulose solvents is rather limited due to some special molecular and supramolecular features of cellulose and thus most of them have been studied comprehensively toward their suitability for cellulosic aerogels preparation. This pertains in particular to N-methylmorpholine-N-oxide (NMMO·H$_2$O) (Innerlohinger et al., 2006a, 2006b; Liebner et al., 2007, 2008, 2009), DMAc/ LiCl (Duchemin et al., 2010), salt hydrate melts, aqueous solutions of calcium thiocyanate (Hoepfner et al., 2008; Jin et al., 2004), NaOH (Gavillon and Budtova, 2008; Sescousse and Budtova, 2009), NaOH/urea (Cai, 2008), NaOH/thiourea, LiOH/urea (Lue and Zhang, 2010), and several ionic liquids (ILs) (Aaltonen and Jauhiainen, 2009; Deng et al., 2009; Tsioptsias et al., 2008). However, even though cellulosic aerogels can be obtained in general from every cellulose solvent, almost all synthetic routes still have some drawbacks either due to strong limitations regarding the DP of the cellulose to be dissolved, toxicological concerns, difficulties with removing the solvent system, with preparing shaped monoliths of complex geometry, undesired side reactions, or laborious and insufficient recovery of expensive solvents. In the following section, some of the solvent systems that have been previously used for preparing aerogels from nonderivatized cellulose will be discussed, including the properties of the obtained materials.

2.2.3.1 Type of Cellulose Solvent

N-Methylmorpholine-N-oxide   The Lyocell process is an environmentally benign, relatively young technology to produce cellulosic materials, mainly fibers, by direct dissolution of cellulose in an organic solvent, N-methylmorpholine-N-oxide

(NMMO). It is used today on an industrial scale with an annual production of about 200,000 tons, and it is an ecological alternative to the viscose process that requires cellulose derivatization (alkalization and xanthation) prior to spinning. Serious advantages of NMMO include its low toxicity in production and further handling, its biodegradability, and its high recovery rate that is >99% in the industrial scale.

The flexibility of the process in terms of cellulosic raw materials is another major advantage. It is commonly agreed upon that the peculiar solvation power of NMMO originates in its ability to disrupt the hydrogen bond network of cellulose and to form solvent complexes by establishing new hydrogen bonds between the macromolecule and the solvent (Harmon et al., 1992; Ioleva et al., 1983). The "chemical" processes during dissolution are thus merely acid–base, that is, donor–acceptor, interactions finally leading to a far-reaching restructuring of the hydrogen bond network (Khanin et al., 1998; Rozhkova et al., 1985).

Besides the anhydrous compound, NMMO forms three hydrates, with monohydrate being that of the highest dissolving potential for cellulose. High shearing provided up to 15% of cellulose that can be dissolved in $NMMO \cdot H_2O$. Cellulose dissolution is commonly accomplished using NMMO of a higher water content that is subsequently evaporated until the stoichiometric percentage of the monohydrate (≈ 13.33 wt%) is achieved.

In addition to its environmental harmlessness, $NMMO.H_2O$ has two further advantages with regard to the preparation of aerogels: It is solid at room temperature, melts at 78°C, and has a very low vapor pressure similar to that of ionic liquids. Therefore, a far-reaching constant solvent/solute ratio can be maintained throughout the cellulose dissolution that is commonly accomplished in a temperature range of 100–120°C. Furthermore, solidification of the Lyocell dope upon cooling allows molding, better handling, and a faster regeneration of the shaped, solid NMMO/cellulose bodies.

However, like most of the known cellulose solvents, the use of $NMMO \cdot H_2O$ also has some drawbacks. First, NMMO is a tertiary amine oxide that has a comparatively high oxidation potential and can thus affect the chemical integrity of the cellulose macromolecules especially at longer dissolving times. Under rather harsh processing conditions and even more pronounced in the presence of acidic groups or heavy metal salts, Lyocell dopes tend toward side reactions that might be reflected in discoloration and cellulose degradation and might also cause severe effects, such as deflagrations or blasts (Rosenau et al., 2001, 2002). Efficient stabilization of cellulose solutions in NMMO means prevention of both homolytic and heterolytic side reactions. The former is commonly accomplished by adding - phenolic antioxidants, mostly propyl gallate, and the latter by providing traps of formaldehyde and NMMO-derived carbenium–iminium ions, which are formed in an autocatalytical process via hydroxymethylation of morpholine and subsequent acid-catalyzed dehydratization. *N*-Benzylmorpholin-*N*-oxide has been recently confirmed to be a sacrificial stabilizer that efficiently converts such Mannich bases into stable, colorless products (Rosenau et al., 2005a, 2005b).

Based on the Lyocell technology, cellulose aerogels from different commercial pulps have been prepared by (i) dissolving cellulose in molten NMMO at about 100–

120°C, (ii) molding, (iii) extracting the solidified castings with an organic cellulose antisolvent—mostly ethanol or acetone—to initiate cellulose aggregation and to replace the NMMO, and (iv) drying of the obtained lyogels by supercritical carbon dioxide (Liebner et al., 2008, 2009).

According to this approach, lightweight cellulosic aerogels were prepared from Lyocell dopes containing 3% and 6% of different commercial hardwood and softwood pulps. The density of the obtained aerogels was in the range of 0.046–0.069 g/cm^3 (3% cellulose) and 0.106–0.137 g/cm^3 (6% cellulose). Scanning electron microscopy (SEM) pictures and nitrogen sorption experiments at 77K revealed a largely uniform, mesoporous structure with an average pore size of 9–12 nm and a specific surface area of 190–310 m^2/g (Liebner et al., 2009). Doubling the cellulose content of Lyocell dopes increased the Young's modulus almost by one order of magnitude and increased slightly the BET surface areas. Similar values were reported by Innerlohinger et al. (2006b) who prepared a large set of aerogels from 0.5 to 13 wt% cellulose containing NMMO dopes. The two papers underline that intensive shrinking of the cellulosic bodies is a major drawback of the Lyocell route that was about 10% for regeneration and 20% for scCO$_2$ drying in case of 3 wt% cellulose containing dopes (Innerlohinger et al., 2006a, 2006b; Liebner et al., 2009).

Shrinking can be caused by several reasons. As it will be discussed in a subsequent section, the presence of water in cellulose aerogels can cause a far-reaching collapsing of the fragile cellulose network structure. That is why polar, organic solvents are used instead of water for initiating cellulose regeneration and performing NMMO extraction. However, due to the high affinity of NMMO to cellulose and its strong tendency to form hydrogen bonds via the exocyclic oxygen atom—the reason why NMMO dissolves cellulose—several subsequent washing steps are necessary to largely get rid of the tertiary amine oxide. However, NMMO eluation profiles that were recorded by GC/MS after reducing the N-oxide with sodium sulfite evidence that a considerable amount of NMMO remains confined to fibrillary interstices of the lyogel. As NMMO is highly hygroscopic, remaining traces can thus absorb considerable amounts of water that finally may lead to the observed pore collapsing and shrinkage of the aerogels.

The crystallization behavior of NMMO is another disadvantage that potentially contributes to shrinking. It has been observed that it is nearly impossible to prepare homogeneous disks from Lyocell dopes even at very slow cooling rates for cellulose contents of up to 6 wt%. This effect has been studied at the Donghua University, Shanghai. Liu et al. (2001a) found that up to a cellulose content of at least 6 wt%, the densities of respective dopes increase linearly even if the solution temperature was much lower than the melting point of the mixture 73–78°C. Only at temperatures below 30°C when crystallization sets in the dope densities increased sharply which was more pronounced for the dopes of lower cellulose content.

For aerogels of comparable cellulose content, the mechanical response profiles upon application of uniaxial compression stress have been found to be very similar and largely independent of the (anti)solvents used for cellulose dissolution and regeneration. The stress–strain curves of different aerogels obtained from 3% and 6% cellulose containing Lyocell dopes is shown in Figure 2.3. After a short initial stage

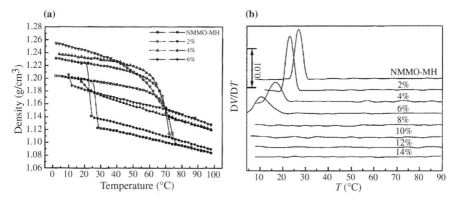

FIGURE 2.2 (a) Density of NMMO and cellulose/NMMO solutions as a function of temperature (open symbols: heating; solid symbols: cooling). (b) Differential curves of the first derivative of the specific solution volume with respect to temperature as a function of temperature. Reproduced from Liu et al. (2001) with permission from Springer Science + Business Media.

($\leq 1\%$ strain) and before the flow limit (2.5–5.5% strain) is reached, the curves are characterized by a largely linear slope that can be described by the coefficient of elasticity (Liebner et al., 2009). Since an exact definition of the yield point is difficult for most cellulosic aerogels, it was set at 0.2% offset strain, a common practice for metallic materials (refer to insert of Figure 2.3). After reaching the flow limit, the pore walls are subjected to progressive elastic buckling that eventually leads to collapsing of an increasing number of pores in the course of which the gels

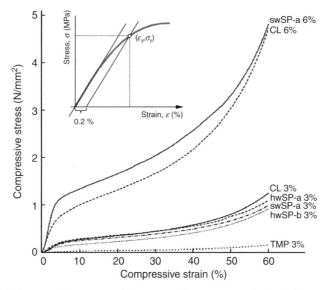

FIGURE 2.3 Compressive stress–strain curves for aerogels obtained from 3% and 6% cellulose containing Lyocell dopes. *Insert:* Scheme showing the determining of the yield point.

are irreversibly damaged. Within this plateau region (5.5–40% strain), the two processes—elastic buckling and pore collapsing—allow the material to absorb a considerable amount of energy. The region beyond 40% strain is characterized by increasing densification when opposing cell walls touch each other, broken fragments pack together, and further deformation starts to compress the wall material itself (Sescousse et al., 2010). Once collapsed, the porous network structure does not recover its shape and the originally porous aerogel structure becomes increasingly compact. This effect causes a significantly steeper slope after reaching a compressive strain of approximately 40–50%. In case of cylindrical cellulose aerogels with an apparent density ranging from 50 to 80 mg/cm^3, the test specimen were compressed from 10 to 2 mm without sample buckling, that is, the cross-sectional area of the aerogel did not change. Furthermore, all aerogels did not suffer a sudden load drop as it is often observed in case of brittle foam plastics (Liebner et al., 2009).

Young's moduli E describing the stiffness of isotropic elastic materials are comparatively low for aerogels from nonderivatized cellulose and highly dependent on the aerogel density, that is, on the cellulose content of the cellulose lyogel prepared beforehand. While aerogels obtained from 3% cellulose containing Lyocell dopes had Young's moduli between 5 and 10 N/mm^2, doubling the cellulose content increased the E values by about one order of magnitude (Liebner et al., 2009). This is due to the increased resistance to cell wall bending and pore collapsing (higher plateau stress). On the other hand, an increasing cellulose content reduces the strain at which densification begins as the cell walls can touch each other at lower stress due to the presence of smaller pore radii (Sescousse et al., 2010). A similar trend was found for cellulose aerogels that were obtained by regenerating Avicell PH-101 micro-crystalline cellulose (MCC) (DP 180) from corresponding solutions in aqueous sodium hydroxide and EMIM acetate (Sescousse et al., 2010).

Ionic Liquids   Ionic liquids have been increasingly moved into the focus of cellulose and lignocelluloses research as they facilitate direct dissolution (Swatloski, 2002; Zhu et al., 2006) and homogeneous derivatization (Barthel and Heinze, 2006; Heinze et al., 2005, 2008; Schöbitz et al., 2009; Wu et al., 2004). They are considered to be potent competitors to NMMO as they are able to dissolve up to 20 wt% of cellulose, are not oxidizing, largely stable, chemically inert, have no vapor pressure, and dissolve cellulose at much lower temperatures. Especially the 1-butyl-3-methyl-imidazolium (BMIM) and 1-ethyl-3-methyl-imidazolium (EMIM) chlorides and acetates have found wide utilization in cellulose and polysaccharide chemistry as they are less expensive than other ILs (Liebner et al., 2010b).

In addition to cellulose fibers (Cai et al., 2010; Quan et al., 2010; Wendler et al., 2009), films (Turner et al., 2004), or beads (Lin et al., 2009), ionic liquids have also been used for producing monolithic aerogels (Aaltonen and Jauhiainen, 2009; Deng et al., 2009; Sescousse et al., 2010) and composite materials (Zhao et al., 2009).

Deng et al. (2009) reported the preparation of highly nanoporous (99% porosity) cellulose foams from BMIM chloride. Following dissolution at room temperature, cellulose was coagulated in water and subsequently subjected to rapid freeze-drying using liquid N_2. The resulting open fibrillar network structure consisted of cellulose II crystalline structure and had a specific surface area of up to $186\,m^2/g$. The same IL was used by Aaltonen and Jauhiainen (2009) who succeeded to obtain nanofibrillar aerogels from cellulose, spruce wood, and mixtures of cellulose, lignin, and xylan. In contrast to the procedure described by Deng et al., coagulation of biopolymers was accomplished with ethanol. The material obtained from a mixture of cellulose (54 wt%, softwood Kraft pulp), xylan (27 wt%, birch, Sigma), and Protobind 1000 sodalignin (19 wt%; Granite SA, Switzerland) was rather brittle, whereas those aerogels obtained from spruce wood (Finnish Picea Abies) exhibited much more structural strength but had a BET surface area of only about $100\,m^2/g$ (Aaltonen and Jauhiainen, 2009). Budtova's group at Mines ParisTech compared the regeneration kinetics of cellulose from solutions in EMIMAc and BMIMCl with those from solutions in NMMO and aqueous sodium hydroxide, using water as antisolvent (Sescousse et al., 2010). SEM pictures revealed the presence of interconnected macropores in all studied materials with diameters covering the range of a few hundreds of nanometers to a few microns. The mesopore fraction ($<50\,nm$) peaked at around 10 nm, which is in good agreement with the data published by Liebner et al. (2009) for aerogels that were obtained from Lyocell dopes (3 wt% pulp of different origin) by regenerating cellulose with ethanol and subsequently converting the alcogels into aerogels by $scCO_2$ drying. BET surface areas were also of comparable dimension, with slightly higher values for the aerogels obtained via the Lyocell route ($200–300\,m^2/g$).

Like other cellulose solvents, ILs also have some drawbacks. For economic considerations, a nearly quantitative IL recovery rate would be required to use them in technical applications. However, actual recovery rates are far away from being quantitative, which is mainly due to their virtually not existing vapor pressure that excludes distillation as measure for IL separation. Purification of ILs is another related issue that amplifies by the good solubility of lower molecular polysaccharide degradation products. In addition, it has been recently reported that aldopyranoses as well as cellulose react at their reducing end with C-2 of imidazolium-based ILs under formation of a carbon–carbon bond (Ebner et al., 2008; Schrems et al., 2010). This type of reactions occur rather slowly in pure ILs, but is strongly catalyzed by bases, namely, imidazol, N-alkylimidazoles, and dimeric compounds that contain two imidazole moieties linked by a methylene bridge. Such compounds were shown to be formed at elevated temperatures from 1-alkyl-3-methylimidazolium ILs (as acetate or chloride) (Liebner et al., 2010b).

ILs can also react with aldehyde/hemiacetal functions other than at the reducing end, such as those introduced by aging processes along the cellulose chain or by deliberate cellulose oxidation (TEMPO, periodate) (Liebner et al., 2010b; Schrems et al., 2010). Therefore, considerable amounts of ILs can be accumulated in the cellulose by covalent binding that not only contribute to the above-mentioned insufficient recovery rates but also prevent such cellulosic fabrics from being used in biomedical, cosmetic, or biological applications, where even minor impurities might induce adverse effects.

The anions of ILs can be separately involved in side reactions, mainly upon cellulose derivatization (Schrems et al., 2010). Thus, IL acetates cannot be recommended in (organic and inorganic) etherifications of cellulose with acid chlorides and anhydrides, as the anion is converted into strongly acetylating species affording (partially) acetylated cellulose rather than the desired organic or inorganic esters. Cellulose methylation in B(E)MIM chloride and B(E)MIM acetate does not proceed, since the anion consumes the commonly used methylating agents methyl iodide and dimethyl sulfate by conversion into methyl chloride and methyl acetate, respectively, which largely exhibits inferior or no methylation behavior reactions (Schrems et al., 2010).

However, the above-mentioned side reactions can be easily avoided if the ILs are either composed of the appropriate anion for a particular reaction or carry an alkyl group at C-2 position if the cation is imidazolium based (Schrems et al., 2010). Considering their application potential—ILs have also been reported to accelerate hydrolysis and condensation of silica monomers and thus interfere in the gelation process (Dai et al., 2000; Karout and Pierre, 2007, 2009)—it is expected that this class of solvents will gain even more importance in the near future.

Salt Hydrate Melts, Aqueous Salt Solutions, and Aqueous Mixtures of Alkali Hydroxides with (Thio)urea Colloidal solutions of cellulose can be obtained by dissolving the latter in salt hydrate melts (Fischer, 2003). As the water content of such salt hydrates corresponds to the coordination number of the respective cation, the resulting system features a larger portion of ionic interactions than that between the water molecules. If such a salt hydrate penetrates the cellulosic fibers, it is assumed that former hydrogen bonds between the cellulose chains can be reorientated in such a way that new hydrogen bonds are formed between one cellulose molecule and the OH groups of the salt hydrate. The cellulose fibrils in the salt hydrate melt perform Brownian motion that results in contacts and connections of the fibrils forming clusters or a fibrillar felt. On cooling the hot cellulose solution down to about 80°C, the cluster size increases until a gel is formed (Hoepfner et al., 2008).

Monolithic aerogels and cryogels with densities of 10–60 mg/cm^3 and surface areas of 200–220 m^2/g were prepared from calcium thiocyanate tetrahydrate melts by regenerating cellulose with ethanol and thoroughly rinsing the alkogel until the salt concentration was negligibly low (Hoepfner et al., 2008). Regarding the final drying step, the authors underline that scCO$_2$ drying is superior to freeze-drying as the highly porous cellulose network structure is much better preserved. Thus, scCO$_2$-dried gels from calcium thiocyanate were reported to have BET surface areas of up to 250 m^2/g, whereas their freeze-dried counterparts feature BET areas of only 160 m^2/g (Hoepfner et al., 2008).

Instead of using the melt, Jin et al. (2004) prepared gels by dissolving Whatman CF11 cellulose powder in an aqueous saturated solution of Ca(SCN)$_2$·4H$_2$O (59 wt%) at 120–140°C (10–20 min). Shaping of the solution has to be performed in hot state as gelation starts below 80°C. Freeze-dried gels that were subjected to a solvent exchange from water to methanol to *tert*.-butylalcohol were found to feature higher BET surface values (160–190 m^2/g) compared to freeze-dried samples (Jin et al., 2004).

Aqueous solutions of sodium hydroxide or alkali hydroxide/(thio)urea mixtures can also be used for dissolving cellulose, and hence for the preparation of aerogels (Cai et al., 2008; Lue and Zhang, 2010). In Budtova's group, a larger set of aerogels was obtained by dissolving different types of cellulose (DP from 180 to 950, preactivated at 5°C) in precooled (-6°C, 2 h of stirring), 12 wt% aqueous NaOH solution (Gavillon and Budtova, 2008; Sescousse et al., 2010). The resulting 8 wt% aqueous NaOH/cellulose solutions were then kept for a few hours above the gel point (25°C, 24 h; 50°C, 2 h) to ensure complete gelation. After irreversibly gelling, cellulose was regenerated and thoroughly rinsed with water (Gavillon and Budtova, 2008). Depending on the type of cellulose, densities between 120 and 140 mg/cm^3, BET surface areas ranging from 240 to 280 m^2/g, and mean pore diameters of about 0.9 µm were obtained for 5 wt% cellulose containing gels.

However, Gavillon and Budtova (2008) underlined that aqueous NaOH is not a very good cellulose solvent as the system cellulose/NaOH/water is increasingly "destabilized" in terms of microphase separation and subsequent syneresis. As the intrinsic viscosity decreases with increasing temperature (Egal, 2006; Roy et al., 2003), preferential cellulose–cellulose interaction and thus gelation in semi-diluted solution are faster at elevated temperatures (Gavillon and Budtova, 2008).

Zhang's group at Wuhan University, PR China, found that precooled aqueous solutions consisting of 7 wt% NaOH and 12 wt% of urea can rapidly dissolve cellulose as long as its molecular weight is less than 1.2×10^5 Da. It is supposed that at low temperature, NaOH "hydrates" such as $[OH(H_2O)_n]^-$ Na$^+$ are formed by dynamic supramolecular assembly, which in turn are able to reorientate and cleave inter- and intramolecular hydrogen bonds between anhydroglucose repeating units. Urea is assumed to interact not directly with the polymer/hydrate/hydrogen bond network but with self-assembly at the surface of the chains forming an water-soluble inclusion complex with cellulose/NaOH hydrate encaged (Lue and Zhang, 2010). This has been confirmed by topochemical studies using SEM-EDX where a homogeneous distribution of the (thio)urea molecules around the cellulose fibers was found (Yan et al., 2007).

In addition to fibers (Cai et al., 2004, 2007; Fialkowski et al., 2006) and transparent films (Qi et al., 2009a, 2009b), cellulose microspheres exhibiting a sensitive magnetic response have recently been obtained by *in situ* generation of uniformly dispersed Fe$_3$O$_4$ nanoparticles, immobilized inside the pore network (Luo et al., 2009).

An even better dissolution behavior for cellulose has been achieved by replacing urea by thiourea and NaOH by LiOH. The optimal temperature for dissolving cellulose in the systems 9.5 wt% NaOH + 4.5 wt% thiourea and 5 wt% LiOH + 12 wt% urea is -5°C and -12°C, respectively (Lue and Zhang, 2010). However, even though these solvent systems bear a fascinating potential for many applications, their limitations regarding the achievable cellulose content of the solutions and the processable molecular weight of cellulose are weak points that disfavor them to some extent for aerogel preparation.

2.2.3.2 Cellulose Regeneration

Regeneration of cellulose II from solution state using a cellulose antisolvent is a diffusion-controlled process at least for cellulose solutions in NMMO monohydrate (Biganska and Navard, 2005), ionic liquids

(BMIMCl, EMIMAc) (Sescousse et al., 2010), and aqueous NaOH (Gavillon and Budtova, 2007; Sescousse and Budtova, 2009).

As a result of a comparative study on the preparation and properties of aerogels from cellulose solutions in BMIMCl, EMIMAc, NMMO, and aqueous NaOH, Sescousse et al. (2010) reported that at a given cellulose concentration, the diffusion coefficients of ILs in water are about twice lower than that of NMMO and four–five times lower compared to aqueous NaOH. Contrary to the aqueous sodium hydroxide solution, the diffusion coefficients for NMMO and the studied ILs showed nearly no dependence on the polymer concentration. The authors explained this as follows: Regeneration of cellulose from aqueous solutions of NaOH follows another mechanism. Gelation for this system is accompanied by a microphase separation that leads to agglomeration of cellulose molecules and the formation of macroscopic network forming particles before any antisolvent is added. Different from cellulose solutions in ILs and NMMO, the comparatively low concentrated NaOH freely diffuses inside this network but is increasingly hampered at higher cellulose content. For NMMO or ionic liquids, phase separation starts at the moment when the antisolvent is added and proceeds slowly into deeper layers of the bodies under regeneration so that only a weak dependence on the cellulose content is plausible.

Another factor influencing the time period required for quantitative regeneration is viscosity. When water is used for regenerating cellulose from 3% cellulose containing solutions, the concentration gradient and thus the viscosity gradient throughout the entire regeneration step are rather small for the 8 wt% aqueous NaOH system (initial content 7.8%). In contrast, the initial percentages of IL and NMMO are 97% and 82%, respectively, so that the diffusion rates distinctly increase with regeneration time.

Effective diffusion coefficients are of considerable value as they allow for a better evaluation of the time periods necessary for quantitative cellulose regeneration and solvent replacement. For NMMO, D_{eff} can be calculated from the increasing NMMO concentration in the regeneration bath using Equation 2.1 (Crank's analytical solution of Fick's second law). The NMMO concentration can be followed *in situ* using an UV/V is probe head.

$$\frac{M(t)}{M} = 1 - \sum_{n=0}^{2} \frac{8}{(2n+1)^2 \pi^2} \exp\left(\frac{-D_{eff}\pi(2n+1)^2}{l^2}\right). \qquad (2.1)$$

Figure 2.4 shows the time-dependent NMMO release from a Lyocell dope containing 3 wt% of *Beech* sulfite pulp (TCF bleached, CCOA 24.3 µmol/g C=O, FDAM 13.9 µmol/g COOH, MW 303.7 kg/mol) using ethanol for cellulose regeneration (25°C). The calculated $D(NMMO)_{eff}$ was found to be 1.32×10^{-10} m^2/s that is only slightly higher than for regeneration with water where D_{eff} lies between 1.30×10^{-10} and 1.25×10^{-10} m^2/s (Biganska and Navard, 2005; Gavillon and Budtova, 2007).

Therefore, the use of ethanol instead of water for regenerating cellulose from Lyocell dopes does not significantly prolong the entire process and it will be shown in the next section that ethanol is superior to water with respect to the preservation of the pore features of lyogel.

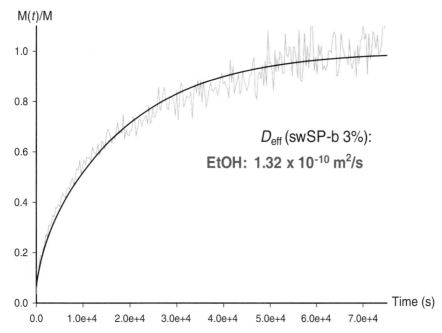

FIGURE 2.4 NMMO release from 3% cellulose (beech sulfite pulp swSP-b) containing Lyocell dope into the ethanolic regeneration bath, indirectly measured by the release of the antioxidant *n*-propyl gallate ($\lambda_{max} = 309$ nm).

An interesting difference regarding the morphology of the regenerated cellulose II polymorph has been reported by Gavillon and Budtova (2008). Although a globular structure is obtained when regenerating cellulose from solutions in ILs (BMIMAc and EMIMAc) or molten NMMO, a net of cellulose strands is formed from solidified Lyocell dopes and gelled cellulose in aqueous NaOH. The authors explain this effect by phase separation (free solvent/cellulose-bound solvent) taking place in the two latter cellulose solvent systems mentioned above prior to regeneration. The addition of water than first removes the free solvent before the cellulose-bound solvent is replaced, which in turn effects the regeneration of preaggregated cellulose structures. In ILs and molten NMMO, cellulose is spatially homogeneously distributed and phase separation occurs in one step, via spinodal decomposition (Biganska and Navard, 2009), creating regular small spheres (Gavillon and Budtova, 2007, 2008; Sescousse et al., 2010).

2.2.3.3 Drying of Cellulose Aerogels The conversion of lyogels into aerogels—commonly referred to as drying—is the last step in the preparation sequence of aerogels. Several techniques have been applied, adapted, and advanced to fit the demands of the highly porous soft matter networks and to preserve their pore characteristics throughout this last step.

Thermal drying of cellulose aerogels is considered to be the worst choice as the resulting xerogels have no interconnected pore network anymore as the majority of

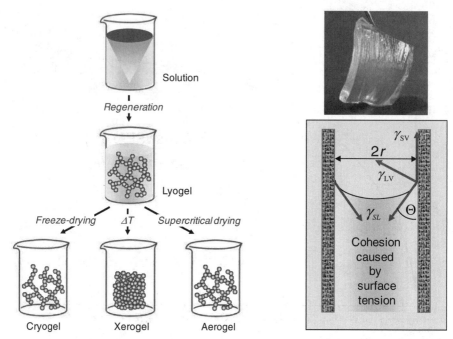

FIGURE 2.5 Schematic representation of the different approaches for converting lyogels into aerogels (left) and of the capillary forces thereof, the occurrence of which should be avoided when drying cellulose aerogels (right).

pores collapse. This can also happen upon vacuum drying or even supercritical drying (see below) when the system is depressurized before the entire amount of solvent is removed. The main reasons for this effect are inward forces alongside the capillary walls adjacent to the solvent menisci that arise from differences in the specific energies of the solid–liquid (γ_{SL}) and liquid–gas phase transitions (γ_{LV}), the same forces that cause liquids to rise in a capillary or spread out onto a solid surface (see Figure 2.5) (Smirnova, 2002). The balance of tensions is given by Equation 2.2 with θ being the contact angle that gives the curvature of a droplet on a plain or a solvent meniscus inside a capillary.

$$\gamma_{SV} = \gamma_{SL} + \gamma_{LV}\cos(\theta). \qquad (2.2)$$

In case of an open porous solid, the liquid rises and causes replacement of the solid–vapor interface with a solid–liquid interface. For cylindrical pores, the gained energy is described by Equation 2.3, where r is the capillary radius and h is the height up to which the liquid rises (Smirnova, 2002).

$$\Delta E_{cap} = 2\pi rh(\gamma_{SV}-\gamma_{SL}) = P_{cap}\pi r^2 h. \qquad (2.3)$$

The work performed by the liquid to move against gravity is equal to the product of the capillary pressure (P_{cap}) and the volume of liquid moved, $\Delta V = \pi h r^2$. Taking into account Equation 2.4, the capillary pressure can be expressed as follows (Smirnova, 2002):

$$P_{cap} = P_g - P_l = \frac{2(\gamma_{SV} - \gamma_{SL})}{r} = \frac{2\gamma_{LV}\cos(\theta)}{r}. \quad (2.4)$$

By introducing a constant and converting Equation 2.2 into Equation 2.3, it becomes evident that for those systems that do not feature any liquid–volatile transition, the capillary pressure must be zero.

$$P_{cap} = \gamma_{LV}K.$$

This pertains to both freeze-drying and supercritical drying as in the ideal case, the liquid and the volatile phase do not coexist. Freeze-drying is based on sublimation of a frozen liquid inside the capillaries, that is, direct transition from solid into gas state. The cryogels (see scheme in Figure 2.5) obtained this way still feature a largely unaltered pore system; however, collapsing of a certain percentage of pores and the formation of cracks have been reported for cellulose aerogels that are mainly due to forces related to the freezing process (Hoepfner et al., 2008). To minimize these effects, problematic solvents like water are commonly replaced by solvents of a lower thermal expansion coefficient and higher sublimation pressure such as ethanol.

Supercritical drying, in particular using scCO$_2$, is considered to be a very "soft" drying method that allows a far-reaching preservation of the polymorphic cellulose II network structure. Supercritical carbon dioxide is furthermore accepted to be a "green" solvent as it is natural, abundant, cheap, and easily recyclable. Most attracting features are the comparatively low supercritical point (31.2°C, 7.38 MPa) and its far-reaching chemical inertness, as these properties allow a gentle processing of thermally, chemically, and mechanically sensitive materials.

Full miscibility of the solvent to be removed and CO$_2$ under supercritical conditions is a prerequisite to scCO$_2$ drying. Ethanol, for example, and carbon dioxide are completely miscible at pressures beyond the critical pressure at a given temperature. At 40°C, full miscibility is achieved at pressures approximately beyond 8 MPa.

The supercritical drying process is schematically presented in Figure 2.6. It starts with a porous gel fully filled with pure ethanol (point A) (Brunner, 2005). Upon pressurization of the system with carbon dioxide, the amount of solved CO$_2$ within the liquid phase increases that leads to an increase of the liquid-phase volume (CO$_2$-expanded liquid phase). For supercritical drying, the system has to be pressurized until a pressure above the mixture critical pressure for the given temperature (point B). After reaching full miscibility, the mixture of ethanol and CO$_2$ is removed from the porous matrix by flushing the autoclave with pure CO$_2$. This flushing step has to be performed until the ethanol content within the pores is low enough, so that no two-phase region occurs upon depressurization (point C, Figure 2.6). Finally, depressurization to atmospheric pressure leads to the final product (point D).

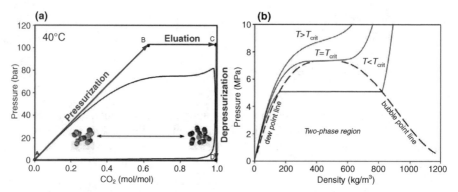

FIGURE 2.6 (a) Binary phase diagram (ethanol/CO_2) showing the main steps of scCO_2 drying. (b) Phase diagram of CO_2 illustrating the strong changes in density in the critical point region.

By comparing different approaches for converting cellulose lyogels into aerogels, it can be concluded that scCO_2 drying is that technology which largely allows preservation of the fragile, highly open porous structure of most cellulosic lyogels.

Ratkes group, for example, applied both rapid freeze-drying and scCO_2 drying to convert lyogels from $Ca(SCN)_2 \cdot 4H_2O$ melts into aerogels (Hoepfner et al., 2008). Rapid freeze-drying was accomplished by dipping the aquogels into liquid nitrogen to get ice crystals as small as possible to avoid mechanical alteration of the pore wall structure. However, conventional freeze-drying of the cellulose gels was finally evaluated to be difficult as the obtained cryogels were reported to have cracks and were more brittle. Furthermore, the distinctly higher BET pore surface area of the scCO_2-dried aerogels ($250\,m^2/g$) unambiguously confirmed the superiority of scCO_2 drying to freeze-drying where the surface area was only $160\,m^2/g$ (Hoepfner et al., 2008).

Besides eluation time necessary to get rid of a sufficiently high amount of expanded solvent–CO_2 phase, other factors related to scCO_2 drying can also decisively affect the quality of an aerogel. This mainly pertains to the presence of water, as the latter—even when present in traces—can lead to considerable pore collapsing due to the above-described formation of capillary forces (Liebner et al., 2009, 2010a). Fast depressurization can also harm the pore structure. As shown in Figure 2.6b, slight changes in pressure are intertwined with considerable changes in reduced volume and density, respectively especially within a certain range around the critical point. Thus, the rapidly expanding CO_2 can tear pore walls if the nearly dry aerogel is depressurized too fast. This effect can be amplified by the well-known Joule–Thompson effect, as rapid cooling would further increase the rate at which the supercritical point is crossed. Changing solvent polarity upon replacing the organic solvent (lyogel) by scCO_2 and subsequent complete removal of the expanded CO_2/ organic solvent phase (aerogel) and hence changing attractive/repulsive interactions between the different solvents and the polymer have been discussed as another, not yet fully understood reason for shrinking of gels upon $scCO_2$ drying (Zhang et al., 2004).

2.2.4 Cellulose I Nanofibers

Cellulose aerogels with a high portion of the thermodynamically meta stable crystal phase I can be obtained from both bacterial cellulose (enriched in I_α, triclinic P1) and plant cellu-lose (mainly I_β, monoclinic $P2_1$). However, in contrast to aerogels from regenerated cellulose, which entirely consist of the stable cellulose II polymorph, the preservation of the cellulose crystal phase I excludes any dissolving step throughout the preparation.

2.2.4.1 Bacterial Cellulose Bacterial cellulose is of increasing interest for many potential aerogel applications as it features some outstanding properties. In partic-ular, purity, average molecular weight, fiber strength, and degree of hydratization are normally significantly higher compared to plant cellulose. Especially its hydratization/dehydratization behavior and its high purity are the main reasons for the permanently increasing number of bacterial cellulose-based applications in the medical or cosmetic sector (Barud et al., 2008b; Sheridan et al., 2002).

There are several techniques for growing bacterial cellulose. It can be obtained either in (semi)continuous mode (Kralisch et al., 2008, 2010) or in discontinuous batch mode (Bungay and Serafica, 1997; Geyer et al., 1994). In the batch mode, static and dynamic cultivation techniques are employed that can be distinguished by the way the nutrient supply is realized.

Acetobacter xylinum is one of the most frequently used strains in commercial bacterial cellulose production even though there are a number of cellulose-forming bacteria such as *Rhizobium, Acrobacterium,* or *Acetobacter.* Its cultivation in static batch mode is commonly performed in (agitated) aquarium-type glass tanks that are filled with the steam-sterilized culture medium such as HS medium that consists of 20 g/L glucose, 5 g/L peptone, 5 g/L yeast extract, 1.15 g/L citric acid monohydrate, and 6.8 g/L $Na_2HPO_4 \cdot 12H_2O$ (Hestrin and Schramm, 1954). After inoculating the culture medium with a suspension of *Gluconacetobacter xylinum* and a certain lag phase, extracellular expression of cellulose I on the surface of the growth medium set in. At a medium temperature of 30°C, most of the glucose is consumed within 30 days and the thickness of the cellulose layer is in the range of about 3–4 cm. The harvested cellulose is then cut into pieces or sheets of desired size and subjected to boiling briefly, repeated alkaline treatment (0.1 M aqueous NaOH at 90°C, 20 min, three times), and several washing steps (deionized water, 24 h) to eliminate growth medium residues and the protein fraction. Prior to its use in medical or cosmetic applications, the materials are typically sterilized.

The first report on bacterial cellulose aerogels is found in the *Japanese Journal of Polymer Science and Technology*, 2006. It was Shoichiro Yano's group at Nihon University, Tokyo, who converted bacterial cellulose aquogels into the corresponding aerogels using supercritical ethanol (6.38 MPa, 243°C) (Maeda et al., 2006b). The obtained ultralightweight materials had densities down to 6 mg/cm^3, a porosity of more than 99%, and consisted of 20–60 nm thick network-forming microfibrils. However, with regard to chemical integrity, it is assumed that significant hornification, fundamental changes in the hydrogen bond network, and loss of structural water must have occurred due to the harsh drying conditions.

It is well known that thermal modification and degradation of cellulose takes place in three steps: (i) removal of physically bound water (up to 150°C), (ii) release of chemically bound water from hydroxyl groups (180–240°C), and (iii) pyrolytic cleavage of the polymer chain (240–400°C) (Tang and Bacon, 1964). Rhee and Yim (1975) who studied graphitization of Ryon under nitrogen atmosphere found that intensive dehydratization and polymer degradation starts at about 240°C and 270°C, respectively. Cheng et al. (2009) reported that structural alteration of dried bacterial cellulose from *A. xylinum* upon heating starts at about 180°C and proceeds in two steps. Yang and Chen (2005) illustrated that the initial weight loss at lower temperatures (200–360°C) is attributed to the removal of small molecular fragments such as hydroxyl and hydroxymethyl groups, whereas depolymerization of bacterial cellulose starts at about 360°C. These data were confirmed by Scheirs et al. (2001), who reported intra- and interring dehydratization of cellulose at temperatures above 220°C, and Lampke (2001), who found that cotton linters start to dehydrate at about 200°C, whereas depolymerization starts at around 325°C.

The application of supercritical carbon dioxide instead of scEtOH for drying of bacterial cellulose lyogels is considered to be superior with respect to the preservation of the chemical integrity of the cellulose due to the low critical point of CO_2 (31.1°C, 7.38 MPa). Hornification and depolymerization are hardly likely under these conditions and the extent of dehydratization or changes in the I_α/I_β ratio are distinctly reduced. The conversion of bacterial cellulose aquogels into aerogels by $scCO_2$ drying has been described by Maeda (2006) who obtained aerogels with a density down to 5.4 mg/cm^3 and a porosity more than 99.5% when he dried the samples at 60°C and 20 MPa. Liebner et al. (2010a) succeeded to obtain aerogels of a similar density under somewhat milder conditions (40°C, 10 MPa) and showed that higher pressure is rather disadvantageous regarding the final density of the aerogels (Liebner et al., 2009). Thorough replacement of water by aprotic organic solvents such as ethanol or acetone is an ultimate precondition to preserve the highly fragile pore network structure of these ultralightweight materials throughout the drying procedure as even the presence of trace amounts of water can effect significant pore collapsing (refer to section on cellulose II aerogels).

Properties of Bacterial Cellulose Aerogels   The extremely low density is obviously the most intriguing property of bacterial cellulose aerogels. ScCO$_2$-dried cellulose aquogels that were obtained from *G. xylinum* AX5 wild-type strain in Hestrin–Schramm's medium had an average density of 8.25 ± 0.7 mg/cm^3 ($n = 10$) (Liebner et al., 2009). With other or mixed strains and different growth media (nata de coco), even lower densities of about 5 mg/cm^3 can be achieved and, thus, electrostatic attraction suffices to hold an ultralightweight aerogel in suspense (see Figure 2.7b).

Cellulose aerogels of comparable density have never been obtained from regenerated cellulose so far. Aerogels obtained by solvent exchange (ethanol) and

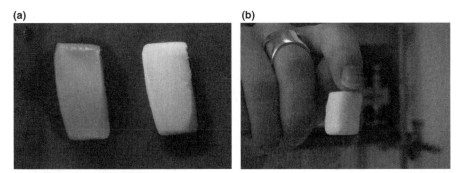

FIGURE 2.7 Different gel states of bacterial cellulose obtained from *Gluconacetobacter xylinum*. (a) Aquogel versus aerogel. (b) Electrostatic attraction suffices to hold an ultralight-weight aerogel in suspense. Reproduced from Liebner et al. (2010a) with permission from Wiley–VCH Verlag GmbH & Co. KGaA

subsequent $scCO_2$ drying from 1% cellulose (Solucell; commercially available eucalyptus pulp) containing Lyocell dopes, for example, were reported to have densities of around $40\,mg/cm^3$ as the gels suffer extensive shrinking during preparation and drying (50–75%) (Innerlohinger et al., 2006b). Even Lyocell dopes consisting of about the fivefold amount of plant cellulose (commercial pulps) showed an average shrinking of about 30% upon extracting the solidified Lyocell dopes with ethanol and subsequent $scCO_2$ under comparable conditions (Liebner et al., 2009). In contrast, bacterial cellulose gels were found to feature a much higher dimensional stability (very little shrinkage) throughout solvent exchange and subsequent $scCO_2$ drying (see Figure 2.7a). This is supposed to be due to the high portion of crystalline domains present in bacterial cellulose and their comparatively high molecular weight.

Mechanical Stability Sample preparation and determination of mechanical parameters of cellulosic aerogels is a challenging task as the materials are highly prone to mechanical forces. On top of this comes the fact that other than aerogels from regenerated cellulose, bacterial cellulose grown under static conditions can be anisotropic with regard to mechanical strength.

Figure 2.8 shows the mechanical response profiles of a bacterial cellulose aerogel obtained by $scCO_2$ drying of the corresponding alcogel at 40°C and 10 MPa in the different directions in space. The stress–strain curves reveal a distinctly lower strength in vertical growth direction of the bacterial cellulose. The profiles are characterized by a short initial stage ($\leq 0.5\%$ strain in horizontal direction, $\leq 2\%$ strain in growth direction) and a largely linear slope before the flow limit is reached (10–12% strain). Once the material yields, the pore structure collapses increasingly and the gels are irreversibly damaged. Interestingly, only the curves recorded perpendicularly to the growth direction resembled those curves of comparable aerogels from regenerated cellulose, that is, the flow limit is followed by a plateau region where the stress–strain curves rise only slightly. In contrast, compression in growth direction requires increasing stress, although at lower absolute level

(a)   **(b)**

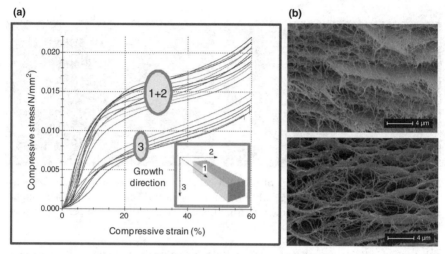

FIGURE 2.8 (a) Compressive stress–strain curves of scCO$_2$-dried aerogels showing the anisotropy of bacterial cellulose grown under static conditions. (b) SEM pictures (kindly provided by FZMB GmbH, Langensalza. Germany) of bacterial cellulose taken in growth direction (from an angle) (top) and perpendicular to it (bottom).

compared to the other two directions. For all directions, no sudden load drop was observed as it is often the case for brittle foam plastics.

The Young's modulus calculated from the linear slope of the elastic deformation was found to be in the range of 140–150 kPa for the two directions perpendicular to the growth direction, and 57 kPa in growth direction. Stress measured at 60% strain was 15 kPa for the former that is in good agreement with data from Maeda et al. (2006b) who reported a σ (60%) value of 20 kPa for bacterial cellulose aquogels. However, after converting the aquogels into the corresponding aerogels (scEtOH, 6.38 MPa, 243°C), a significantly increased σ (60%) value of 54 kPa was found by the same author indicating the occurrence of hornification reactions under these rather harsh drying conditions. Hornification is known to take place in lignocellulosic materials upon drying or removal of water leading to stiffening of the polymer structure (Diniz et al., 2004). Thus, it can be concluded that scCO$_2$ drying (31.1°C, 7.38 MPa) is advantageous with respect to a full preservation of bacterial cellulose properties.

Porosity The spread porosity of cellulosic aerogels commonly includes all the three IUPAC groups, that is, micro- (< 2 nm), meso- (2–50 nm), and macropores (> 50 nm). The characterization of such multiscale pore size distributions requires the application of multiple, complementary methods such as scanning electron microscopy (> 10 nm), environmental scanning electron microscopy (ESEM) (> 10 nm), small-angle X-ray scattering (SAXS) (2–100 nm), nitrogen sorption at 77K (0.35– 50 nm), and mercury intrusion (2 nm to about 1 mm) (Pinnow et al., 2008). However, different from most inorganic and organic aerogels, qualitative and quantitative characterization of the porosity of ultralightweight aerogels from nonderivatized plant or bacterial cellulose is hampered by the low mechanical stability of these materials.

While SAXS and nitrogen sorption can be applied without larger adaptation, sample preparation plays a key role in the quality and informational value of SEM and ESEM pictures, especially in the case of bacterial cellulose aerogels. Scanning electron micrographs of the inner structure have been obtained from aerogels that were either deep frozen (Nge et al., 2010) and subsequently cut or dissected by notching and subsequent application of tension and flexural stress (Liebner et al., 2009). The quantification of macropores and thus the determination of the multiscale pore size distribution are obviously the most challenging analytical tasks with cellulosic aerogels, in particular with those from bacterial cellulose due to the fragility of the ultralightweight, highly porous materials. For example, mercury intrusion—the standard method for studying macropores—cannot be applied as the aerogel structure completely collapses under the applied pressure. This is the case not only for BC aerogels but also for aerogels from plant cellulose (Sescousse et al., 2010), cellulose acetate (Fischer et al., 2006), polyurethane (Pirard et al., 2003), and silica (Majling et al., 1995; Scherer et al., 1995).

Thermoporosimetry (TPM) has been recently advanced by Nedelec and Baba as a technique that is considered to have the potential for characterizing soft matter of multiscale porosity (e.g., polymers and gels) in one single experiment as the impact of compressive stress exerted on the material by the used solvent is almost negligible compared to other methods (Baba et al., 2003; Nedelec and Baba, 2004). The first calibration for the use of TPM in the macropore range of soft matter has been recently accomplished with hybrid silica gels of hierarchical porosity. Thermoporosimetry is based on the principles of differential scanning calorimetry. For a given aerogel whose pores are filled with a certain solvent (lyogel), two exothermic peaks can be observed in the DSC curve upon cooling down the solvent slowly. While the first peak at higher temperature can be assigned to the melting point of the "free" solvent, the second signal peak at lower temperature is caused by the so-called confined solvent. As both shift and shape of the "confined solvent peak" are largely affected by the pore diameter and the width of the pore size distribution, these parameters can be determined within a wide range of pore diameters. On the macropore side the method is only limited by the transient from the "confined" to the "free" solvent state that also depends on the molecular features of the constituents of aerogel, their solvent affinity, and to a certain extent on the pore geometry of the materials. However, the solvent used in TPM needs to be carefully chosen to avoid underestimation of macropores due to pore contraction (Scherer et al., 1995).

The application of TPM (with o-xylene as solvent) to bacterial cellulose aerogels revealed a comparatively broad pore size distribution that peaks at around 70–150 nm.

2.2.4.2 Microfibrillated Cellulose
Microfibrillation has been recently advanced as a technique that allows the preparation of cellulose I nanocrystals from both plant and bacterial celluloses. Depending on the applied method, micelle cleavage preferably in amorphous regions and deaggregation can vary to a large extent.

Hydrolysis using strong mineral acids gives colloidal suspensions of rod-like, highly crystalline fibril aggregates of low aspect ratio (Battista, 1950; Ono et al., 2004). Extended hydrolysis in combination with sonication effects further disaggregation to cellulose whiskers that form mechanically rather weak networks clung together by hydrogen bonds (Araki et al., 1998; Fleming et al., 2001). Due to their rod-like character and surface charges, they can form liquid crystalline solutions (Ikkala et al., 2009). Johnsy et al. (2010) recently demonstrated that microfibrillation of bacterial cellulose using HCl affords the formation of nanofibrils that significantly improved the thermal stability and mechanical properties of PVA nanocomposites for food packaging applications.

Using the classical approach, that is, disintegration of cellulose under application of strong shear forces, a distinct higher mechanical stability of the microfibrillated cellulose (MFC) is obtained. This is due to high entanglement of the formed interconnected fibrils and fibril aggregates (diameters of 10–100 nm) and their comparatively high aspect ratio (Herrick et al., 1983). Pääkö et al. (2007) extended the classical approach by introducing a subsequent mild enzymatic hydrolysis that allows an improved control of the particle homogeneity. The aspect ratio of the resulting mixture of dominantly cellulose I fibrils (about 5 nm thickness) and fibril aggregates (about 10–20 nm thickness) was confirmed to be still sufficient for strong entanglement.

MFC obtained according to the above-mentioned approach has been recently demonstrated to be an interesting alternative source for aerogels of mainly cellulose I crystal structure (Pääkkö et al., 2008). In contrast to aquogels from regenerated cellulose, MFC aquogels can be directly converted into the corresponding aerogels by freeze-drying without suffering extensive pore collapsing. This is mainly due to their high robustness toward mechanical stress that overmatches that of comparable aquogels from acidic hydrolyzed cellulose by up to two orders of magnitude (for 1 wt% MFC containing aquogels) (Pääkkö et al., 2007). Laborious cross-linking, solvent exchange, or supercritical drying all aiming at a far-reaching preservation of the lyogel's pore features can thus be omitted (see Section 2.2.3).

MFC aerogels that were obtained from bleached sulfite softwood pulp by consecutive mechanical refining (Escher Wyss refiner), enzymatic treatment with endoglucanase (Novozym 476; Novozym A/S), repeated refining, and homogenization using z-shaped chambers were reported to have densities of about 20 mg/cm^3 and a porosity of approximately 98% (Pääkkö et al., 2008). The high maximum compressive strain of the obtained aerogels (about 70%) suggests a pronounced flexibility that distinguishes them from many other aerogels.

The above-described properties of highly entangled MFC scaffolds allow a facile modification and subsequent conversion into the corresponding aerogels. Pääkkö et al. (2008) reported on the preparation of electrically conducting flexible MFC aerogels with relatively high conductivity of around 1×10^{-2} S/cm that had been obtained by dipping the aerogels in an polyaniline–surfactant solution, subsequent rinsing off the unbound conducting polymer, and freeze-drying.

Bacterial cellulose nanocrystals obtained by HCl hydrolysis were used for preparing polyvinyl alcohol (PVA) films that had significantly improved thermal and mechanical stability compared to the reference material consisting of pure PVA (George et al., 2010).

2.2.5 Reinforcing Cellulosic Aerogels and Controlling Their Pore Size

2.2.5.1 Reinforcement

Composite Materials   Cellulosic aerogels and in particular BC aerogels are comparatively prone to mechanical stress that is considered a drawback for a couple of applications. Depending on the envisaged type of utilization and the required mechanical stability, different reinforcing approaches are generally applicable such as the preparation of *all*-cellulose composite materials, cross-linking, insertion of an interpenetrating network consisting of a second polymer, or the incorporation of network stiffening inorganic or organic particles. However, all of the aforementioned techniques are still in their infancies regarding their adaptation to cellulose aerogels as controlling the mechanical properties under preserving the interconnected, high porosity—certainly the most valuable feature of aerogels—is a challenging task.

All-cellulose composite materials are aiming at a homomacromolecular reinforcement of cellulosic aerogels and can be accomplished either by heterogeneous modification of a single cellulose source or by joint processing of cellulosic materials of different morphology.

All-cellulose-reinforced composite materials of high tensile strength comparable to those of conventional glass–fiber-reinforced composites can be prepared from microcrystalline cellulose as the only substrate (Abbott and Bismarck, 2010; Duchemin et al., 2010; Gindl and Keckes, 2005, 2007). Gindl et al. (2006) obtained optically transparent, self-reinforced, *all*-cellulose films by incompletely dissolving commercial MCC in LiCl/DMAc and subsequent regeneration of the dissolved cellulose portion with water. During dissolution, large MCC crystals and fiber fragments are progressively split into thinner crystals and cellulose fibrils (Duchemin et al., 2010). By drawing the mixture of undissolved MCC and regenerated cellulose in wet state and subsequent drying, self-reinforced cellulose films with preferred cellulose orientation can be obtained (Gindl and Keckes, 2007). Abbott and Bismarck (2010) controlled the level of reinforcement by tailoring the crystallinity of cellulose, which was accomplished by the extent of MCC dissolution in DMAc/LiCl. In addition to structural alterations, considerable modification of physical and thermal properties and of the moisture uptake has been reported. Duchemin et al. (2010) employed the same technique for preparing *all*-cellulose-reinforced aerogels. After regenerating the dissolved portion of MCC using water, the composite materials were freeze-dried. The density of *all*-cellulose-reinforced aerogels ranged from 116 up to 350 mg/cm^3 depending on the dissolved cellulose content (5–20 wt%). Flexural strength of the materials reached up to 8.1 MPa and their stiffness was as high as 280 MPa.

A heterogeneous approach using cellulosic materials of different morphology was studied by Nishino and coworkers (Nishino and Arimoto, 2007; Nishino et al., 2004) who prepared *all*-cellulose composites from pure cellulose and ramie fiber in DMAc/

LiCl. The obtained material exhibited excellent tensile strength of 480 MPa at 25°C, which is comparable to those of conventional glass–fiber-reinforced composites (Nishino et al., 2004).

Zhao et al. (2009) communicated recently the preparation of an *all*-cellulose ecocomposite material from Toyo filter paper and dewaxed rice husk (boiling ethanol/toluene, 2 h) that followed the above-described basic approach using BMIM chloride instead of DMAc/LiCl for partial cellulose dissolution. Although some of the obtained films also featured enhanced mechanical strength, the morphology of the silicates—present in considerable amounts in rice husk—seems largely to interfere with the mechanical properties of the products.

Interpenetrating Networks   The preparation of interpenetrating networks consisting of a second, mechanically more stable polymer and optional partial cross-linking between the two polymers is another approach for reinforcing cellulose aerogels. Although not being widely used so far for aerogels, numerous examples can be found in the literature for both inorganic (Aljaberi et al., 2009; Gill et al., 2005; Gonçalves et al., 2008; Tanaka and Kozuka, 2004) and organic interpenetrating polymers such as cellulose derivatives (Kamath et al., 1996), PLA, PLLA (Pei et al., 2010), or chitin (Karaaslan et al., 2010).

Modification of cellulose with silica is one of the approaches to improve the mechanical properties of cellulosic materials (Aljaberi et al., 2009; Gill et al., 2005; Gonçalves et al., 2008; Tanaka and Kozuka, 2004). Studies that had been dedicated to hybrid silica/cellulose xerogels and fibers evidenced the possible improvement of thermal and mechanical properties by silica modification (Amarasekara and Owereh, 2009; Barud et al., 2008a; Gonçalves et al., 2008; Hou et al., 2009; Kulpinski, 2005; Love et al., 2008; Maeda et al., 2006a; Pinto et al., 2008; Sequeira et al., 2007, 2009; Tanaka and Kozuka, 2004; Xie et al., 2009; Yano et al., 2008). Acid-catalyzed hydrolysis of tetraalkoxysilanes, in particular of tetraethoxysilane, and subsequent condensation of the intermediary formed silanols is obviously the most commonly used approach for silica modification of aerogels (Maeda et al., 2006a; Sequeira et al., 2007; Yano et al., 1998).

Introduction of silica-based reinforcing structures furthermore allows the controlled tuning of the material properties by adding certain functional groups. Gonçalves et al. (2008), for example, used fluorosiloxanes as silica precursor to get superhydrophobic surfaces. Reinforced shaped aerogels from cotton linters consisting of two interpenetrating networks of cellulose and silica were recently prepared from shaped Lyocell dopes by (i) regenerating cellulose with ethanol, (ii) subjecting the obtained shaped alcogels to sol-gel condensation with tetraethoxysilane as the principal network-forming compound, and (iii) drying the reinforced cellulose bodies with supercritical carbon dioxide (Litschauer et al., 2010). Acidic hydrolysis (0.1 M HCl) and condensation of TEOS, as sometimes recommended for silica modification of cellulose xerogels, was found to be not a suitable method to prepare homogeneously modified cellulosic aerogels because of three disadvantages: (i) the formation of a crust surrounding the aerogel, (ii) significant cellulose degradation and thus the decreased degree of polymerization, and (iii) reduced pore volumes of up to 15% (Litschauer et al., 2010). Silica modification under less acidic conditions (0.1 mM acetic acid) was found to maintain both cellulose integrity and open porous network structure of the cellulose aerogels to a large extent. Cohydrolysis and condensation of TEOS with 3-chloropropyl-trimethoxysilane (CPTMS) gave rise to very homogeneous samples that featured slightly smaller pore volumes compared to their unmodified counterparts. The best results in terms of high surface homogeneity, largely preserved porosity (pore volume: $0.65 \, cm^3/g$, specific surface area: $265 \, m^2/g$), and cellulose integrity were obtained when CPTMS was used as the only precursor. CPTMS, in addition, is a suitable donor to introduce anchor groups for further functionalization (Litschauer et al., 2010).

Interpenetrating networks have been also obtained for bacterial cellulose. Heßler and Klemm (2009) reported that cationic starch (2-hydroxy-3-trimethylammonium-propyl starch chloride, TMAP starch) added to the growth medium of *G. xylinum* forms stabilized double-network composites. In contrast to BC composites containing carboxymethyl cellulose (CMC) or methyl cellulose (MC) that have been reported by Seifert et al. (2004), the cationic starch derivative TMAP were shown to be incorporated into the wide-mashed BC prepolymer formed during the first 2 days of incubation (Heßler and Klemm, 2009).

Bacterial cellulose nanocomposites of up to 50% silica were obtained by Yano et al. (2008) by adding silica sol (Snowtex ST 0: pH 2–4 or Snowtex ST 20: pH 9.5–10.0) to the incubation medium for *G. xylinum*. Enhanced elastic modulus was observed for silica contents below 4% (ST 20) and 8.7% (ST 0), respectively, whereas higher silica contents led to reduced strength and modulus of the aerogels. X-ray diffraction measurements revealed that the silica particles disturbed the formation of ribbon-shaped fibrils and affected the preferential orientation of the (110) plane (Yano et al., 2008).

Interestingly, the addition of microstructuring organic sources such as multiwalled carbon nanotubes (MWNTs) in the growth medium of *A. xylinum* results in the formation of more rigid cellulosic pore walls, the latter having different crystal structures, cellulose I_α content, crystallinity index (CrI), and crystalline size compared to BC grown without the addition of MWNTs (Yan et al., 2008). A similar result was observed when wax spheres were added to the growth medium for controlling the pore size of the BC aquogels.

Cross-Linking   Cross-linking of both cellulose homopolymers or interpenetrating networks of cellulose and a second different polymer is another general approach for reinforcing cellulosic aerogels. In addition to hydroxyl groups that are present in high abundance, other anchor groups such as carbonyl or carboxyl groups, introduced by TEMPO (2,2,6,6-tetramethylpiperidine-1-oxyl) (Nge and Sugiyama, 2007; Saito and Isogai, 2004) or periodate oxidation (Kim et al., 2000; Malaprade, 1928), respectively, can also be used.

The preparation of dimensionally stable, highly porous cellulosic scaffolds for tissue engineering has been recently communicated by Nge et al. (2010) who cross-linked never-dried, microfibrillated, TEMPO-oxidized bacterial cellulose with chitosan (CTS) using 1-ethyl-3-(3-dimethylaminopropyl)carbodiimide (EDC)/

N-hydroxysuccinimide (NHS) as cross-linking mediator. After cross-linking at room temperature in slightly acidic aqueous medium (pH 5.5–6) (Araki et al., 2002; Nge et al., 2010), subsequent dialysis (removal of excess of EDC and NHS), and concentrating the suspension by immersing the dialysis tubing in aqueous polyethylene, the BC/CTS slurry was degassed and cast into moulds. The samples were deep frozen (e.g., −30°C) to solidify the solvent and to induce liquid–solid phase separation. The freeze-dried lyogels feature a macroporosity that seems to be well suited for tissue engineering applications. BC/CTS aerogels containing 60% of microfibrillated BC, for example, had an average pore diameter of $284 \pm 32\,\mu m$, which is more than three orders of magnitude larger than that of native bacterial cellulose and meets the requirements of cell scaffolding materials (Puppi et al., 2010). The Young's modulus of this particular sample was $702 \pm 142\,kPa$. Similar composites consisting of interpenetrating cellulose–chitosan networks have been prepared by Cai and Kim (2009) for electroactive actuator applications. They reported an optimum cellulose/chitosan ratio of 60/40 with regard to density and bending performance of the aerogels. Cellulose electroactive paper (EAPap) is an attractive material for constructing biomimetic actuators and microelectromechanical systems (MEMS) as it is lightweight, biodegradable, and features large displacement and low actuation voltage (Mahadeva et al., 2009).

2.2.5.2 Adjusting Pore Size There are numerous potential applications (see below) for cellulose aerogels that require certain pore features. Highly porous materials composed of micropores or nanopores are desired if the material is intended to find use for thermal or sound insulation purposes, as a filter or gas adsorbing material, catalyst support, or as a matrix for controlled release of bioactive compounds. On the other hand, their use as a biocompatible scaffolding material would require both small pores to provide sufficient stability and—equally important—the presence of large pores (≥ 100–$400\,\mu m$) for cell ingrowth, *in vivo* neovascularization, cell shaping, propagation and reorganization, and gene expression.

Several techniques have been developed for tailoring the pore size distribution of aerogels from both regenerated cellulose and bacterial cellulose. The addition of surfactants is one approach that has been proven to affect the porosity of cellulosic gels to some extent (Gavillon and Budtova, 2008). Budtova and coworkers, for example, added small amounts (0.1–1 wt%) of foaming agent Simusol® SL8 (alkyl polyglycosid; Seppic Inc., Fairfield, NC) to a solution of variable amounts of cellulose in 8 wt% aqueous NaOH. After stirring the mixture at 1000 rpm for 5 min at +5°C, stable air bubbles were formed. Immediate subsequent gelling of the mixture at 50°C (2 h) afforded aerogels consisting of distinctly larger pores (Sescousse et al., 2010).

Controlling the porosity of bacterial cellulose aquogels is another challenging task that recently gained considerable dedication of the scientific community as bacterial cellulose is considered to be a suitable, biocompatible material for cartilage repair or bone replacement (Heßler and Klemm, 2009; Huang et al., 2010).

First attempts aiming at the preparation of bacterial cellulose with tailored physical properties—the latter being primarily influenced by the ultrastructure of cellulose and by cellulose ribbon size in particular—have been made by Yamanaka et al. (1989). Later on it was found that the addition of xylan inhibits ribbon formation and decreases crystal size and ratio of cellulose I_α, while the xylan sticks on the surface of individual microfibrils (Iijima, 1991). Therefore, it was concluded that assembly and crystallization of glucan chains can be affected by adding water-soluble agents into the culture medium (Huang et al., 2010; Tokoh et al., 2002; Yan et al., 2008).

Heßler and Klemm (2009), for example, were able to tune the pore size of bacterial cellulose using different types of ethylene glycol (PEG). Thus, the addition of PEG 4000 reduced the average pore diameter, whereas PEG 400 or β-cyclodextrin caused the adverse effect. Interestingly, PEG is not incorporated into the BC network during biosynthesis as it has been reported to be the case for many other cosubstrates such as carboxymethyl cellulose, methyl cellulose, and hemicellulose (Heßler and Klemm, 2009).

The preparation of honeycomb-patterned bacterial cellulose networks has been described by Uraki et al. (2007). Based on the fact that cellulose-producing bacteria are able to move along linear microgrooves of a stripe-patterned cellulosic scaffold, they successfully attempted to control the bacterial movement using an agarose film scaffold with honeycomb-patterned grooves (concave type).

Another approach for tailoring the porosity of scaffolds makes use of porogens such as salts (e.g., NaCl) (Mikos et al., 1994), carbon nanotubes (Yan et al., 2008), paraffin (Ma and Choi, 2001), ice (Chen et al., 2001), natural hydrogels like gelatin (Zhou et al., 2005), and synthetic hydrogels such as poly(ethylene glycol) sebacic acid diacrylate (Kim et al., 2009). Some of the porogens have also been applied to control the porosity of bacterial aerogels. Currently, wax spheres that had been first applied by Ma and Choi (2001) as a new tool for preparing cell scaffolds from PLLA and PLGA gels are considered to be one of the most promising porogens for preparing BC aquogels or aerogels of well-defined porosity (Andersson et al., 2010; Bäckdahl et al., 2008). Paraffin spheres of defined size can be easily prepared by vigorous stirring of molten paraffin in hot PVA, rapid cooling, and sieve fractionation. Subsequently, the spheres are slightly compacted to ensure that the final BC network does not consist of confined bubbles but of interconnected pores. The conglomerate of wax spheres is then placed into the growth medium inoculated with the cellulose-producing bacteria strain. In the final step, the wax spheres are removed from the BC gel using a nonpolar solvent such as n-heptane (Ma and Choi, 2001). The formation of naturally reinforced pore walls is an additional effect reported by Andersson et al. (2010) that is obviously due to the formation of biofilms on the wax surface.

Cross-linking of TEMPO-oxidized, microfibrillated bacterial cellulose with chitosan using 1-ethyl-3-(3-dimethylaminopropyl)-carbodiimide hydrochloride (EDC) as a cross-linking mediator has been confirmed to be another measure for microstructuring (porosity of 120–280 µm) and tailoring the mechanical properties of scaffolds for tissue engineering (Zimmermann et al., 2009).

2.2.6 Potential Applications for Cellulosic Aerogels

The huge application potential of silica and inorganic aerogels has been comprehensively described by several authors (Akimov, 2003; Pajonk, 2003; Rolison and Dunn, 2001; Schmidt and Schwertfeger, 1998). Cellulosic aerogels can be used in similar applications, but they have the potential of entering new fields due to the intriguing properties of this renewable biomaterial. However, these fields can be different for cellulose of different origin. Aerogels from plant cellulose, for example, are expected to preferably find use in technical applications, as plant cellulose is much more plentiful and cheaper than bacterial cellulose. Furthermore, most technical applications have less requirements regarding cellulose purity, which renders plant cellulose an ideal raw material for applications such as thermal or acoustic insulation, shock adsorption, highly effective filter systems, catalyst carrier, or precursor material for highly porous carbogels.

The mechanical properties of cellulose microfibrils are of considerable interest for many applications. Bacterial cellulose microfibrils, for example, have a density of 1.6 g/cm^3, Young's modulus of about 138 GPa (Nishino et al., 1995), and a tensile strength of up to 2 GPa, which are on the same order of magnitude as those of aramide fibers such as Kevlar® (Yano et al., 2005).

Cellulose is obviously one of the oldest insulation materials and is also used these days in various application forms such as loose-fill, wet-spray, stabilized cellulose or low-dust cellulose. The thermal conductivity of loose-fill cellulose ranges from about 35 to 50 mW/(m K) (Eurima, 2002, 2004; Nicolajsen, 2005), but it is expected that cellulosic aerogels can reach a thermal conductivity that is as low as that of silica aerogels (14 mW/(m K); Spaceloft®) that are sold, for example, by Aspen Aerogels, Inc. (2010b). The thermal expansion coefficient of bacterial cellulose is about 1×10^{-7}°C^{-1}, which is similar to that of glass (Yano et al., 2005). Due to its high interconnected porosity, cellulosic aerogels are also excellent sound insulating materials.

Compared to classical "cone paper" (CP), sheets from bacterial cellulose are superior with respect to their use as vibration membranes for ear and speaker phones, respectively (Tabuchi, 2007). This is due to their high flexural rigidity with Young's moduli E of more than 30 GPa (CP: 1.5 GPa), sound propagation velocity C of around 5000 m/s (CP: 1600 m/s), and a comparatively high internal loss tan δ (CP: 0.001) (Nishi et al., 1990).

The interconnected pore system and high pore surface area render cellulosic aerogels excellent matrices for controlled deposition of finely spread nanoparticles of noble metals or metal oxides aiming at their use in electronic, optical, sensor, medical, or catalysis applications (Cai et al., 2009). Cellulosic aerogels furnished with silver, gold, and platinum nanoparticles have been obtained by dissolving cellulose in precooled (-10°C) aqueous, 4.6 wt% LiOH and 15 wt% urea-containing solution, gel formation using ethanol, and subsequent hydrothermal reduction using either cellulose or NaBH$_4$ as a reductant. Drying of the metal-carrying gel using scCO$_2$ yields the corresponding cellulosic aerogels that were shown to feature high transmittance, porosity, surface area, moderate thermal stability, and good mechanical strength (Cai et al., 2009).

A similar approach has been described by Olsson et al. (2010) who prepared flexible magnetic aerogels and stiff magnetic nanopaper from bacterial cellulose nanofibrils of 20–70 nm thickness. The latter were reported to act as templates for nonagglomerated growth of ferromagnetic cobalt ferrite nanoparticles (diameter 40–120 nm) that are obtained from aqueous solutions of $FeSO_4$ and $CoCl_2$ by controlling temperature and pH. The freeze-dried aerogels can be actuated by a small magnet, they adsorb water and release it upon compression.

For applications that require a higher mechanical toughness, wood nanofibrils can be used instead of plant fibers (Henriksson et al., 2008). Compounding cellulose with other polymers such as polylactic acid (Suryanegara et al., 2009, 2010) or chitosan (Cai and Kim, 2009) is another measure for tuning the mechanical properties of aerogels. Cellulose/chitosan composite materials consisting of two interpenetrating networks, for example, were shown to feature high bending displacement upon applying an electrical field that renders them promising materials for electroactive actuator applications (Cai and Kim, 2009). Different from earlier approaches with underivatized or polypyrrol/polyaniline-coated cellulose, the introduction of chitosan was found to reduce significantly the sensitivity of the materials toward humidity (Cai and Kim, 2009).

In addition to pore volume, average pore diameter, pore size distribution, and pore surface area (refer to Section 2.2.5.2), other application-related properties of the aerogels such as their sensitivity toward water vapor (adsorption, swelling, and aging) or nonpolar lipophilic solvents can also be controlled. Chemical vapor deposition (CVD) of perfluorodecyltrichlorosilane (PFDTS), for example, was used to uniformly coat aerogels from nanofibrillated cellulose to tune their oleophobicity, that is, their wetting properties toward nonpolar liquids (Aulin et al., 2010). Increased water sorption on the other hand can be obtained for cellulose/starch nanocomposites (Svagan et al., 2009). Chemical derivatization, that is, the introduction of hydrophilic or hydrophobic substituents, would be another approach in this respect.

Bacterial Cellulose in Cosmetic and Biomedical Applications The purity of bacterial cellulose along with its special features, that is, comparatively high crystallinity, mechanical stability (Bodin et al., 2007a), and biocompatibility (Helenius et al., 2006; Jonas and Farah, 1998; Yamanaka et al., 1989) allows its use in cosmetic and medical applications. Cosmetic applications comprise mainly the different fields of skin care where BC is used as light scatterer for sunscreen applications (Tabuchi, 2007) or as a stimulant for skin tissue regeneration (Sutherland, 1998).

In the field of biomedicine, bacterial cellulose is increasingly considered for artificial blood vessels (Klemm et al., 2001; Schumann et al., 2009), matrices for slow release applications of bioactive compounds (Haimer et al., 2010), and wound dressings for patients with burns, chronic skin ulcers, or other extensive loss of tissue (Czaja et al., 2006). BC acts as a temporary skin substitute with high mechanical strength in wet state. The high water capacity of the oxygen-permeable BC along with the high abundance of surface hydroxyl groups seems to stimulate regrowth of skin tissue, limiting infection at the same time (Sutherland, 1998). In this respect, bacterial cellulose has been confirmed to be one of the best materials to promote wound healing from second and third degree burns. In addition to significantly reducing scar tissue formation and pain sensitivity, bacterial cellulose promotes cell interaction and tissue regrowth. Furthermore, wound care materials can be easily and safely released from the burn site during treatment.

Reinforced, yet biodegradable, bacterial cellulose aerogels of appropriate porosity are considered as a potential alternative to organ or tissue transplantation as these materials are able to direct tissue generation from culture cells (Nge et al., 2010). Svensson et al. (2005) reported a significantly increased growth of bovine chondrocytes on bacterial scaffolds compared to tissue culture plastics and confirmed that BC scaffolds support the proliferation of chondrocytes. As BC does not furthermore induce significant activation of proinflammatory cytokine production during *in vitro* macrophage screening, bacterial cellulose scaffolds appear to be promising materials for tissue engineering of cartilage (Svensson et al., 2005). Accordingly, the first meniscus implants from bacterial cellulose have been prepared recently, and good implantation properties with regard to handling and fixation have been reported (Bodin et al., 2007b). Improving the fixation to the bone at the anterior and posterior meniscus horn of the implant by inducing bone formation, enhancing the compression strength by manipulating the fiber direction in radial direction, and modifying the porosity in the outer third of the cellulose implant are subject to further studies (Bodin et al., 2007b).

Cellulosic Scaffolds for Bone Replacement   Bones are fascinating, highly porous materials that are mainly composed of a collagen matrix that is reinforced with hydroxyapatite (HAp) nanocrystals (Hutchens et al., 2006). Although bone is a living tissue that has a great capacity of spontaneous regeneration (Puppi et al., 2010), it is the second most transplanted tissue after blood with a number of bone graft procedures that exceed one million cases per year (Zaborowska et al., 2010). Motivated by the limited availability of natural replacement tissue, including the risks for graft rejection and pathogen transmission, and stimulated by the recent advances in material science, considerable efforts have been made during the past decades to develop artificial bone graft materials. Tissue engineering offers almost unlimited possibilities for providing bone tissue replacements (Zaborowska et al., 2010). Artificial cell scaffolds, such as those for bone replacement, have considerable demands regarding biocompatibility, mechanical properties, porosity, and surface geometry to provide structural support for cell attachment, spreading, migration, proliferation, and differentiation. An interconnected, spread pore network with a highly porous, microstructured surface is required for diffusion of physiological nutrients and gases to cells, the removal of metabolic by-products from cells, *in vitro* cell adhesion, ingrowth, and *in vivo* neovascularization (Puppi et al., 2010). The size, distribution, and shape of pores furthermore control cell survival, signaling, growth, propagation, reorganization, cell shaping, and gene expression.

The utilization of highly open porous inorganic materials as potential scaffolding material for bone repair or even bone replacement has been an exciting topic in material research for about 20 years. One of the obviously most fascinating inorganic materials for bone repair is bioactive sol–gel glass of the CaO-P_2O_5-SiO_2 type that was first prepared in 1991 (Li et al., 1991). Much research has been conducted in this field since then and led to numerous improved bioglass compositions such as the clinically applied Bioglass 45S5® (Hench, 2006) or the more recently developed quaternary system SiO_2-CaO-P_2O_5-MgO. The latter was shown to support the growth of human fetal osteoblastic cells to be nontoxic and compatible in segmental defects in the goat model *in vivo* (Saboori et al., 2009). Biomineralization, that is, the formation of a nanocrystalline carbonate apatite-like phase (CHAp) is initialized by leaching of Na^+ and Ca^{2+} cations to the surrounding fluids, followed by the formation of silanols(Si–OH) at the glass surface. These reactions lead to the formation of a silica-gel layer. The latter is slowly transformed to an amorphous calcium phosphate layer as calcium and phosphorous atoms from the surrounding solution are incorporated into the network structure. Simultaneous incorporation of carbonates eventually leads to crystallization of CHAp (Hench and Andersson, 1993). However, their brittleness and comparatively low mechanical stability are considered to be serious drawbacks of bioceramics and bioglasses, respectively.

Among the organic materials that come into consideration for tissue engineering, natural polymers such as collagen, fibrinogen, starch, chitosan, and, in particular, cellulose are superior for both their high natural abundance and their low immunogenic potential (Zaborowska et al., 2010). Cellulosic aerogels and in particular bacterial cellulose have been recently moved into the limelight of artificial scaffold research. The preparation of highly porous cellulose/hydroxyapatite composite materials is one major strategy in this respect as hydroxyapatite $Ca_{10}(PO_4)_6(OH)_2$ is osteoconductive, biocompatible, and bioactive. Model studies have shown that the presence of hydroxyapatite increases the expression of mRNA encoding the bone matrix proteins osteocalcin, osteopontin, and bone sialoprotein (Fang et al., 2009; Liu et al., 2009). Last but not least, hydroxyapatite considerably increases the mechanical strength of cellulosic scaffolds, thus providing improved conditions for cell growth, osteogenic differentiation, and bone grafting.

A facile, rapid microwave-assisted synthesis of highly porous cellulosic matrices containing homogeneously distributed HAp nanorods has been described by Ma et al. (2010) who codissolved microcrystalline cellulose with $CaCl_2$ and NaH_2PO_4 in *N,N*-dimethylacetamide (DMAc) and subjected the mixture to a microwave-assisted thermal treatment at 150°C.

In addition to plant cellulose, bacterial cellulose has been also confirmed to permit the formation of calcium-deficient hydroxyapatite (cdHAp) nanocrystallites under aqueous conditions at ambient pH and temperature (Hutchens et al., 2009). As the porosity of naturally grown bacterial cellulose is considered to be too low for tissue engineering, several approaches have been studied aiming at the preparation of materials that feature a hierarchical pore system with a large portion of macropores, preferably in the range of 50–400 µm. The utilization of wax spheres as easily removable porogens (refer to Section 2.2.5.2) seems to be the most promising approach to date in this respect, as this approach has been confirmed to allow the preparation of materials with interconnected porosity. Correspondingly, it leads to larger cell concentration and cluster formation within the pores compared to previous studies (Zaborowska et al., 2010).

The introduction of surface charges to cellulose has been known for a while to increase both biodegradation and molecular recognition (Hayashi, 1994; Zaborowska et al., 2010). Periodate oxidation of bacterial cellulose, for example, was shown to retain porosity and ability for the formation of calcium-deficient hydroxyapatite, but renders the composite degradable and thus more suitable for bone regeneration (Hutchens et al., 2009). Furthermore, negative charges on the cellulose surface have been confirmed to initiate nucleation of calcium-deficient hydroxyapatite (Zimmermann et al., 2009). Bacterial cellulose with various surface morphologies (pellicles and tubes) was negatively charged by adsorption of carboxymethyl cellulose, and cdHAp was grown *in vitro* via dynamic simulated body fluid (SBF) treatments over a one week period. The obtained BC/cdHAp composite materials were found to be superior with regard to cell attachment to neat bacterial cellulose scaffolds.

However, despite the above-described advances regarding cell attachment, growth, and osteogenic differentiation, the major problem with cellulose aerogels as bone graft material is the insufficient binding and crystallization tendency of hydroxyapatite on the cellulose matrix (Wan et al., 2006, 2007). Even if cdHAp nanoparticles are homogeneously formed and deposited within the pore system of the cellulosic aerogel, there is no chemical bond between cellulose as the "organic" and hydroxyapatite as the "inorganic" part, so that a complete biomineralization and osseointegration are considered to be impossible (Granja et al., 2001a).

Grafting of negatively charged phosphorous-containing groups onto cellulose has been confirmed to be a promising technique for improving the mineralizing properties of cellulose (Wan et al., 2007). As with phosphorylated synthetic polymers such as poly(ethylene terephthalate) (Kato et al., 1996) or biopolymers such as chitin (Yokogawa et al., 1997), phosphorylated cotton linters (Mucalo et al., 1995), Avicell® PH-101 (Granja et al., 2001a), and bacterial cellulose (Wan et al., 2006, 2007) have been found to induce the formation of cdHAp in simulated body fluid. Phosphorylated Avicell® PH-101 was furthermore confirmed to be nontoxic in cultured human osteoblasts and fibroblasts (Granja et al., 2001b).

Wan et al. (2006) demonstrated that bacterial cellulose/hydroxyapatite composites are formed only after covalently grafting phosphorous groups onto the cellulose. The required cellulose phosphates were obtained by activating BC pellicles in a solution of urea in DMF at 110°C and subsequent phosphorylation using a solution of 98 wt% H_3PO_4 in DMF (136°C, 1 h). The formation of cdHAp layer was accomplished by immersing the phosphorylated BC (DS 0.067) in an aqueous solution of $CaCl_2$ (37°C, 3 days) and subsequent soaking in 1.5 simulated body fluid at 37°C for 7–14 days (Wan et al., 2007).

Similar results have been reported by Granja et al. (2001a, 2001b) who studied cdHAp formation on the surface of regenerated cellulose xanthogenate disks after phosphorylating the samples according to Touey and Kingsport (1956) using a mixture of H_3PO_4, P_4O_{10}, and triethyl phosphate in hexanol. It was also shown that

Hap formation is suppressed for water-soluble phosphates of high DS or when the phosphates were not pretreated with $CaCl_2$ to form the corresponding salts. Furthermore, the degree of phosphorylation was found to influence the extent of surface mineralization, that is, formation of cdHAp with the highest values at moderate degrees of surface phosphorylation (Granja et al., 2001a). Similarly, better attachment and proliferation of cultured human bone marrow stromal cells (HBMSC) were observed on unmodified cellulosic scaffolds compared to highly negatively charged, hydrophilic samples having a high degree of phosphorylation. This result may illustrate the well-known fact that there is still insufficient knowledge regarding the mechanisms that promote cell attachment, growth, proliferation, or differentiation on foreign porous materials (Granja et al., 2006).

Our studies confirmed good hemocompatibility of cellulose phosphate aquogels and aerogels that were prepared from phosphorylated cotton linters and hardwood prehydrolysis Kraft pulp of low DS (≤ 0.5), as blood platelet activation, coagulation (hemostasis), and complement activation on the alternative pathway (inflammatory response) were low. Calcium saturation of the phosphate groups (aqueous $CaCl_2$ solution) was demonstrated to render the aerogels promising cell scaffolding materials with regard to osteointegration as blood platelets and growth factors were simulatenously activated without inducing an inflammatory response. Cell culture experiments employing human mesenchymal stromal cells showed a robust growth and spontaneous osteogenic differentiation on the studied cellulose phosphates.

The studied ultralightweight, highly porous cellulose phosphate aquogels and aerogels were prepared via the Lyocell route by (i) dissolving cellulose phosphate in stabilized N-methylmorpholine-N-oxide monohydrate, (ii) shaping the Lyocell dope by casting, (iii) regenerating cellulose phosphate with ethanol (aerogels) or water (aquogels), and (iv) converting alcogels into the corresponding aerogels by $scCO_2$ drying at 40°C and 10 MPa.

Processing of cellulose derivatives containing large amounts of acidic groups such as cellulose phosphates, sulfates, or oxidized cellulose according to the Lyocell procedure was hitherto considered not to be viable as acidic groups are potent inducers of Polonowski-type reactions, and hence increase these side reactions considerably (Rosenau et al., 2001). However, using the sacrifice stabilizer N-benzylmorpholine-N-oxide that was recently developed in our group (Rosenau et al., 2005b) as a very efficient scavenger for highly reactive carbonium–iminium ions, cellulose phosphates can also be safely processed under Lyocell conditions.

Cellulose-Based Carbon Aerogels Similar to aerogels from silica, metals, oxides, and synthetic and natural organic polymers, carbogels of controlled porosity are expected to find use in an increasing field of applications. Carbon aerogels of tailored porosity have been confirmed to be promising materials for gas separation and adsorption (Carrasco-Marín et al., 2009; Maldonado-Hòdar et al., 2000), catalysis (Gomes et al., 2004; Maillard et al., 2008; Maldonado-Hòdar et al., 2000; Smirnova et al., 2005), hydrogen storage (Babel and Jurewicz, 2008; Jordá-Beneyto et al., 2007; Schimmel et al., 2004), and electrochemical applications where carbogels have been used as electrode material for fuel cells (Guilminot et al., 2007, 2008; Marie et al., 2009), batteries (Béguin and Frackowiak, 2006), or in electrical double-layer capacitors (EDLCs) (Chmiola et al., 2006a, 2006b; Frackowiak and Béguin, 2002).

Carbogels are commonly prepared by pyrolysis of appropriate organic aerogels at temperatures up to $> 1000°C$. Suitable precursors are polymers with a high percentage of aromatic or heteroaromatic moieties, with functionalities that undergo radical reaction upon pyrolysis, or moieties that prevent parts of the polymer molecules from thermal degradation up to a certain temperature level. Nanoporous carbon aerogels can be obtained, for example, by pyrolysis of resorcinol/formaldehyde gels, which could be cost-effectively manufactured from RF wet gels by a modified ambient drying technique using acetone exchange/controlled evaporation instead of conventional supercritical drying (Hwang and Hyun, 2004).

Carbogels are promising materials as electrodes for capacitive deionization (GDI) units and electrical double-layer capacitors. The latter are also referred to as supercapacitors that store energy via separation of charges across a polarized electrode/ electrolyte interface (Zheng et al., 1997). Supercapacitors are a special type of capacitors that bridge the gap between batteries (accumulators) and conventional capacitors. They are able (i) to store more energy than conventional capacitors, (ii) to release a higher voltage than batteries, (iii)to store electrical energy almost lossless for a long time period, and (iv) they can be charged and discharged very quickly. Potential applications of supercapacitors are uninterruptible power supplies for bridging electrical power outage, short-term supply of high electrical power such as for starting up industrial machinery, and storage of relatively short energy impulses.

In addition to voltage, the surface of the interface between electroconductive solid and surrounding electrolyte is the main criterion determining charge storage. Suitable electrolytes in combination with carbogels are aqueous KOH, H_2SO_4, or polymeric electrolytes. Although carbogels have been prepared from a multitude of synthetic organic precursors such as resorcinol/formaldehyde resins, resol/melamin cocondensates, PVC, polyurethanes, cellulose acetates, and so on, tailoring the pores of carbon aerogels throughout the gel formation, drying, modification, and pyrolysis remains a very challenging issue. The latter is even more demanding as different electrochemical applications require different pore and pore surface properties. Thus, the porosity of electrodes for EDCL applications should be in the micropore range, whereas proton exchange membrane fuel cells (PEMFCs) require a high mesoporosity. Macroporous materials are required for supercapacitor applications especially if polymer electrolytes are used.

Mesoporous cellulose-based carbon aerogels with pore surface areas of 117– 165 m^2/g for EDLC applications have been recently obtained by pyrolysis of crosslinked cellulose acetate under nitrogen atmosphere ($T_{max} = 1000°C$, 4°C/min) (Grzyb et al., 2010). Cross-linking was performed with polymeric diphenylmethane diisocyanate in dry acetone, catalyzed by 1.4-diazabicyclo[2.2.2]octane (DABCO® TMR). Subsequent consecutive surface treatment with 4N HNO_3 and ammonia at 400°C (3 h) was shown to further increase the pore surface area by establishing a higher microporosity.

Cellulose-based carbon aerogels doped by platinum nanoparticles have been reported by several authors to be promising substrates for oxygen reduction electrocatalysis as their properties can compete with state-of-the-art Pt/CB (carbon black) electrocatalysts (Guilminot et al., 2007, 2008; Rooke et al., 2010). Carbogels from microcrystalline cellulose (Avicell Ph-101), for example, were obtained by dissolving the cellulose in precooled aqueous NaOH ($-6°C$), gelling at $50°C$, regenerating cellulose with water, replacing water by aceton, scCO$_2$ drying, and converting the aerogels into carbogels by pyrolysis at $830°C$ ($1050°C$) in nitrogen atmosphere (Rooke et al., 2010). Subsequently, the carbogels were doped with platinum particles by (i) thermal activation in CO$_2$ atmosphere, (ii) impregnation with H$_2$PtCl$_6$, and (iii) platinum salt reduction using either hydrogen ($300–400°C$) (Rooke et al., 2010) or NaBH$_4$ (Guilminot et al., 2008; Rooke et al., 2010). Similar materials featuring pore surface areas of $400–450 \, m^2/g$ have been obtained from cellulose acetate aerogels after pyrolysis at $830°C$, preactivation of the carbogels at $800°C$ in CO$_2$ atmosphere, subsequent impregnation with H$_6$PtCl$_6$, and chemical reduction using NaBH$_4$ (Guilminot et al., 2007).

Highly crystalline native cellulose (e.g., bacterial and algal celluloses and ramie fibers) seems to be suitable raw material also for the preparation of carbon aerogels, as Kim et al. (2001) were able to show that the ultrastructure of the parent material and in particular their microfibrillar structure are largely retained throughout the carbonization ($\geq 500°C$) and subsequent graphitization ($\geq 2000°C$) processes.

Polymer/carbon aerogel hybrid materials have been obtained by surface coating of the carbogel with ultrathin (< 15 nm), conformal, electroactive polymer coatings based on the self-limiting electropolymerization of o-methoxyaniline and related arylamine monomers. The studied coating approach allows a far-reaching retention of the pore features of the parent carbon aerogel and strongly increases the energy storing capacities of the resulting hybrid structures due to faradaic pseudocapacitance reactions (Long et al., 2004).

2.3 SUMMARY

Within only about half a decade, the international research community in the fields of polysaccharide chemistry, material research, and process engineering has succeeded to add a new rapidly growing branch—*cellulosic aerogels*—to the field of tailored, highly porous functional materials.

Cellulosic aerogels were obtained from both plant and bacterial celluloses. Different approaches have been studied that allow the preparation of monolithic bodies that consist of either predominantly cellulose I_α or I_β or (regenerated) cellulose II. Surface modification, cross-linking, improvement of nanofiber and fibril entanglement, preparation of interpenetrating networks with additional inorganic or organic polymers, and modification of the growth media used in bacterial cellulose production are some of the measures that have been comprehensively studied for reinforcing cellulosic aerogels aiming at a far-reaching retention of their pore features. Hydrophilization, oleophilization, or introduction of anchor groups are further techniques that have been adapted to extend the application potential of the novel functional materials. The preparation of cellulosic aerogels with tailored macroporosity, their use in biomedical applications, and the development of reliable techniques for a multiscale characterization of porosity are only few subjects of current research projects.

2.4 FUTURE PERSPECTIVES

In view of the recent efforts and advances in the biorefinery sector and the permanently increasing public awareness of renewable sources, it can be assumed that at least the abundant biopolymers—among them cellulose—will continue to move increasingly into the limelight of biomaterial research. Cellulosic aerogels as very lightweight, highly porous functional materials are considered to have a large application potential due to their fascinating properties. The current number of research projects in this field spread over many countries around the world, responsible and increasing funding of such projects by national and international authorities, and the awakening curiosity and interest of the respective industry branches are strong indicators of this development and important signals toward new resource-saving concepts.

REFERENCES

Aaltonen, O., and O. Jauhiainen (2009). The preparation of lignocellulosic aerogels from ionic liquid solutions. *Carbohydr. Polym.* **75**(1):125–129.

Abbott, A., and A. Bismarck (2010). Self-reinforced cellulose nanocomposites. *Cellulose* **17**:779–791.

Akimov, Y. K. (2002) Fields of applications of aerogels. *Instrum. Exp. Tech.* **46**:287–299.

Akimov, Y. K. (2003). Fields of application of aerogels (Review). *Instrum. Exp. Tech.* **46**(3):287–299.

Aljaberi, A., A. Chatterji, N. H. Shah, and H. K. Sandhu (2009). Functional performance of silicified microcrystalline cellulose versus microcrystalline cellulose: a case study. *Drug Dev. Ind. Pharm.* **35**(9):1066–1071.

Amarasekara, A. S., and O. S. Owereh (2009). Homogeneous phase synthesis of cellulose carbamate silica hybrid materials using 1-*n*-butyl-3-methylimidazolium chloride ionic liquid medium. *Carbohydr. Polym.* **78**(3):635–638.

Andersson, J., H. Stenhamre, H. Bäckdahl, and P. Gatenholm (2010). Behavior of human chondrocytes in engineered porous bacterial cellulose scaffolds. *J. Biomed. Mater. Res. A* **94A**(4):1124–1132.

Araki, J., M. Wada, S. Kuga, and T. Okano (1998). Flow properties of microcrystalline cellulose suspension prepared by acid treatment of native cellulose. *Colloids Surf. A* **142**(1):75–82.

Araki, J., S. Kuga, and J. Magoshi (2002). Influence of reagent addition on carbodiimide-mediated amidation for poly(ethylene glycol) grafting. *J. Appl. Polym. Sci.* **85**(6): 1349–1352.

Aspen Aerogels, Inc. (2010a). BASF venture capital invests in Aspen Aerogels. Available at http://press.aerogel.com/index.php?s=118&item=267. Accessed Nov. 27, 2010.

Aspen Aerogels, Inc. (2010b). Spaceloft. Available at http://www.aerogel.com/products/pdf/ Spaceloft_DS_GERMAN.pdf. Accessed Sept. 24, 2010. http://press.aerogel.com/index .php?s=25881&item=66360

Aulin, C., J. Netrval, L. Wågberg, and T. Lindström (2010). Aerogels from nanofibrillated cellulose with tunable oleophobicity. *Soft Matter* **6**(14):3298–3305.

Baba, M., J. M. Nedelec, J. Lacoste, J. L. Gardette, and M. Morel (2003). Crosslinking of elastomers resulting from ageing: use of thermoporosimetry to characterise the polymeric network with *n*-heptane as condensate. *Polym. Degrad. Stab.* **80**(2):305–313.

Babel, K., and K. Jurewicz (2008). KOH activated lignin based nanostructured carbon exhibiting high hydrogen electrosorption. *Carbon* **46**(14):1948–1956.

Bäckdahl, H., M. Esguerra, D. Delbro, B. Risberg, and P. Gatenholm (2008). Engineering microporosity in bacterial cellulose scaffolds. *J. Tissue Eng. Regen. Med.* **2**(6):320–330.

Bag, S., P. N. Trikalitis, P. J. Chupas, G. S. Armatas, and M. G. Kanatzidis (2007). Porous semiconducting gels and aerogels from chalcogenide clusters. *Science* **317**:490–493.

Bag, S., A. F. Gaudette, M. E. Bussell, and M. G. Kanatzidis (2009). Spongy chalcogels of non-platinum metals act as effective hydrodesulfurization catalysts. *Nat. Chem.* **1**(3):217–224.

Barthel, S., and T. Heinze (2006). Acylation and carbanilation of cellulose in ionic liquids. *Green Chem.* **8**:301.

Barud, H. S., R. M. N. Assuncao, M. A. U. Martines, J. Dexpert-Ghys, R. F. C. Marques, Y. M., and S. J. L. Ribeiro (2008a). Bacterial cellulose–silica organic–inorganic hybrids. *J. Sol-Gel Sci. Technol.* **46**:363–367.

Barud, H. S., C. Barrios, T. Regiani, R. F. C. Marques, M. Verelst, J. Dexpert-Ghys, Y. Messaddeq, and S. J. L. Ribeiro (2008b). Self-supported silver nanoparticles containing bacterial cellulose membranes. *Mater. Sci. Eng. C* **28**:515–518.

Basta, N. (1985). Supercritical fluids: still seeking acceptance. *Chem. Eng.* **92**(3):14.

Battista, O. A. (1950). Hydrolysis and crystallization of cellulose. *Ind. Eng. Chem.* **42**(3):502–507.

Béguin, F., and E. Frackowiak (2006). Nanotextured carbons for electrochemical energy storage. *Nanomaterials Handbook*, CRC Press.

Biesmans, G., A. Mertens, L. Duffours, T. Woigner, and J. Phalippou (1998). Polyurethane based organic aerogels and their transformation into carbon aerogels. *J. Non-Cryst. Solids* **225**:64–68.

Biganska, O., and P. Navard (2005). Kinetics of precipitation of cellulose from cellulose–NMMO–water solutions. *Biomacromolecules* **6**:1948–1953.

Biganska, O., and P. Navard (2009). Morphology of cellulose objects regenerated from cellulose–*N*-methylmorpholine *N*-oxide–water solutions. *Cellulose* **16**(2):179–188.

Bodin, A., H. Bäckdahl, H. Fink, L. Gustafsson, B. Risberg, and P. Gatenholm (2007a). Influence of cultivation conditions on mechanical and morphological properties of bacterial cellulose tubes. *Biotechnol. Bioeng.* **97**(2):425–434.

Bodin, A., S. Concaro, M. Brittberg, and P. Gatenholm (2007b). Bacterial cellulose as a potential meniscus implant. *J. Tissue Eng. Regen. Med.* **1**(5):406–408.

Brunner, G.,editor, (2005). *Supercritical Fluids as Solvents and Reaction Media*. Elsevier Science & Technology.

Bungay, H. R., and G. C. Serafica (1997). Production of microbial cellulose. US patent 6,071,727.

Burchell, M. J., R. Thomson, and H. Yano (1999). Capture of hypervelocity particles in aerogel in ground laboratory and low earth orbit. *Planet. Space Sci.* **47**:189–204.

Cai, Z., and J. Kim (2009). Cellulose–chitosan interpenetrating polymer network for electro-active paper actuator. *J. Appl. Polym. Sci.* **114**(1):288–297.

Cai, J., L. Zhang, J. Zhou, H. Li, H. Chen, and H. Jin (2004). Novel fibers prepared from cellulose in NaOH/urea aqueous solution. *Macromol. Rapid Commun.* **25**(17): 1558–1562.

Cai, J., L. Zhang, J. Zhou, H. Qi, H. Chen, T. Kondo, X. Chen, and B. Chu (2007). Multifilament fibers based on dissolution of cellulose in NaOH/urea aqueous solution: structure and properties. *Adv. Mater.* **19**(6):821–825.

Cai, J., S. Kimura, M. Wada, S. Kuga, and L. Zhang (2008). Cellulose aerogels from aqueous alkali hydroxide–urea solution. *ChemSusChem* **1**:149–154.

Cai, T., H. Zhang, Q. Guo, H. Shao, and X. Hu (2010). Structure and properties of cellulose fibers from ionic liquids. *J. Appl. Polym. Sci.* **115**(2):1047–1053.

Carrasco-Marín, F., D. Fairén-Jiménez, and C. Moreno-Castilla (2009). Carbon aerogels from gallic acid-resorcinol mixtures as adsorbents of benzene, toluene and xylenes from dry and wet air under dynamic conditions. *Carbon* **47**(2):463–469.

Chen, G., T. Ushida, and T. Tateishi (2001). Development of biodegradable porous scaffolds for tissue engineering. *Mater. Sci. Eng. C* **17**(1–2):63–69.

Cheng, K.-C., J. M. Catchmark, and A. Demirci (2009). Enhanced production of bacterial cellulose by using a biofilm reactor and its material property analysis. *J. Biol. Eng.* **3**:12.

Cherepy, N. J., A. F. Jankowski, T. Tillotson, and K. Fiet (2006). Carbon aerogel and xerogel fuels for fuel cells and batteries. US patent WO/2006/025993.

Chmiola, J., G. Yushin, R. Dash, and Y. Gogotsi (2006a). Effect of pore size and surface area of carbide derived carbons on specific capacitance. *J. Power Sources* **158**(1):765–772.

Chmiola, J., G. Yushin, Y. Gogotsi, C. Portet, P. Simon, and P. L. Taberna (2006b). Anomalous increase in carbon capacitance at pore sizes less than 1 nanometer. *Science* **313**(5794):1760–1763.

Czaja, W., A. Krystynowicz, S. Bielecki, and J. R. M. Brown (2006). Microbial cellulose: the natural power to heal wounds. *Biomaterials* **27**(2):145–151.

Dai, S., Y. H. Ju, H. J. Gao, J. S. Lin, S. J. Pennycook, and C. E. Barnes (2000). Preparation of silica aerogel using ionic liquids as solvents. *Chem. Commun.* **3**:243–244.

DECHEMA, G. f. C. T. u. B. 2007. Entwicklung neuartiger Schutzschichtsysteme für extrem korrosive Hochtemperaturumgebungen: Schlussbericht für den Zeitraum 01.11.04 bis 28.02.2007, Frankfurt, p. 51.

Deng, M., Q. Zhou, A. Du, J. van Kasteren, and Y. Wang (2009). Preparation of nanoporous cellulose foams from cellulose–ionic liquid solutions. *Mater. Lett.* **63**:1851–1854.

Diniz, J. M. B. F., M. H. Gil, and J. A. A. M. Castro (2004). Hornification: its origin and interpretation in wood pulps. *Wood Sci. Technol.* **37**:489–494.

Domínguez, G., A. J. Westphal, M. L. F. Phillips, and S. M. Jones (2003). A fluorescent aerogel for capture and identification of interplanetary and interstellar dust. *Astrophys. J.* **592**:631–635.

Duchemin, B. J. C., M. P. Staiger, N. Tucker, and R. H. Newman (2010). Aerocellulose based on all-cellulose composites. *J. Appl. Polym. Sci.* **115**(1):216–221.

Ebner, G., S. Schiehser, A. Potthast, and T. Rosenau (2008). Side reaction of cellulose with common 1-alkyl-3-methylimidazolium-based ionic liquids. *Tetrahedron Lett.* **49**(51): 7322–7324.

Egal, M. (2006). Thèse de doctorat. Ecole Nationale Supérieure des Mines de Paris, Sophia-Antipolis.

Eurima (2002) The Contribution of Mineral Wool and Other Thermal Insulation Materials to Energy Savings and Climate Protection in Europe, ECOFYS, Cologne, p. 36.

Eurima (2004) Cellulose Fibre Insulation, Fact Sheet, p. 7.

Fang, B., Y.-Z. Wan, T.-T. Tang, C. Gao, and K.-R. Dai, (2009). Proliferation and osteoblastic differentiation of human bone marrow stromal cells on hydroxyapatite/bacterial cellulose nanocomposite scaffolds. *Tissue Eng. A* **15**(5):1091–1098.

Farmer, J. C., D. Fix, G. V. Mack, R. W. Pekala, and J. F. Poco (1996). Capacitive deionization of NaCl and $NaNO_3$ solutions with carbon aerogel electrodes. *J. Electrochem. Soc.* **143**:159–169.

Feldmann, M., and P. Desrochers (2003). Research universities and local economic development: lessons from the history of the Johns Hopkins University. *Ind. Innov.* **10**(1):5–24.

Fialkowski, M., K. J. M. Bishop, R. Klajn, S. K. Smoukov, C. J. Campbell, and B. A. Grzybowski (2006). Principles and implementations of dissipative (dynamic) self-assembly. *J. Phys. Chem. B* **110**(6):2482–2496.

Fischer, S. (2003). Anorganische Salzschmelzen- ein unkonventionelles Löse- und Reaktionsmedium für Cellulose Habilitation. Ph.D. thesis, TU Bergakademie Freiberg, Freiberg, Germany.

Fischer, F., A. Rigacci, R. Pirard, Berthon-Fabry, S., and P. Achard (2006). Cellulose-based aerogels. *Polymer* **47**:7636–7645.

Fleming, K., D. G. Gray, and S. Matthews (2001). Cellulose crystallites. *Chemistry* **7**(9): 1831–1836.

Frackowiak, E., and F. Béguin (2002). Electrochemical storage of energy in carbon nanotubes and nanostructured carbons. *Carbon* **40**(10):1775–1787.

Fricke, J. (1993). The unbeatable lightness of aerogels. *New Sci.* **137**:31–34.

Fricke, J., and A. Emmerling (1992). Aerogels: preparation, properties, applications. *Struct. Bonding* **77**:37–87.

Gavillon, R., and T. Budtova (2007). Kinetics of cellulose regeneration from cellulose-NaOH water gels and comparison with cellulose-NMMO-water solutions. *Biomacromolecules* **8**:424–432.

Gavillon, R., and T. Budtova (2008). Aerocellulose: new highly porous cellulose prepared from cellulose-NaOH aqueous solutions. *Biomacromolecules* **9**:269–277.

Geyer, U., T. Heinze, A. Stein, D. Klemm, S. Marsch, D. Schumann, and H. P. Schmauder (1994). Formation, derivatization and applications of bacterial cellulose. *Int. J. Biol. Macromol.* **16**(6):343–347.

Gill, R. S., M. Marquez, and G. Larsen (2005). Molecular imprinting of a cellulose/silica composite with caffeine and its characterization. *Microporous Mesoporous Mater.* **85**(1–2):129–135.

Gindl, W., and J. Keckes (2005). All-cellulose nanocomposite. *Polymer* **46**(23):10221–10225.

Gindl, W., and J. Keckes (2007). Drawing of self-reinforced cellulose films. *J. Appl. Polym. Sci.* **103**(4):2703–2708.

Gindl, W., K. J. Martinschitz, P. Boesecke, and J. Keckes (2006). Structural changes during tensile testing of an all-cellulose composite by *in situ* synchrotron X-ray diffraction. *Compos. Sci. Technol.* **66**(15):2639–2647.

Gomes, H. T., P. V. Samant, P. Serp, P. Kalck, J. L. Figueiredo, and J. L. Faria (2004). Carbon nanotubes and xerogels as supports of well-dispersed Pt catalysts for environmental applications. *Appl. Catal. B* **54**(3):175–182.

Gonçalves, G., P. A. A. P. Marques, T. Trindade, C. P. Neto, and A. Gandini (2008). Superhydrophobic cellulose nanocomposites. *J. Colloid Interface Sci.* **324**(1–2):42–46.

Granja, P. L., L. Pouysegu, B. De Jeso, F. Rouais, C. Baquey, and M. A. Barbosa (2001a). Cellulose phosphates as biomaterials: mineralisation of chemically modified regenerated cellulose hydrogels. *J. Mater. Sci.* **36**:2163–2172.

Granja, P. L., L. Pouységu, M. Pétraud, B. De Jéso, C. Baquey, and M. A. Barbosa (2001b). Cellulose phosphates as biomaterials. I. Synthesis and characterisation of highly phosphorylated cellulose gels. *J. Appl. Polym. Sci.* **82**:3341–3353.

Granja, P. L., B. De Jeso, R. Bareille, F. Rouais, C. Baquey, and M. A. Barbosa (2006). Cellulose phosphates as biomaterials: *in vitro* biocompatibility studies. *React. Funct. Polym.* **66**:728–739.

Grzyb, B., C. Hildenbrand, S. Berthon-Fabry, D. Bégin, N. Job, A. Rigacci, and P. Achard (2010). Functionalisation and chemical characterisation of cellulose-derived carbon aerogels. *Carbon* **48**(8):2297–2307.

Guilminot, E., F. Fischer, M. Chatenet, A. Rigacci, S. Berthon-Fabry, P. Achard, and E. Chainet (2007). Use of cellulose-based carbon aerogels as catalyst support for PEM fuel cell electrodes: electrochemical characterization. *J. Power Sources* **166**(1):104–111.

Guilminot, E., R. Gavillon, M. Chatenet, S. Berthon-Fabry, A. Rigacci, and T. Budtova (2008). New nanostructured carbons based on porous cellulose: elaboration, pyrolysis and use as platinum nanoparticles substrate for oxygen reduction electrocatalysis. *J. Power Sources* **185**(2):717–726.

Haimer, E., M. Wendland, K. Schlufter, K. Frankenfeld, P. Miethe, A. Potthast, T. Rosenau, and F. Liebner (2010). Loading of bacterial cellulose aerogels with bioactive compounds by antisolvent precipitation with supercritical carbon dioxide. *Macromol. Symp.* **294**(2):64–74.

Harmon, K. M., A. C. Akin, P. K. Keefer, and B. L. Snider (1992). Hydrogen bonding Part 45. Thermodynamic and IR study of the hydrates of N-methylmorpholine oxide and quinuclidine oxide: effect of hydrate stoichiometry on strength of H–O–H–O–N hydrogen bonds: implications for the dissolution of cellulose in amine oxide solvents. *J. Mol. Struct.* **269**(1–2):109–121.

Hayashi, T. (1994). Biodegradable polymers for biomedical uses. *Prog. Polym. Sci.* **19**:663–702.

Heßler, N., and D. Klemm (2009). Alteration of bacterial nanocellulose structure by *in situ* modification using polyethylene glycol and carbohydrate additives. *Cellulose* **16**:899–910.

Heinze, T., K. Schwikal, and S. Barthel (2005). Ionic liquids as reaction medium in cellulose functionalization. *Macromol. Biosci.* **5**(6):520–525.

Heinze, T., S. Dorn, M. Schöbitz, T. Liebert, S. Köhler, and F. Meister (2008). Interactions of ionic liquids with polysaccharides: 2: cellulose. *Macromol. Symp.* **262**(1):8–22.

Helenius, G., H. Bäckdahl, A. Bodin, U. Nannmark, P. Gatenholm, and B. Risberg (2006). *In vivo* biocompatibility of bacterial cellulose. *J. Biomed. Mater. Res. A* **76**(2):431–438.

Hench, L. L. (2006). The story of Bioglass. *J. Mater. Sci. Mater. Med.* **17**:967–78.

Hench, L. L., and O. Andersson (1993). In L. L. H. a. J. Wilson, editor, *Bioactive Glasses: An Introduction to Bioceramics*. World Scientific Publishing, Singapore, p. 41.

Henriksson, M., L. A. Berglund, P. Isaksson, T. Lindström, and T. Nishino (2008). Cellulose nanopaper structures of high toughness. *Biomacromolecules* 9(6):1579–1585.

Herrick, F. W., R. L. Casebier, J. K. Hamilton, and K. R. Sandberg (1983). Microfibrillated cellulose: morphology and accessibility. *J. Appl. Polym. Sci.* 37:797–813.

Hestrin, S., and M. Schramm (1954). Synthesis of cellulose by *Acetobacter xylinum*. 2. Preparation of freeze-dried cells capable of polymerizing glucose to cellulose. *Biochem. J.* 58:345–352.

Hoepfner, S., L. Ratke, and B. Milow (2008). Synthesis and characterisation of nanofibrillar cellulose aerogels. *Cellulose* 15(1):121–129.

Hou, A., Y. Shi, and Y. Yu (2009). Preparation of the cellulose/silica hybrid containing cationic group by sol-gel crosslinking process and its dyeing properties. *Carbohydr. Polym.* 77(2):201–205.

Hrubesh, L. W., and R. W. Pekala (1994). Thermal properties of organic and inorganic aerogels. *J. Mater. Sci.* 9:731–738.

Huang, H.-C., L.-C. Chen, S.-B. Lin, C.-P. Hsu, and H.-H. Chen (2010). *In situ* modification of bacterial cellulose network structure by adding interfering substances during fermentation. *Bioresour. Technol.* 101:6084–6091.

Hunt, A., and M. Ayers (2000). The history of silica aerogels. Available at http://eetd.lbl.gov/ecs/aerogels/kistler-early.html. Accessed Sept. 2010.

Hutchens, S. A., R. S. Benson, B. R. Evans, H. M. O'Neill, and C. J. Rawn (2006). Biomimetic synthesis of calcium-deficient hydroxyapatite in a natural hydrogel. *Biomaterials* 27(26):4661–4670.

Hutchens, S. A., R. S. Benson, B. R. Evans, C. J. Rawn, and H. O'Neill (2009). A resorbable calcium-deficient hydroxyapatite hydrogel composite for osseous regeneration. *Cellulose* 2009(16):887–898.

Hwang, S.-W., and S.-H. Hyun (2004). Capacitance control of carbon aerogel electrodes. *J. Non-Cryst. Solids* 347:238–245.

Iijima, S. (1991). Helical microtubules of graphitic carbon. *Nature* 354(6348):56–58.

Ikkala, O., R. H. A. Ras, N. Houbenov, J. Ruokolainen, M. Paakko, J. Laine, M. Leskela, L. A. Berglund, T. Lindstrom, G. Ten Brinke, et al. (2009). Solid state nanofibers based on self-assemblies: from cleaving from self-assemblies to multilevel hierarchical constructs. *Faraday Discuss.* 143:95–107.

Innerlohinger, J., H. K. Weber, and G. Kraft (2006a). Aerocell: aerogels from cellulosic materials. *Lenzinger Ber.* 86:137–143.

Innerlohinger, J., H. K. Weber, and G. Kraft (2006b). Aerocellulose: aerogels and aerogel-like materials made from cellulose. *Macromol. Symp.* 244(1):126–135.

Loleva, M. M., A. S. Goikhman, S. I. Banduryan, and S. P. Papkov (1983). Characteristics of the interaction of cellulose with *N*-methylmorpholine-*N*-oxide. *Vysokomol. Soedin. B* 25(11):803–804.

Jin, H., Nishiyama, M. Wada, and S. Kuga (2004). Nanofibrillar cellulose aerogels. *Colloids Surf. A* 240:63–67.

George, J., S. Amarinder, and S. Bawa (2010). Synthesis and characterization of bacterial cellulose nanocrystals and their PVA nanocomposites. *Adv. Mater. Res.* 123–125:383–386.

Jonas, R., and L. F. Farah (1998). Production and application of microbial cellulose. *Polym. Degrad. Stab.* 59(1–3):101–106.

Jordá-Beneyto, M., F. Suárez-García, D. Lozano-Castelló, D. Cazorla-Amorós, and A. Linares-Solano (2007). Hydrogen storage on chemically activated carbons and carbon nanomaterials at high pressures. *Carbon* **45**(2):293–303.

Kamath, M., J. Kincaid, and B. K. Mandal (1996). Interpenetrating polymer networks of photocrosslinkable cellulose derivatives. *J. Appl. Polym. Sci.* **59**(1):45–50.

Karaaslan, A. M., M. A. Tshabalala, and G. Buschle-Diller (2010). Wood hemicellulose/ chitosan-based semi-interpenetrating network hydrogels: mechanical swelling and controlled drug release properties. *Bioresources* **5**(2):1036–1054.

Karout, A., and A. C. Pierre (2007). Silica xerogels and aerogels synthesized with ionic liquids. *J. Non-Cryst. Solids* **353**(30–31):2900–2909.

Karout, A., and A. C. Pierre (2009). Silica gelation catalysis by ionic liquids. *Catal. Commun.* **10**(4):359–361.

Kato, K., Y. Eika, and Y. Ikada (1996). Deposition of a hydroxyapatite thin layer onto a polymer surface carrying grafted phosphate polymer chains. *J. Biomed. Mater. Res.* **32**(4):687–691.

Khanin, V. A., A. V. Bandura, and N. P. Novoselov (1998). Barriers to rotation of bridging bonds of cellulose molecule in its interaction with *N*-methylmorpholine *N*-oxide. *Russ. J. Gen. Chem.* **68**(2):305–308.

Kim, U.-J., S. Kuga, M. Wada, T. Okano, and T. Kondo (2000). Periodate oxidation of crystalline cellulose. *Biomacromolecules* **1**(3):488–492.

Kim, D.-Y., Y. Nishiyama, M. Wada, and S. Kuga (2001). Graphitization of highly crystalline cellulose. *Carbon* **39**(7):1051–1056.

Kim, J., M. J. Yaszemski, and L. Lu (2009). Three-dimensional porous biodegradable polymeric scaffolds fabricated with biodegradable hydrogel porogens. *Tissue Eng. C* **15**(4):583–594.

Klemm, D., D. Schumann, U. Udhardt, and S. Marsch (2001). Bacterial synthesized cellulose: artificial blood vessels for microsurgery. *Prog. Polym. Sci.* **26**(9):1561–1603.

Kralisch, D., N. Hessler, and D. Klemm (2008). Kontinuierliches Verfahren zur Darstellung von bakteriell synthetisierter Cellulose in flächiger Form. German Patent DE 10 2008 046 644.1

Kralisch, D., N. Hessler, D. Klemm, R. Erdmann, and W. Schmidt (2010). White biotechnology for cellulose manufacturing: the HoLiR concept. *Biotechnol. Bioeng.* **105**(4): 740–747.

Kulpinski, P. (2005). Cellulose fibers modified by silicon dioxide nanoparticles. *J. Appl. Polym. Sci.* **98**(4):1793–1798.

Lampke, T. (2001). Beitrag zur Charakterisierung naturfaserverstärkter Verbundwerkstoffe mit hochpolymerer matrix. Ph.D. thesis, TU Chemnitz, Chemnitz, Germany, p. 58.

Li, R., A. E. Clark, and L. L. Hench (1991). An investigation of bioactive glass powders by sol- gel processing. *J. Appl. Biomat.* **2**:231–239.

Liebner, F., A. Potthast, T. Rosenau, E. Haimer, and M. Wendland (2007). Ultralight-weight cellulose aerogels from NBnMO-stabilized Lyocell dopes. *Res. Lett. Mater. Sci.* doi: 10.1155/2007/73724.

Liebner, F., A. Potthast, T. Rosenau, E. Haimer, and M. Wendland (2008). Cellulose aerogels: highly porous, ultra-lightweight biomaterials. *Holzforschung* **62**:129–135.

Liebner, F., E. Haimer, A. Potthast, D. Loidl, S. Tschegg, M.-A. Neouze, M. Wendland, and T. Rosenau (2009). Cellulosic aerogels as ultra-lightweight materials. Part II: synthesis and properties. *Holzforschung* **63**(1):3–11.

Liebner, F., E. Haimer, M. Wendland, M.-A. Neouze, K. Schlufter, P. Miethe, T. Heinze, A. Potthast, and T. Rosenau (2010a). Aerogels from unaltered bacterial cellulose: application of scCO$_2$ drying for the preparation of shaped, ultra-lightweight cellulosic aerogels. *Macromol. Biosci.* **10**(4):349–352.

Liebner, F., I. Patel, G. Ebner, E. Becker, M. Horix, A. Potthast, and T. Rosenau (2010b). Thermal aging of 1-alkyl-3-methylimidazolium ionic liquids and its effect on dissolved cellulose. *Holzforschung* **64**:161–166.

Lin, C. X., H. Y. Zhan, M. H. Liu, S. Y. Fu, and L. A. Lucia (2009). Novel preparation and characterization of cellulose microparticles functionalized in ionic liquids. *Langmuir* **25**(17):10116–10120.

Litschauer, M., M.-A. Neouze, E. Haimer, U. Henniges, A. Potthast, T. Rosenau, and F. Liebner (2010). Silica modified cellulosic aerogels. *Cellulose* **18**(1):143–149.

Liu, R.-G., Y.Y. Shen, H.-L. Shao, C.-X. Wu, and X.-C. Hu (2001). An analysis of Lyocell fiber formation as a melt-spinning process. *Cellulose* **8**(1):13–21.

Liu, X., L. A. Smith, J. Hu, and P. X. Ma (2009). Biomimetic nanofibrous gelatin/apatite composite scaffolds for bone tissue engineering. *Biomaterials* **30**(12):2252–2258.

Long, J. W., B. M. Dening, T. M. McEvoy, and D. R. Rolison (2004). Carbon aerogels with ultrathin, electroactive poly(o-methoxyaniline) coatings for high-performance electrochemical capacitors. *J. Non-Cryst. Solids* **350**:97–106.

Love, K. T., B. K. Nicholson, J. A. Lloyd, R. A. Franich, R. P. Kibblewhite, and S. D. Mansfield (2008). Modification of kraft wood pulp fibre with silica for surface functionalisation. *Composites A* **39**(12):1815–1821.

Lue, A., and L. Zhang (2010). Advances in aqueous cellulose solvents. *Cellulose Solvents: For Analysis, Shaping and Chemical Modification.* American Chemical Society, pp. 67–89.

Luo, X., S. Liu, J. Zhou, and L. Zhang (2009). In situ synthesis of Fe$_3$O$_4$/cellulose microspheres with magnetic-induced protein delivery. *J. Mater. Chem.* **19**:3538–3545.

Ma, P. X., and J. W. Choi (2001). Biodegradable polymer scaffolds with well-defined interconnected spherical pore network. *Tissue. Eng.* **7**(1):23–33.

Ma, M.-G., J.-F. Zhu, N. Jia, S.-M. Li, R.-C. Sun, S.-W. Cao, and F. Chen (2010). Rapid microwave-assisted synthesis and characterization of cellulose-hydroxyapatite nanocomposites in N,N-dimethylacetamide solvent. *Carbohydr. Res.* **345**(8):1046–1050.

Maeda, H. (2006). Preparation and properties of bacterial cellulose aerogel and its application. *Cell Commun.* **13**(4):169–172.

Maeda, H., M. Nakajima, T. Hagiwara, T. Sawaguchi, and S. Yano (2006a). Bacterial cellulose/silica hybrid fabricated by mimicking biocomposites. *J. Mater. Sci.* **41**(17):5646–5656.

Maeda, H., M. Nakajima, T. Hagiwara, T. Sawaguchi, and S. Yano (2006b). Preparation and properties of bacterial cellulose aerogel. *Jpn J. Polym. Sci. Technol.* **63**:135–137.

Mahadeva, S. K., S. Yun, and J. Kim (2009). Dry electroactive paper actuator based on cellulose/poly(ethylene oxide): poly(ethylene glycol) microcomposite. *J. Intell. Mater. Syst. Struct.* **20**(10):1141–1146.

Maillard, F., P. A. Simonov, and E. R. Savinova (2008). *Carbon Materials as Supports for Fuel Cell Electrocatalysts*, John Wiley & Sons, Inc., pp. 429–480.

Majling, J., S. Komarneni, and V. S. Fajnor (1995). Mercury porosimeter as a means to measure mechanical properties of aerogels. *J. Porous Mater.* **1**(1):91–95.

Malaprade, L. (1928). Action of polyalcohols on periodic acid. *Bull. Soc. Chim. Fr.* **43**:683.

Maldonado-Hòdar, F. J., C. Moreno-Castilla, J. Rivera-Utrilla, and M. A. Ferro-Garcla (2000). Metal-carbon aerogels as catalysts and catalyst supports. In: V. M. S. M. Avelino Corma, G. F. José Luis,editors, *Studies in Surface Science and Catalysis*, Elsevier, pp. 1007–1012.

Marie, J., R. Chenitz, M. Chatenet, S. Berthon-Fabry, N. Cornet, and P. Achard (2009). Highly porous PEM fuel cell cathodes based on low density carbon aerogels as Pt-support: experimental study of the mass-transport losses. *J. Power Sources* **190**(2):423–434.

Mikos, A. G., A. J. Thorsen, L. A. Czerwonka, Y. Bao, R. Langer, D. N. Winslow, and J. P. Vacanti (1994). Preparation and characterization of poly(L-lactic acid) foams. *Polymer* **35**(5):1068–1077.

Mucalo, M. R., Y. Yokogawa, T. Suzuki, Y. Kawamoto, F. Nagata, and K. Nishizawa (1995). Further studies of calcium phosphate growth on phosphorylated cotton fibres. *J. Mater. Sci.* **6**(11):658–669.

Nedelec, J.-M., and M. Baba (2004). On the use of monolithic sol-gel derived mesoporous silica for the calibration of thermoporisemetry using various solvents. *J. Sol-Gel Sci. Technol.* **31**(1):169–173.

Nge, T. T., and J. Sugiyama (2007). Surface functional group dependent apatite formation on bacterial cellulose microfibrils network in a simulated body fluid. *J. Biomed. Mater. Res. A* **81A**(1):124–134.

Nge, T. T., M. Nogi, H. Yano, and J. Sugiyama (2010). Microstructure and mechanical properties of bacterial cellulose/chitosan porous scaffold. *Cellulose* **17**(2):349–363.

Nicolajsen, A. (2005). Thermal transmittance of a cellulose loose-fill insulation material. *Build. Environ.* **40**(7):907–914.

Nilsson, O., V. Bock, R. Caps, and J. Fricke (1994). High temperature thermal properties of carbon aerogels. *Therm. Conduct.* **22**:878–887.

Nishi, Y., M. Uryu, S. Yamanaka, K. Watanabe, N. Kitamura, M. Iguchi, and S. Mitsuhashi (1990). The structure and mechanical properties of sheets prepared from bacterial cellulose. *J. Mater. Sci.* **25**(6):2997–3001.

Nishino, T., and N. Arimoto (2007). All-cellulose composite prepared by selective dissolving of fiber surface. *Biomacromolecules* **8**(9):2712–2716.

Nishino, T., K. Takano, and K. Nakamae (1995). Elastic modulus of the crystalline regions of cellulose polymorphs. *J. Polym. Sci. B* **33**(11):1647–1651.

Nishino, T., I. Matsuda, and K. Hirao (2004). All-cellulose composite. *Macromolecules* **37**(20):7683–7687.

Olsson, R. T., M. A. S. Azizi Samir, G. Salazar Alvarez, L. Belova, V. Strom, L. A. Berglund, O. Ikkala, J. Nogues, and U. W. Gedde (2010). Making flexible magnetic aerogels and stiff magnetic nanopaper using cellulose nanofibrils as templates. *Nat. Nanotechnol.* **5**(8):584–588.

Ono, H., Y. Shimaya, K. Sato, and T. Hongo (2004). 1H spin–spin relaxation time of water and rheological properties of cellulose nanofiber dispersion, transparent cellulose hydrogel (TCG). *Polym. J.* **36**(9):684–694.

Pääkkö, M., M. Ankerfors, H. Kosonen, A. Nykänen, S. Ahola, M. Österberg, J. Ruokolainen, J. Laine, P. T. Larsson, O. Ikkala et al. (2007). Enzymatic hydrolysis combined with mechanical shearing and high-pressure homogenization for nanoscale cellulose fibrils and strong gels. *Biomacromolecules* **8**(6):1934–1941.

Pääkko, M., J. Vapaavuori, R. Silvennoinen, H. Kosonen, M. Ankerfors, T. Lindstrom, L. A. Berglund, and O. Ikkala (2008). Long and entangled native cellulose I nanofibers allow flexible aerogels and hierarchically porous templates for functionalities. *Soft Matter* **4**(12):2492–2499.

Pajonk, G. M. (2003). Some applications of silica aerogels. *Colloid Polym. Sci.* **281**(7):637–651.

Pei, A., Q. Zhou, and L. A. Berglund (2010). Functionalized cellulose nanocrystals as biobased nucleation agents in poly(L-lactide) (PLLA): crystallization and mechanical property effects. *Compos. Sci. Technol.* **70**(5):815–821.

Pekala, R. W. (1989). Organic aerogels from the polycondensation of resorcinol with formaldehyde. *J. Mater. Sci.* **24**:3221–3227.

Pekala, R. W., C. T. Alviso, X. Lu, J. Gross, and J. Fricke (1995). New organic aerogels based upon a phenolic-furfural reaction. *J. Non-Cryst. Solids* **188**:34–40.

Pierre, A. C., and G. M. Pajonk (2002). Chemistry of aerogels and their applications. *Chem. Rev.* **102**:4243–4266.

Pinnow, M., H. P. Fink, C. Fanter, and J. Kunze (2008). Characterization of highly porous materials from cellulose carbamate. *Macromol. Symp.* **262**(1):129–139.

Pinto, R. J. B., P. A. A. P. Marques, A. M. Barros-Timmons, T. Trindade, and C. P. Neto (2008). Novel SiO$_2$/cellulose nanocomposites obtained by *in situ* synthesis and via polyelectrolytes assembly. *Compos. Sci. Technol.* **68**(3–4):1088–1093.

Pirard, R., A. Rigacci, J. C. Maréchal, D. Quenard, B. Chevalier, P. Achard, and J. P. Pirard (2003). Characterization of hyperporous polyurethane-based gels by non-intrusive mercury porosimetry. *Polymer* **44**(17):4881–4887.

Puppi, D., F. Chiellini, A. M. Piras, and E. Chiellini (2010). Polymeric materials for bone and cartilage repair. *Prog. Polym. Sci.* **35**(4):403–440.

Qi, H., J. Cai, L. Zhang, and S. Kuga (2009a). Properties of films composed of cellulose nanowhiskers and a cellulose matrix regenerated from alkali/urea solution. *Biomacromolecules* **10**:1597–1602.

Qi, H., C. Chang, and L. Zhang (2009b). Properties and applications of biodegradable transparent and photoluminescent cellulose films prepared via a green process. *Green Chem.* **11**(2):177–184.

Quan, S.-L., S.-G. Kang, and I.-J. Chin (2010). Characterization of cellulose fibers electrospun using ionic liquid. *Cellulose* **17**(2):223–230.

Reynolds, G. A. M., A. W. P. Fung, Z. H. Wang, M. S. Dresselhaus, and R. W. Pekala (1995). The effects of external conditions on the internal structure of carbon aerogels. *J. Non-Cryst. Solids* **188**:27–33.

Rhee, B., and H. B. Yim (1975). Optimierung des kontinuierlichen thermischen Abbaus zur Graphitierung von Endlos-Zellulose (Trans.). *Hwahak Konghak* **13**(5): 261–268.

Rolison, D. R., and B. Dunn (2001). Electrically conductive oxide aerogels: new materials in electrochemistry. *J. Mater. Chem.* **11**(4):963–980.

Rooke, J., C. Matos, M. Chatenet, R. Sescousse, T. Budtova, S. Berthon-Fabry, R. Mosdale, and F. Maillard (2010). Elaboration and characterizations of platinum nanoparticles supported on cellulose-based carbon aerogel. *ECS Trans.* **33**(1):447–459.

Rosenau, T., A. Potthast, H. Sixta, and P. Kosma (2001). The chemistry of side reactions and by-product formation in the system NMMO/cellulose (Lyocell process). *Prog. Polym. Sci.* **26**:1763–1837.

Rosenau, T., A. Potthast, I. Adorjan, A. Hofinger, H. Sixta, H. Firgo, and P. Kosma (2002). Cellulose solutions in N-methylmorpholine-N-oxide (NMMO): degradation processes and stabilizers. *Cellulose* 9:283–291.

Rosenau, T., A. Potthast, P. Schmid, and P. Kosma (2005a). On the non-classical course of Polonowski reactions of N-benzylmorpholine-N-oxide (NBnMO). *Tetrahedron* 61:3483–3487.

Rosenau, T., P. Schmid, A. Potthast, and P. Kosma (2005b). Stabilization of cellulose solutions in N-methylmorpholine-N-oxide (Lyocell dopes) by addition of an N-oxide as sacrificial substrate. *Holzforschung* 59:503–506.

Roy, C., T. Budtova, and P. Navard (2003). Rheological properties and gelation of aqueous cellulose−NaOH solutions. *Biomacromolecules* 4(2):259–264.

Rozhkova, O. V., V. V. Myasoedova, and G. A. Krestov (1985). Effect of donor–acceptor interactions on solubility of cellulose in methylmorpholine N-oxide-based systems. *Khim Drev* 2:26–29.

Saboori, A., M. Rabiee, F. Mutarzadeh, M. Sheikhi, M. Tahriri, and M. Karimi (2009). *Mater. Sci. Eng. C* 29:335–340.

Saito, T., and A. Isogai (2004). TEMPO-mediated oxidation of native cellulose. the effect of oxidation conditions on chemical and crystal structures of the water-insoluble fractions. *Biomacromolecules* 5(5):1983–1989.

Scheirs, J., G. Camino, and W. Tumiatti (2001). Overview of water evolution during the thermal degradation of cellulose. *Eur. Polym. J.* 37(5):933–942.

Scherer, G. W., D. M. Smith, X. Qiu, and J. M. Anderson (1995). Compression of aerogels. *J. Non-Cryst. Solids* 186:316–320.

Schimmel, H. G., G. Nijkamp, G. J. Kearley, A. Rivera, K. P. de Jong, and F. M. Mulder (2004). Hydrogen adsorption in carbon nanostructures compared. *Mater. Sci. Eng. B* 108(1–2):124–129.

Schmidt, M., and F. Schwertfeger (1998). Applications for silica aerogel products. *J. Non-Cryst. Solids* 225(1):364–368.

Schöbitz, M., F. Meister, and T. Heinze (2009). Unconventional reactivity of cellulose dissolved in ionic liquids. *Macromol. Symp.* 280(1):102–111.

Schrems, M., G. Ebner, F. Liebner, E. Becker, A. Potthast, and T. Rosenau (2010). Side reactions in the system cellulose/1-alkyl-3-methyl-imidazolium ionic liquid. *Cellulose Solvents: For Analysis, Shaping and Chemical Modification*, American Chemical Society, pp. 149–164.

Schumann, D., J. Wippermann, D. Klemm, F. Kramer, D. Koth, H. Kosmehl, T. Wahlers, and S. Salehi-Gelani (2009). Artificial vascular implants from bacterial cellulose: preliminary results of small arterial substitutes. *Cellulose* 16(5):877–885.

Seifert, M., S. Hesse, V. Kabrelian, and D. Klemm (2004). Controlling the water content of never dried and reswollen bacterial cellulose by the addition of water-soluble polymers to the culture medium. *J. Polym. Sci. A* 42(3):463–470.

Sequeira, S., D. V. Evtuguin, I. Portugal, and A. P. Esculcas (2007). Synthesis and characterisation of cellulose/silica hybrids obtained by heteropoly acid catalysed sol-gel process. *Mater. Sci. Eng. C* 27(1):172–179.

Sequeira, S., D. V. Evtuguin, and I. Portugal (2009). Preparation and properties of cellulose/silica hybrid composites. *Polym. Compos.* 30(9):1275–1282.

Sescousse, R., and T. Budtova (2009). Influence of processing parameters on regeneration kinetics and morphology of porous cellulose from cellulose–NaOH–water solutions. *Cellulose* **16**(3):417–426.

Sescousse, R., R. Gavillon, and T. Budtova (2010). Aerocellulose from cellulose-ionic liquid solutions: preparation, properties and comparison with cellulose-NaOH and cellulose-NMMO routes. *Carbohydr. Polym.* **83**(4):1766–1774.

Sheridan, R. L., J. R. Morgan, and R. Mohamed (2002). Biomaterials in burn, wound dressings. In: D. Severian,editor, *Handbook of Polymeric Biomaterials*, Marcel Dekker, New York.

Smirnova, I. (2002). Synthesis of silica aerogels and their application as a drug delivery system. Ph.D. thesis, Technical University of Berlin, Berlin.

Smirnova, A., X. Dong, H. Hara, A. Vasiliev, and N. Sammes (2005). Novel carbon aerogel-supported catalysts for PEM fuel cell application. *Int. J. Hydrogen Energy* **30**(2):149–158.

Suryanegara, L., A. N. Nakagaito, and H. Yano (2009). The effect of crystallization of PLA on the thermal and mechanical properties of microfibrillated cellulose-reinforced PLA composites. *Compos. Sci. Technol.* **69**(7–8):1187–1192.

Suryanegara, L., A. Nakagaito, and H. Yano, (2010). Thermo-mechanical properties of microfibrillated cellulose-reinforced partially crystallized PLA composites. *Cellulose* **17**(4):771–778.

Sutherland, I. W. (1998). Novel and established applications of microbial polysaccharides. *Trends Biotechnol.* **16**(1):41–46.

Svagan, A. J., M. S. Hedenqvist, and L. Berglund (2009). Reduced water vapour sorption in cellulose nanocomposites with starch matrix. *Compos. Sci. Technol.* **69**(3–4):500–506.

Svensson, A., E. Nicklasson, T. Harrah, B. Panilaitis, D. L. Kaplan, M. Brittberg, and P. Gatenholm (2005). Bacterial cellulose as a potential scaffold for tissue engineering of cartilage. *Biomaterials* **26**(4):419–431.

Swatloski, R. P. (2002). Dissolution of cellose with ionic liquids. *JACS* **124**(18):4974–4975.

Tabuchi, M. (2007). Nanobiotech versus synthetic nanotech? *Nat. Biotechnol.* **25**(4):389–390.

Tan, C., M. Fung, J. K. Newman, and C. Vu (2001). Organic aerogels with very high impact strength. *Adv Mater.* **13**:644–646.

Tanaka, K., and H. Kozuka (2004). Sol-gel preparation and mechanical properties of machinable cellulose/silica and polyvinylpyrrolidone/silica composites. *J. Sol-Gel Sci. Technol.* **32**(1):73–77.

Tang, M. M., and R. Bacon (1964). Carbonization of cellulose fibers: I. Low temperature pyrolysis. *Carbon* **2**(3):211–214, 215–220.

Tokoh, C., K. Takabe, J. Sugiyama, and M. Fujita, (2002). CPMAS 13C NMR and electron diffraction study of bacterial cellulose structure affected by cell wall polysaccharides. *Cellulose* **9**(3):351–360.

Touey, G. P., and T. Kingsport (1956) Preparation of cellulose phosphates. US patent 2,759,924, 1956.

Tsioptsias, C., A. Stefopoulos, I. Kokkinomalis, L. Papadopoulou, and C. Panayiotou (2008). Development of micro- and nano-porous composite materials by processing cellulose with ionic liquids and supercritical CO_2. *Green Chem.* **10**(9):965–971.

Turner, M. B., S. K. Spear, J. D. Holbrey, and R. D. Rogers (2004). Production of bioactive cellulose films reconstituted from ionic liquids. *Biomacromolecules* **5**:1379–1384.

Uraki, Y., J. Nemoto, H. Otsuka, Y. Tamai, J. Sugiyama, T. Kishimoto, M. Ubukata, H. Yabu, M. Tanaka, and M. Shimomura (2007). Honeycomb-like architecture produced by living bacteria, *Gluconacetobacter xylinus*. *Carbohydr. Polym.* **69**(1):1–6.

Wan, Y. Z., L. Hong, S. R. Jia, Y. Huang, Y. Zhu, Y. L. Wang, and H. J. Jiang (2006). Synthesis and characterization of hydroxyapatite-bacterial cellulose nanocomposites. *Compos. Sci. Technol.* **66**(11–12):1825–1832.

Wan, Y. Z., Y. Huang, C. D. Yuan, S. Raman, Y. Zhu, H. J. Jiang, F. He, and C. Gao (2007). Biomimetic synthesis of hydroxyapatite/bacterial cellulose nanocomposites for biomedical applications. *Mater. Sci. Eng. C* **27**(4):855–864.

Wendler, F., B. Kosan, M. Krieg, and F. Meister (2009). Possibilities for the physical modification of cellulose shapes using ionic liquids. *Macromol. Symp.* **280**(1):112–122.

Wu, J., J. Zhang, H. Zhang, J. He, Q. Ren, and M. Guo (2004). Homogeneous acetylation of cellulose in a new ionic liquid. *Biomacromolecules* **5**(2):266–268.

Xie, K., Y. Yu, and Y. Shi (2009). Synthesis and characterization of cellulose/silica hybrid materials with chemical crosslinking. *Carbohydr. Polym.* **78**:799–805.

Yamanaka, S., K. Watanabe, N. Kitamura, M. Iguchi, S. Mitsuhashi, Y. Nishi, and M. Uryu (1989). The structure and mechanical properties of sheets prepared from bacterial cellulose. *J. Mater. Sci.* **24**(9):3141–3145.

Yan, L., J. Chen, and P. R. Bangal (2007). *Macromol. Biosci.* **7**:11391148.

Yan, Z., S. Chen, H. Wang, B. Wang, C. Wang, and J. Jiang (2008). Cellulose synthesized by *Acetobacter xylinum* in the presence of multi-walled carbon nanotubes. *Carbohydr. Res.* **343**(1):73–80.

Yang, C. M., and C. Y. Chen (2005). Synthesis, characterization and properties of polyanilines containing transition metal ions. *Synth. Met.* **153**:133–136.

Yano, S., K. Iwata, and K. Kurita (1998). Biomimetic materials, sensors and systems. *Mater. Sci. Eng. C* **6**:75–90.

Yano, H., J. Sugiyama, A. N. Nakagaito, M. Nogi, T. Matsuura, M. Hikita, and K. Handa (2005). Optically transparent composites reinforced with networks of bacterial nanofibers. *Adv. Mater.* **17**(2):153–155.

Yano, S., H. Maeda, M. Nakajima, T. Hagiwara, and T. Sawaguchi (2008). Preparation and mechanical properties of bacterial cellulose nanocomposites loaded with silica nanoparticles. *Cellulose* **15**(1):111–120.

Yokogawa, Y., J. Paz Reyes, M. R. Mucalo, M. Toriyama, Y. Kawamoto, T. Suzuki, K. Nishizawa, F. Nagata, and T. Kamayama (1997). Growth of calcium phosphate on phosphorylated chitin fibres. *J. Mater. Sci.* **8**(7):407–412.

Zaborowska, M., A. Bodin, H. Bäckdahl, J. Popp, A. Goldstein, and P. Gatenholm (2010). Microporous bacterial cellulose as a potential scaffold for bone regeneration. *Acta Biomater.* **6**(7):2540–2547.

Zhang, R., Y. Xue, Q. Meng, L. Zhan, K. Li, D. Wu, L. Ling, J. Wang, H. Zhao, and B. Dong (2004). Small angle X-ray scattering study of microstructure changes of organic hydrogels from supercritical carbon dioxide drying. *J. Supercrit. Fluids* **28**:263–276.

Zhao, Q., R. Yam, B. Zhang, Y. Yang, X. Cheng, and R. Li (2009). Novel all-cellulose ecocomposites prepared in ionic liquids. *Cellulose* **16**(2):217–226.

Zheng, J. P., J. Huang, and T. R. Jow (1997). The limitation of energy density for electrochemical capacitors. *J. Electrochem. Soc.* **144**:2026.

Zhou, Q., Y. Gong, and C. Gao (2005). Microstructure and mechanical properties of poly(L-lactide) scaffolds fabricated by gelatin particle leaching method. *J. Appl. Polym. Sci.* **98**(3):1373–1379.

Zhu, S., Y. Wu, Q. Chen, Z. Yu, C. Wang, S. Jin, Y. Dinga, and G. Wuc (2006). Dissolution of cellulose with ionic liquids and its application: a mini-review. *Green Chem.* **8**:325–327.

Zimmermann, K. A., J. M. LeBlanc, K. T. Sheets, R. W. Fox, and P. Gatenholm (2009). Biomimetic design of a bacterial cellulose/hydroxyapatite nanocomposite for bone healing applications. *Mater. Sci. Eng. C* **31**(1):43–49.

3

NANOCELLULOSES: EMERGING BUILDING BLOCKS FOR RENEWABLE MATERIALS

Youssef Habibi and Lucian A. Lucia

3.1 INTRODUCTION

Cellulose constitutes the most abundant renewable polymeric resource available today worldwide having an annual biosynthetic production estimated to be over 7.5×10^{10} tons. This fascinating and almost inexhaustible sustainable polymeric raw material has been used in the form of (macro)fibers or derivatives for thousands of years for a wide range of materials and products in diurnal applications. Only now are the remarkable chemical and physical attributes of cellulose being "re-discovered" through the singular emergence of nanosized cellulosic substrates known as *nanocelluloses*. These substrates have recently gained, in the materials community, a tremendous level of attention that does not appear to be waning. In fact, nanocelluloses have been the subject of wide array of research efforts for different applications due to their low cost, availability, renewability, lightweight, nanoscale dimensionality, unique morphology as well as their unsurpassed quintessential physical and chemical properties. Several reviews have been published recently attesting to the importance of these emerging raw materials as building blocks for renewable materials (Eichhorn et al., 2010; Gatenholm and Klemm, 2010; Habibi and Dufresne, 2010; Habibi et al., 2010; Siró and Plackett, 2010; Eichhorn, 2011; Spence et al., 2011).

The aim of this chapter, therefore, is to collocate vital current knowledge in the research, development, and applications of nanocelluloses. These building blocks can not only be obtained from numerous sources, but also with different means

Polysaccharide Building Blocks: A Sustainable Approach to the Development of Renewable Biomaterials, First Edition. Edited by Youssef Habibi and Lucian A. Lucia.

leading to distinctive morphologies and characteristics thus implying the potential for diverse functionalities and uses as will become evident in this chapter.

3.2 STRUCTURAL AND MORPHOLOGICAL FEATURES OF CELLULOSE

Cellulose is widely distributed in the biosphere, principally in higher plants, but also in several marine animals (for example, tunicates), and to a lesser degree in algae, fungi, bacteria, and invertebrates. In general, cellulose is a fibrous, mechanically tough, water-insoluble biomaterial, which plays an essential role in maintaining the structure of plant cell walls while providing the key internal load resistance to facilitate uninterrupted transport mechanisms within a plant organism. Regardless of its origin, cellulose may be typically characterized as a semicrystalline high molecular weight homopolymer of β-1,4 linked anhydroglucose (Figure 3.1) (Fengel and Wegener, 1983). In plant cell walls, the exquisite architectural arrangement of cellulose microfibrils results from the combined action of biopolymerization spinning and

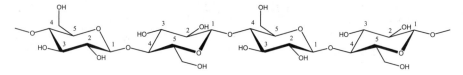

FIGURE 3.1 Chemical structure of cellulose.

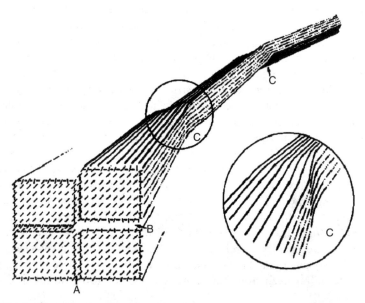

FIGURE 3.2 Schematic representation of cellulosic fibers containing elementary fibrils and illustrating the microstructure of the elementary fibril and strain-distorted tilt and twist regions (defects) Adapted from Rowland and Roberts (1972).

crystallization. All these events are orchestrated by specific enzymatic terminal complexes (TC) that adopt a rosette configuration behaving as precise biological spinnerets that are credited with synthesizing up to 36 glucan chains simultaneously and in close proximity to each another. The architectural precision of the complex itself templates the growth of glucan chains to co-crystallize to microfibrils that adopt a linear and rigid conformation in lieu of an amorphous array of β-glucans. On one hand, the polymer chains are assembled through van der Waals forces and both intra- and inter-molecular hydrogen bonds in a hierarchical order to form elementary nanofibrils having a cross-dimensional thickness of 2–5 nm that in turn aggregates laterally to larger macrofibers (Hon and Shiraishi, 1991; Ding and Himmel, 2006).

On the other hand, if the TCs are not perturbed, they can generate an interminable number of microfibrils having only a limited number of defects or amorphous regions (Brown, 1996, 2004). These regions are distributed on segments of the elementary fibril, which are distorted by internal strain in the fiber to undergo tilt and twist (Figure 3.2) (Rowland and Roberts, 1972).

3.3 PREPARATION OF NANOCELLULOSES

According to their morphology, cellulose fibers can be dissociated transversally at the amorphous parts present along their axis to yield defect-free rod-like nanoparticles referred hereafter as cellulose nanocrystals (CNs), or laterally yielding to bundles of their elementary protofibrils known as nanofibrillated celluloses (NFCs). The latter biomaterial can be also biosynthesized through microorganisms and is well known as bacterial cellulose (BC).

3.3.1 Preparation of Nanofibrillated Cellulose

3.3.1.1 Top-down Deconstruction The top-down deconstruction process for iso- lating NFC consists of the disintegration of natural cellulose fibers along the long axis. They include simple mechanical methods or a combination of enzymatic or chemical pretreatments in conjunction with mechanical processing. Different tech- nical approaches consist of using, for example, a Manton Gaulin Homogenizer or a Microfluidizer, Microgrinder, for the mechanical treatment. NFC was first developed by Turbak et al. (1983) using purified cellulose fibers from wood pulp as starting materials, after high pressure mechanical homogenization. Cellulosic fibers were disintegrated into their sub-structural fibrils and microfibrils that displayed lengths on the micrometers scale and widths ranging from ten to few hundred nanometers depending on the nature of the plant cell walls. The resulting aqueous suspensions exhibit gel-like properties in water with pseudoplastic and thixotropic properties even at low solid content. The major obstacle, for industrial production, has been related to the very high-energy consumption involved in processing pure cellulosic fibers. In fact, homogenization, utilized most popularly in the food processing industry, particularly to homogenize milk, is estimated to consume approximately 4000 kJ of energy per kilogram of NFC (Spence et al., 2010b). Materials are defibrillated utilizing rapid pressure drops inducing high shear provoked by impact forces against a valve and an impact ring. In comparison, microfluidization, utilized most commonly in the cosmetic and pharmaceutical industries, consumes approximately 200, 390, and 630 kJ/kg at processing pressures of 10, 20, and 30 kpsi, respectively. Specifically, a large pressure drop and an interaction chamber that imposes shear and impact forces on the fibers induce a low level of defibrillation that can only be augmented by multiple passes to achieve the final NFC product (Nakagaito and Yano, 2004). Unfortunately, the compromising issue with the processing methods mentioned (*vide supra*) is that the wood fibers tend to tangle that leads to equipment pluggage. Pretreatments are required for these methods to decrease the initial fiber length and to reduce energy consumption. Hydrolytic pretreatment methods have been introduced in the past that attempt to overcome the obstacles encountered; however, these approaches tend to diminish the aspect ratio and hence mechanical properties of the crystals, which although useful for chiral nematic phase preparations in aqueous media, nonetheless are not idea for reinforcement in composites (Fleming et al., 2001). Alternative successful approaches have been pursued; for example, work done by Lindström's group demonstrated that a combination of high pressure shear forces and mild enzymatic hydrolysis represents an efficient approach to preparing MFC with a well-controlled diameter in the nanometer range while concomitantly maintaining a high aspect ratio (Henriksson et al., 2007; Lindström et al., 2007; Pääkkö et al., 2007). Another method, microgrinding, is typically used in food processing. It is estimated to consume approximately 620 kJ/kg of energy to process NFC. The application of this method involves forcing wood pulp through a gap between rotary and stator disks having patented bursts and grooves that contact the fibers (Nakagaito and Yano, 2004).

As already has been mentioned, chemical and/or enzymatic pretreatments have been introduced to facilitate the efficient production of NFCs. An additionally exciting approach that represents a cost-effective chemical pretreatment before mechanical shearing involves oxidizing cellulose fibers using TEMPO-mediated oxidation which creates carboxyl groups on the fiber and microfibril surfaces. The TEMPO-oxidized cellulose fibers can be converted, utilizing mechanical shearing, to transparent and highly viscous dispersions in water, consisting of highly crystalline individual nanofibers (Saito et al., 2006, 2007; Fukuzumi et al., 2008). It was shown that the optimal conditions for the processing were attained at pH 10, giving cellulose nanofibers with 3–4 nm in width and a few microns in length. Carboxymethylation has also been successfully used to chemically pre-treat cellulose fibers before mechanical processing to generate NFC (Aulin et al., 2008; Wågberg et al., 2008). The processing of cellulosic materials extracted from primary cells such as parenchyma cells from sugar beet pulps (Dinand et al., 1996, 1999), potato pulp (Dufresne and Vignon, 1998), banana rachis (Zuluaga et al., 2009), cactus cladodes (Malainine et al., 2005), and fruits (Habibi et al., 2009) has been shown to be more facile to mechanical process without any enzymatic or chemical pretreatment.

3.3.1.2 Bottom-up Construction Besides being a dominant component of cell walls in plants, cellulose is also secreted extracellularly as synthesized cellulose fibers by several bacterial species (Gatenholm and Klemm, 2010). In particular, cellulose production has been verified for *G. xylinus, Agrobacterium tumefaciens, Rhizobium leguminosarum* bv. *trifolii, Sarcina ventriculi,* and recently for the enterobacteriaceae *Salmonella* spp., *Escherichia coli, Klebsiella pneumoniae,* and cyanobacteria (Deinema and Zevenhuizen, 1971; Napoli et al., 1975; Ross et al., 1991; Matthysse et al., 1995; Ausmees et al., 1999; Nobles et al., 2001; Zogaj et al., 2001). Instead of being obtained by fibrillation of fibers as already described (*vide supra*), (BC) is produced by bacteria in a reverse manner, namely, by synthesizing cellulose and building up bundles of microfibrils. A cellulose-producing enzyme has been identified in all the above-mentioned bacteria in addition to the Gram-positive bacterium *S. ventriculi*. BC can be produced by *Acetobacter* species *G. xylinus* by cultivation in aqueous culture media containing carbon and nitrogen sources during a time period of days up to 2 weeks. The resulting cellulosic network structure is in the form of a pellicle made up of a random assembly of ribbon-shaped fibrils, less than 100 nm wide, composed of a bundle of much finer microfibrils, 2–4 nm in diameter. These bundles are relatively straight, continuous, and dimensionally uniform.

3.3.2 Preparation of Cellulose Nanocrystals

One of the most accepted and traditional methodologies for the isolation of (CNs) from cellulose fibers is based on acid hydrolysis. Disordered (noncrystalline) and para-crystalline regions of cellulose are preferentially hydrolyzed, whereas mostly crystalline regions by virtue of their tight packing display a higher resistance to acid attack. Thus, following an acid treatment that hydrolyzes the cellulose (leading to removal of the microfibrils at the structural defects), true cellulose rod-like nano-crystals are produced. The obtained CNs have a morphology and crystallinity similar to the original cellulose fibers. The actual occurrence of the acid cleavage event is attributed to differences in the kinetics of hydrolysis between amorphous and crystalline domains. In general, acid hydrolysis of native cellulose induces a rapid decrease in its degree of polymerization (DP), to the so-called leveling-off DP (LODP). The DP subsequently decreases much more slowly, even during prolonged hydrolysis times (Battista et al., 1956; Sharples, 1958; Martin, 1971, 1974; Yachi et al., 1983; Håkansson and Ahlgren, 2005). LODP has been thought to correlate with crystal sizes along the longitudinal direction of cellulose chains present in the original celluloses before the acid hydrolysis. The value of LODP has been shown to depend on the cellulose origin, with typical values of 250 being recorded for hydrolyzed cotton (Battista, 1950), 300 for ramie fibers (Nishiyama et al., 2003), 140–200 for bleached wood pulp (Battista et al., 1956), and up to 6000 for the highly crystalline *Valonia* cellulose (Kai, 1976).

Typical procedures currently employed for the production of CNs consist of subjecting pure cellulosic material to strong acid hydrolysis under strictly controlled conditions of temperature, agitation/sonication, and time. The nature of the acid and the acid-to-pulp ratio are also critical parameters that affect the ultimate chemical character and efficiency of production of CNs. A typical procedure is described:

A suspension resulting from an acid hydrolysis is diluted with water and washed with a series of centrifugations. Dialysis against distilled water is then performed to remove any residual acid molecules remaining in the dispersion. Additional steps such as filtration (Elazzouzi-Hafraoui et al., 2008), differential centrifugation (Bai et al., 2009), or ultracentrifugation (using a saccharose gradient) (de Souza Lima and Borsali, 2002) have all been reported.

Sulfuric and hydrochloric acids have been extensively used for CN preparation, but phosphoric (Koshizawa, 1960; Usuda et al., 1967; Okano et al., 1999; Ono et al., 1999) acid has also been reported. The concentration of sulfuric acid in hydrolysis reactions to obtain CNs does not vary much from a typical value of about 65% (wt%); however, the temperature can range from room temperature up to 70 °C and the corresponding hydrolysis time can be tuned from 30 min to overnight depending on the temperature. In the case of hydrochloric acid-catalyzed hydrolysis, the reaction is usually carried out at a reflux temperature at an acid concentration between 2.5 and 4 N with variable time of reaction depending on the source of the cellulosic material. By increasing the temperature or by prolonging hydrolysis times, shorter nanocrystals, but with narrow size polydispersity can be obtained; however, no clear influence on the width of the crystal has to this point been revealed (Dong et al., 1998; Beck-Candanedo et al., 2005; Bondeson et al., 2006a, 2006b; Elazzouzi-Hafraoui et al., 2008).

If the CNs are prepared by the action of hydrochloric acid, their ability to deaggregate is limited and the colloidal dispersion generated in water tends to flocculate (Araki et al., 1998b). On the other hand, when sulfuric acid is used as the hydrolyzing agent, its reaction with the surface hydroxyl groups of cellulose tends to yield negatively charged surface sulfate ester functionalities that promote Coulombic-driven homogeneous dispersions in water. At a certain solids content, typically between 1 and 10%, this homogeneous suspension self-organizes spontaneously into spectacular liquid crystalline orders such as witnessed in the development of the anisotropic chiral nematic phase (Revol et al., 1992). However, the introduction of charged sulfate groups compromises the thermostability of such nanocrystals (Roman and Winter, 2004).

3.4 MORPHOLOGY OF NANOCELLULOSES

The origin of the cellulose fibers, mainly related to the nature of the plant cell wall, for example, primary or secondary, determines the morphology of the generated NFCs. The type of the NFC-developing pretreatment may also affect the morphological properties of the NFCs. Commonly, they are elongated nanoparticles with widths ranging from 3 to 20 nm and lengths of few microns. NFCs from primary cell walls are generally thinner and longer and have been much easier to produce compared to those extracted from the secondary walls of plants such as wood. Figure 3.3 shows an example of NFCs extracted from primary or secondary cell walls. Bacterial cellulose, on the other hand, is manufactured in the form of a pellicle made up of a random assembly of ribbon-shaped fibrils, less than 100 nm wide, which are in turn composed of bundles of much finer microfibrils, 2–4 nm in diameter (Figure 3.4).

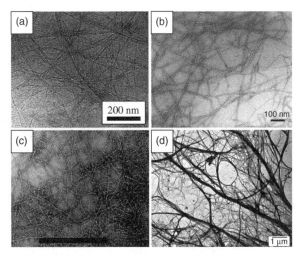

FIGURE 3.3 Transmission electron micrographs of a dilute suspension of NFCs obtained from wood fibers by mechanical processing that was employed in conjunction with (a) enzymatic treatment (Pääkkö et al., 2007); (b) TEMPO-mediated oxidation (Saito et al., 2007); (c) carboxylmethylation pretreatment (Wågberg et al., 2008) (a, b and c reprinted with permissions. Copyright 2007/2008 American Chemical Society); (d) NFCs extracted from Opuntia ficus-indica (Malainine et al., 2005) (Reprinted with permission, Elsevier).

In general, CNs can theoretically be prepared from any botanical source containing cellulose. The literature is replete with the various cellulosic furnishes that have been used. Regardless of the source, CNs occur as elongated nanoparticles. Each nanocrystal may be considered as a cellulosic crystal that does not possess any ostensible defect. The dimensions of the CNs depend on several factors, but mainly on the source of the cellulose, although the hydrolysis conditions and ionic strength must be considered.

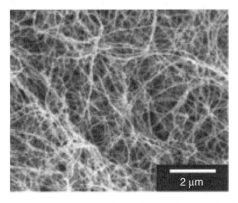

FIGURE 3.4 Scanning electron micrographs (micro-scale order) of a bacterial cellulose pellicle (Nakagaito et al., 2005) with kind permission from Springer Science & Business Media.

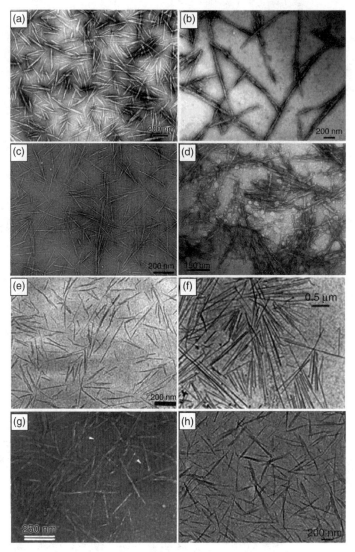

FIGURE 3.5 Transmission electron micrographs from dilute suspension of cellulose nanocrystals from (a) ramie (Habibi et al., 2008), (b) bacterial (Grunnert and Winter, 2002), (c) sisal (Garcia de Rodriguez et al., 2006), (d) microcrystalline cellulose (Kvien et al., 2005), (e) sugar beet pulp (Azizi Samir et al., 2004a), (f) tunicin (Anglès and Dufresne, 2000), (g) wheat straw (Helbert et al., 1996), and (h) cotton (Fleming et al., 2000). (a: *Reproduced by permission of The Royal Society of Chemistry; b, c*: with kind permission from Springer Science & Business Media; d, e, f, h: reprinted with permission, Copyright 2000, 2004 & 2005, American Chemical Society; g: reprinted with permission from John Wiley and Sons.

Examples and the typical geometrical characteristics for CNs derived from different species are depicted in Figure 3.5 and collected in Table 3.1. Although, they may be composed of a few laterally bound elementary crystallites that are not

TABLE 3.1 Geometrical Characteristics of Cellulose Nanocrystals from Various Sources: Length (L) and Cross Section (D)

Source	L (nm)	D (nm)	References
Algal (*Valonia*)	> 1000	10–20	Revol (1982), Hanley et al. (1992)
Bacterial	100–several 1000	5–10 × 30–50	Tokoh et al. (1998), Grunnert and Winter (2002), Roman and Winter (2004)
Cladophora	-	20 × 20	Kim et al. (2000)
Cotton	100–300	5–10	Fengel and Wegener (1983), Dong et al. (1998), Ebeling et al. (1999), Araki et al. (2000), Podsiadlo et al. (2005)
MCC	150–300	3–7	Kvien et al. (2005)
Ramie	200–300	10–15	Habibi et al. (2007, 2008)
Sisal	100–500	3–5	Garcia de Rodriguez et al. (2006)
Tunicin	100–several 1000	10–20	Favier et al. (1995a)
Wood	100–300	3–5	Fengel and Wegener (1983), Araki et al. (1998a), Araki et al. (1999), Beck-Candanedo et al. (2005)

separated by conventional acid hydrolysis and sonication process (Elazzouzi-Hafraoui et al., 2008), the length and width of cellulose nanocrystals derived from hydrolyzed furnishes are generally in the order of few hundred nanometers and few nanometers, respectively.

3.5 NANOCELLULOSE-BASED MATERIALS

3.5.1 Foams and Aerogels

Aerogels are materials that have been known since the 1930s when silica aerogels became well-established first generation aero-materials. In general, they are highly porous, extremely lightweight materials having only 1–15% of solid material that exhibit a very high specific surface area of up to $1600 \, m^2/g$, extremely low thermal conductivity, and good strength and dimensional stability. Aerogels are composed of materials having diameters of a few nanometers that can be linked to each other to form stable three-dimensional networks. Aerogels have received extensive attention as storage media for gases, catalysts, insulation, or as sorbents mainly because of their high porosity and the accessible and connected open pore system. In addition to the above-noted benefits, cellulosic aerogels are particularly exemplary aerogels; they also display the additional advantages of being a renewable, biodegradable, sustainable, and eco-friendly biomaterial. Nanocellulose-based foams are currently being studied mainly for packaging applications to replace petroleum-based polymeric foams. Nanocelluloses, especially nanofibrillated cellulose, alone or incorporated in composite formulations, can be used to prepare foams/aerogels. As thin as the cells in starch foam, nanofibrillated cellulose demonstrated ability to enhance the mechanical performance of starch-based foams prepared with successive freeze-drying techniques (Svagan et al., 2008). By controlling the density and interaction of the nanofibril obtained by varying the concentration of the nanocellulose in aqueous suspensions before freeze-drying, Sehaqui et al. (2010) prepared tough and ultra-high foams with tuned porosity. Similarly Aulin et al. (2010) prepared structured porous aerogels from nanocellulose by freeze-drying. The resulting aerogels were further chemically modified with fluorinated silane to uniformly coat them and thus tune their wetting properties toward nonpolar liquids and oils. The authors demonstrated that it is possible to switch the wettability behavior of the aerogels between super-wetting and super-repellent using different aerogel textures.

Cellulose nanofibrils derived from *Gluconobacter* strains of bacteria have been freeze-dried and further impregnated with metalhydroxide/oxide precursors to recently make lightweight porous magnetic aerogels (Olsson et al., 2010). These porous magnetic foams can be used in number of novel applications such as magnetic super sponges absorbing 1 g of water within only 60 mg of aeorogel foam. More interestingly, they can be also compacted into a stiff string magnetic nanopaper.

Recently cellulose nanocrystal-based aerogels have been prepared through the self-assembly of cellulose nanowhiskers in a benign manner (Heath and Thielemans, 2010). Preparation of these aerogels only requires sonication in water to form a hydrogel, solvent exchange with ethanol, followed by supercritical carbon dioxide drying resulting in very limited shrinkage of the hydrogel. These aerogels display very low densities down to $78 \, mg/cm^3$ with high specific surface areas up to $605 \, m^2/g$.

3.5.2 Films and Nanopapers

The incorporation of a limited amount of TEMPO-mediated oxidized NFCs has been shown to improve to some extent the wet strength of paper made with cellulose fibers (Saito and Isogai, 2005, 2006, 2007). Yet the first films, with a high content of nanocelluloses, were prepared by Nakagaito et al. by impregnating dried sheets made out of BC (Nakagaito et al., 2005) or wood-sourced NFC (Nakagaito and Yano, 2005) with phenolic resins. The resulting films displayed a dense paper-like organization with nano-order-scale-interconnected network structures. They were porous and exhibited outstanding mechanical properties, but were brittle. Similarly nanocellulose-based films were impregnated with melamine formaldehyde resins displaying comparable properties (Henriksson and Berglund, 2007). Full nanocellulose-based paper sheets have been prepared by drying suspensions of NFCs in various media other than water. These suspensions were obtained by solvent exchanging water with different media such as methanol, ethanol, and acetone. The resulting porous films obtained from carboxymethylated NCFs show remarkably high toughness in relation to their large strain-to-failure that is as high as 10%. Other studies have shown that NFC-based films were found to meet the requirements for packaging applications because at a $35 \, g/m^2$ basis weight, NFC films were found to have the requisite mechanical properties: tensile index of $146 \, Nm/g$, elongation of 8.6%, an elastic

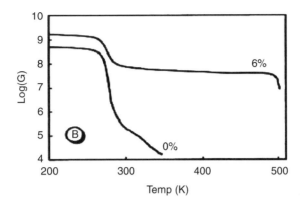

FIGURE 3.6 Logarithm of storage shear modulus versus temperature for poly(S-*co*-BuA) nanocomposite reinforced by a 6% weight fraction of tunicate cellulose nanocrystals (Favier et al., 1995b) (Reprinted with permission. Copyright 1995, American Chemical Society).

modulus of 17.5 GPa, low oxygen transmission rates of 17 mL/m^2 day, which were comparable to synthetic packaging derived from oriented polyester ethylene vinyl alcohol (Syverud and Stenius, 2009). These properties were further enhanced when these films were produced from NFCs containing residual lignin (Spence et al., 2010a).

3.5.3 Polymer-Based Composites

Since the first publication related to the use of CNs as reinforcing fillers in poly-(styrene-*co*-butyl acrylate) (poly(S-*co*-BuA))-based nanocomposites by Favier et al. (1995a, 1995b) who demonstrated a spectacular improvement in the storage modulus, as measured by dynamic mechanical analysis, above the glass–rubber transition temperature range, even at low CNs loading (see Figure 3.6), nanocelluloses have attracted a great deal of interest in the nanocomposites field. The use of nanocelluloses especially CNs as fillers in polymeric nanocomposites materials is among the most studied fields in biomaterials science. For more detailed information, the cited reviews are recommended (Azizi Samir et al., 2005a; Habibi et al., 2010; Siró and Plackett, 2010).

The nanoscale dimensions, low density, and extremely attractive mechanical properties of CNs make them ideal candidates to improve the mechanical properties of polymeric materials. They comprise a generic class of biomaterials that display mechanical strengths that are approximately on the order of the binding forces of adjacent atoms. The tensile strength properties of CNs are far in excess of current high-volume content reinforcement materials and allow access among the highest attainable composite strengths. In fact, the axial Young's modulus of a CN is theoretically stronger than stainless steel and similar to that of Kevlar. For CNs, the theoretical value of Young's modulus for the native cellulose perfect crystal has been estimated to be 167.5 GPa (Tashiro and Kobayashi, 1991). Recently, Raman spectroscopy was applied to measure the elastic modulus of native cellulose crystals from tunicate and cotton to yield values of 143 GPa (Šturcova et al., 2005) and

105 GPa (Rusli and Eichhorn, 2008), respectively. In fact, this technique was also recently applied by Hsieh et al. to determine the Young's modulus of a single filament of BC. Their measurements provided a value of 114 GPa which is not as high as values by Šturcova et al. (although the authors offer reasoning based on crystallinity and structural differences), but nevertheless represents a great advancement in our ability to probe the overall physical properties of these unique nanocelluloses (Hsieh et al., 2008).

Due to their hydrophilic nature, nanocelluloses have been easily incorporated in waterborne systems containing hydrophilic or hydro-dispersible and latex-based polymer matrices. These composite systems are usually processed by casting-evaporation techniques, but alternative processing methods such as freeze-drying combined with hot-pressing or extrusion/hot pressing have also been reported. Fiber spinning and electrospinning are techniques that are now seeing much more applications.

Examples of polymer-based latex include, but are not limited poly-(styrene-*co*-butyl acrylate) (Favier et al., 1995a), poly(hydroxyoctanoate) (Dubief et al., 1999; Dufresne, 2000), poly(vinyl acetate) (Garcia de Rodriguez et al., 2006; Roohani et al., 2008), polyethylene-vinyl acetate (EVA) (Chauve et al., 2005), poly(vinyl chloride) (Chazeau et al., 1999a, 1999b, 1999c, 2000), waterborne epoxy (Matos Ruiz et al., 2001), natural rubber (NR) (Bendahou et al., 2009, 2010; Siqueira et al., 2010), poly(styrene-*co*-hexyl-acrylate) (Ben Elmabrouk et al., 2009). Several hydrosoluble or hydro-dispersible polymers are reported where CNs were incorporated such as poly (oxyethylene) (Azizi Samir et al., 2004b, 2004d, 2005b, 2005c), carboxymethyl cellulose (Choi and Simonsen, 2006), polyvinyl alcohol (PVA) (Zimmerman et al., 2004, Zimmerman et al., 2005; Lu et al., 2008; Paralikar et al., 2008) waterborne polyurethane (Cao et al., 2009), hydroxypropyl cellulose (HPC) (Zimmerman et al., 2004, 2005), starch (Anglès and Dufresne, 2001; Orts et al., 2004; Lu et al., 2005, 2006; Kvien et al., 2007; Cao et al., 2008a, 2008b), soy protein (Wang et al., 2006), chitosan (Li et al., 2009), or regenerated cellulose (Qi et al., 2009).

However, the main challenge in incorporating nanocellulose particles in the most common nonpolar polymeric matrices is interfacial compatibility issues that control homogeneous dispersibility within the matrix. Recently, it has been demonstrated that tunicin CNs could be dispersed in DMF without additives or surface modifications (Azizi Samir et al., 2004c). The clear message from this result is the possibility of using hydrophobic polymers as a matrix in addition to permitting additional chemical modifications of CNs especially those incompatible in water. Different strategies such as surfactant surface coating, covalent, or noncovalent chemical modifications (acetylation, esterification, silylation, polymers grafting, etc.) have been pursued to improve the dispersibility/compatibility of nanocelluloses within nonpolar media. Following these strategies, nanocelluloses have been incorporated into a wide range of nonpolymer matrices, including thermoplastics, thermosets, and urethanes. Examples of such polymers include: acrylic-based resins (Nogi et al., 2006; Ifuku et al., 2007), siloxanes (Grunnert and Winter, 2000) polysulfonates (Noorani et al., 2006), poly(caprolactone) (Habibi and Dufresne, 2008; Habibi et al., 2008; Goffin et al., 2011), cellulose acetate butyrate (Grunnert and Winter, 2002; Petersson et al., 2009), polyethylene (Junior de Menezes et al., 2009), polypropylene (Bonini, 2000), polyurethane (Marcovich et al., 2006), polylactic acid (Oksman et al., 2006; Bondeson and Oksman, 2007a, 2007b; Petersson et al., 2007), and polyhydroxybutyrates (Jiang et al., 2008).

3.6 CONCLUSION AND FUTURE PERSPECTIVES

Nanocelluloses, including NFCs and (CNs), are currently the object of intense scientific curiosity and study within the materials community. They not only represent a unique biomaterial of significant abundance and sustainable production but they also permit the pursuit of novel applications in a diverse range of fields especially in the nanocomposite field that can be ascribed to their inherently attractive chemical (polyhydroxyl functionalities) and physical (MOEs up to 145 GPa) functionality. They can be obtained from a number of cellulosic furnishes that ultimately provide CNs that display various aspect ratios (length/width), and can be modified/compatibilized with substrates for inclusion as reinforcement materials. The present contribution intends to provide further knowledge to the nascent material research area of nanocellulosics by examining their basic unit structure, their chemical origin (top-down, bottom-up), their morphologies, their applications particularly as foams, fillers, films, nanopapers, and polymer composites. Their contribution to physical improvements is outstanding and nearly unsurpassable within the materials community. Clearly, increased attention is needed to enhance their uses in nonpolar matrices (focusing on compatibility issues), but the obvious advantages and accrued benefits are extremely high. Increased performance of the CNs-reinforced composite material and reduced cost are strong drivers to the growth of this area although there are still significant scientific and technological challenges to address including scale up, uniformity of manufacture, and chemical compatibilization for composite applications.

ACKNOWLEDGMENTS

We are especially grateful for receiving support from a number of US federal agencies that allowed us to contribute to the development of this chapter, including the Department of Energy (Cooperative Agreement DE-FC36-04GO14308) and the Department of Agriculture (Cooperative Agreement 2006-38411-17035). We are also thankful to the members of the Laboratory of Soft Materials and Green Chemistry for their active contributions.

REFERENCES

Anglès, M. N. and A. Dufresne (2000). Plasticized starch/tunicin whiskers nanocomposites. 1. Structural analysis. *Macromolecules* **33**:8344–8353.

Anglès, M. N. and A. Dufresne (2001). Plasticized starch/tunicin whiskers nanocomposite materials. 2. Mechanical behavior. *Macromolecules* **34**:2921–2931.

Araki, J., M. Wada, S. Kuga, and T. Okano (1998a). Flow properties of microcrystalline cellulose suspension prepared by acid treatment of native cellulose. *Colloids Surf. A* **142**:75–82.

Araki, J., M. Wada, S. Kuga, and T. Okano (1998b). Flow properties of microcrystalline cellulose suspension prepared by acid treatment of native cellulose. *Colloids Surf. A* **142**:75–82.

Araki, J., M. Wada, S. Kuga, and T. Okano (1999). Influence of surface charge on viscosity behavior of cellulose microcrystal suspension. *J. Wood Sci.* **45**:258–261.

Araki, J., M. Wada, S. Kuga, and T. Okano (2000). Birefringent glassy phase of a cellulose microcrystal suspension. *Langmuir* **16**:2413–2415.

Aulin, C., J. Netrval, L. Wagberg, and T. Lindstrom (2010). Aerogels from nanofibrillated cellulose with tunable oleophobicity. *Soft Matter* **6**:3298–3305.

Aulin, C., I. Varga, P. M. Claesson, L. Wågberg, and T. Lindström (2008). Buildup of polyelectrolyte multilayers of polyethyleneimine and microfibrillated cellulose studied by *in situ* dual-polarization interferometry and quartz crystal microbalance with dissipation. *Langmuir* **24**:2509–2518.

Ausmees, N., H. Jonsson, S. Höglund, H. Ljunggren, and M. Lindberg (1999). Structural and putative regulatory genes involved in cellulose synthesis in *Rhizobium leguminosarum* bv. *trifolii*. *Microbiology* **145**:1253–1262.

Azizi Samir, M. A. S., F. Alloin, and A. Dufresne, (2005a). Review of recent research into cellulosic whiskers, their properties and their application in nanocomposite field. *Biomacromolecules* **6**:612–626.

Azizi Samir, M. A. S., F. Alloin, M. Paillet, and A. Dufresne (2004a). Tangling effect in fibrillated cellulose reinforced nanocomposites. *Macromolecules* **37**:4313–4316.

Azizi Samir, M. A. S., F. Alloin, J.-Y. Sanchez, and A. Dufresne (2004b). Cellulose nanocrystals reinforced poly(oxyethylene). *Polymer* **45**:4149–4157.

Azizi Samir, M. A. S., F. Alloin, J.-Y. Sanchez, and A. Dufresne (2005b). Nanocomposite polymer electrolytes based on poly(oxyethylene) and cellulose whiskers. *Polim. Cienc. Tecnol.* **15**:109–113.

Azizi Samir, M. A. S., F. Alloin, J.-Y. Sanchez, N. El Kissi, and A. Dufresne (2004c). Preparation of cellulose whiskers reinforced nanocomposites from an organic medium suspension. *Macromolecules* **37**:1386–1393.

Azizi Samir, M. A. S., L. Chazeau, F. Alloin, J. Y. Cavaille, A. Dufresne, and J. Y. Sanchez (2005c). POE-based nanocomposite polymer electrolytes reinforced with cellulose whiskers. *Electrochim. Acta* **50**:3897–3903.

Azizi Samir, M. A. S., A. M. Mateos, F. Alloin, J.-Y. Sanchez, and A. Dufresne (2004d). Plasticized nanocomposite polymer electrolytes based on poly(oxyethylene) and cellulose whiskers. *Electrochim. Acta* **49**:4667–4677.

Bai, W., J. Holbery, and K. Li (2009). A technique for production of nanocrystalline cellulose with a narrow size distribution. *Cellulose* **16**:455–465.

Battista, O. A. (1950). Hydrolysis and crystallization of cellulose. *Ind. Eng. Chem.* **42**:502–507.

Battista, O. A., S. Coppick, J. A. Howsmon, F. F. Morehead, and W. A. Sisson (1956). Level-off degree of polymerization. Relation to polyphase structure of cellulose fibers. *Ind. Eng. Chem. Res.* **48**:333–335.

Beck-Candanedo, S., M. Roman, and D. G. Gray (2005). Effect of reaction conditions on the properties and behavior of wood cellulose nanocrystal suspensions. *Biomacromolecules* **6**:1048–1054.

Ben Elmabrouk, A., T. Wim, A. Dufresne, and S. Boufi (2009). Preparation of poly(styrene-*co*-hexylacrylate)/cellulose whiskers nanocomposites via miniemulsion polymerization. *J. Appl. Polym. Sci.* **114**:2946–2955.

Bendahou, A., Y. Habibi, H. Kaddami, and A. Dufresne (2009). Physico-chemical characterization of palm from *Phoenix dactylifera*-L., preparation of cellulose whiskers and natural rubber-based nanocomposites. *J. Biobased Mater. Bioener.* **3**:81–90.

Bendahou, A., H. Kaddami, and A. Dufresne (2010). Investigation on the effect of cellulosic nanoparticles' morphology on the properties of natural rubber based nanocomposites. *Eur. Polym. J.* **46**:609–620.

Bondeson, D., I. Kvien, and K. Oksman (2006a). Strategies for preparation of cellulose whiskers from microcrystalline cellulose as reinforcement in nanocomposites. In: K. Oksman, and M. Sain, editors, *Cellulose Nanocomposites: Processing, Characterization, and Properties*, ACS Symposium Series, 938, Washington, DC: American Chemical Society, pp. 10–25.

Bondeson, D., A. Mathew, and K. Oksman (2006b). Optimization of the isolation of nanocrystals from microcrystalline cellulose by acid hydrolysis. *Cellulose* **13**:171–180.

Bondeson, D., and K. Oksman (2007a). Dispersion and characteristics of surfactant modified cellulose whiskers nanocomposites. *Compos. Interfaces* **14**:617–630.

Bondeson, D. and K. Oksman (2007b). Polylactic acid/cellulose whisker nanocomposites modified by polyvinyl alcohol. *Compos. Part A Appl. Sci. Manuf.* **38A**:2486–2492.

Bonini, C. (2000). *Mise en évidence du rôle des interactions fibre/fibre et fibre/matrice dans des nanocomposites à renfort cellulosique et matrice apolaire (atactique et isotactique).* Grenoble, France: Joseph Fourier University.

Brown, R. M. J. (1996). The biosynthesis of cellulose. *J. Macromol. Sci. Pure Appl. Chem.* **A33**:1345–1373.

Brown, R. M. J. (2004). Cellulose structure and biosynthesis: what is in store for the 21st century? *J. Polym. Sci., Polym. Chem.* **42**:487–495.

Cao, X., Y. Chen, P. R. Chang, A. D. Muir, and G. Falk (2008a). Starch-based nanocomposites reinforced with flax cellulose nanocrystals. *eXPRESS Polym. Lett.* **2**:502–510.

Cao, X., Y. Chen, P. R. Chang, M. Stumborg, and M. A. Huneault (2008b). Green composites reinforced with hemp nanocrystals in plasticized starch. *J. Appl. Polym. Sci.* **109**:3804–3810.

Cao, X., Y. Habibi, and L. A. Lucia (2009). One-pot polymerization, surface grafting, and processing of waterborne polyurethane-cellulose nanocrystal nanocomposites. *J. Mater. Chem.* **19**:7137–7145.

Chauve, G., L. Heux, R. Arouini, and K. Mazeau (2005). Cellulose poly(ethylene-*co*-vinyl acetate) nanocomposites studied by molecular modeling and mechanical spectroscopy. *Biomacromolecules* **6**:2025–2031.

Chazeau, L., J. Y. Cavaillé, G. Canova, R. Dendievel, and B. Boutherin (1999a). Viscoelastic properties of plasticized PVC reinforced with cellulose whiskers. *J. Appl. Polym. Sci.* **71**:1797–1808.

Chazeau, L., J. Y. Cavaille, and J. Perez (2000). Plasticized PVC reinforced with cellulose whiskers. II. Plastic behavior. *J. Polym. Sci. Part A Polym. Phys.* **38**:383–392.

Chazeau, L., J. Y. Cavaillé, and P. Terech (1999b). Mechanical behaviour above Tg of a plasticised PVC reinforced with cellulose whiskers: a SANS structural study. *Polymer* **40**:5333–5344.

Chazeau, L., M. Paillet, and J. Y. Cavaillé (1999c). Plasticized PVC reinforced with cellulose whiskers. I. Linear viscoelastic behavior analyzed through the quasi-point defect theory. *J. Polym. Sci. Part A Polym. Phys.* **37**:2151–2164.

Choi, Y., and J. Simonsen (2006). Cellulose nanocrystal-filled carboxymethyl cellulose nanocomposites. *J. Nanosci. Nanotechnol.* **6**:633–639.

de Souza Lima, M. M., and R. Borsali (2002). Static and dynamic light scattering from polyelectrolyte microcrystal cellulose. *Langmuir* **18**:992–996.

Deinema, M. H., and L. P. T. M. Zevenhuizen (1971). Formation of cellulose fibrils by gram-negative bacteria and their role in bacterial flocculation. *Arch. Microbiol.* **78**:42–57.

Dinand, E., H. Chanzy, and M. R. Vignon (1996). Parenchymal cell cellulose from sugar beet pulp: preparation and properties. *Cellulose* **3**:183–188.

Dinand, E., H. Chanzy, and R. Vignon (1999). Suspensions of cellulose microfibrils from sugar beet pulp. *Food Hydrocolloids* **13**:275–283.

Dong, X. M., J.-F. Revol, and D. G. Gray (1998). Effect of microcrystallite preparation conditions on the formation of colloid crystals of cellulose. *Cellulose* **5**:19–32.

Dubief, D., E. Samain, and A. Dufresne (1999). Polysaccharide microcrystals reinforced amorphous poly(beta-hydroxyoctanoate) nanocomposite materials. *Macromolecules* **32**:5765–5771.

Dufresne, A. (2000). Dynamic mechanical analysis of the interphase in bacterial polyester/ cellulose whiskers natural composites. *Compos. Interfaces* **7**:53–67.

Dufresne, A., and M. R. Vignon (1998). Improvement of starch film performances using cellulose microfibrils. *Macromolecules* **31**:2693–2696.

Ebeling, T., M. Paillet, R. Borsali, O. Diat, A. Dufresne, J. Y. Cavaillé, and H. Chanzy (1999). Shear-induced orientation phenomena in suspensions of cellulose microcrystals, revealed by small angle x-ray scattering. *Langmuir* **15**:6123–6126.

Eichhorn, S. J. (2011). Cellulose nanowhiskers: promising materials for advanced applications. *Soft Matter* **7**:303–315.

Eichhorn, S., A. Dufresne, M. Aranguren, N. Marcovich, J. Capadona, S. Rowan, C. Weder, W. Thielemans, M. Roman, S. Renneckar, W. Gindl, S. Veigel, J. Keckes, H. Yano, K. Abe, M. Nogi, A. Nakagaito, A. Mangalam, J. Simonsen, A. Benight, A. Bismarck, L. Berglund, and T. Peijs (2010). Review: current international research into cellulose nanofibres and nanocomposites. *J. Mater. Sci.* **45**:1–33.

Elazzouzi-Hafraoui, S., Y. Nishiyama, J.-L. Putaux, L. Heux, F. Dubreuil, and C. Rochas (2008). The shape and size distribution of crystalline nanoparticles prepared by acid hydrolysis of native cellulose. *Biomacromolecules* **9**:57–65.

Favier, V., G. R. Canova, J. Y. Cavaille, H. Chanzy, A. Dufreshne, and C. Gauthier (1995a). Nanocomposite materials from latex and cellulose whiskers. *Polym. Adv. Technol.* **6**:351–355.

Favier, V., H. Chanzy, and J. Y. Cavaillé (1995b). Polymer nanocomposites reinforced by cellulose whiskers. *Macromolecules* **28**:6365–6367.

Fengel, D., and G. Wegener (1983). *Wood, Chemistry, Ultrastructure, Reactions.* New York: Walter de Gruyter.

Fleming, K., D. G. Gray, and S. Matthews (2001). Cellulose crystallites. *Chem. Eur. J.* **7**:1831–1836.

Fleming, K., D. G. Gray, S. Prasannan, and S. Matthews (2000). Cellullose nanocrystals: a new and robust liquid crystalline medium for the measurement of residual dipolar couplings. *J. Am. Chem. Soc.* **122**:5224–5225.

Fukuzumi, H., T. Saito, T. Iwata, Y. Kumamoto, and A. Isogai (2008). Transparent and high gas barrier films of cellulose nanofibers prepared by TEMPO-mediated oxidation. *Biomacromolecules* **10**:162–165.

Garcia de Rodriguez, N. L., W. Thielemans, and A. Dufresne (2006). Sisal cellulose whiskers reinforced polyvinyl acetate nanocomposites. *Cellulose* **13**:261–270.

Gatenholm, P., and D. Klemm (2010). Bacterial nanocellulose as a renewable material for biomedical applications. *MRS Bull.* **35**:208–213.

Goffin, A. L., J. M. Raquez, E. Duquesne, G. Siqueira, Y. Habibi, A. Dufresne, and P. Dubois (2011). Poly(ε-caprolactone) based nanocomposites reinforced by surface-grafted cellulose nanowhiskers via extrusion processing: morphology, rheology, and thermo-mechanical properties. *Polymer* **52**:1532–1538.

Grunnert, M., and W. T. Winter (2000). Progress in the development of cellulose reinforced nanocomposites. *Polym. Mater. Sci. Eng.* **82**:232.

Grunnert, M., and W. T. Winter (2002). Nanocomposites of cellulose acetate butyrate reinforced with cellulose nanocrystals. *J. Polym. Environ.* **10**:27–30.

Habibi, Y., and A. Dufresne (2008). Highly filled bionanocomposites from functionalized polysaccharide nanocrystals. *Biomacromolecules* **9**:1974–1980.

Habibi, Y., and A. Dufresne (2010). Nanocrystals from natural polysaccharides. In: K. D. Sattler, editor, *Handbook of Nanophysics*, Nanoparticles and Quantum Dots, Boca Raton, FL: CRC Press. pp. 10/11–10/15.

Habibi, Y., L. Foulon, V. Aguié-Béghin, M. Molinari, and R. Douillard (2007). Langmuir–Blodgett films of cellulose nanocrystals: preparation and characterization. *J. Colloid Interface Sci.* **316**:388–397.

Habibi, Y., A.-L. Goffin, N. Schiltz, E. Duquesne, P. Dubois, and A. Dufresne (2008). Bionanocomposites based on poly(ε-caprolactone)-grafted cellulose nanocrystals by ring opening polymerization. *J. Mater. Chem.* **18**:5002–5010.

Habibi, Y., L. A. Lucia, and O. J. Rojas (2010). Cellulose nanocrystals: chemistry, self-assembly, and applications. *Chem. Rev.* **110**:3479–3500.

Habibi, Y., M. Mahrouz, and M. Vignon (2009). Microfibrillated cellulose from the peel of prickly pear fruits. *Food Chem.* **115**:423–429.

Håkansson, H., and P. Ahlgren (2005). Acid hydrolysis of some industrial pulps: effect of hydrolysis conditions and raw material. *Cellulose* **12**:177–183.

Hanley, S. J., J. Giasson, J. F. Revol, and D. G. Gray (1992). Atomic force microscopy of cellulose microfibrils—comparison with transmission electron-microscopy. *Polymer* **33**:4639–4642.

Heath, L., and W. Thielemans (2010). Cellulose nanowhisker aerogels. *Green Chem.* **12**:1448–1453.

Helbert, W., J. Y. Cavaille, and A. Dufresne (1996). Thermoplastic nanocomposites filled with wheat straw cellulose whiskers. Part I: processing and mechanical behavior. *Polym. Compos.* **17**:604–611.

Henriksson, M., and L. A. Berglund (2007). Structure and properties of cellulose nanocomposite films containing melamine formaldehyde. *J. Appl. Polym. Sci.* **106**:2817–2824.

Henriksson, M., G. Henriksson, L. A. Berglund, and T. Lindström (2007). An environmentally friendly method for enzyme-assisted preparation of microfibrillated cellulose (MFC) nanofibers. *Eur. Polym. J.* **43**:3434–3441.

Hsieh, Y. C., H. Yano, M. Nogi, and S. Eichhorn (2008). An estimation of the Young's modulus of bacterial cellulose filaments. *Cellulose* **15**:507–513.

Hon, D. N.-S. and N. Shiraishi (1991). Wood and cellulosic chemistry. New York, Marcel Dekker, Inc.

Ifuku, S., M. Nogi, K. Abe, K. Handa, F. Nakatsubo, and H. Yano (2007). Surface modification of bacterial cellulose nanofibers for property enhancement of optically transparent composites: dependence on acetyl-group DS. *Biomacromolecules* **8**:1973–1978.

Jiang, L., E. Morelius, J. Zhang, M. Wolcott, and J. Holbery (2008). Study of the poly(3-hydroxybutyrate-*co*-3-hydroxyvalerate)/cellulose nanowhisker composites prepared by solution casting and melt processing. *J. Compos. Mater.* **42**:2629–2645.

Junior de Menezes, A., G. Siqueira, A. A. S. Curvelo, and A. Dufresne (2009). Extrusion and characterization of functionalized cellulose whiskers reinforced polyethylene nanocomposites. *Polymer* **50**:4552–4563.

Kai, A. (1976). The fine structure of *Valonia microfibril*. Gel permeation chromatographic studies of *Valonia* cellulose *Sen-i Gakkaishi* **32**:T326–T334.

Kim, U. J., S. Kuga, M. Wada, T. Okano, and T. Kondo (2000). Periodate oxidation of crystalline cellulose. *Biomacromolecules* **1**:488–492.

Koshizawa, T. (1960). Degradation of wood cellulose and cotton linters in phosphoric acid. *Kami Pa Gikyoshi* **14**:455–458.

Kvien, I., J. Sugiyama, M. Votrubec, and K. Oksman (2007). Characterization of starch based nanocomposites. *J. Mater. Sci.* **42**:8163–8171.

Kvien, I., B. S. Tanem, and K. Oksman (2005). Characterization of cellulose whiskers and their nanocomposites by atomic force and electron microscopy. *Biomacromolecules* **6**:3160–3165.

Li, Q., J. Zhou, and L. Zhang (2009). Structure and properties of the nanocomposite films of chitosan reinforced with cellulose whiskers. *J. Polym. Sci. Part A Polym. Phys.* **47**:1069–1077.

Lindström, T., M. Ankerfors, and G. Henriksson, (2007). Method for treating chemical pulp for manufacturing microfibrillated cellulose WO Patent 2007091942, STFI-Packforsk AB. 14pp.

Lu, J., T. Wang, and L. T. Drzal (2008). Preparation and properties of microfibrillated cellulose polyvinyl alcohol composite materials. *Compos. Part A Appl. Sci. Manuf.* **39A**:738–746.

Lu, Y., L. Weng, and X. Cao (2005). Biocomposites of plasticized starch reinforced with cellulose crystallites from cottonseed linter. *Macromol. Biosci.* **5**:1101–1107.

Lu, Y., L. Weng, and X. Cao (2006). Morphological, thermal and mechanical properties of ramie crystallites-reinforced plasticized starch biocomposites. *Carbohydr. Polym.* **63**:198–204.

Malainine, M. E., M. Mahrouz, and A. Dufresne (2005). Thermoplastic nanocomposites based on cellulose microfibrils from *Opuntia ficus-indica* parenchyma cell. *Compos. Sci. Technol.* **65**:1520–1526.

Marcovich, N. E., M. L. Auad, N. E. Bellesi, S. R. Nutt, and M. I. Aranguren (2006). Cellulose micro/nanocrystals reinforced polyurethane. *J. Mater. Res.* **21**:870–881.

Martin, C. (1971). Folding chain model and annealing of cellulose. *J. Polym. Sci. Part C Polym. Sym.* **36**:343–362.

Martin, M. Y. C. (1974). Crystallite structure of cellulose. *J. Polym. Sci. Part A Polym. Chem.* **12**:1349–1374.

Matos Ruiz, M., J. Y. Cavaillé, A. Dufresne, C. Graillat, and J.-F. Gerard (2001). New waterborne epoxy coatings based on cellulose nanofillers. *Macromol. Symp.* **169**:211–222.

Matthysse, A., S. White, and R. Lightfoot (1995). Genes required for cellulose synthesis in *Agrobacterium tumefaciens. J. Bacteriol.* **177**:1069–1075.

Nakagaito, A. N., S. Iwamoto, and H. Yano (2005). Bacterial cellulose: the ultimate nano-scalar cellulose morphology for the production of high-strength composites. *Appl. Phys. A Mater. Sci. Process.* **80**:93–97.

Nakagaito, A. N., and H. Yano (2004). The effect of morphological changes from pulp fiber towards nano-scale fibrillated cellulose on the mechanical properties of high-strength plant fiber based composites. *Appl. Phys. A Mater. Sci. Process.* **78**:547–552.

Nakagaito, A. N., and H. Yano (2005). Novel high-strength biocomposites based on micro-fibrillated cellulose having nano-order-unit web-like network structure. *Appl. Phys. A Mater. Sci. Process.* **80**:155–159.

Napoli, C., F. Dazzo, and D. Hubbell (1975). Production of cellulose microfibrils by *Rhizobium. Appl. Microbiol.* **30**:123–131.

Nishiyama, Y., U. J. Kim, D. Y. Kim, K. S. Katsumata, R. P. May, and P. Langan (2003). Periodic disorder along Ramie cellulose microfibrils. *Biomacromolecules* **4**:1013–1017.

Nobles, D. R., D. K. Romanovicz, and R. M. Brown (2001). Cellulose in cyanobacteria. Origin of vascular plant cellulose synthase? *Plant Physiol.* **127**:529–542.

Nogi, M., K. Abe, K. Handa, F. Nakatsubo, S. Ifuku, and H. Yano (2006). Property enhancement of optically transparent bionanofiber composites by acetylation. *Appl. Phys. Lett.* **89**:233123/233121–233123/233123.

Noorani, S., J. Simonsen, and S. Atre (2006). Polysulfone-cellulose nanocomposites. In: K. Oksman, and M. Sain, editors, *Cellulose Nanocomposites: Processing, Characterization and Properties*, ACS Symposium Series, 938, Washington, DC: American Chemical Society.

Okano, T., S. Kuga, M. Wada, J. Araki, and J. Ikuina, (1999). Fine cellulose particle and its production. JP Patent, JP11343301, NISSHIN OIL MILLS LTD.

Oksman, K., A. P. Mathew, D. Bondeson, and I. Kvien (2006). Manufacturing process of cellulose whiskers/polylactic acid nanocomposites. *Compos. Sci. Technol.* **66**:2776–2784.

Olsson, R. T., M. A. S. Azizi Samir, G. Salazar Alvarez, L Belova, V Strom, L. A. Berglund, O. Ikkala, J. Nogues, and U. W. Gedde (2010). Making flexible magnetic aerogels and stiff magnetic nanopaper using cellulose nanofibrils as templates. *Nat. Nanotechnol.* **5**:584–588.

Ono, H., T. Matsui, and I. Miyamato, (1999). Cellulose dispersion, WO Patent 9928350, ASAHI CHEMICAL IND.

Orts, W. J., S. H. Imam, J. Shey, G. M. Glenn, M. K. Inglesby, M. E. Guttman, and A. Nguyen (2004). Effect of fiber source on cellulose reinforced polymer nanocomposites. 62nd Annual Technical Conference—Society of Plastics Engineers 2427–2431.

Pääkkö, M., M. Ankerfors, H. Kosonen, A. Nykänen, S. Ahola, M. Österberg, J. Ruokolainen, J. Laine, P. T. Larsson, O. Ikkala, and T. Lindström (2007). Enzymatic hydrolysis combined with mechanical shearing and high-pressure homogenization for nanoscale cellulose fibrils and strong gels. *Biomacromolecules* **8**:1934–1941.

Paralikar, S. A., J. Simonsen, and J. Lombardi (2008). Poly(vinyl alcohol)/cellulose nano-crystal barrier membranes. *J. Membr. Sci.* **320**:248–258.

Petersson, L., I. Kvien, and K. Oksman (2007). Structure and thermal properties of poly(lactic acid)/cellulose whiskers nanocomposite materials. *Compos. Sci. Technol.* **67**:2535–2544.

Petersson, L., A. P. Mathew, and K. Oksman (2009). Dispersion and properties of cellulose nanowhiskers and layered silicates in cellulose acetate butyrate nanocomposites. *J. Appl. Polym. Sci.* **112**:2001–2009.

Podsiadlo, P., S.-Y. Choi, B. Shim, J. Lee, M. Cuddihy, and N. A. Kotov (2005). Molecularly engineered nanocomposites: layer-by-layer assembly of cellulose nanocrystals. *Biomacromolecules* **6**:2914–2918.

Qi, H., J. Cai, L. Zhang, and S. Kuga (2009). Properties of films composed of cellulose nanowhiskers and a cellulose matrix regenerated from alkali/urea solution. *Biomacromolecules* **10**:1597–1602.

Revol J. F. (1982). On the cross-sectional shape of cellulose crystallites in *Valonia ventricosa Carbohydr. Polym.* **2**:123–134.

Revol J. F., H. Bradford, J. Giasson, R. H. Marchessault, and D. G. Gray (1992). Helicoidal self-ordering of cellulose microfibrils in aqueous suspension. *Int. J. Biol. Macromol.* **14**:170–172.

Roman, M. and W. T. Winter (2004). Effect of sulfate groups from sulfuric acid hydrolysis on the thermal degradation behavior of bacterial cellulose. *Biomacromolecules* **5**:1671–1677.

Roohani, M., Y. Habibi, N. M. Belgacem, G. Ebrahim, A. N. Karimi, and A. Dufresne (2008). Cellulose whiskers reinforced polyvinyl alcohol copolymers nanocomposites. *Eur. Polym. J.* **44**:2489–2498.

Ross, P., R. Mayer, and M. Benziman (1991). Cellulose biosynthesis and function in bacteria. *Microbiol. Mol. Biol. Rev.* **55**:35–58.

Rowland, S. P., and E. J. Roberts (1972). The nature of accessible surfaces in the microstructure of cotton cellulose. *J. Polym. Sci. Part A Polym. Chem.* **10**:2447–2461.

Rusli, R., and S. J. Eichhorn (2008). Determination of the stiffness of cellulose nanowhiskers and the fiber–matrix interface in a nanocomposite using Raman spectroscopy. *Appl. Phys. Lett.* **93**:033111/033111–033111/033113.

Saito, T., and A. Isogai (2005). A novel method to improve wet strength of paper. *TAPPI J.* **4**:3–8.

Saito, T., and A. Isogai (2006). Wet strength improvement of TEMPO-oxidized cellulose sheets prepared with cationic polymers. *Ind. Eng. Chem. Res.* **46**:773–780.

Saito, T., and A. Isogai (2007). Wet strength improvement of TEMPO-oxidized cellulose sheets prepared with cationic polymers. *Ind. Eng. Chem. Res.* **46**:773–780.

Saito, T., S. Kimura, Y. Nishiyama, and A. Isogai (2007). Cellulose nanofibers prepared by TEMPO-mediated oxidation of native cellulose. *Biomacromolecules* **8**:2485–2491.

Saito, T., Y. Nishiyama, J. Putaux, M. R. Vignon, and A. Isogai (2006). Homogeneous suspensions of individualized microfibrils from TEMPO-catalyzed oxidation of native cellulose. *Biomacromolecules* **7**:1687–1691.

Sehaqui, H., M. Salajkova, Q. Zhou, and L. A. Berglund (2010). Mechanical performance tailoring of tough ultra-high porosity foams prepared from cellulose I nanofiber suspensions. *Soft Matter* **6**:1824–1832.

Sharples, A. (1958). The hydrolysis of cellulose and its relation to structure. *Trans. Faraday Soc.* **54**:913–917.

Siqueira, G., H. Abdillahi, J. Bras, and A. Dufresne (2010). High reinforcing capability cellulose nanocrystals extracted from *Syngonanthus nitens* (Capim Dourado). *Cellulose* **17**:289–298.

Siró, I., and D. Plackett (2010). Microfibrillated cellulose and new nanocomposite materials: a review. *Cellulose* **17**:459–494.

Spence, K., Y. Habibi, and A. Dufresne (2011). Nanocellulose-based composites. In: S. Kalia, B. S. Kaith, and I. Kaur, editors, *Cellulose Fibers: Bio- and Nano-Polymer Composites*, Berlin, Heidelberg: Springer, pp. 179–213.

Spence, K., R. Venditti, O. Rojas, Y. Habibi, and J. Pawlak (2010a). The effect of chemical composition on microfibrillar cellulose films from wood pulps: water interactions and physical properties for packaging applications. *Cellulose* **17**:835–848.

Spence, K., R. A. Venditti, O. J. Rojas, Y. Habibi, and J. Pawlak (2010b). A comparative study of energy consumption and physical properties of microfibrillated cellulose produced by different processing methods. *Cellulose* **2011**, 18, (4), 1097–1111.

Šturcova, A., G. R. Davies, and S. J. Eichhorn (2005). Elastic modulus and stress-transfer properties of tunicate cellulose whiskers. *Biomacromolecules* **6**:1055–1061.

Svagan, A. J., M. A. S. A. Samir, and L. A. Berglund (2008). Biomimetic foams of high mechanical performance based on nanostructured cell walls reinforced by native cellulose nanofibrils. *Adv. Mater.* **20**:1263–1269.

Syverud, K., and P. Stenius (2009). Strength and barrier properties of MFC films. *Cellulose* **16**:75–85.

Tashiro, K., and M. Kobayashi (1991). Theoretical evaluation of three-dimensional elastic constants of native and regenerated celluloses: role of hydrogen bonds *Polymer* **32**:1516–1526.

Tokoh, C., K. Takabe, M. Fujita, and H. Saiki (1998). Cellulose synthesized by *Acetobacter xylinum* in the presence of Acetyl glucomannan. *Cellulose* **5**:249–261.

Turbak, A., F. Snyder, and K. Sandberg (1983). Microfibrillated cellulose: a new cellulose product: properties, uses, and commercial potential. *J. Appl. Polym. Sci. Appl. Polym. Symp.* **37**:815–827.

Usuda, M., O. Suzuki, J. Nakano, and N. Migita (1967). Acid hydrolysis of cellulose in concentrated phosphoric acid: effects of modified groups of cellulose on the rate of hydrolysis. *Kogyo Kagaku Zasshi* **70**:349–352.

Wågberg, L., G. Decher, M. Norgren, T. Lindström, M. Ankerfors, and K. Axnäs (2008). The build-up of polyelectrolyte multilayers of microfibrillated cellulose and cationic polyelectrolytes. *Langmuir* **24**:784–795.

Wang, Y., X. Cao, and L. Zhang (2006). Effects of cellulose whiskers on properties of soy protein thermoplastics. *Macromol. Biosci.* **6**:524–531.

Yachi, T., J. Hayashi, M. Takai, and Y. Shimizu (1983). Supermolecular structures of cellulose: stepwise decrease in LODP and particle size of cellulose hydrolyzed after chemical treatment. *J. Appl. Polym. Sci. Appl. Polym. Symp.* **37**:325–343.

Zimmerman, T., E. Poehler, and T. Geiger (2004). Cellulose fibrils for polymer reinforcement. *Adv. Eng. Mater.* **6**:754–761.

Zimmermann, T., E. Pöhler, and P. Schwaller (2005). Mechanical and morphological properties of cellulose fibril reinforced nanocomposites. *Adv. Eng. Mater.* **7**:1156–1161.

Zogaj, X., M. Nimtz, M. Rohde, W. Bokranz, and U. Römling (2001). The multicellular morphotypes of *Salmonella typhimurium* and *Escherichia coli* produce cellulose as the second component of the extracellular matrix. *Mol. Microbiol.* **39**:1452–1463.

Zuluaga, R., J. L. Putaux, J. Cruz, J. Vélez, I. Mondragon, and P. Gañán (2009). Cellulose microfibrils from banana rachis: effect of alkaline treatments on structural and morphological features. *Carbohydr. Polym.* **76**:51–59.

4

INTERACTIONS OF CHITOSAN WITH METALS FOR WATER PURIFICATION

Mohammed Rhazi, Abdelouhad Tolaimate, and Youssef Habibi

4.1 INTRODUCTION

Increased environmental concerns have made the control and treatment of liquid effluents, mainly wastewaters, discharged by various industries such as mining, food processing, petrochemicals, and so on, an integral part of operational systems. These wastewaters may contain organic and inorganic contaminants such as suspended solid particles, dyes, pesticides, and heavy metals. The potential of these contaminants to negatively impact aquatic ecosystems and drinking water resources necessitates that wastewaters either be recycled back to processing circuits for reuse or treated prior to disposal or release to the environment. Thus, the contaminants must be effectively removed to meet increasingly stringent environmental quality standards. Some natural biopolymers such as chitosan are being recognized as efficient materials used for wastewater treatment.

Chitin is the second most widely produced polymer in the biosphere after cellulose and is composed of N-acetylglucosamine monomer units linked by a β $(1 \rightarrow 4)$ glycosidic bond. It is commonly found in the exoskeletons or cuticles of many invertebrates and in the cell walls of most fungi and some algae (Ruiz-Herrera, 1978; Jeuniaux, 1982). Chitosan is a deacetylated derivative of chitin (Muzzarelli, 1977; Roberts, 1992; Rinaudo, 2006). However, in almost all cases, monomers of chitin are not actually all acetylated, and likewise for the chitosan, the monomers are not all deacetylated. In fact, chitin and chitosan correspond to copolymers consisting of a sequence of N-acetyl-D-glucosamine and D-glucosamine units (Figure 4.1).

Polysaccharide Building Blocks: A Sustainable Approach to the Development of Renewable Biomaterials, First Edition. Edited by Youssef Habibi and Lucian A. Lucia.

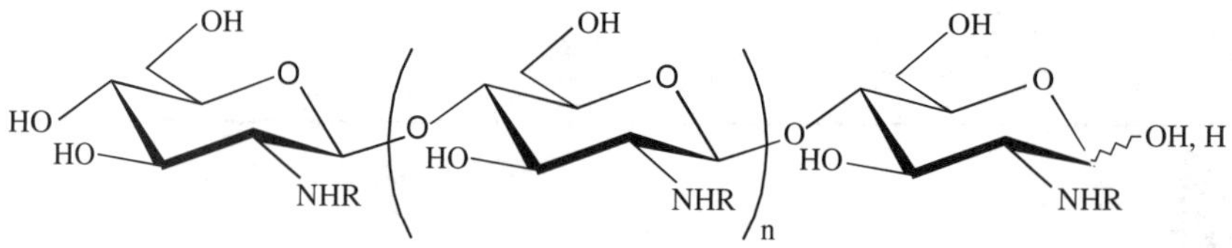

Chitin R = COCH$_3$ Chitosan R = H

FIGURE 4.1 Chemical structure of chitin and chitosan.

They differ from each other by a medium degree of acetylation (DA), which reflects the relative proportion of monomers of *N*-acetyl-glucosamine present in the copolymer. Generally, when the DA exceeds 50%, the copolymer is practically insoluble in dilute acids and corresponds to the chitin, whereas when the DA is less than 50%, the copolymer is more soluble in acidic solutions, and therefore the copolymer is assigned as chitosan (Rinaudo et al., 1993).

These polymers have many properties that allow numerous potential applications in a wide range of areas such as the fields of cosmetics, agriculture, food, biomedicine, textile, or the treatment of industrial effluents (Allan et al., 1984; Gordon, 1984; Nagai et al., 1984; Pangburn et al., 1984; Rutherford and Dunson, 1984; Spreen et al., 1984; Brzeski, 1987; Ember, 1997; Witoon, 2005; Rinaudo, 2006). Yet, for each of these uses, it is necessary to control the various parameters influencing the physicochemical properties of chitin-based materials so that they can be produced according to the performances required for the final uses.

The ability of chitosan to form complexes with metal ions has attracted the interest of researchers for many years (Muzzarelli, 1973; Hauer, 1978; Onsosyen and Skaugrud, 1990; Guibal, 2004). Although, any comparison between the different studies is difficult because of the divergence in preparation and operating conditions, the chitosan–metal ions interactions are far from being completely understood mainly because of the diversity on chelation mechanisms.

In this chapter, some aspects of the interaction of chitosan from various sources with metal ions were highlighted. Chitin with different crystallographic structure types were extracted from numerous sea sources and investigated in regard to the deacetylation process to access chitosan. Indeed, deacetylation was carried out and the role of the process parameters and the nature of the chitin on the physicochemical properties of the obtained chitosan were studied. Finally, the influence of the nature of the metal ion on chitosan–metal interactions was examined.

4.2 EXTRACTION OF CHITIN

The main issue met in the chemistry of chitin relies upon its extraction and preparation with limited degradation. Chitin isolated from crab and shrimp shells have been studied extensively owing to their easy accessibility. These chitins have the α-crystallographic structure where polymer chains are arranged in an anti-parallel fashion with strong intermolecular hydrogen bonding (Minke and Blackwell, 1978; Rinaudo, 2006). Chitin may also be obtained from squid pens (Kurita et al., 1993b), but it adopts the β-crystallographic structure characterized by a parallel arrangement of the polymer chains with relatively weak intermolecular forces (Gardner and Blackwell, 1975). So far, only limited attention has been paid to β-chitin, and its chemistry has been barely exploited, primarily because of its low accessibility.

β-Chitin is considered of interest owing to some specific properties. It shows higher solubility and swelling ability than α-chitin (Lee, 1974; Austin et al., 1989) mainly due to much weaker intermolecular hydrogen bonding ascribable to the parallel arrangement of the chains. Also, it shows enhanced reactivity compared to α-chitin (Sannan et al., 1976). For example, β-chitin shows higher reactivity than α-chitin during deacetylation (Kurita et al., 1993b) with limited degradation through acetolysis (Kurita et al., 1993a).

The main extraction procedure of chitin is based on the use of combined low concentrated acidic (HCl 0.55 M) and alkaline (NaOH 0.3 M) solutions at hot temperatures (Rhazi et al., 2000; Tolaimate et al., 2000). These conditions are very moderate compared to other methods reported in the literature (Hackman, 1954; BeMiller and Whistler, 1962; Broussignac, 1968; Hackman and Goldberg, 1974; Madhavan and Ramachandran Nair, 1974; Moorjani et al., 1975; Muzzarelli et al., 1980). However, this extraction procedure has to be adapted according to the source of chitin and its chemical composition particularly the content of mineral, which is of about 1.7% in the case of the pen of squid and 18.90% for the squilla and 34% for lobster (see Table 4.1). The ratios of the solid raw material to the extracting solvent as well as the extraction time are the more important parameters that affect the yield and quality of extracted chitin (Rhazi et al., 2000; Tolaimate et al., 2000) as shown by the work of Kurita et al. (1993b) in which chitin was extracted from the pen of squid and shrimp and the conditions were scrutinized and compared. The application of this extraction process to isolate chitin from wastes of several crustaceans and cephalopods allowed the preparation of pure, colorless, and fully acetylated chitin (Table 4.1).

The residual mineral content of different extracted chitin is very low and it is in the order of several tens of ppm (micrograms per gram), whereas it is approximately of 1.5% for chitin prepared by other methods (Tolaimate et al., 2003). The examination of extracted chitin with NMR/CPMAS shows its high degree of acetylation (Figure 4.2) that is in most cases around 100%. Such results are particularly interesting and to our knowledge, such values of DA were never obtained since published results report mainly DA values ranging between 80 and 90% (Roberts, 1992; Kurita et al., 1993b). As such, this high DA was achieved with this extraction procedure without affecting any structural features particularly in relation to the molecular weight or the original crystalline structure (α-chitin or β-chitin) (Tolaimate et al., 2003).

4.3 PREPARATION OF CHITOSAN

Chitosan, a deacetylated form of chitin, occurs rarely in nature. It is found in small amount in the exoskeletons of some insects like the queens of the termites and in the

TABLE 4.1 Extraction Conditions and Properties of Chitin from Different Sea Sources (Rhazi et al., 2000; Tolaimate et al., 2000)

Chitin Source	Mineral Content (%)	Acid Bath 0.55 M HCl Number	Alkaline Bath 0.3 M NaOH Number	H_2O_2 Treatment	Chitin Content (%)	Mineral Residual (%)	Type of Chitin	DA (%)
Barnacle	–	2	4	Yes	07	–	α	100
Marbled crab	31.13	5	3	Yes	10	0.025	α	98
Red crab	–	3	3	Yes	10	–	α	99
Spider crab	25.96	3	3	Yes	16	0.014	α	96
Lobster	33.99	3	3	Yes	17	0.062	α	–
Locust lobster	–	3	7	Yes	25	0.168	α	100
Spiny lobster	–	2	3	Yes	32	–	α	–
Crayfish	–	2	3	Yes	36	–	α	–
Pink shrimp	21.5	3	3	Yes	22	0.019	α	100
Gray shrimp	12.95	2	2	Yes	24	0.02	α	100
Squilla	18.89	3	3	Yes	24	0.014	α	100
Cuttlefish	–	3	3	No	20	–	β	–
Squid	1.70	2	2	No	42	0.017	β	100

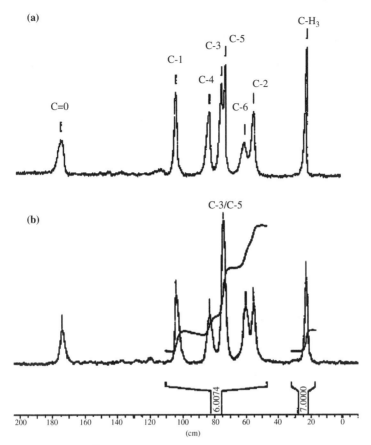

FIGURE 4.2 Solid-state ^{13}C NMR spectra of chitinβ-chitin extracted from cephalopod and α-chitin extracted from crustacean (Rhazi et al., 2000; Tolaimate et al., 2000).

cell walls of zygomycetes that form a particular class of fungi (Seng, 1988). Hence, chitosan is mainly produced by thermochemical alkaline deacetylation of chitin. Chitosan is characterized by its (DA) and this influences not only its physicochemical properties (Sannan et al., 1976; Wang et al., 1991; Errington et al., 1993; Rinaudo et al., 1993; Vårum et al., 1994; Zydowicz et al., 1996) but also its biodegradability (Shigemasa et al., 1994; Hutadilok et al., 1995; Nordtveit et al., 1996) and immunological activity (Peluso et al., 1994) among others.

Regeneration of amine functions from acetamidodeoxy carbohydrates, can be performed in acid and basic conditions (Hanessian, 1972), but unfavorable steric effects frequently hinder the reaction (Thompson and Wolfrom, 1963). Despite numerous attempts by the means of acid-catalyzed procedures, the N-acetyl groups could not be removed without inducing the hydrolysis of the polysaccharide backbone. In the presence of alkali, polysaccharide chains were found to undergo degradation because of the high concentration of reagents and prolonged reaction times required to obtain a complete deacetylation. The low reactivity of chitin against the deacetylation reaction was ascribed to the *trans*-arrangement of acetamido groups in the monomeric unit with respect to the hydroxyl group OH-3 (Muzzarelli, 1977). Several methods of deacetylation of chitin are described in the literature; however there is no regular process allowing the preparation of chitosan. Nevertheless, two processes for instance are the most employed (Rhazi et al., 2000; Tolaimate et al., 2000). Applied by Broussignac (1968) to the deacetylation of chitin extracted from different sources, this process uses a potassium hydroxide (50 w/w%)as deacetylation reagent, dissolved in a mixture of ethanol (25 w/w%) and mono-ethyleneglycol (25 w/w%), which is nearly an anhydrous reaction medium. The second process published by Kurita et al. (1993a, 1993b) consists of a treatment with an aqueous sodium hydroxide solution 40% (wt) at high temperature (e.g., 80°C) under nitrogen.

Since physicochemical properties of chitosan are usually influenced by the parameters of its preparation, the reaction time, concentration, and nature of the alkaline reagent were investigated. This study allowed the determination of the optimal conditions to reach the adequate molecular weight and degree of acetylation for desired properties.

4.3.1 Role of Source Chitin

Under the same experimental conditions, chitosan with different physicochemical characteristics was obtained by the deacetylation of chitin from different sources (Rhazi et al., 2000). The molecular weights of the obtained samples ranged from 8300 g/mol for chitin from cuttlefish up to 125,000 g/mol for chitin from shrimps. The DA also varied between 0.5% (squid) and 16% (Locust lobster) (Table 4.2).

TABLE 4.2   Influence of the Source of Chitin on the Properties of Chitosan Obtained Under the Same Conditions (KOH (50 w/w%)/Ethanol (25 w/w%)/Monoethyleneglycol (25 w/w%), 120 °C)

Source of Chitin	Type of Chitin	Yield of Chitosan (%)	Degree of Acetylation (%)	Average Molecular Weight (M_v, g/mol)
Barnacle	α	78	13	66,000
Marbled crab	α	65	04	55,000
Red crab	α	76	09	56,000
Spider crab	α	75	03	45,000
Lobster	α	77	10	72,000
Locust lobster	α	77	16	98,000
Spiny lobster	α	78	12	82,000
Crayfish	α	79	08	83,000
Shrimp	α	78	10	125,000
Squilla	α	67	04	84,000
Cuttlefish	β	64	03	8300
Squid	β	70	00.5	17,000

4.4 INFLUENCE OF THE N-DEACETYLATION METHOD

For the preparation of chitosan, two methods were studied according to, respectively, Broussignac's method, using KOH solution under anhydrous conditions, and Kurita's method, using aqueous NaOH solution. In the case of β-chitin (squid), the N-deacetylation with aqueous NaOH under mild conditions (NaOH 40% wt, T 80 °C) was conducted more easily and chitosan samples with low DA and medium to high molecular weights were obtained. However, in the case of α-chitin (crustaceans), the method using KOH in anhydrous medium has the advantage of providing good quality chitosan in a single step and relatively short treatment time (2 h for shrimp, 4 h for squilla, and 8 h for spider crab), whereas no deacetylation occurred under the same conditions particularly the base concentration, for example, 50% and temperature, for example, 120°C, with the method using NaOH. The reaction time has to be prolonged for the deacetylation to occur and the resulting chitosan was too degraded and colorful enough (Tolaimate et al., 2000, 2003).

4.5 ROLE OF REPEATED TREATMENT AND THE ADDITION OF NaBH₄

When the deacetylation was carried out in aqueous caustic soda solution, the combined effects of the nature and mode of repeated treatments and the addition of $NaBH_4$ improved the quality of the chitosan obtained (Table 4.3). When the deacetylation is carried out with caustic potash solution in the alcohol mixture, the

TABLE 4.3 Influence of Repeated Treatment and the Addition of NaBH₄ on the Degree of Acetylation and the Average Molecular Weight of Chitosan

Chitin	Deacetylation Process				Degree of Acetylation (DA; %)	Average Molar Mass (M_v; g/mol)
	NaOH (w/w%)	T (°C)	Time (h)	Addition NaBH₄		
β-Chitin DA: 100%	40	80	9	−	17	298,000
	40	80	3 h × 3	−	1	500,000
	40	80	3 h × 3	+	0	644,000
α-Chitin DA: 100%	50	120	12	−	3	45,000
	50	120	6 h × 2	−	3.9	84,000
	50	120	6 h × 2	+	1.8	153,000
	50	120	4 h × 3	+	1	147,000
	50	120	3 h × 4	+	0	149,000
	50	120	3 h × 3	+	1	190,000

addition of NaBH$_4$ did not improve the molecular weight of chitosan. The observed effect of such addition tends to be very negative, especially at the DA of chitosan obtained (Tolaimate et al., 2000, 2003).

The experimental conditions that allow for fully deacetylated chitosan (DA: 0–1%) with remarkably high molar mass (600,000 to over 1 million g/mol) are particularly interesting (Tolaimate et al., 2003, 2008). They can remedy satisfactorily with the dilemma involved in the preparation of chitosan that when an advanced deacetylation, highly desirable, is obtained, it is often accompanied by a degradation of the polymer highly undesirable.

4.6 INTERACTION OF CHITOSAN WITH METALS

Chitosan presents a large capacity to fix molecules such as pesticides (Van Daele and Thomé, 1986; Thomé et al., 1997), proteins (Yoshida et al., 1995), and dyes (Kim et al., 1997). The free amine function of chitosan gives it a better ability to chelate ions of transition metals (Koshijima et al., 1973; Muzzarelli, 1973; Muzzarelli and Rocchetti, 1974) compared to other natural compounds such as cellulose derivatives (Masri et al., 1974). These chelating properties are of great interest for water treatment and particularly to recover metals present in wastewaters or seawaters. The mechanisms of the chelation seem to be very complex, and not fully understood which stimulate vigorous discussion to the present day.

Recovery of metals from metal-bearing effluents may be performed using different treatment processes (Maghami and Roberts, 1988; Zhu and Sen Gupta, 1992). Besides mechanical treatments of wastewaters (sedimentation) or biological ones (activated muds), some chemical treatments are used to eliminate these metals. Most current processes are the precipitation by hydroxides or sulfides, the oxidation–reduction, ions exchange, liquid–solid separation by decanting–flotation, and the separation using membranes. However, the major drawback of these treatments is the formation of muds that have to be confined. Hence alternative processes, more economical, were developed that are based on the use of natural materials or polymers, such as chitosan, to recover metallic ions. Adsorption on active coal was widely studied but new adsorbent materials are considered to be more efficient and less expensive. Thus, the adsorption by living organisms (bacteria, fungi, seaweeds, etc.) or by chemicals from these organisms was studied. Chitosan is one of these natural products.

The capacity of chitosan to form complexes with metallic ions gained the interest of researchers for many years (Muzzarelli, 1973; Guibal et al., 1997, 1999; Guibal, 2004). However, any apple-to-apple comparison between different published studies would be conjectural due to the large discrepancy of experimental conditions. It was shown that the chelation process and the stability of the metal–chitosan complex may be influenced by the stirring (mechanical or ultrasound) (Muzzarelli, 1977). The chelation may depend on the physical state of the chitosan (powder, gel, fiber, film, etc.) (Guibal et al., 1997). However, the prominent parameter in the complexation process seems to be the (DA). Yaku and Koshijima (1978) have found that a film of chitin obtained by N-acetylation of chitosan does not form any complex with metallic ions such as

Cu(II), Co(II), or Ni(II). This was confirmed by an interesting work from Micera et al. (1985) on monomers from which it was concluded that N-acetyl-D-glucosamine units do not interact with Cu(II). Moreover Kurita et al. (1979) have confirmed that complexation was easier when the chitosan has a high degree of deacetylation.

Different models were proposed to elucidate the mechanism of coordination implied in the formation of complexes (Ogawa et al., 1993). The first one, called "bridge model," suggests that the metallic ion is bound with several nitrogen atoms from the same molecular chain or from different chains (Schlick, 1986). The second one known as "pendant model" considers that the metallic ion is bound to the amino group as a pendant (Ogawa et al., 1984, 1993; Nieto et al., 1992). This model was also suggested and confirmed by Domard (1987). Using potentiometric and dichroic measurements it was concluded that fully deacetylated chitosan may form with the cupric ions, in a giving pH range (pH < 6.1), only one complex for which the assumed structure is [CuNH$_2$(OH)$_2$]. The fourth site could be occupied by a water molecule or by a hydroxyl group linked on C-3.

Other authors (Monteiro and Airoldi, 1999) have suggested another structure for the Cu–chitosan complex in which the metal ion Cu is likely to be bonded to three oxygen and one nitrogen ligands in a square planar or a tetrahedral geometry. Within this structure, the two bonded oxygen atoms and the nitrogen atom are believed to emerge from a monosaccharide unit and it happens that two units are involved to ultimately form the metal coordination sphere.

Studies by Rhazi et al. (2002a) identified an optimal pH, which ranges between 5 and 7, where the highest fixation of cupric ions by chitosan occurs. These authors have suggested two different complexes. From potentiometric and spectrophotometric studies and taking into account the repartition of the species in solution, it is suggested that the chitosan–Cu(II) complexes have structures as $[Cu(-NH_2)^{2+}, 2OH^-, H_2O]$ and $[Cu(-NH_2)_2^{2+}, 2OH^-]$. In the 5–5.8 pH range, the first complex is the more stable and for pH values larger than 5.8 the second structure is predominant. It should be noted that the complexation of ions by the polymer was carried outside the pH range where the chitosan precipitation befalls. Under these conditions, the system remains homogeneous allowing a thriving study of the complex formation since in such conditions a monochain fixation of metallic ions would be favored (Rhazi et al., 2002a).

These interactions between the chitosan and the copper ions may be considered as a competition between the following equilibrium:

$$Cu^{2+} + -NH_3^+ + H_2O \rightleftharpoons [Cu(-NH_2)]^{2+} + H_3O^+$$

$$Cu^{2+} + -NH_2 \rightleftharpoons [Cu(-NH_2)]^{2+}$$

$$[Cu(-NH_2)]^{2+} + -NH_3^+ + H_2O \rightleftharpoons [Cu(-NH_2)_2]^{2+} + H_3O^+$$

$$[Cu(-NH_2)]^{2+} + -NH_2 \rightleftharpoons [Cu(-NH_2)_2]^{2+}$$

Regarding the interaction of Fe ions with chitosan, it was found that the complex formed is the result of binding of Fe(III) with two moles of amino groups and four

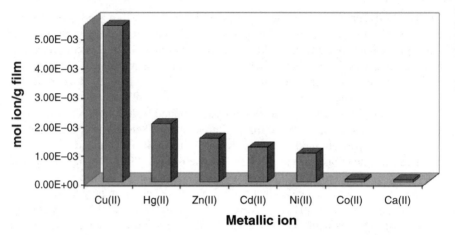

FIGURE 4.3 Ion displacement by addition of metallic chloride to chitosan solution.

moles of oxygen atoms (Nieto et al., 1992). It was suggested that the ligand around Fe ion is either penta or hexa coordinated with N/O ligands including two amino groups and three or four oxygen atoms, respectively (Bhatia and Ravi, 2000).

In another work (Rhazi et al., 2002b), the selectivity of chitosan for metallic ions was studied according to their ionic charge. The study was performed on solid chitosan, for example, film or powder or using chitosan solution. The results showed that chitosan films have a higher affinity for copper and mercury (Figure 4.3).

The mobilization depends strongly on the nature of the ionic metal added to the polymer solution; better ionic displacement is observed with copper and mercury ions showing the greatest affinity of chitosan for these ions. Ions zinc and cadmium show a smaller attraction and only a slight change in pH was detected with cobalt and calcium ions (Figure 4.4).

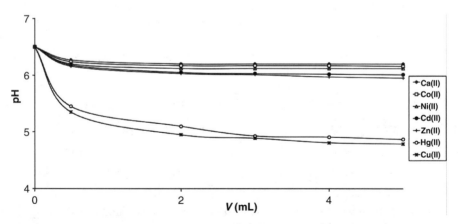

FIGURE 4.4 Quantity of divalent metallic ions fixed per gram of chitosan film.

TABLE 4.4 Quantity of the Metallic Ion Fixed at Different pH per One Mole of Chitosan Determined from Microanalysis and Weighing

Metal	Solution (Equilibrium pH)	Microanalysis (mmol ion/gram of Film)	Weighing (mol ion/1 mol of Chitosan)	Microanalysis (mol ion/1 mol of Chitosan)
Cu	5.70	4.38	0.863	0.704
Hg	7.20	1.91	0.318	0.309
Zn	5.88	1.23	0.248	0.193
Cd	7.47	0.85	0.198	0.137
Ni	7.93	0.61	0.157	0.097
Co	7.87	0.07	0.018	0.011
Ca	7.94	0.03	0.013	0.005

A summary of the results obtained with different experimental techniques is gathered in Table 4.4. The affinity of chitosan in film form or in solution can be represented by the following sequence:

$$Cu(II) \geqslant Hg(II) > Zn(II) > Cd(II) > Ni(II) > Co(II), Ca(II)$$

As mentioned previously, the ability of chitosan to complex metallic ions is one of its major potentialities. This polymer has shown selectivity according to the cation. It was determined that fixation capacities of metal per gram of chitosan vary from 0.02 mmol (Co^{2+}, Ca^{2+}) up to 1.2 mmol (Cu^{2+}) in the same experimental conditions (Rhazi et al., 2002b).

The quantity of ions fixed on the polymer is more important in the case of the film than the powder. Besides the effect of the structure, mainly related to DA, physical parameters such as its morphology could alter the accessibility of metal ions to the active sites of the polymer (Rhazi et al., 2008).

4.7 CONCLUSION

Numerous sea sources of chitin were explored in order to produce chitosan. Studies showed that the proper adjustment of the process conditions will make possible to produce pure chitin very similar to their native form, and to prepare chitosan with well-controlled properties.

The capacity of chitosan to complex metallic ions is one of its most important potentialities. Several factors affect the formation of chitosan–metallic ion complexes from which some are intrinsic to chitosan such as the degree of polymerization (DP), degree of acetylation and its physical state and others depend on the operational conditions mainly the pH and the nature of the ion and cons-ion. The complex begins to form when DP is greater than 6 and it is optimal with low degree of acetylation. The nature of the ion and cons-ion also plays an important role in the interaction between units of glucosamine and the cation. Although the new mechanisms of chitosan–ion interactions that drive their chelation were suggested, these mechanisms are still not fully elucidated. Finally, these chitosan-based materials were successfully tested to depolluted wastewaters from Marrakesh region (Rhazi et al., 2008). This study would provide new venues to add value to waste products from Moroccan fishing industries.

REFERENCES

Allan, G. G., L. C. Altman, R. E. Bensinger, D. K. Ghosh, Y. Hirabayashi, A. N. Neogi, and S. Neogi (1984). Biomedical applications of chitin and chitosan. In: J. P. Zikakis,editor, *Chitin Chitosan and Related Enzymes*, New York: Academic Press, pp. 119–133.

Austin, P. E., J. E. Castle, and C. J. Albisetti (1989). Beta-chitin from squid: new solvents and plasticizers. In: G. Skjak-Barck, T. Anthosen, and P. Sandford,editors, *Chitin and Chitosan: Sources, Chemistry, Biochemistry, Physical Properties and Applications*, Essex: Elsevier Applied Science, pp. 749–755.

BeMiller, J. N. and R. L. Whistler (1962). Alkaline degradation of amino sugars. *J. Org. Chem.* **27**:1161–1164.

Bhatia, S. C. and N. Ravi (2000). A magnetic study of an Fe–chitosan complex and its relevance to other biomolecules. *Biomacromolecules* **1**:413–417.

Broussignac, P. (1968). Chitosan: a natural polymer not well known by the industry. *Chim. Ind. Genie Chim.* **99**:1241–1247.

Brzeski, M. M. (1987). Chitin and chitosan-puting waste to good use. *Infofish Inter.* **87**:38–40.

Domard, A. (1987). pH and CD measurements on fully deacetylated chitosan: application to Cu II—polymer interactions. *Int. J. Biol. Macromol.* **9**:98–104.

Ember, L. (1997). Detoxifying nerve agents: academic, army, and scientists join forces on enzyme anchored in foams and fiber. *Chem. Eng. News* **15**:26–29.

Errington, N., S. E. Harding, K. M. Vårum, and L. Illum (1993). Hydrodynamic characterization of chitosans varying in degree of acetylation. *Int. J. Biol. Macromol.* **15**:113–117.

Gardner, K. H., and J. Blackwell (1975). Refinement of the structure of β-chitin. *Biopolymers* **14**:1581–1595.

Gordon, D. T. (1984). Action of amino acids on iron status, gut morphology, and cholesterol levels in the rat. In: J. P. Zikakis,editor, *Chitin Chitosan and Related Enzymes*, New York: Academic Press, pp. 97–117.

Guibal, E. (2004). Interactions of metal ions with chitosan-based sorbents: a review. *Sep. Purif. Technol.* **38**:43–74.

Guibal, E., C. Milot, O. Eterradossi, C. Gauffier, and A. Domard (1999). Study of molybdate ion sorption on chitosan gel beads by different spectrometric analyses. *Int. J. Biol. Macromol.* **24**:49–59.

Guibal, E., C. Milot, and J. Roussy (1997). Chitosan gel beads for metal ion recovery. In: R. Muzzarelli, and M. G. Peter,editors, *Chitin Handbook*, Grottamare: European Chitin Society, pp. 423–429.

Hackman, R. H. (1954). Studies on chitin. 1. Enzymic degradation of chitin and chitin esters. *Aust. J. Biol. Sci.* **7**:168–178.

Hackman, R. H., and M. Goldberg (1974). Light-scattering and infrared spectrophotometric studies of chitin and chitin derivatives. *Carbohydr. Res.* **38**:35–45.

Hanessian, S. (1972). *Methods Carbohydr. Chem.* **6**:208.

Hauer, H. (1978). *The Chelating Properties of Kytex H Chitosan. First International Conference on Chitin/Chitosan.* Cambridge, MA: MIT Sea Grant Program.

Hutadilok, N., T. Mochimasu, H. Hisamori, K.-i. Hayashi, H. Tachibana, T. Ishii, and S. Hirano (1995). The effect of N-substitution on the hydrolysis of chitosan by an endo-chitosanase. *Carbohydr. Res.* **268**:143–149.

Jeuniaux, C. (1982). Composition chimique comparée des formations squelettiques chez lophophoriens et endoproctes. *Bull. Soc. Zool. Fr.* **107**:233–249.

Kim, C. Y., H.-M. Choi, and H. T. Cho (1997). Effect of deacetylation on sorption of dyes and chromium on chitin. *J. Appl. Polym. Sci.* **63**:725–736.

Koshijima, T., R. Tanaka, E. Muraki, A. Yamada, and F. Yaku (1973). Chelating polymers derived from cellulose and chitin. *Cellul. Chem. Technol.* **7**:197–208.

Kurita, K., T. Sannan, and Y. Iwakura (1979). Studies on chitin. VI. Binding of metal cations. *J. Appl. Polym. Sci.* **23**:511–515.

Kurita, K., K. Tomita, S. Ishii, S.-I. Nishimura, and K. Shimoda (1993a). β-Chitin as a convenient starting material for acetolysis for efficient preparation of *N*-acetylchitooligosaccharides. *J. Polym. Sci. Part A Polym. Chem.* **31**:2393–2395.

Kurita, K., K. Tomita, T. Tada, S. Ishii, S.-I. Nishimura, and K. Shimoda (1993b). Squid chitin as a potential alternative chitin source: deacetylation behavior and characteristic properties. *J. Polym. Sci. Part A Polym. Chem.* **31**:485–491.

Lee, V. F. P. (1974). *Solution and Shear Properties of Chitin and Chitosan.* Seattle, WA: University of Washington. PhD dissertation.

Madhavan, P. and K. G. Ramachandran Nair (1974). Utilization of prawn waste-isolation of chitin and its conversion to chitosan. *Fish. Technol.* **11**:50–53.

Maghami, G. G., and G. A. F. Roberts (1988). Studies on the adsorption of anionic dyes on chitosan. *Makromol. Chem.* **189**:2239–2243.

Masri, M. S., F. W. Reuter, and M. Friedman (1974). Binding of metal cations by natural substances. *J. Appl. Polym. Sci.* **18**:675–681.

Micera, G., S. Deiana, A. Dessi, P. Decock, B. Dubois, and H. Kozlowski (1985). Copper(II) complexation by D-glucosamine. Spectroscopic and potentiometric studies. *Inorg. Chim. Acta* **107**:45–48.

Minke, R., and J. Blackwell (1978). The structure of [alpha]-chitin. *J. Mol. Biol.* **120**:167–181.

Monteiro, O. A. C., and C. Airoldi (1999). Some thermodynamic data on copper–chitin and copper–chitosan biopolymer interactions. *J. Colloid Interface Sci.* **212**:212–219.

Moorjani, M. N., V. Achutha, and D. I. Khasim (1975). Parameters affecting the viscosity of chitosan from prawn waste. *J. Food Sci. Technol.* **12**:187–189.

Muzzarelli, R. A. A. (1973). *Natural Chelating Polymers.* Oxford: Pergamon Press.

Muzzarelli, R. A. A. (1977). *Chitin.* New York: Pergamon Press.

Muzzarelli, R. A. A., and R. Rocchetti (1974). Enhanced capacity of chitosan for transition metal ions in sulphuric acid solutions. *Talanta* **21**:1173–1143.

Muzzarelli, R. A. A., F. Tanfani, M. Emanuelli, and S. Gentile (1980). The chelation of cupric ions by chitosan membranes. *J. Appl. Biochem.* **2**:380–389.

Nagai, T., Y. Sawayanagi, and N. Nambu (1984). Application of chitin and chitosan to pharmaceutical preparations. In: J. P. Zikakis, editor, *Chitin Chitosan and Related Enzymes*, New York: Academic Press, pp. 21–39.

Nieto, J. M., C. Peniche-Covas, and J. Del Bosque (1992). Preparation and characterization of a chitosan–Fe(III) complex. *Carbohydr. Polym.* **18**:221–224.

Nordtveit, R. J., K. M. Vårum, and O. Smidsrød (1996). Degradation of partially N-acetylated chitosans with hen egg white and human lysozyme. *Carbohydr. Polym.* **29**:163–167.

Ogawa, K., T. Miganiski, and S. Hirano (1984). X-ray diffraction study on chitosan–metal complexes. In: J. P. Zikakis,editor, *Advances in Chitin, Chitosan and Related Enzymes*, Orlando: Academic Press Inc., pp. 327–345.

Ogawa, K., K. Oka, and T. Yui (1993). X-ray study of chitosan-transition metal complexes. *Chem. Mater.* **5**:726–728.

Onsosyen, E., and O. Skaugrud (1990). Metal recovery using chitosan. *J. Chem. Technol. Biotechnol.* **49**:395–404.

Pangburn, S. H., P. V. Trescony, and J. Heller (1984). Partially deacetylated chitin: its use in self-regulated drug delivery systems. In: J. P. Zikakis,editor, *Chitin Chitosan and Related Enzymes*, New York: Academic Press, pp. 3–19.

Peluso, G., O. Petillo, M. Ranieri, M. Santin, L. Ambrosic, D. Calabró, B. Avallone, and G. Balsamo (1994). Chitosan-mediated stimulation of macrophage function. *Biomaterials* **15**:1215–1220.

Rhazi, M., J. Desbrières, A. Tolaimate, A. Alagui, and P. Vottero (2000). Investigation of different natural sources of chitin: influence of the source and deacetylation process on the physicochemical characteristics of chitosan. *Polym. Int.* **49**:337–344.

Rhazi, M., J. Desbrières, A. Tolaimate, M. Rinaudo, P. Vottero, and A. Alagui (2002a). Contribution to the study of the complexation of copper by chitosan and oligomers. *Polymer* **43**:1267–1276.

Rhazi, M., J. Desbrieres, A. Tolaimate, M. Rinaudo, P. Vottero, A. Alagui, and M. El Meray (2002b). Influence of the nature of the metal ions on the complexation with chitosan: application to the treatment of liquid waste. *Eur. Polym. J.* **38**:1523–1530.

Rhazi, M., A. Tolaimate, A. Alagui, J. Desbrières, and M. Rinaudo (2008). Interaction of chitosan with metallic ions. Applications on the metallic depollution of waste water from Marrakech. *Phys. Chem. News* **39**:136–141.

Rinaudo, M. (2006). Chitin and chitosan: properties and applications. *Prog. Polym. Sci.* **31**:603–632.

Rinaudo, M., M. Milas, and P. L. Dung (1993). Characterization of chitosan. Influence of ionic strength and degree of acetylation on chain expansion. *Int. J. Biol. Macromol.* **15**:281–285.

Roberts, G. A. F. (1992). *Chitin Chemistry*. London: MacMillan Press.

Ruiz-Herrera, J. (1978). *First International Conference on Chitin and Chitosan*. Cambridge, MA: MIT Sea Grant Program.

Rutherford, F. A., and W. A. Dunson (1984). The permeability of chitin films to water and solutes. In: J. P. Zikakis,editor, *Chitin Chitosan and Related Enzymes*, New York: Academic Press pp. 135–143.

Sannan, T., K. Kurita, and Y. Iwakura (1976). Studies on chitin. 2. Effect of deacetylation on solubility. *Makromol. Chem.* **177**: pp. 3589–3600.

Schlick, S. (1986). Binding sites of Cu^{2+} in chitin and chitosan. An electron spin resonance study. *Macromolecules* **19**:192–195.

Seng, J. M. (1988). Chitine, chitosane et dérivés: de nouvelles perspectives pour l'industrie. *Biofutur* **71**:40–44.

Shigemasa, Y., K. Saito, H. Sashiwa, and H. Saimoto (1994). Enzymatic degradation of chitins and partially deacetylated chitins. *Int. J. Biol. Macromol.* **16**:43–49.

Spreen, K. A., J. P. Zikakis, and P. R. Austin (1984). The effect of chitinous materials on the intestinal microflora and the utilization of whey in monogastric animals. In: J. P. Zikakis, editor, *Chitin Chitosan and Related Enzymes*, New York: Academic Press pp. 57–75.

Thomé, J. P., C. Jeuniaux, and M. Weltrowski (1997). Applications of chitosan for the elimination of organochlorine xenobiotics from wastewater. In: M. F. A. Goosen,editor, *Applications of Chitin and Chitosan*, Lancaster, PA: Technomic, pp. 309–331.

Thompson, A., and M. L. Wolfrom (1963). *Methods Carbohydr. Chem.* **2**:215.

Tolaimate, A., J. Desbrieres, M. Rhazi, and A. Alagui (2003). Contribution to the preparation of chitins and chitosans with controlled physico-chemical properties. *Polymer* **44**:7939–7952.

Tolaimate, A., J. Desbrières, M. Rhazi, A. Alagui, M. Vincendon, and P. Vottero (2000). On the influence of deacetylation process on the physicochemical characteristics of chitosan from squid chitin. *Polymer* **41**:2463–2469.

Tolaimate, A., M. Rhazi, A. Alagui, J. Desbrières, and M. Rinaudo (2008). Valorization of waste products from fishing industry by production of the chitin and chitosan. *Phys. Chem. News* **42**:120–127.

Van Daele, Y., and J. P. Thomé (1986). Purification of PCB contaminated water by chitosan: a biological test of efficiency using the common barbel, *Barbus barbus*. *Bull. Environ. Contam. Toxicol.* **37**:858–865.

Vårum, K. M., M. H. Ottøy, and O. Smidsrød (1994). Water-solubility of partially *N*-acetylated chitosans as a function of pH: effect of chemical composition and depolymerisation. *Carbohydr. Polym.* **25**:65–70.

Wang, W., S. Bo, S. Li, and W. Qin (1991). Determination of the Mark–Houwink equation for chitosans with different degrees of deacetylation. *Int. J. Biol. Macromol.* **13**:281–285.

Witoon, H. K. N. (2005). Treatment of wastewaters with the biopolymer chitosan. In: K. J. Yarema,editor, *Handbook of Carbohydrate Engineering*, Boca Raton, FL: Taylor & Francis Group, LLC, pp. 535–562.

Yaku, F., and T. Koshijima (1978). *First International Conference on Chitin/Chitosan.* Cambridge, MA: MIT Sea Grant Program.

Yoshida, H., N. Kishimoto, and T. Kataoka (1995). Adsorption of glutamic acid on poly-aminated highly porous chitosan: equilibria. *Ind. Eng. Chem. Res.* **34**:347–355.

Zhu, Y., and A. K. Sen Gupta (1992). Sorption enhancement of some hydrophilic organic solutes through polymeric ligand exchange. *Environ. Sci. Technol.* **26**:1990–1998.

Zydowicz, N., L. Vachoud, and A. Domard (1996). In: C. Jeuniaux, R. Muzzarelli, A. Domard, and G. Roberts,editors, *Advances in Chitin Science*, Vol. I, Lyon: Jacques André. pp. 262–270.

5

RECENT DEVELOPMENTS IN CHITIN AND CHITOSAN BIO-BASED MATERIALS USED FOR FOOD PRESERVATION

Véronique Coma

5.1 INTRODUCTION

The growing shift of interest to renewable resources has reached impressive proportions in terms of the number of scientific publications and books, international symposia, and patents, although the output of concrete results is still modest (Gandini, 2008).

Food packaging is a particular domain in which there is growing interest in using biomaterials in place of their petroleum-based counterparts. Indeed, with the recent increase in ecological consciousness, research has turned toward finding bio-based materials and often biodegradable materials. The main identified objectives are (1) to decrease our impact on the environment and (2) to overcome the petroleum resource limitations.

Cellulose, chitin, and starch are three heavyweights among natural polymers due to their abundance and importance. Despite the fact that their basic building blocks are not so different (Figure 5.1), they display very varied properties, including crystallinity, solubility, and aptitude for chemical modification, and hence quite varied applications as macromolecular materials (Gandini, 2008). Most commonly, chitin is a linear chain molecule typically composed of several hundred 1 → 4-linked 2-acetamido-2-deoxy-β-D-glucopyranose units. Chitin and its derivatives such as chitosan are among the most promising materials derived from renewable resources

Polysaccharide Building Blocks: A Sustainable Approach to the Development of Renewable Biomaterials, First Edition. Edited by Youssef Habibi and Lucian A. Lucia.
© 2012 John Wiley & Sons, Inc. Published 2012 by John Wiley & Sons, Inc.

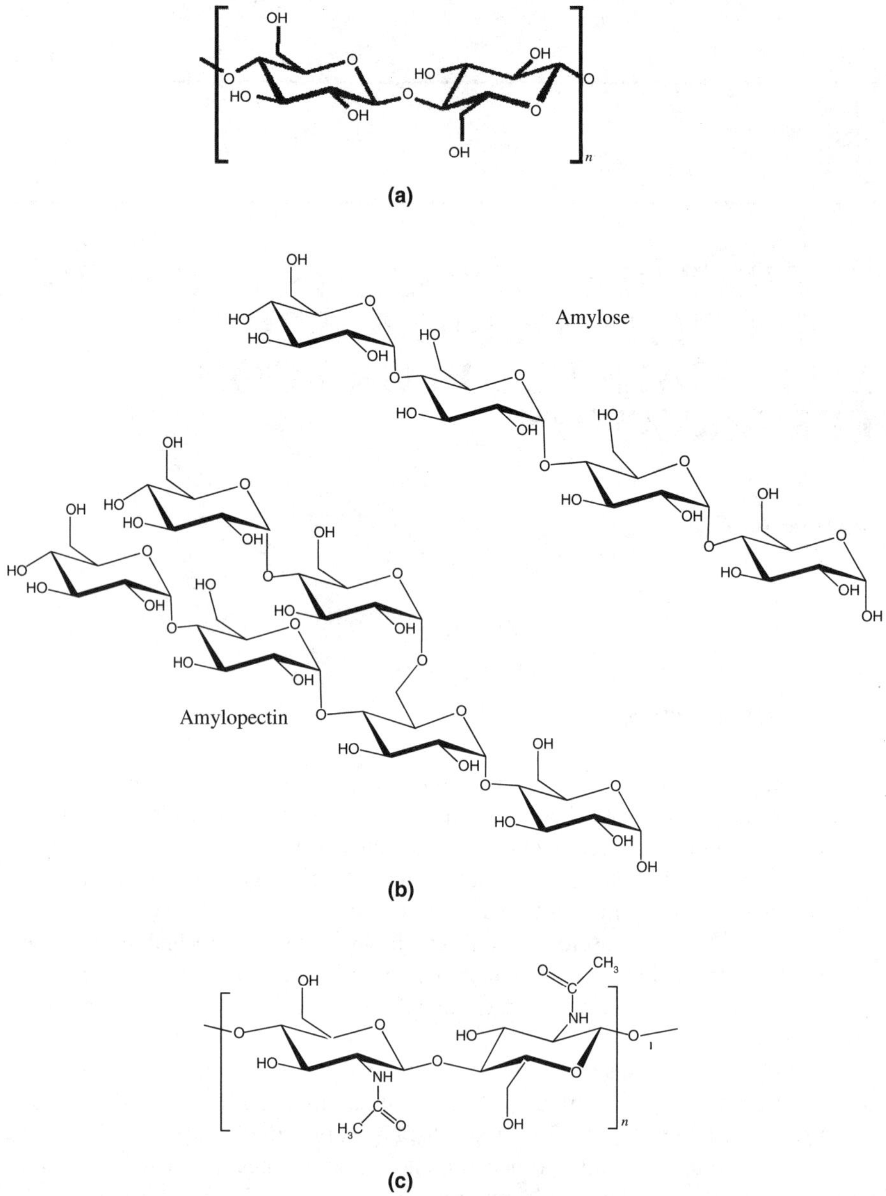

(a)

(b)

(c)

FIGURE 5.1 Chemical structures of (a) cellulose, (b) starch, and (c) chitin.

even though the interest in these polysaccharides as materials is relatively recent, compared with the age-old exploitation of cellulose and starch.

Today, the main industrial source of chitin is crustacean shells. However, its conventional extraction requires harsh solvents and high-temperature treatments and has seasonal supplies as well as geographical limitations. Knowing that some fungal chitinous biopolymers, or glycosaminoglycans, consist mainly of chitin and chitosan (Wu et al., 2004), mycelia of various fungi have been suggested as alternative chitosan sources to crustaceans. *Agaricus bisporus*, one of the most consumed mushrooms in the United States, is particularly rich in chitinous biopolymers. All these fungal strains synthesize chitin following a common pathway that ends with the polymerization of *N*-acetylglucosamine from an activated precursor (UDP-GlcNAc). The regulation, biochemistry, and genetics of chitin synthases are well known (Ravi Kumar et al., 2004). It is estimated that crustaceans, mollusks, insects, and fungi synthesize about 100 billion tons of chitin annually (Tharanathan and Kittur, 2003). At least 10 Gtons (1×10^{13} kg) of chitin is constantly present in the biosphere (Jeuniaux and Voss-Foucart, 1991).

In chitin, the degree of acetylation is typically 0.90, indicating the presence of some amino groups, but as some deacetylation might take place during extraction, chitin may also contain about 5–15% amino groups (Pillai et al., 2009).

Depending on the sources, this polymer has been found in three polymorphic crystal structures: α-, β-, and γ-chitin (Figure 5.2); the main forms are α- and β-chitin. The polymer molecules can be aligned in antiparallel fashion (α-chitin) to yield highly crystalline, durable structures. This form is predominantly found in crustacean shells and fungal cell walls (Kjartansson et al., 2006). In the β-form, obtained from squid, all chains are aligned in a parallel manner (Ravi Kumar et al., 2004). The molecular order of chitin depends on the physiological role and tissue characteristics.

A specific character of chitin is its low solubility, which is remarkably less than that of cellulose, because of the high crystallinity of chitin due to hydrogen bonds mainly through the acetamido group (Ravi Kumar et al., 2004).

More soluble, especially in dilute aqueous acid solutions, chitosan is more interesting from a biomaterial point of view.

Chitosan is a family of linear copolymers of 2-acetamido-2-deoxy-β-D-glucopyranose (GlcNAc) and 2-amino-2-deoxy-β-D-glucopyranose (GlcN) connected by $(1 \rightarrow 4)$ linkages (Figure 5.3). The presence of a prevailing number of

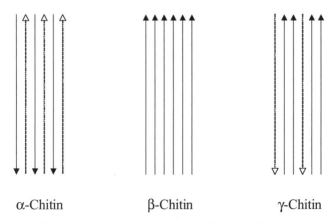

α-Chitin β-Chitin γ-Chitin

FIGURE 5.2 Scheme of the three chitin crystalline forms.

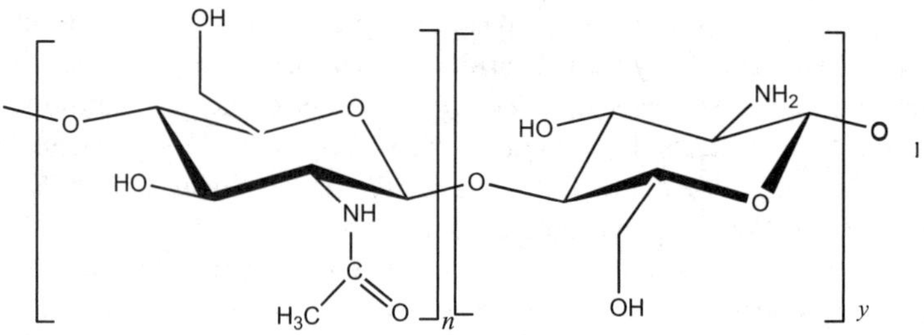

FIGURE 5.3 Chemical structure of chitosan.

2-amino-2-deoxyglucose units in a chitosan permits bringing the polymer into solution by salt formation. Due to its semicrystalline nature, derived mainly from inter- and intramolecular hydrogen bonds, chitosan is water soluble only at pH <6. Indeed, its amino groups can be partially protonated in acidic conditions, resulting in repulsion between positively charged macrochains, thereby allowing diffusion of water molecules and subsequent solvation of macromolecules (Sogias et al., 2010).

It is important to note that the term "chitosan" does not refer to a single well-defined structure, and chitosans can differ in molecular weight (MW), degree of deacetylation (DD), and sequence (i.e., whether the acetylated residues are distributed along the backbone in a random or an organized manner) (Yi et al., 2005). Commercial chitosan is available with DD $> 85\%$ deacetylated units, and MW between 100 and 1000 kDa. There is no specific standard to define molecular weight, but it is accepted that low MW corresponds to MW < 50 kDa, medium MW to 50–150 kDa, and high MW > 150 kDa (Goy et al., 2009). The fully deacetylated product is rarely obtained due to the risks of secondary reactions and chain depolymerization. Generally, chitosans have a heterogeneous distribution of acetyl groups along the chains and this distribution is very important in controlling the solution's and material's properties.

Chitosan also shows polymorphism depending on its physical state. It crystallizes in the orthorhombic system, like α-chitin, and two types of chitosan can be differentiated. Chitosan I, in salt form, with a weak DD, is more disordered than chitosan II, which has a high DD and which is in the form of free amine (Crini et al., 2009). Depending on the origin of the polymer and its treatment during extraction from raw materials, the residual crystallinity may vary considerably.

Some specific properties of chitosan, for example, capability of forming films, and intrinsic antioxidant and antimicrobial activities, make it suitable for active bio-packaging for food applications. Indeed, due to the numerous outbreaks linked to contaminated food products, and since minimal processing is being implemented for a broad range of foodstuffs, food preservation requires the use of integral solutions where the packaging plays a more active role (Fernandez et al., 2008).

The objective of this chapter is to present the recent developments and the future trends in chitin- and chitosan-based matrices as antimicrobial materials.

The structure of this chapter is as follows. In Section 5.2, a background is presented, consisting of general considerations concerning food preservation and current concepts of antimicrobial matrices. The main properties of interest of chitin and chitosan are discussed from the point of view of functional material elaboration. Section 5.3 is related to the state of the art of chitin- and chitosan-based materials, followed by a critical discussion. In the final section, future trends are described, taking into account the new processes/concepts that have recently emerged.

5.2 BACKGROUND

5.2.1 General Considerations of Food Safety and Active Packaging

Food-borne diseases have always topped the list of food safety concerns for most governments around the world (Wallace et al., 2000; Jones, 2002), and pathogenic bacteria are responsible for the majority of food-related outbreaks in developed countries. Estimates from the Center for Disease Control and Prevention in the United States suggest that there are 5000 deaths each year in the United States from food-borne diseases. The report traces a large majority of the microbial disease outbreaks to food service establishments and homes (Table 5.1).

Among pathogen strains that contribute significantly to these concerns, we can find bacteria such as *Salmonella* Typhimurium, *Listeria monocytogenes*, *Campylobacter jejuni*, and *Escherichia coli* 0157. Salmonellosis and listeriosis could represent, respectively, 31% and 28% of total food-related deaths (Aymerich et al., 2006; Mead et al., 1999). In addition, some fungal strains belong to the pathogen world. Indeed, a few fungal strains, under the right conditions, can produce mycotoxins, which are naturally occurring toxic metabolites such as aflatoxins, ochratoxin A, and fumonisins. These compounds are health hazards that particularly contaminate a wide variety of crops. Human and animal health can be at risk due to mycotoxin contamination. Mycotoxicoses are examples of "poisoning by natural means" and

TABLE 5.1 Impact of Food-Borne Disease and Identification of the Contamination Sources in the United States According to the Center for Disease Control and Prevention Quoted in Jones (2002)

Estimates of impact of food-borne disease in the United States	
	Number of cases per year
Total illnesses	76 million
Identified pathogenic microorganism	*14 million*
Total hospitalizations	32,5000
Identified pathogenic microorganism	*60,000*
Total deaths	5000
Identified pathogenic microorganism	*1800*

Main sources of contamination in the United States	
	Microbial disease outbreaks (%)
Food service establishments	77
Homes	20
Food processing plants	3

thus could be analogous to the pathologies caused by exposure to pesticides or heavy metal residues. The majority of mycotoxicoses result from eating contaminated foods (Bennett and Klich, 2003). In 1985, the World Health Organization estimated that approximately 25% of the world's grains were contaminated by mycotoxins, of which the most notorious are aflatoxins. This figure has certainly grown since then due to an increase in global import and export of grains and cereals and the changing environmental and weather patterns.

Aside from disease-causing bacteria and fungi, spoilage microorganisms also lead to decreased food quality and to significant economic losses (Wolffs and Radstrom, 2006).

To prevent the negative consequences of food contamination, active packaging materials, and particularly those that have antimicrobial properties, could help minimize the risk of contamination. Active packaging material is defined as "a type of packaging that changes the storage conditions to extend shelf life or improve safety or sensory properties while maintaining the quality of the food" (Anonymous, 2001). According to Lopez-Rubio et al. (2006), there is a difference between active and bioactive packaging technologies: active packaging primarily deals with maintaining or increasing quality and safety of packaged foods, and bioactive packaging has a direct impact on the health of the consumer by generating healthier packaged foods. This latter system can include the possible migration of functional compounds such as probiotics, phytochemicals, marine oils, encapsulated vitamins, and bioavailable flavonoids, allowing a more efficient way to provide foods with improved impact on human health.

Chitosan and chitin may be used to prepare antimicrobial films for food preservation. This type of material, compared to direct incorporation of antimicrobial agents into the food composition, or other processes frequently used for preservation, maintains the minimal inhibitory concentration of biocides on the surface of the food where the microbial growth is generally found, with a lower preservative concentration per product mass (Figure 5.4).

It is important to note that the use of antimicrobial packaging is not meant to be a substitute for good handling practices, but it should enhance the safety of food as an additional impediment to the growth of pathogenic and/or spoilage microorganisms (Cooksey, 2005).

We can, however, note that the application of antimicrobial packaging systems to food is currently limited due to the regulatory concerns and the lack of availability of suitable antimicrobials, new polymer materials, and appropriate testing methods (Jin and Zhang, 2008).

5.2.2 Current Concepts of Antimicrobial Materials and Legislation

Different approaches may be used for the creation of antimicrobial material for food preservation (Coma, 2008a, 2010) (Figure 5.5):

1. *Direct incorporation* of active agents into a matrix (e.g., blends, multilayer systems). This category of materials can release the antimicrobial agents onto

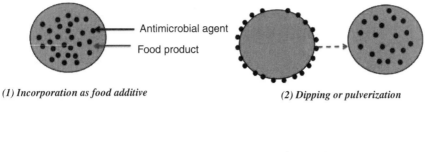

(1) Incorporation as food additive　　　　　　　*(2) Dipping or pulverization*

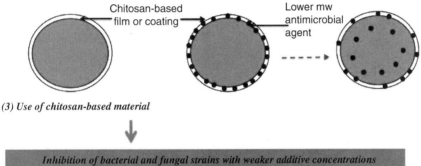

(3) Use of chitosan-based material

Inhibition of bacterial and fungal strains with weaker additive concentrations

FIGURE 5.4 Schematic representation of the advantages of chitosan-based films whether or not associated with other antimicrobial agents.

the surface on which antimicrobial action is needed. The active matrix can be used directly as a packaging material or can be coated on a packaging surface, where it acts as a carrier for the active agent.

2. *Chemical modification or biocide immobilization* on polymers. For this system, the release of the biocide agent is generally not required, and can even be prohibited in the case of a potentially toxic active agent. In some cases, such as pH-sensitive or temperature-sensitive bonds, chemical links are created to allow a slow release of the active agent system.

3. *Use of an inherently antimicrobial biopolymer*, such as some cationic amino-polysaccharides.

In terms of legislation, the safety of food packaging materials is generally based on the lack of potential toxic substances and the absence of migration of such substances (Galić et al., 2011). EU Directive 2002/72/EC lays down limits with respect to the concentration of some substances in the packaging and of migrants into foodstuffs or food simulants. This directive stipulates that a maximum of $10 \, mg/dm^2$ or $60 \, mg/kg$ of migrants may transfer from the packaging to the food. Taking into account these specifications, the question is: What about active and intelligent materials? In Europe, food packages are regulated by Regulation EC 1935/2004, with respect to materials and articles intended to come into contact with foodstuffs. This regulation contains general provisions on the safety of active and intelligent

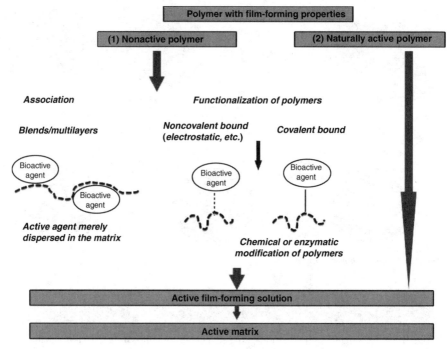

FIGURE 5.5 Schematic representation of the different concepts of active matrices from polymers having film-forming properties.

packaging and sets the framework for the safety evaluation process. In addition, a new regulation published in 2009 (Regulation EC 450/2009) establishes the specific rules for this type of material, with additional requirements to ensure their safe use.

5.2.3 Chitin and Chitosan as Functional Bio-Based Materials: Properties of Interest for the Creation of Antimicrobial Material

Most chitin and chitosan properties are variable, depending on their origin and their production processes. It is well known that their properties are highly connected to their chemical and physicochemical characters (e.g., polymerization and acetylation degree, crystallinity, solubility). However, some general properties related to their use as biomaterials can be observed.

5.2.3.1 Film-Forming Ability   The film-forming ability of chitosan is an important aspect that cannot be found with cellulose, for example. In addition, both α- and β-forms of chitin are insoluble in all the usual solvents, despite natural variations in crystallinity (Rinaudo, 2006). The insolubility in relatively safe solvents is a major problem for biomaterial development. Cast transparent chitin film with good mechanical properties was obtained from an alkali chitin after dissolution in NaOH solution (Einbu et al., 2004), but it was suggested that chitin deacetylation plays an important role in these film-forming properties. Chitin-based membranes were also prepared from α- or β-chitin hydrogels (Jayakumar et al., 2007), but in general chitin-based films have been obtained after chemical modification of the biopolymer (Jayakumar et al., 2009) or, more recently, by the use of ionic solvent (Takegawa et al., 2010).

Due to these good film-forming properties, chitosan-based films are particularly interesting for food packaging. Because chitosan can be dissolved under mildly acidic aqueous conditions, it can be readily cast into membranes and films that can be converted into insoluble networks by neutralization.

5.2.3.2 Antimicrobial Properties Early research describing the antimicrobial potential of chitin, chitosan, and their derivatives dates from the 1980 to the 1990s (Hadwiger et al., 1981; Papineau et al., 1991). Chitosan is better known for its bioactivity than chitin and has been investigated as an antimicrobial biopolymer against a wide range of target organisms such as algae, bacteria, yeasts, and fungi in experiments involving *in vivo* and *in vitro* interactions with chitosan in different forms, for example, solutions, films, and composites (Goy et al., 2009). Despite the large number of publications related to antimicrobial properties of chitosan and its derivatives, there is still a strong controversy regarding the phenomenology and mechanisms of its activity (Fernandez-Saiz et al., 2009a, 2009b). The exact mechanism of antimicrobial action of chitin and chitosan is not fully understood and several factors may contribute to this action. As mentioned by Rabea et al. (2009), it is recognized that the biological activity of chitosan depends on its MW, DD, chemical modification, degree of substitution, pH, length and position of substituents on the glucosamine units of chitosan, and, of course, the target organism.

The mode of action of chitosan certainly involves cell lysis, breakdown of the cytoplasmic membrane barrier, and selective chelating of trace metal cations (Je and Kim, 2006; Coma, 2008b). However, due to the numerous factors influencing antimicrobial activity, Rabea et al. (2009) suggested two different mechanisms of chitosan and target microorganism interaction:

- the first is the adsorption of the chitosan to cell walls, leading to cell wall covering and membrane disruption;
- the second is the penetration of chitosan into living cells, leading to the inhibition of various enzymes and interference with the synthesis of proteins.

The sequence of elementary events in the direct bacteriostatic or lethal action of the cationic active agents may be considered as follows (Figure 5.6):

1. Adsorption onto the bacterial cell surface. This interaction is mediated by the electrostatic forces between the charged NH_3^+ groups and the negative residues on the membrane surface, presumably by competing with Ca^{2+} for electronegative sites (Young and Kauss, 1983; Goy et al., 2009). Indeed, it is commonly accepted that polycationic chitosan can bind with negatively

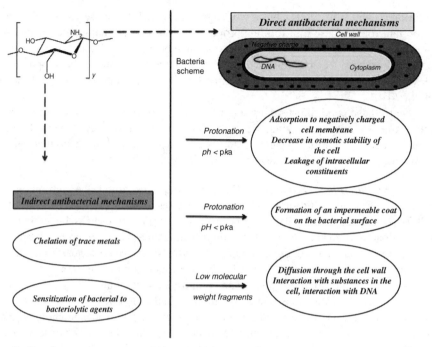

FIGURE 5.6 Schematic representation of the suggested mechanism of antibacterial activity of chitosan.

charged cell membranes (Rabea et al., 2003; Devlieghere et al., 2004). Liu et al. (2004), in a study on *E. coli* and *Staphylococcus aureus*, showed interaction between chitosan-NH$_3^+$ and phosphoryl groups of phospholipid components of cell membranes.

Due to the different structure of their cell walls, interaction may be different on Gram-positive and Gram-negative bacteria. Some authors reported a positive correlation between the molecular weight of chitosan and activity against Gram-positive bacteria, such as *S. aureus*, whereas the correlation is negative for Gram-negative bacteria (Zheng and Zhu, 2003; Másson et al., 2008).

These electrostatic interactions have two main consequences depending on the molecular weight of chitosan or chitosan fragments.

2. In the case of high MW chitosan: change in membrane wall permeability. Interactions lead to a decrease in the cell osmotic stability, followed by subsequent leakage of intracellular constituents (Rabea et al., 2003; Devlieghere et al., 2004). Liu et al. (2004) showed on a Gram-positive strain that chitosan effectively increased the permeability of the bacterial outer membrane and inner membrane, and ultimately disrupted the bacterial cell membranes, with the release of cellular contents.

3. In the case of low MW chitosan: possible diffusion through the cell wall. If diffusion is possible because the low MW chitosan fragments are able to pass

through the bacterial cell wall, specific interactions can occur, such as binding of chitosan with microbial DNA, which leads to the inhibition of the mRNA and protein synthesis (Shahidi et al., 1999).

4. Inhibition or death of the cell. All this damage leads to inhibition of the bacterial growth and can lead to cell death.

Other actions may be possible based on indirect mechanisms, such as chelating of metals, suppression of spore elements, and binding to nutrients that are essential for microbial growth, but they do not seem to have a significant role.

Relatively recent data in the literature generally characterize chitosan as a bacteriostatic agent rather than a bactericidal one (Raafat et al., 2008).

The polycationic nature of chitosan is also the key to its antifungal properties. An additional explanation includes the possible effect that chitosan might have on the synthesis or on the activity of certain fungal enzymes (Ziani et al., 2009; Muhizi et al., 2008).

5.2.3.3 Mechanical and Barrier Properties
The main development of chitin film being in medical and pharmaceutical applications, only chitosan properties will be discussed here.

Chitosan application in biopackaging for food preservation has been limited due to its brittle behavior. An important parameter, which influences mechanical properties, is the process of matrix formation. Indeed, longer film formation time tends to favor molecular rearrangements as the drying process occurs, resulting in the formation of a more ordered structure and increasing crystallinity, thus raising the maximum tension. It is important to point out that some processing parameters, that is, solution pH, solvent drying profile, and use of plasticizers, will interact to influence the overall mechanical properties. The nature of chitosan, such as its DD and MW, also has a significant influence on the different potential interactions. This explains the wide variation found in the literature in tensile strength and tensile elongation ranging from 1.4 to 57.2 MPa and from 3.5% to 115%, respectively (Costa-Júnior et al., 2009).

In most cases, water barrier materials are desirable in order to retard the surface dehydration of fresh or frozen products. Unfortunately, chitin and, much more, chitosan are water-sensitive biopolymers, which lead to materials exhibiting low moisture barrier properties. The control of gas exchange, particularly oxygen, allows to better limit the degradation process in foods (fruit ripening), aerobic contamination, or to significantly reduce the oxidation of oxygen-sensitive foods and the rancidity of polyunsaturated fats (Dutta et al., 2009). Organic vapor transfer also has to be controlled for certain foods, particularly those exhibiting an aromatic profile. Chitosan has generally good gas and organic compound barrier properties at low relative humidity, but the barrier is dramatically reduced as humidity increases. Water sensitivity can be decreased by the association with less hydrophilic molecules or by chemical modification, but it seems that this is still a critical point because moisture sensitivity/properties of polysaccharide-based material are still not comparable to those of petroleum-based materials.

5.2.3.4 Natural Origin/Biodegradability/Edibility   Although the main source of chitosan allows valorization of marine subproducts, we can predict a problematic shell crustaceous dependence. In addition, according to Tayel et al. (2010), uniform deacetylation, which is difficult to obtain from the marine chitosan, is a requirement to specific industrial applications. The potential key for these restrictions might be the microbial chitosan. Chitosan can be found in the cell wall of certain groups of fungi, particularly zygomycetes. With advances in fermentation technology, chitosan preparation from fungal cell walls becomes an alternative route for the production of this polymer in an eco-friendly pathway.

Chitin and chitosan are biodegraded by various microorganisms. Chitinases and chitosanases and some other enzymes are able to transform chitin and chitosan to lower molecular weight products (monomers, dimers, trimers, . . ., oligosaccharides) by hydrolysis of the $\beta(1 \rightarrow 4)$ glycosidic bonds (Kumar et al., 2004). Concerning chitin, while in bacterial genome databases on average only between 2 and 4 chitinases can be found, the genomes of filamentous fungi typically contain between 10 and 25 different chitinases. Chitinolytic enzymes can be divided into N-acetylglucosaminidases and chitinases. N-Acetylglucosaminidases catalyze the release of terminal, nonreducing N-acetylglucosamine residues from chitin, but in general they have the highest affinity for the dimer N,N'-diacetylchitobiose and convert it into two monomers. Furthermore, chitinases can be divided into endo- and exochitinases. Endochitinases degrade chitin from any point along the polymer chain forming random-size length products while exochitinases cleave from the nonreducing chain end and the released product is the dimer (Seidl, 2008). In the case of chitosan, total depolymerization releases N-acetyl-β-glucosamine and β-glucosamine (Figure 5.7) but, in general, both rate and extent of chitosan enzymatic hydrolysis are dependent on the DD (Kean and Thanou, 2010). Most microbial chitosanases catalyze an endo-type cleavage reaction that randomly hydrolyzes β-1,4-linkages in chitosan or its oligomers. The ability to hydrolyze β-glucosaminidic and N-acetyl-β-glucosaminidic chitosan linkages with different degrees of deacetylation depends on the microorganisms from which the chitosanase is produced (Lin et al., 2009). Sashiwa et al. (1990) studied the relative rates of degradation of six chitosans varying in degree of deacetylation (45%, 66%, 70%, 84%, 91%, and 95%) and found that 70% deacetylated chitosan is degraded most quickly.

The edibility of chitosan is also an interesting characteristic in terms of biopackaging. The use of an edible coating is one of the most effective methods of maintaining food quality and safety.

As a result, chitosan is thought to be degraded in vertebrates predominantly by lysozyme and by bacterial enzymes in the colon.

5.3 STATE OF THE ART OF APPLICATIONS OF CHITOSAN IN BIOMATRICES

Its insolubility in ordinary solvents makes chitin difficult to characterize and to process (Rinaudo, 2006). This section is thus mainly devoted to chitosan.

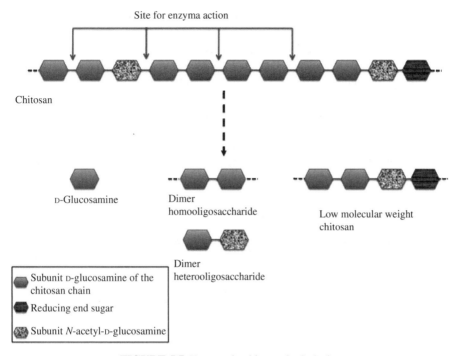

FIGURE 5.7 Enzymatic chitosan hydrolysis.

5.3.1 Preparation of Chitosan-Based Antimicrobial Matrices

Different techniques are employed to prepare chitosan films. Chitosan cannot be melted and processed as a typical thermoplastic because of the large number of inter- and intramolecular bonds that are responsible for its rigid, semicrystalline structure. Nevertheless, it can be dissolved in weakly acidic aqueous solutions and converted into hydrogels (Murray and Dutcher, 2006), before being converted into films.

The solution casting method is one of the most popular. This technique involves the drying of a complex colloidal solution, which is composed of the polymer, a solvent, and sometimes some additives, such as plasticizers. In the case of chitosan, aqueous film-forming solutions (e.g., acetic acid solutions) are prepared and generally poured onto an appropriate support. The cast solution is then set to air-dry in a laminar flow hood or under specific conditions of temperature and relative humidity. The films are then left to dry for 24–48 h until all the solvent has evaporated to give solid chitosan-based films. After drying, the films are mechanically removed. Films can be neutralized or not, depending on the potential application.

Uniform thin chitosan-based films can also be prepared by spin coating solutions of chitosan dissolved in dilute acetic acid onto appropriate supports (Murray and Dutcher, 2006).

Most recently, Tripathi et al. (2008) have created chitosan-based films using supercritical carbon dioxide and microwave techniques. Extrusion can also be used for chitosan-based active film preparation, a method widely used in the packaging film preparation industry (Aider, 2010).

We have to note that the drying conditions (e.g., temperature, moisture) strongly affect the film properties, particularly in bio-based film preparation. The effect of a specific drying condition depends on various characteristics of the raw materials, and various phenomena may occur, such as transition from a rubbery to a vitreous phase, phase separation, or crystallization. Unfortunately, the influence of the drying process is still neglected and poorly understood (Denavi et al., 2009). Mayachiew and Devahastin (2008) investigated the influence of different drying methods and conditions on the drying kinetics and various properties of chitosan films, such as color, elongation and tensile strength, water vapor permeability, and crystallinity.

5.3.2 Applications of Chitosan as Antimicrobial Matrices

Chitosan can be used to produce active matrices by the development of antimicrobial surface system that exhibits contact bioactive properties. Moreover, chitosan-based edible coatings or films can, for example, reduce or replace barrier layers in traditional food packaging systems and also make recycling of plastics easier.

Chung and Chen (2008) reported that chitosan exhibited different antibacterial activity. The inhibition order *Enterobacter aerogenes* > *Salmonella* Typhimurium > *S. aureus* > *E. coli* can be different, depending on the authors. This could be due to the differences in the methods used to determine the antimicrobial properties, the nature and age of the target microbial strain, the chitosan nature and state, and the composition of chitosan film-forming solutions (e.g., solvent and counter ion, additives). Whatever the efficiency order, the antimicrobial activity of chitosan-based films was observed against a wide variety of food bacteria (e.g., in addition to the bacteria listed before, *Pseudomonas aeruginosa*, *L. monocytogenes*, *Bacillus cereus*, *Bacillus subtilis*, *Serratia liquefaciens*) and fungi (e.g., *Aspergillus niger*, *Aspergillus ochraceus*, *Fusarium proliferatum*).

From an antimicrobial packaging point of view, a sufficient number of carboxylate groups would be needed in the film or coating formulation. Lagaron et al. (2007) showed that the coating should be kept under dry conditions prior to its use to avoid losses of the biocide groups by organic acid evaporation in the presence of relative humidity. The results of this work suggest that, in a humid environment, the bioactive carboxylate groups that form when chitosan is cast from acetic acid solutions continuously evaporate from the film (in the form of acetic acid). This could decrease the antimicrobial properties of the system.

Some chemical modifications of chitosan have been carried out to enhance antimicrobial properties and open various ways to utilize this amino-polysaccharide (Figure 5.8). Further derivatization of the amine function may be interesting and relatively easy due to the high amino group reactivity. Other approaches have also been tried to exploit the reactivity of hydroxyl groups or both-site derivatization.

As an efficient solution to improve its solubility and antibacterial activity in water, quaternization of the amine group has been reported. For a polymeric quaternary ammonium biocide, the hydrophilic–lipophilic balance influences antimicrobial

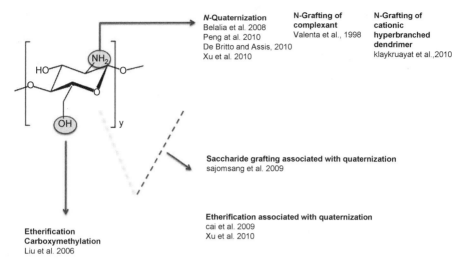

N-Quaternization
Belalia et al. 2008
Peng at al. 2010
De Britto and Assis, 2010
Xu et al. 2010

N-Grafting of complexant
Valenta et al., 1998

N-Grafting of cationic hyperbranched dendrimer
klaykruayat et al.,2010

Saccharide grafting associated with quaternization
sajomsang et al. 2009

Etherification associated with quaternization
cai et al. 2009
Xu et al. 2010

Etherification Carboxymethylation
Liu et al. 2006

FIGURE 5.8 Chemical modifications conducted to improve the antibacterial or antifungal properties of chitosan.

properties by affecting the mode of interaction with the cytoplasmic membrane. Grafting alkyl chains on chitosan generally leads to higher antimicrobial activity. De Britto and Assis (2010) synthesized chemically modified chitosans by carrying out reductive alkylation and quaternization reactions. Quaternized chitosans were the most hydrophilic derivatives. They also exhibited the highest swelling tendency. Probably due to agglomeration of polymeric chains, the alkyl quaternary salts unfortunately showed an irregular topography with undulations and mountain-like formations, compared to the homogeneous chitosan-based films. Belalia et al. (2008) prepared N-alkyl chitosan derivatives by introducing alkyl groups into the amino groups of chitosan via a Schiff's base intermediate. N,N,N-Trimethyl chitosan (TMC) showed a higher listerial strain inhibition compared to the nonchemically modified chitosan. Xu et al. (2010) synthesized O-methyl-free TMC by first treating chitosan with formic acid and formaldehyde, followed by methylation with CH_3I. TMC exhibited stronger activity than raw chitosan at both pH 5.5 and 7.2. Interestingly, TMC's activity decreased as the degree of substitution increased at pH 5.5, while at pH 7.2 the structure–activity relationship was reversed. Both chitosan's and TMC's antibacterial activity increased as the pH value decreased. However, TMC was more efficient than chitosan at pH 5.5, while at pH 3.5 it was less efficient.

Quaternized chitosan was also synthesized by Peng et al. (2010), by reacting chitosan with glycidyl trimethylammonium chloride to produce hydroxypropyl-trimethyl ammonium chloride chitosan (Figure 5.9) with different degrees of substitution (DS 6%, 18%, and 44%) of quaternary ammonium. Higher antibacterial activity was observed with 18% or 44% substitution against *S. aureus*, methicillin-resistant *S. aureus*, and *Staphylococcus epidermidis*, with no cytotoxicity for 18% substitution. The solubility of derivatives increased with the DS due to the steric hindrance and good

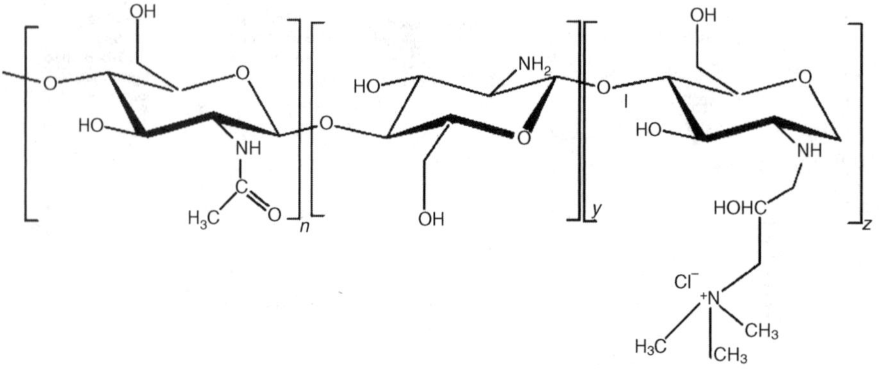

FIGURE 5.9 Structure of hydroxypropyltrimethyl ammonium chloride chitosan according to Peng et al. (2010).

hydration capacity of quaternary ammonium groups, which greatly reduced the intermolecular and intramolecular hydrogen bonds of chitosan. The authors also showed a pH dependence of the water solubility of chitosan derivatives. The differences in solubility between chitosan and hydroxypropyltrimethyl ammonium chloride chitosan were particularly significant at pH higher than 8. Solubility can play a role in bioactive properties and can enhance antibacterial activity.

It seems that the best antimicrobial properties of quaternized chitosan are connected to more parameters than only the permanent positive charge. Chitosan *N*-betainates have been prepared with various degrees of substitution by Holappaa et al. (2006). The inhibitory activity of water-soluble chitosan *N*-betainate chloride was measured against *E. coli* and *S. aureus* at pH 7.2 and 5.5. The antimicrobial activity of chitosan derivatives increased with a decreasing degree of substitution under acidic conditions, which suggests that the positive charge has to be located on the amino group in the chitosan backbone if efficient antimicrobial activity is to be achieved. Besides the positive charge of quaternary ammonium groups, the hydrophobic character introduced by the hydrocarbon chains of the different alkyl chains on the amino group might help elevate the antibacterial activities (Badawy, 2010; Vallapa et al., 2010).

To improve antimicrobial properties, another approach based on the grafting of substances exhibiting chelating properties may be used. Ethylenediaminetetraacetic acid (EDTA) grafted onto chitosan increases the antibacterial activity of chitosan by complexing magnesium that, under normal circumstances, stabilizes the outer membrane of Gram-negative bacteria (Valenta et al., 1998).

The grafting of carboxylic functions has been regarded as a way to increase the antimicrobial and metal ion sorption properties of chitosan (Alves and Mano, 2008). According to Liu et al. (2006), *O*-carboxymethyl chitosan (*O*-CMCS) exhibits increased activity against *E. coli*. This could be due to a higher content of NH_3^+ groups in *O*-CMCS than in raw chitosan. The authors suggested that COOH groups in *O*-CMCS may react (intra- or intermolecular reaction) with NH_2 groups and thus charge them. *O*-CMCS also makes essential transition metal ions unavailable for bacteria. In addition, derivatization by introducing small functional groups to the chitosan structure, such as alkyl or carboxymethyl groups, can drastically increase the solubility of chitosan at neutral and alkaline pH values without affecting its cationic character (Alves and Mano, 2008), and the solubility can have an impact on the antimicrobial character.

Synergistic effects that associate various chemical modifications have been looked for. Sajomsang et al. (2009) successfully synthesized quaternary ammonium chitosans containing monosaccharide or disaccharide moieties by reductive *N*-alkylation, followed by quaternization using *N*-(3-chloro-2-hydroxypropyl)-trimethylammonium chloride. Unfortunately, they observed that quaternary ammonium mono- and disaccharide chitosan derivatives had very high minimal inhibitory concentration values, in the range of 32 to $>256\,\mu g/mL$, against target bacteria (*E. coli* and *S. aureus*). Also, it was found that the antibacterial activity decreased with increasing DS. These authors suggested that this was due to the increased hydrophilicity caused by mono- or disaccharide moieties. Against both of these bacterial strains, Yang et al. (2005) concluded that, although the disaccharide chitosan derivatives showed less antibacterial activity than the native chitosan at pH 6.0, the derivatives exhibited a higher activity than native chitosan at pH 7.0.

It was also shown that the enhanced antibacterial activity of quaternized *N*, *O*-(2-carboxyethyl) chitosan was the result of synergy between the carboxyalkyl group and the quaternary ammonium group (Cai et al., 2009). The quaternized chitosan derivative showed much stronger antimicrobial activity, which increased with increasing chain length of the alkyl group in the quaternary ammonium groups. Xu et al. (2010) synthesized *N,N,N*-trimethyl-*O*-carboxymethyl chitosan. They first synthesized TMC, which was further carboxymethylated by monochloroacetic acid. *N,N,N*-Trimethyl-*O*-carboxymethyl chitosan acted less strongly than TMC and they showed that carboxymethylation did not enhance the antibacterial activity directly.

Other approaches can be found in the literature. Chemical modification of chitosan with cationic hyperbranched denditric polyamidoamine carried out by Klaykruayat et al. (2010) led to an improvement of antimicrobial activity against *S. aureus* strain, even at a near-neutral pH, compared to that obtained with unmodified chitosan.

5.3.3 Applications of Chitosan for Active Agent Immobilization

Bioactive agents can be easily immobilized onto the chitosan backbone. This functionalization may be done by simple superficial adsorption of active substances onto the biopolymer (e.g., from electrostatic interactions) or by anchoring molecules by covalent bonds, which can potentially be broken to release the active agent (e.g., pH-sensitive, temperature-sensitive, or enzyme-sensitive bond) (Figure 5.10).

5.3.3.1 Noncovalent Immobilization The adsorption of a biocide onto the surface of packaging material could provide several advantages, such as the maintenance of a high concentration of the active agent directly on the food surface, the decrease of the risk of active substance inactivation by food constituents, and the avoidance of the use

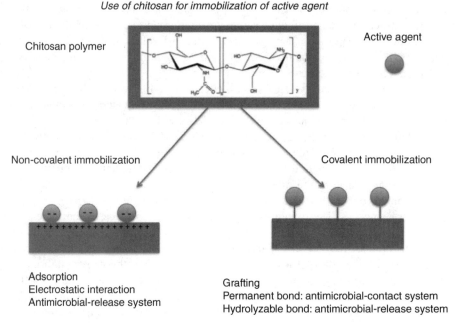

Use of chitosan for immobilization of active agent

FIGURE 5.10 Potential application of chitosan for active agent immobilization.

of this substance as a food additive (Coma, 2008b). Noncovalent immobilization is generally connected to electrostatic interaction.

The positive charges of chitosan can attract negatively charged substances, such as some active proteins. Hoven et al. (2007) have demonstrated that positive or negative charges can be successfully introduced to the surface of chitosan films using heterogeneous chemical reactions. This leads to a remarkable impact on the way the chitosan film responds to charged proteins in terms of adsorbed quantity and selectivity. This approach could be used to fix negatively charged active agents onto a chitosan matrix, in order to control their release in foods.

5.3.3.2 Covalent Immobilization   As specified above, chitosan has two types of reactive groups that can be grafted. First, the free amino groups on deacetylated units, and second, the hydroxyl groups on the C_3 and C_6 carbons on acetylated or deacetylated units. Reactivity of the chitosan amine function was often exploited to covalently conjugate functional agents to the polymer backbone. Chemical or enzymatic catalyzed reactions can be used for active compound immobilization, especially on chitosan. There are several potential advantages for the use of biocatalytic complexes in polymer synthesis and modification. With respect to health and safety, enzymes offer the potential for eliminating the hazards associated with reactive reagents. An environmental benefit for using enzymes is that their selectivity may be exploited to eliminate the need for full protection and deprotection steps. Finally, enzyme specificity may offer the potential for precisely modifying macromolecular structure to better control polymer function (Alves and Mano, 2008).

Certain naturally occurring proteins or peptides have long been recognized for their antimicrobial properties. In addition, chitosan can be easily cross-linked by reagents, such as glutaraldehyde, to form rigid aquagels. In fact, chitosan has been shown to be a superior polymer for protein or peptide immobilization, compared with polysaccharides such as alginate (Cetinus and Oztop, 2000). Lysozyme has been immobilized onto chitosan without using any intermediate reagent by Muzzarelli et al. (2004). The immobilization of antimicrobial enzymes (such as lactoperoxidase, lactoferrin, or glucose oxidase) or antimicrobial peptides (such as bacteriocins, magainins, cecropins, defensins, or lysozyme) could be very useful for the development of active chitosan-based materials.

Mushroom tyrosinases (EC 1.14.18.1) are oxidative enzymes capable of converting low molecular weight phenols and accessible tyrosyl residues of proteins, like gelatin, into quinones. It is well known that these quinones are reactive and undergo nonenzymatic reactions with a variety of nucleophiles. Thus, tyrosinase can catalyze reactions enabling proteins to be covalently tethered to a three-dimensional chitosan gel network. If the protein exhibits antimicrobial properties, this novel hybrid biomaterial can become a promising material for on-demand active packaging.

Knowing that antioxidant properties may be connected to antimicrobial activity, particularly on fungal strains, phenolic compound immobilization can be a good way to create antimicrobial chitosan-based matrices with high performance. Curcio et al. (2009) synthesized gallic acid–chitosan and catechin–chitosan conjugates by adopting a free radical-induced grafting procedure. A biocompatible and water-soluble system (an ascorbic acid/hydrogen peroxide pair) was chosen as the redox initiator system. The interaction mechanism between the two components of the redox pair involves the oxidation of ascorbic acid by H_2O_2 at room temperature with the formation of ascorbate and hydroxyl radicals, which initiate the reaction. These authors showed that the rapidity of the reaction, together with the absence of toxic reaction products, makes this procedure very useful to enhance the biological properties of chitosan. The antioxidant activity of chitosan was strengthened after its functionalization with the antioxidant molecules.

Sousa et al. (2009) grafted chitosan fibers with flavonoids using tyrosinase to produce reactive O-quinones, which subsequently reacted with primary amino groups of the chitosan. Oxidation reactions were conducted at 20 °C for 20 h in a phosphate buffer. The grafted flavonoid led to an enhanced antimicrobial activity against different bacteria, more effectively against Gram-positive than Gram-negative bacteria.

5.3.4 Applications of Chitosan as a Vehicle for Active Agents

Incorporation of active agents, including antimicrobials, into a polymeric matrix has been commercially applied in drug and pesticide delivery, household goods, textiles, surgical implants, and other biomedical devices. Few food-related applications have been commercialized (Appendini and Hotchkiss, 2002). Chitosan can be associated with other antimicrobials, potentially released in foods following a controlled desorption process, and diffusion coefficients. Numerous papers have suggested the use of chitosan to carry organic acids, sorbates, benzoates, bacteriocins, phenolic compounds, spices, and so on (Appendini and Hotchkiss, 2002; Coma, 2008b).

Chitosan can serve simply as a matrix to entrap biologically active components within the film's network (Özmeric et al., 2000). This system can be useful for preventing post-processing contamination. To associate the active component with the chitosan matrix, a blending is generally suggested, but other techniques can be used. As a result, Constantine et al. (2003) reported that layer-by-layer assembly has been utilized to fabricate an ultrathin film of polyelectrolytes. The architecture was composed of chitosan and organophosphorus hydrolase polycations. Chitosan-based microcapsules (Kosaraju et al., 2006; Parize et al., 2008) may also be used as carriers for active molecules.

During the past few years, due to the increased multidrug resistance of many human pathogenic microorganisms, as well as the consumer's demand for natural additives, the investigation of chemical compounds from traditional plants has become desirable. Antimicrobial properties have been reported on a wide range of extracts and natural products, such as essential oils (São Pedro et al., 2009). Chitosan-based films were then frequently studied as an essential oil carrier. Pelissari et al. (2009) showed that chitosan films added with oregano essential oil strongly inhibited the development of some microorganisms such as *B. cereus, E. coli, S. enteritidis*, and *S. aureus*. Portes et al. (2009) elaborated environment-friendly films exhibiting both antibacterial and antioxidative properties from chitosan and tetrahydrocurcuminoids (THCs), which are *Curcumin* derivatives (*Curcuma longa* L.). Two THCs (Figure 5.11), THC1 (5-hydroxy-1,7-bis(4-hydroxy-3-methoxyphenyl)hept-4-en-3-one) and THC2 (5-hydroxy-1,7-bis(4-hydroxy-3,5-dimethoxyphenyl)hept-4-en-3-one), were incorporated into a chitosan film. Results indicated that THC1 was more quickly released than THC2 in methanol. The differences observed in the kinetics and in the amounts of

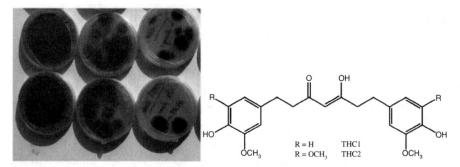

FIGURE 5.11 Chitosan antifungal activity of *Fusarium proliferatum* after 7 days of incubation and structure of THC1 and THC2 tetrahydrocurcuminoids studied by Portes et al. (2009).

both THCs released from the chitosan films are probably due to stronger interactions between protonated chitosan and THC2 than with THC1. Both THCs contain two electron-donating groups (i.e., phenol and methoxy), which increase the density of electrons on the benzene rings. Based on the assumption that the main interactions between chitosan and THCs are governed by electrostatic forces, it is not surprising that THC2 is released less than THC1, due to the two methoxy groups. In addition, chitosan and THC1 showed a significant action against the development of fungal strains such as *Fusarium* species, and also allowed the inhibition of the mycotoxin production.

In another study, interaction between chitosan and polyphenols separated from spruce wood bark was observed. Chitosan and the studied polyphenols formed a stable complex; the release of the polyphenols from the chitosan matrix occurred only in an alkaline medium close to pH 9 (Popa et al., 2000).

Higazya et al. (2010) treated scoured jute fabrics with chitosan (predissolved in 1% acetic acid) and chitosan–metal complexes aiming to impart the jute fabric antimicrobial properties. In this regard, Ag^+, Zn^{2+}, and Zr^{2+} ions were allowed to form a complex with chitosan. Jute fabrics treated with chitosan–metal complexes show better antimicrobial properties than the fabrics treated with either chitosan or metal salt separately. Moreover, the jute fabrics treated with chitosan–Zn complex have higher antimicrobial properties (*S. aureus* and *Candida albicans*), compared with the samples treated with chitosan–Zn or chitosan–Ag complexes.

Nanotechnology in food packaging is expected to grow strongly over the next 5 years as increased globalization sets demands for shelf life enhancing packaging (Vartiainen et al., 2010). There has been growing interest in delivery systems based on chitosan nanoparticles incorporated into a packaging matrix. Preparation by ionic gelation between the protonated amino groups of chitosan and anions, such as tripolyphosphate, is particularly studied because of their very simple and mild preparation conditions and homogeneous particle sizes (Agnihotri et al., 2004).

Due to the well-known guanidine group antimicrobial properties, Sun et al. (2010) prepared chitosan–guanidine complexes by reacting chitosan and polyhexamethylene guanidine hydrochloride. Particles exhibited antibacterial activity against *E. coli* and *S. aureus*. Lee et al. (2010) studied some chitosan-based nanoparticles loaded with *E. splendens* extract, a subclass of the family Labiatae that has, for a long time, been a traditional treatment for a variety of disorders such as fevers and diarrhea in eastern Asia. The authors showed that encapsulation using chitosan nanoparticle is potentially a valuable technique for improving the activity of such natural extracts.

Chitosan microparticles can be used to entrap other essential oils. Particles of chitosan and citronella oil were obtained by Hsieh et al. (2006) with different sizes, ranging from $11 \pm 3 \, \mu m$ to $225 \pm 24 \, \mu m$. The smallest microparticles showed the biggest release rate. This could be due to a larger specific surface area, causing the oil release rate to be faster. *Zanthoxylum limonella* essential oil was encapsulated by Hussain and Maji (2008), using a natural and nontoxic chemical cross-linker (genipin) after complex coacervation between chitosan and gelatin. The release rate of oil from the microcapsules was dependent on the percentage of chitosan.

Banerjee et al. (2010) reported antimicrobial properties against *E. coli* of chitosan–silver nanoparticles. According to the authors, the essential action of the positively charged chitosan matrix was to capture negatively charged bacteria on its surface. After that, small-sized silver nanoparticles created pores on the bacterial wall, thus causing rapid disintegration of the bacteria. Moreover, molecular iodine was then added to the composite in order to improve its bioactive properties, because chitosan, silver nanoparticles, and iodine have antimicrobial properties individually. The authors established the importance of synergy between different antimicrobial agents within a material, taking advantage of the mechanism of action of each individual agent.

5.4 DISCUSSION

5.4.1 General Properties

With respect to creating new bio-based material, chitosan is much easier to process than chitin, but the stability of chitosan materials is generally lower, owing to their greater hydrophilic character and, especially, pH sensitivity (Rinaudo, 2006). Chitosan possesses very significant film-forming properties after solubilization under some acidic conditions; however, as for cellulose, the theoretical melting temperature is above the degradation temperature, and chitosan is not thermoplastic, so cannot be heat sealed (Weber, 2000), which can be a disadvantage for its development in biopackaging. In addition, it is generally noted that to improve concrete commercial results, polysaccharide-based films or coatings should decrease their hydrophilic character in order to improve their moisture barrier properties and reduce the impact of the relative humidity on barrier and mechanical properties. Nevertheless, this seems very difficult if we want to maintain their biodegradable/edible character, and bioactive and film-forming properties. Due to the competition between fossil-based polymers and biopolymers, such as chitosan, it seems that the disadvantages of this bio-based polymer with regard to its direct production costs and water sensitivity can be compensated by other advantages, such as bioactivity and specific functionality, and it can be produced to serve small-volume niche markets. In other words, perhaps it is not necessary to replace all nonbiodegradable petroleum-based materials, especially those exhibiting very good properties in terms of mechanical and moisture resistance. Let us remember that polyolefins can be synthesized from bioethanol.

In future, chitosan could be commercially attractive for the production of bulk products for some market segments, notably due to the customer's preference for bio-based materials over fossil-based materials. In addition, chitin and chitosan could be competitive with respect to their direct and indirect production costs, due to substantial improvements in, for example, process optimization, exploitation of economies of scale, use, and commercialization of by-products and waste.

5.4.2 Variability

The variability of the chitosan structure, in terms of molecular weight, acetylation degree, and crystallinity, has a considerable effect on some biopolymer properties, such as solubility and bioactivity, and on material properties, that is, barrier and mechanical properties. De Britto and Assis (2010) showed, for example, that final chitosan-based film morphology was dependent on the polymer solubility. Depending on the origin of the polymer and its treatment during extraction from raw resources, the residual crystallinity may also vary considerably.

Due to various parameters (MW, DD, and chitosan origin), considerable variations in minimal inhibitory concentrations and/or minimum bactericidal concentrations of chitosan have been described. A significant review summarizes the chitosan MIC values according to recent data (Goy et al., 2009), indicating that the effectiveness of this amino-polysaccharide is also dependent on microbial species. This variability may be a disadvantage for the creation of new biopackaging.

5.4.3 Lack of Applied Research Regarding the Requirements Set by Food Packaging Legislation

It is well known that the key safety objective for traditional materials in contact with foods is to be as inert as possible; that is, there should be a minimum of interaction between food and packaging (Restuccia et al., 2010). Indeed, we need more information about the total migration of generally studied bio-based materials and whether the raw materials used in the formulation are "generally recognized as safe" (GRAS). On this matter, reported by Kean and Thanou (2010), regulatory agencies encounter many difficulties in approving all existing chitosans as GRAS materials. At this time, all chitosan GRAS applications are "At notifier's request, FDA ceased to evaluate the notice." Chitosan's chemical versatility and the variety of formulations confuse researchers and regulatory scientists. This is one reason why a thorough description of the chitosan used in a study should be included.

The basic concept of active and intelligent packaging poses new challenges to the evaluation of its safety as compared to traditional packaging, due to its deliberate interaction with the food and/or its environment. In Europe, both Regulations 1935/2004/EC and 450/2009/EC represent a partial answer to the lack of penetration of active and intelligent packaging in the European market in comparison to Japan, United States, and Australia, where more adequate and flexible regulations permitted technological innovations in the food packaging sector in the past years (Restuccia et al., 2010).

5.4.4 Potential Volume

The potential volumes of chitin and chitosan are generally too small to consider them as real materials to be used in bio-based matrices. Starch polymers and polylactic acid (PLA) are now clearly the most important types of polymers. However, even if by-product chitin is relatively low in terms of production, its microbial production, which is a novel process, could increase its potential since it could be produced in a natural or genetically modified organism.

5.4.5 Chemical Modification Facilities

The advantage of chitosan over other polysaccharides (cellulose, starch, galactomannans, etc.) is that its chemical structure allows specific modifications without too

TABLE 5.2   Obstacles and Promoters of Chitosan to Compete with Traditional Materials in Food Packaging

Obstacles	Promoters
Cost	Renewable resource
Manufacturing (injection, molding, extrusion, etc.)	Biodegradability, edibility
Sensitivity toward water	Water solubility could be an advantage in some cases
Limited availability and variability in chemical structure	Development of chitosan from fungal fermentation, which can bring down costs and decrease the variability
Approval for food contact?	No toxicity
Need applied research work on real food products and not only on model foods	Public concern about the environment, and limited fossil fuel resources
Need studies on life cycle analysis, taking into account use and disposal or reuse in order to fully evaluate their benefits	More international focus given to environmental impact and packaging directives
	Support to rural development

many difficulties in the C_2 position, as described in the review of Rinaudo (2006). Specific groups can be introduced on the amino group to design polymers for selected applications. Unfortunately, if we want to chemically modify this biopolymer while maintaining its active properties, it would be more complicated. In other words, specific O-chemical modifications are more difficult to control.

In conclusion, Table 5.2 summarizes obstacles and drivers for chitosan-based material development.

It is important to note that chemical modifications or association of chitosan with nonbiodegradable substances (to improve its properties) may dramatically decrease its biodegradability and enzyme susceptibility. Due to the development of biofragmentable material (unfortunately frequently presented as biodegradable materials), such as some composites, the risk is the creation of chemical pollution without visible pollution, a phenomenon that is much more dangerous for the future than the well-known visible plastic pollution.

5.5 SUMMARY

Because the ecosystem is substantially disturbed and damaged as a result of the nonbiodegradable plastic materials for disposable items, chitin and chitosan have received much attention in the recent years, notably due to their natural origin. It is well known that microbial contamination and subsequent growth reduces the shelf life of foods and especially increases the risk of food-borne illness. In recent years, antimicrobial packaging has attracted much attention from the food industry because of the increase in consumer demand for food safety and for minimally processed, preservative-free products. As a result, in order to inhibit the development of pathogens or spoilage microorganisms, packaging with antimicrobial properties can be used. New packaging solutions will increasingly focus on food safety notably by controlling microbial growth or delaying oxidation. Natural biopolymers such as chitin and chitosan exhibit very useful properties in terms of active bio-based matrices. This chapter gives a description of the different concepts of antimicrobial packaging and evaluates the potential for development by discussing recent research with respect to the use of chitin and chitosan. Relevant regulations and future trends are also reviewed. Due to governmental encouragement, this bio-based packaging with antimicrobial properties notably based on chitin or chitosan as inherently antimicrobial biopolymer (which can also be used as carriers for food preservatives) will be one of the most investigated trends.

5.6 FUTURE TRENDS

There is a pressing need for scientific institutions, industrial R&D sectors, or appropriate government departments to enhance considerably the implementation of activities aiming at accelerating the development of polymers from renewable resources. Although petroleum, natural gas, and carbon are here to stay, for some time to come, it is likely that their price will remain very high and that their reserves will start dwindling within a few decades (Gandini, 2008). Chitosan or chitin-based materials may offer important contributions by reducing the dependence on fossil fuels and the related environmental impacts.

Nature employs a small number of organic materials as building blocks to construct a diversity of structures that perform a range of functions. These capabilities have sparked interest in fabricating devices using these same biological building blocks or their mimics. The use of biological materials also allows access to enzymes for selective catalysis under benign conditions. The future is, perhaps, in the *enzymatic polymerization* and *enzymatic polymer modification*. As mentioned by Kobayashi and Makino (2009), enzymatic polymerization is defined as the *in vitro* polymerization of artificial substrate monomers catalyzed by an isolated enzyme via nonbiosynthetic (nonmetabolic) pathways. Enzymatic polymer modification denotes a modification reaction of the existing polymers with enzymatic catalysis. This is another effective method to produce new polymeric materials with improved properties.

Enzymatic assembly offers exciting opportunities that will be further broadened to design and build new macromolecular structures from renewable resources. Chitosan's pH-responsive solubility and reactivity suggest that this biopolymer has a particularly bright future for biofabrication of new macromolecular structures (Yi et al., 2005).

According to Kobayashi and Makino (2009), if *in vivo* characteristics, such as high catalytic activity, reactions under mild conditions, and high reaction selectivity, could be realized for *in vitro* enzymatic polymer synthesis, we may expect the following outcomes: (1) better control of polymer structures, (2) creation of polymers with new structures, (3) clean, selective processes without forming by-products, (4) low loading processes with energy savings, and (5) biodegradable properties of the new polymers in many cases. These are indicative of the "green" nature of the *in vitro* enzymatic catalysis for developing new polymeric materials. Polysaccharides, such as chitin and chitosan, have very complicated structures, having many stereo- and regioisomers. Chitin is synthesized *in vivo* by the catalysis of chitin synthase (Merzendorfer, 2006). The first *in vitro* synthesis of chitin was accomplished in 1995 via ring-opening polyaddition of a chitobiose oxazoline derivative monomer catalyzed by chitinase (EC 3.2.1.14), a glycoside hydrolase of chitin (Kobayashi et al., 1996). In the chitinase-catalyzed synthesis of chitin, the oxazoline structure at the reducing end and C_4 hydroxyl group at the nonreducing end are essential for the exclusive formation of $\beta(1 \rightarrow 4)$ glycosidic linkage (Kobayashi and Makino, 2009). This is a very promising way to produce chitinous substrates with less variability.

Besides homobiopolymer, we can also design new chitin- or chitosan-based polymers, such as cellulose–chitin hybrid and chitin–xylan hybrid (Kobayashi et al., 2006).

To create effective chitosan antimicrobial films, it is necessary to select the method of fabrication of the films, the optimal temperature/time conditions for mixing the biopolymer with the functional substances, and the suitable engineering mechanism to attain the desired release rate, taking into account the storage time before consumption (Lopez-Rubio et al., 2006). As mentioned by Mayachiew and Devahastin (2010), many studies have been made on the effects of many parameters on the release of active agents, the effects of drying methods, and conditions used to prepare functional films. Unfortunately, in the majority of research studies, the retention and release characteristics of added active compounds have not been well established, in particular in chitosan films. These authors suggest that drying methods may be used to engineer chitosan films for controlled release in food packaging applications. Concerning the antimicrobial activity of chitin and chitosan, there is a critical need to establish more reliable analytical methods for proper quality control in chitinous polymer production, especially regarding MW and DD, to better understand the mode of action and then increase its activity spectra (No et al., 2007). Furthermore, bacteria in different growth stages or initial microbial populations also seem to be crucial factors in the susceptibility tests, but little information can be found in the literature regarding these particular aspects.

Moreover, consumers favor foods with fewer synthetic additives, but products must also be safe to eat and have a sufficiently long shelf life. Active biopackaging can be an alternative. In order to work for real development of active biopackaging, we have to identify if and how active packaging provides added value to the target food. To find new applications in food preservation, polysaccharide-based materials have to generally improve their performance in terms of water sensitivity, barrier properties, and mechanical properties. According to the majority of research studies, it appears that biopolymer chemical modifications or association with compounds having complementary properties are not particularly efficient. Perhaps the addition of nanostuctured compounds or nanoparticles could be useful. Indeed, nanomaterials, due to a very high aspect ratio, have the capability to change the properties of packaging materials without significant changes in clarity and processing performance. Nevertheless, we have to seriously study the potential impact of these nanostructures on the environment and human or animal health. Antimicrobial biopackaging can provide solutions to producer and consumer problems with respect to microbial development. Consumers become more selective about what qualifies as fresh and safe food and we, as scientists, have to work to decrease our environmental impact and to measurably increase the safety of foods. We need to consider our older methods of preservation, for example, by creating "intelligent" biopackaging that has a "release on command" preservative, which will save the food before it begins to spoil, and with a relatively low preservative content.

Perhaps one of the main objectives for the future in biopackaging will be first the development of (1) environment-friendly processes for biopolymer chemical modification or elaboration of bio-based materials and (2) 100% biodegradable additives commonly used in packaging. Second, we will have to evaluate the global impact (positive and negative) of the biopackaging creation on the environment (life cycle analysis).

To conclude, there is an urgent need to develop materials from renewable resources, which will be of great importance to the materials domain, not only as a solution to the growing environmental threat, but also as a solution to the uncertainty of the petroleum supply. Bio-based polymers are an *emerging field* that is characterized by a number of different developments. One development is that *established chemical companies* are moving into *biotechnology* and engaging in R&D efforts. Main drivers are the biodegradability of the product, the reduction in production costs associated with using carbohydrate feedstocks due to advances in fermentation and aerobic bioprocesses, unique properties of bio-based polymers, and (to a lesser extent) the use of renewable resources (Wolf et al., 2005).

REFERENCES

Agnihotri, S. A., N. N. Mallikarjuna, and T. M. Aminabhavi (2004). Recent advances on chitosan-based micro- and nanoparticles in drug delivery. *J. Control. Release* **100**:5–28.

Aider, M. (2010). Chitosan application for active bio-based films production and potential in the food industry. *LWT Food Sci. Technol.* **43**:837–842.

Alves, N. M., and J. F. Mano (2008). Chitosan derivatives obtained by chemical modifications for biomedical and environmental applications. *Int. J. Biol. Macromol.* **43**:401–414.

Anonymous (2001). U.S. Food and Drug Administration, Center for Food Safety and Applied Nutrition, Office of Premarket Approval. GRAS Notices. Available at http:/vm.cfsan.fda.gov.

Appendini, P., and J. H. Hotchkiss (2002). Review of antimicrobial food packaging. *Innov. Food Sci. Emerg. Technol.* **3**:113–126.

Aymerich, M., J. Garriga, A. Jofre, B. Martin, and J. M. Monfort (2006). The use of bacteriocins against meat-borne pathogens. In: L. M. L. Nollet and F. Toldra, editors, *Advanced Technologies for Meat Processing*, London: CRC Press, pp. 371–399.

Badawy, M. E. I. (2010). Structure and antimicrobial activity relationship of quaternary *N*-alkyl chitosan derivatives against some plant pathogens. *J. Appl. Polym. Sci.* **117**:960–969.

Banerjee, M., S. Mallicks, P. Anumita, A. Chattopadhyay, and S. G. Siddhartha (2010). Heightened reactive oxygen species generation in the antimicrobial activity of a three component iodinated chitosan–silver nanoparticle. *Langmir* **26**:5901–5908.

Belalia, R., S. Grelier, M. Benaissa, and V. Coma (2008). New bioactive biomaterials based on quaternized chitosan. *J. Agric. Food Chem.* **65**:1582–1588.

Bennett, J. W., and Klich M. (2003). Mycotoxins. *Clin. Microbiol. Rev.* **16**:497–516.

Cai, Z. S., Z. Q. Song, C. S. Yang, S. B. Shang, and Y. B. Yin (2009). Synthesis, characterization and antibacterial activity of quaternized *N,O*-(2-carboxyethyl) chitosan. *Polym. Bull.* **62**:445–456.

Cetinus, S. A., and H. N. Oztop (2000). Immobilization of catalase on chitosan film. *Enzyme Microb. Technol.* **26**:497–501.

Chung, Y. C., and C. Y. Chen (2008). Antibacterial characteristics and activity of acid-soluble chitosan. *Bioresour. Technol.* **99**:2806–2814.

Coma V., (2006). Perspectives for the active packaging of meat products. In : L. M. L. Nollet and F. Toldra, editors, *Advances technologies for meat processing.* Boca Raton: CRC Taylor&Francis, pp. 449–472.

Coma, V. (2008a). Bioactive packaging technologies for extended shelf life of meat-based products. *Meat Sci.* **78**:90–103.

Coma V. (2008b). Bioactive chitosan-based substances and films. In: R. Jayakumar and M. Prabaharan,editors, *Current Research and Developments on Chitin and Chitosan in Biomaterials Sciences*, Trivandrum: Research Signpost, pp. 21–51.

Coma, V. (2010). Overview: polysaccharide-based biomaterials with antimicrobial and antioxidant properties. *Polímeros Ciência e Tecnologia* doi:10. 4322/polimeros020ov002.

Constantine, C. A., K. M. Gattas-Asfura, S. V. Mello, G. Crespo, V. Rastogi, T. C. Cheng, J. J. DeFrank, and R. M. Leblanc (2003). Layer-by-layer biosensor assembly incorporating functionalized quantum dots. *Langmuir* **19**:9863–9867.

Cooksey, K. (2005). Effectiveness of antimicrobial food packaging materials. *Food Addit. Contam. A* **22**:980–987.

Costa-Júnior, E. S., E. F. Barbosa-Stanciolib, A. A. P. Mansurc, W. L. Vasconcelosc, and H. S. Mansur (2009). Preparation and characterization of chitosan/poly(vinyl alcohol) chemically crosslinked blends for biomedical applications. *Carbohydr. Polym.* **76**:472–481.

Crini, G., E., Guibal, M., Morcellet, G. Torri, and P. M. Badot (2009). Chitine et chitosane: synthèse, propriétés et principales application. In: G. Crini, P. M. Badot, and E. Guibal, editors, *Chitine et chitosane, du biopolymère à l'application*, Besançon: Presse Universitaires de Franche-Comté, pp. 19–37.

Curcio, M., F. Puoci, F. Iemma, O. I. Parisi, G. Cirillo, U. G. Spizzirri, and N. Picci (2009). Covalent insertion of antioxidant molecules on chitosan by a free radical grafting procedure. *J. Agric. Food Chem.* **57**:5933–5938.

De Britto, D., and B. G. O. Assis (2010). Hydrophilic and morphological aspects of films based on quaternary salts of chitosan for edible applications. *Packag. Technol. Sci.* **23**:111–119.

Denavi, G., D. R. Tapia-Blácido, M. C. Añón, P. J. A. Sobral, A. N. Mauri, and F. C. Menegalli (2009). Effects of drying conditions on some physical properties of soy protein films. *J. Food Eng.* **90**:341–349.

Devlieghere, F., A. Vermeulen, and J. Debevere (2004). Chitosan: antimicrobial activity, interactions with food components and applicability as a coating on fruit and vegetables. *Food Microbiol.* **21**:703–714.

Dutta, P. K., S. Tripathi, G. K. Mehrotra, and J. Dutta (2009). Perspectives for chitosan based antimicrobial films in food applications. *Food Chem.* **114**:1173–1182.

Einbu, A., S. N. Naess, A. Elgsaeter, and K. M. Varum (2004). Solution properties of chitin in alkali. *Biomacromolecules* **5**:2048–2054.

Fernandez, A., D. Cava, M. J. Ocio, and J. M. Lagaron (2008). Perspectives for biocatalysts in food packaging. *Trends Food Sci. Technol.* **19**:198–206.

Fernandez-Saiz, P., M. J. Ocio, and J. M. Lagaron (2009a). The use of chitosan in antimicrobial films for food protection. *Perspect. Agric. Vet. Sci. Nutr. Nat. Resour.* **5**:1–11.

Fernandez-Saiz, P., J. M. Lagaron, and M. J. Ocio (2009b). Optimization of the biocide properties of chitosan for its application in the design of active films of interest in the food area. *Food Hydrocolloids* **23**:913–921.

Galić, K., M. Ščetar, and M. Kurek (2011). The benefits of processing and packaging. *Trends Food Sci. Technol.* **22**:127–137.

Gandini, A. (2008). Polymers from renewable resources: a challenge for the future of macromolecular materials. *Macromolecules* **41**:9491–9504.

Goy, R. C., D. De Britto, and O. B. G. Assis (2009). A review of the antimicrobial activity of chitosan. *Polimeros Ciencia e Technologia* **19**:241–247.

Hadwiger, L. A., D. G. Kendra, B. W. Fristensky, and W. Wagoner (1981). Chitosan both activated genes in plants and inhibits RNA synthesis in fungi. In: R. A. A Muzzarelli, C. Jeuniaux, and G. W. Gooday, editors, *Chitin in Nature and Technology*, New York: Plenum, pp. 209–222.

Higazya, A., M. Hashema, A. ElShafeib, N. Shakerc, and M. A. Hadya (2010). Development of antimicrobial jute packaging using chitosan and chitosan–metal complex. *Carbohydr. Polym.* **79**:867–874.

Holappaa, J., M. Hjálmarsdóttirb, M. Mássonc, Ö. Rúnarssonc, T. Asplundd, P. Soininene, T. Nevalainena, and T. Järvinena (2006). Antimicrobial activity of chitosan *N*-betainates. *Carbohydr. Polym.* **65**:114–118.

Hoven, V. P., V. Tangpasuthadol, Y. Angkitpaiboon, N. Vallapa, and S. Kiatkamjornwong (2007). Surface-charged chitosan: preparation and protein adsorption. *Carbohydr. Polym.* **68**:44–53.

Hsieh, W. C., C. P. Chang, and Y. L. Gao (2006). Controlled release properties of chitosan encapsulated volatile citronella oil microcapsules by thermal treatments. *Colloids Surf. B* **53**:209–214.

Hussain, M. R., and T. K. Maji (2008). Preparation of genipin cross-linked chitosan–gelatin microcapsules for encapsulation of *Zanthoxylum limonella* oil (ZLO) using salting-out method. *J. Microencapsul.* **25**:1–7.

Jayakumar, R., N. Nwe, S. Tokura, and H. Tamura (2007). Sulfated chitin and chitosan as novel biomaterials. *Int. J. Biol. Macromol.* **40**:175–181.

Jayakumar, R., M. Rajkumar, H. Freitas, N. Selvamurugan, S. V. Nair, T. Furuike, and H. Tamura (2009). Preparation, characterization, bioactive and metal uptake studies of alginate/phosphorylated chitin blend film. *Int. J. Biol. Macromol.* **44**:107–111.

Je, J. Y., and S. K. Kim (2006). Antimicrobial action of novel chitin derivative. *Biochim. Biophys. Acta* **1760**, 104–109.

Jin, T., and H. Zhang (2008). Biodegradable polylactic acid polymer with nisin for use in antimicrobial food packaging. *J. Food Sci.* **73**:127–134.

Jones, J. M. (2002). Food safety. In: Z. E. Sikorski,editor, *Chemical and Functional Properties of Food Components*, London: CRC Press, pp. 291–305.

Jeuniaux, C., and M. F. Voss-Foucart (1991). Chitin biomass and production in the marine environment. *Biochem. Syst. Ecol.*, **19**:347–356.

Kean, T., and M. Thanou (2010). Biodegradation, biodistribution and toxicity of chitosan. *Adv. Drug Deliv. Rev.* **62**:3–11.

Kjartansson, G. T., S. Zivanovic, K. Kristbergsson, and J. Weiss (2006). Sonication-assisted extraction of chitin from North Atlantic shrimps (*Pandalus borealis*). *J. Agric. Food Chem.* **54**:5894–5902.

Klaykruayat, B., K. Siralertmukul, and K. Srikulkit (2010). Chemical modification of chitosan with cationic hyperbranched dendritic polyamidoamine and its antimicrobial activity on cotton fabric. *Carbohydr. Polym.* **80**:197–207.

Kobayashi, S., and A. Makino (2009). Enzymatic polymer synthesis: an opportunity for green polymer chemistry. *Chem. Rev.* **109**:5288–5353.

Kobayashi, S., T. Kiyosada, and S. Shoda (1996). Synthesis of artificial chitin: irreversible catalytic behavior of a glycosyl hydrolase through a transition state analogue substrate. *J. Am. Chem. Soc.* **118**:13113–13114.

Kobayashi, S., A. Makino, H. Matsumoto, S. Kunii, M. Ohmae, T. Kiyosada, K. Makiguchi, A. Matsumoto, M. Horie, and S. I. Shoda (2006). Enzymatic polymerization to novel polysaccharides having a glucose *N*-acetylglucosamine repeating unit, a cellulose–chitin hybrid polysaccharide. *Biomacromolecules* **7**:1644–1656.

Kosaraju, S. L., L. D'ath, and A. Lawrence (2006). Preparation and characterisation of chitosan microspheres for antioxidant delivery. *Carbohydr. Polym.* **64**:163–167.

Kumar, V. A. B., M. C. Varadaraj, R. G. Lalitha, and R. N. Tharanathan (2004). Low molecular weight chitosan: preparation with the aid of papain and characterization. *Biochim. Biophys. Acta* **1670**:137–146.

Lagaron, J. M., P. Fernandez-Saiz, and M. J. Ocio (2007). Using ATR-FTIR spectroscopy to design active antimicrobial food packaging structures based on high molecular weight chitosan polysaccharide. *J. Agric. Food Chem.* **55**:2554–2562.

Lee, J. S., G. H. Kim, and H. G. Lee (2010). Characteristics and antioxidant activity of *Elsholtzia splendens* extract-loaded nanoparticles. *J. Agric. Food Chem.* **58**:3316–3321.

Lin, Y. W., Y. C. Hsiao, and B. H. Chiang (2009). Production of high degree polymerized chitooligosaccharides in a membrane reactor using purified chitosanase from *Bacillus cereus*. *Food Res. Int.* **42**:1355–1361.

Liu, H., Y. Du, X. Wang, and L. Sun (2004). Chitosan kills bacteria through cell membrane damage. *Int. J. Food Microbiol.* **95**:147–155.

Liu, X. F., L. Song, L. Li, S. Y. Li, and K. D. Yao (2006). Antibacterial effects of chitosan and its water-soluble derivatives on *E. coli*, plasmids DNA, and mRNA. *J. Appl. Polym. Sci.* **103**:3521–3528.

Lopez-Rubio, A., R. Gavara, and J. M. Lagaron (2006). Bioactive packaging: turning foods into healthier foods through biomaterials. *Trends Food Sci. Technol.* **7**:567–575.

Másson, M., J. Holappa, M. Hjálmarsdóttir, Ö. V. R. Rúnarsson, T. Nevalainen, and T. Järvinen (2008). Antimicrobial activity of piperazine derivatives of chitosan. *Carbohydr. Polym.* **74**:566–571.

Mayachiew, P., and S. Devahastin (2008). Comparative evaluation of physical properties of edible chitosan films prepared by different drying methods. *Drying Technol.* **26**:176–185.

Mayachiew, P., and S. Devahastin (2010). Effects of drying methods and conditions on release characteristics of edible chitosan films enriched with Indian gooseberry extract. *Food Chem.* **118**:594–601.

Mead, P. S., L. Slutsker, V. Dietz, L. F. McCaig, J. S. Bresee, C. Shapiro, P. M. Griffin, and R. V. Tauxe (1999). Food-related illness and death in the United States. *Emerg. Infect. Dis.* **5**:607–625.

Merzendorfer, H. J. (2006). Insect chitin synthases: a review. *Comp. Physiol.* **176**:1–15.

Muhizi, T., V. Coma, and S. Grelier (2008). Synthesis and evaluation of *N*-alkyl-β-D-glucosylamines on the growth of two wood fungi, *Coriolus versicolor* and *Poria placenta*. *Carbohydr. Res.* **343**:2369–2375.

Murray, C. A., and J. R. Dutcher (2006). Effect of changes in relative humidity and temperature on ultrathin chitosan films. *Biomacromolecules* **7**:3460–3465.

Muzzarelli, R. A. A., G. Barontini, and R. Rochetti (2004). Isolation of lysozyme on chitosan. *Biotechnol. Bioeng.* **20**:87–94.

No, H. K., S. P. Meyers, W. Prinyawiwatkul, and Z. Xu (2007). Applications of chitosan for improvement of quality and shelf life of foods: a review. *J. Food Sci.* **72**:87–100.

Özmeric, N., G. Özcan, C. M. Haytac, E. E. Alaaddinoglu, M. F. Sargon, and S. Senel (2000). Chitosan film enriched with an antioxidant agent, taurine, in fenestration defects. *J. Biomed. Mater. Res.* **51**:500–503.

Papineau, A. M., D. G. Hoover, D. Knorr, and D. F. Farkas (1991). Antimicrobial effect of water-soluble chitosans with high hydrostatic pressure. *Food Biotechnol.* **5**:45–57.

Parize, A. L., T. C. Rozone de Souza, I. M. Costa Brighente, V. T. Fávere, M. C. M. Laranjeira, A. Spinelli, and E. Longo (2008). Microencapsulation of the natural urucum pigment with chitosan by spray drying in different solvents. *Afr. J. Biotechnol.* **7**:3107–3114.

Pelissari, F. M., M. V. E. Grossmann, F. Yamashita, and E. A. G. Pineda (2009). Antimicrobial, mechanical, and barrier properties of cassava starch–chitosan films incorporated with oregano essential oil. *J. Agric. Food Chem.* **57**:7499–7504.

Peng, Z.-X., L. Wang, L. Du, S. R. Guo, X. Q. Wang, and T. T. Tang (2010). Adjustment of the antibacterial activity and biocompatibility of hydroxypropyltrimethyl ammonium chloride chitosan by varying the degree of substitution of quaternary ammonium. *Carbohydr. Polym.* **81**:275–283.

Pillai, C. K. S., W. Paul, and C. P. Sharma (2009). Chitin and chitosan polymers: chemistry, solubility and fiber formation. *Prog. Polym. Sci.* **34**:641–678.

Popa, M. I., N. Aelenei, V. I. Popa, and D. Andrei D. (2000). Study of the interactions between polyphenolic compounds and chitosan. *React. Funct. Polym.* **45**:35–43.

Portes, E., C. Gardrat, A. Castellan, and V. Coma (2009). Environmentally friendly films based on chitosan and tetrahydrocurcuminoid derivatives exhibiting antibacterial and antioxidative properties. *Carbohydr. Polym.* **76**:578–584.

Raafat, D., K. Von Bargen, A. Haas, and H. G. Sahl (2008). Insights into the mode of action of chitosan as an antibacterial compound. *Appl. Environ. Microbiol.* **74**:3764–3773.

Rabea, E. I., M. E. T. Badawy, V. Christian, C. V. Stevens, G. Smagghe, and W. Steurbaut (2003). Chitosan as antimicrobial agent: applications and mode of action. *Biomacromolecules* **4**:1454–1465.

Rabea, E. I., M. E. I. Badawy, W. Steurbaut, and C. V. Stevens (2009). *In vitro* assessment of *N*-(benzyl)chitosan derivatives against some plant pathogenic bacteria and fungi. *Eur. Polym. J.* **45**:237–245.

Ravi Kumar, M. N. V., R. A. A. Muzzarelli, C. Muzzarelli, H. Sashiwa, and A. J. Domb (2004). Chitosan chemistry and pharmaceutical perspectives. *Chem. Rev.* **104**:6017–6084.

Restuccia, D., U. G. Spizzirri, O. I. Parisi, G. Cirillo, M. Curcio, F. Iemma, F. Puoci, G. Vinci, and N. Picci (2010). New EU regulation aspects and global market of active and intelligent packaging for food industry applications. *Food Control* **21**:1425–1435.

Rinaudo, M. (2006). Chitin and chitosan: properties and applications. *Prog. Polym. Sci.* **31**:603–632.

Sajomsang, W., P. Gonil, and S. Tantayanon (2009). Antibacterial activity of quaternary ammonium chitosan containing mono or disaccharide moieties: preparation and characterization. *Int. J. Biol. Macromol.* **44**:419–427.

São Pedro, A., E. Cabral-Albuquerque, D. Ferreira, and B. Sarmento (2009). Chitosan: an option for development of essential oil delivery systems for oral cavity care? *Carbohydr. Polym.* **76**:501–508.

Sashiwa, H., H. Saimoto, Y. Shigemasa, R. Ogawa, and S. Tokura (1990). Lysozyme susceptibility of partially deacetylated chitin. *Int. J. Biol. Macromol.* **12**:295–296.

Seidl, V. (2008). Chitinases of filamentous fungi: a large group of diverse proteins with multiple physiological functions. *Fungal Biol. Rev.* **22**:36–42.

Shahidi, F., J. K. V. Arachchi, and Y. J. Jeon (1999). Food applications of chitin and chitosans. *Trends Food Sci. Technol.* **10**:37–51.

Sogias, I. A., V. V. Khutoryanskiy, and A. C. Williams (2010). Exploring the factors affecting the solubility of chitosan in water. *Macromol. Chem. Phys.* **211**:426–433.

Sousa, F., G. M. Guebitz, and V. Kokol (2009). Antimicrobial and antioxidant properties of chitosan enzymatically functionalized with flavonoids. *Process Biochem.* **44**:749–756.

Sun, S., Q. An, X. Li, L. Qian, B. He, and H. Xiao (2010). Synergistic effects of chitosan–guanidine complexes on enhancing antimicrobial activity and wet-strength of paper. *Bioresour. Technol.* **101**:5693–5700.

Takegawa, A., M. Murakami, Y. Kaneko, and J. Kadokawa (2010). Preparation of chitin/cellulose composite gels and films with ionic liquids. *Carbohydr. Polym.* **79**:85–90.

Tayel, A. A., S. Moussa, K. Opwis, D. Knittel, E. Schollmeyer, and A. Nickisch-Hartfiel (2010). Inhibition of microbial pathogens by fungal chitosan. *Int. J. Biol. Macromol.* **47**:10–14.

Tharanathan, R. N., and F. S. Kittur (2003). Chitin: the undisputed biomolecule of great potential. *Crit. Rev. Food Sci. Nutr.* **43**:61–87.

Tripathi, S., G. K. Mehrotra, C. K. M. Tripathi, B. Banerjee, A. K. Joshi, and P. K. Dutta (2008). Chitosan based bioactive film: functional properties towards biotechnological needs. *Asian Chitin J.* **4**:29–36.

Valenta, C., B. Christen, and A. Bernkop-Schnürch (1998). Chitosan–EDTA conjugate: a novel polymer for topical gels. *J. Pharm. Pharmacol.* **50**:445–452.

Vallapa, N., O. Wiarachai, N. Thongchul, J. Pan, V. Tangpasuthadol, S. Kiatkamjornwong, and V. P. Hoven (2010). Enhancing antibacterial activity of chitosan surface by heterogeneous quaternization. *Carbohydr. Polym.* **83**:868–875.

Vartiainen, J., M. Tuominen, and K. Nättinen (2010). Bio-hybrid nanocomposite coatings from sonicated chitosan and nanoclay. *J. Appl. Polym. Sci.* **116**:3638–3647.

Wallace, D., T. V. Gilder, S. Shallow, S. Fiorentino, S. Segler, K. Smith, B. Shiferaw, R. Etzel, W. Garthright, and F. Angulo (2000). Incidence of food-borne illnesses reported by the food-borne diseases active surveillance network. *J. Food Prot.* **63**:807–809.

Weber, C.,editor (2000). *Biobased Packaging Materials for the Food Industry: Status and Perspectives*, Frederiksberg: KVL Department of Dairy and Food Science, pp. 1–136.

Wolf, O., M. Crank, F. Marscheider-Weidemann, J. Schleich, B. Hüsing, and G. Angerer (2005). Techno-economic feasibility of large-scale production of bio-based polymers in Europe. Technical Report EUR 22103 EN, Brussels, Belgium: European Commission, Joint Research Center, pp. 1–260.

Wolffs, P., and P. Radstrom (2006). Real-time PCR for the detection of pathogens in meat. In: L. M. L. Nollet and F. Toldra,editors, *Advanced Technologies for Meat Processing*, London: CRC Press, pp. 131–153.

Wu, Y. B., S. H. Yu, F. L. Mi, C. W. Wu, S. S. Shyu, C. K. Peng, and A. C. Chao (2004). Preparation and characterization on mechanical and antibacterial properties of chitosan/cellulose blends. *Carbohydr. Polym.* **57**:435–440.

Xu, T., M. Xin, M. Li, H. Huang, and S. Zhou (2010). Synthesis, characteristic and antibacterial activity of *N,N,N*-trimethyl chitosan and its carboxymethyl derivatives *Carbohydr. Polym.* **81**:931–936.

Yang, T. C., C. C. Chou, and C. F. Li (2005). Antibacterial activity of *N*-alkylated disaccharide chitosan derivatives. *Int. J. Food Microbiol.* **97**:237–245.

Yi, H., L. Q. Wu, W. E. Bentley, R. Ghodssi, G. W. Rubloff, J. N. Culver, and G. F. Payne (2005). Biofabrication with chitosan. *Biomacromolecules* **6**:2881–2894.

Young, D. H., and H. Kauss (1983). Release of calcium from suspension-cultured glycine max cells by chitosan, other polycations, and polyamines in relation to effects on membrane permeability. *Plant Physiol.* **73**:698–702.

Zheng, L. Y., and J. A. F. Zhu (2003). Study on antimicrobial activity of chitosan with different molecular weights. *Carbohydr. Polym.* **54**:527–530.

Ziani, K., I. Fernández-Pan, M. Royo, and J. I. Maté (2009). Antifungal activity of films and solutions based on chitosan against typical seed fungi. *Food Hydrocolloids* **23**:2309–2314.

6

CHITIN AND CHITOSAN AS BIOMATERIAL BUILDING BLOCKS

José F. Louvier-Hernández and Ram B. Gupta

6.1 INTRODUCTION

In the past two decades, chitin and chitosan biopolymers have attracted growing interest for their use as biomaterials. Chitin (CTN) is not only the second most common polysaccharide in world but it also has a combination of individual properties that makes it very attractive for use in several high-valued applications. Chitin occurs as a component of crustacean shells, insect cuticles, certain fungi, and the cell walls of specific plants. It is found in crustacean exoskeletons in association with proteins and minerals such as calcium carbonate. Different sources of chitin differ somewhat in their structure and percentage of chitin content. For example, α-chitin, from crab and shrimp, is a highly crystalline polymorph in which polymer chains are tightly packed in antiparallel arrangement forming an orthorhombic crystal structure, making this polymorph very difficult to dissolve. In the other hand, β-chitin, from squid pen, has a less crystalline structure, in which polymer chains are arranged in parallel, forming a monoclinic crystal structure. β-Chitin has weaker intermolecular hydrogen bonding than α-chitin, and dissolution is not as difficult; however, β-chitin could revert to the α-form when precipitated from solution. Chemically, chitin is a naturally occurring polymer formed primarily of $\beta-(1 \rightarrow 4)$-2-acetamido-2-deoxy-D-glucose repeating units. Only a few known solvents can dissolve chitin, such as mixtures of N,N-dimethylacetamide (DMAc) with 5 wt% LiCl, or 1,1,1,3,3,3-hexafluoro-2-propanol (HFIP). Chitosan (CTS) is

Polysaccharide Building Blocks: A Sustainable Approach to the Development of Renewable Biomaterials, First Edition. Edited by Youssef Habibi and Lucian A. Lucia. © 2012 John Wiley & Sons, Inc. Published 2012 by John Wiley & Sons, Inc.

obtained upon deacetylation of chitin, it is soluble in aqueous dilute organic acids, and for this reason is more widely used than chitin.

6.2 BACKGROUND

6.2.1 History and Definitions

Chitin was first discovered in mushrooms by Henri Braconnot, a Natural History professor, in 1811; he named this material "fungine." In 1823, Odier isolated the same material from insects and plants and named it "chitine" (Prajapati, ; Khoushab and Yamabhai, 2010), derived from the Greek *chiton* which means "tunic" or "envelope" because of the protective character of chitin in insects and arthropods. Discovery of chitosan is attributed to Rouget in 1859, when he found that boiling chitin in potassium hydroxide yielded an acid soluble chitin. In 1894, Hoppe-Seyler named it chitosan; however, its molecular structure was not resolved until 1950. The existence of chitosan in nature was unknown before 1954, when it was discovered in the yeast *Phycomyces blakesleeanus*. Chitosan occurs as the major structural component of the cell walls of certain fungi, mainly of the Zygomycetes species, but it is not as abundant as chitin; however, it is easily produced by chemical deacetylation of chitin (Saburo et al., 2006).

Chitin, found in nature as ordered crystalline microfibers, is a unique biopolymer based on the *N*-acetyl-glucosamine monomer. It is a highly insoluble material resembling cellulose in its solubility and low chemical reactivity. It may be regarded as cellulose with hydroxyl at position C-2 replaced by an acetamido group. Chitin is a white, hard, inelastic, nitrogenous polysaccharide (Ravi Kumar, 2000a).

There is no agreement for a sharp nomenclature regarding degree of N-deacetylation between chitin and chitosan (Ravi Kumar, 2000a). Chitin ideal molecular structure should be a linear polysaccharide of β-(1 → 4)-2-acetamido-2-deoxy-D-glucopyranose where all units are comprised entirely of *N*-acetyl-D-glucosamine. Ideal molecular structure of chitosan, a deacetylated form of chitin, is a linear polysaccharide of β-(1 → 4)-2-amino-2-deoxy-D-glucopyranose where all units are comprised entirely of D-glucosamine. However, in most natural forms, this biopolymer exists as a random copolymer of D-glucosamine and *N*-acetyl-D-glucosamine units (see Figure 6.1). There are two different rules proposed for naming them: (a) when the number of acetamido groups is more than 50% (more commonly 70–90%) the biopolymer is termed chitin (Khor, 2001; Khor and Lim, 2003), in contrast, when the number of acetamido groups is less than 50% (more commonly 10–30%) it is termed chitosan; (b) the second one is based on the solubility in aqueous acetic acid, that is, chitosan as soluble and chitin as insoluble (Roberts, 1992).

6.2.2 Sources

Chitin occurs as component of crustacean shells, insect cuticles, certain fungi, and the cells walls of specific plants. It is found in crustacean exoskeletons in association with proteins and minerals such as calcium carbonate. Chitin is produced from the processing waste of shellfish, krill, clams, oysters, and fungi and is the second most

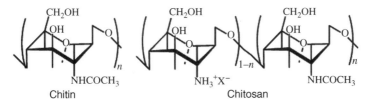

FIGURE 6.1 Chemical structures for chitin and chitosan (in its protonated form). Figure taken from Pavinatto et al. (2010). Reprinted from Biomcromolecules. Pavinatto, F.J., Caseli, L. and O.N. Oliveira, Jr. Chitosan in nanostructured thin films. 11:1897–1908. Copyright 2010 American Chemical Society.

abundant polysaccharide in nature after cellulose. Its total annual production by arthropods has been estimated to be more than one billion ton per year (Jeuniaux and Voss-Foucart, 1991; Cauchie, 2002). Chitin is the raw material for all commercial production of chitosan and glucosamine, with an estimated annual production of 2000 and 4000 ton, respectively (Sandford, 2003). Chitin and chitosan are also of commercial interest due to their relatively high nitrogen content (6.89%). Many different sources of chitin have been reported: chitin in the α- and β-forms from marine crustacean from the Arabian Gulf (Sagheer et al., 2009), the North Atlantic (Kjartansson et al., 2006), or from six different local sources in Egypt (Abdou et al., 2008).

6.2.3 Solubilization

Solubilization and reactivity of chitin is significantly limited due to its strong hydrogen-bond network. For chitin-based copolymers with less than 70% deacetylation, only a few solvents, such as N,N-dimethylacetamide containing 5–10% LiCl, hexafluoroacetone, and hexafluoro-2-propanol can dissolve the copolymer without altering the chemical structure (Louvier-Hernández et al., 2005; Phongying et al., 2007). Chitosan, on the other hand, can be easily solubilized in acidic media, due to the protonation of the amino groups. The most common solvents include aqueous solutions of organic acids such as acetic, formic, and lactic acids.

6.2.4 Properties

Following properties of chitin and chitosan make them suitable as building blocks for biomaterials.

6.2.4.1 Biodegradable Chitin and chitosan are biodegradable biopolymers. Enzymes including chitinase, chitosanase, and lysozyme present in human body, break them down into oligomers. In humans, chitin monomers are precursors of the disaccharide units in glycosaminoglycans (such as hyaluronic acid, chondroitin sulfate, and keratin sulfate), which are necessary to repair and maintain healthy cartilage and joint functions (Zhao et al., 2010).

6.2.4.2 Biocompatible Chitin and chitosan are natural biopolymers. They have no known antigenic properties, and thus are perfectly compatible with living tissue.

Their antithrombogenic and hemostatic properties make them very suitable for use in various fields of biology and medicine.

6.2.4.3 Hemostasis The process of blood clotting and the subsequent dissolution of the clot, following repair of the injured tissue is called hemostasis. Chitin and chitosan enhance platelet adhesion and aggregation (Kim and Jung, 2010). Hemostasis is promoted when chitin or chitosan dressings are applied to wounds, activating platelets with superior performance compared with known hemostatic materials (Muzzarelli, 2009).

6.2.4.4 Antimicrobial Activity Because of its positively charged amino group, chitosan interacts with microbial cell membranes, causing leakage of proteins and intracellular material (Shaidi et al., 1999). Molecular weight and degree of deacetylation have great effect in antimicrobial activity of chitosan. Antimicrobial activity increases with a decrease in the molecular weight and increase in the deacetylation (Dutta and Dutta, 2010).

6.2.4.5 Anticholesterolemic Agent Upon insolubilization, due to pH, in the digestive tract, chitosan can trap lipids. When administered to rats, chitosan has shown to considerably reduce the level of cholesterol in the blood. (Ravi Kumar et al., 2004)

6.2.4.6 Anti-inflammatory Activity The anti-inflammatory activity of chitin and chitosan are not fully established, yet. However, some experiments have shown that chitosan can inhibit the activity of mast cells with a potential to reduce the allergic inflammatory response (Kim et al., 2004; Kim and Kim, 2010).

6.2.4.7 Chelation Agent Chitin and its derivatives are remarkable chelation agents. Applications include chromatography medium, removing heavy metals, and water treatment. Chitosan has a strong positive charge, which allows it to bind with negatively charged surfaces or materials, including metals, skin, and macromolecules such as proteins, with a high potential for biomaterial. The charge density allows chitosan to form insoluble ionic complexes or complex coacervates with a wide variety of water-soluble anionic polymers (Francis Suh and Matthew, 2000).

Due to the above interesting properties, a wide number of scholarly research publications have intensively discussed several aspects of biomedical applications of chitosan and chitin for tissue engineering, drug delivery, bone regeneration, scaffolds, and wound dressing material.

6.2.5 State of the Art

Rinaudo (2008) reviewed properties of polysaccharides for their use as biomaterials. Because chitin is insoluble in usual solvents, it is difficult to determine the molecular weight. Also, there is not many available data on its physical properties, mainly due to its low solubility. Rinaudo (2008) reports Mark–Houwink parameters for chitin and chitosan. Processing of chitosan is easier than that of chitin. Several chemically modified forms of chitin and chitosan have been produced including carboxymethylchitin 6-O-carboxymethylchitin, hydroxypropylchitin, phosphoryl-chitin, O-carboxymethylchitosan, n-carboxymethylchitosan, phosphorylated chitosan, dibutyryl chitin, hydroxybutylchitosan, and so on (Kurita, 2001; Onishi et al., 2001; Tokura and Tamura, 2001; Ravi Kumar et al., 2004; Muzzarelli et al., 2005; Vikhoreva et al., 2005). Chitin and chitosan blends have been prepared including chitosan/collagen, chitosan/PEG, chitosan/PLLA, chitosan/PLA, chitosan/starch, gelatin/chitosan, alginate/chitosan, chitin/polystyrene, and so on (Kurita et al., 1996; Lee et al., 2002; Zhou et al., 2006; Liji Sobhana et al., 2009; Bama et al., 2010; Saboktakin et al., 2010; Watthanaphanit et al., 2010; Fernandes et al., 2011). Also, composites of chitosan with inorganic materials have been examined including chitosan–hydroxyapatite, chitosan–chitin–hydroxyapatite, chitosan–calcium phosphates, mainly for tissue engineering and orthopedic applications (Muzzarelli and Muzzarelli, 2002; Tigh et al., 2009; Peniche et al., 2010).

6.2.5.1 Tissue Engineering   Tissue engineering involves a scaffold made of a biocompatible, biodegradable polymer, living cells capable of differentiation and growth factors. The cells can be stem cells or come from living tissue. Multiplication of the cells fills the scaffold, yielding a three-dimensional tissue. Once implanted in the body, cells recreate their proper tissue functions, blood vessels grow into the new tissue, the scaffold degrades, and lab-grown tissue merges with natural tissue. Matrices used need to be biocompatible and have good mechanical characteristics (Vacanti and Langer, 1999). Francis Suh and Matthew (2000) have reviewed the application of chitosan-based biomaterials in cartilage tissue engineering. The review describes the cartilage tissue engineering concepts and the benefits of chitosan. Degradation kinetics of scaffolds is inversely related to the degree of crystallinity, which in turns depends on the degree of acetylation. Maximum crystallinity values would correspond to 0% deacetylated chitin or 100% deacetylated chitosan. Minimum crystallinity is achieved at intermediate degrees of acetylation. Highly deacetylated chitin (i.e., >85%) exhibits high crystallinity and, therefore, the lowest degradation rates and could last several months *in vivo*.

A more recent review is provided by Mano et al. (2007), where different classes of natural polymers are analyzed. Due to similarities with extracellular matrix (ECM), polysaccharides are among the most attractive options, since they may avoid the stimulation of chronic inflammation or immunological reactions and toxicity. Cellulose, starch, arabinogalactan (larch gum), alginic acid, agar, carragenan, hyaluronic acid, dextran, gellan gum, pullulan, and chitin are reviewed. In particular, various processing methods for tissue engineering scaffolds are discussed including particle leaching, freeze-drying, phase separation, fiber meshes and fiber bonding, melt processing, batch foaming, electrospinning, and rapid prototyping. Different processing methods are proposed for natural macromolecules, mainly because they degrade before melting (no observable melting point) and they are difficult to dissolve in common solvents. It is desirable to obtain repeatable porous structures with controlled porous morphology, and therefore, new innovative processing methods need to be developed.

6.2.5.2 Drug Delivery The use of chitin and chitosan is reviewed by Ravi Kumar et al. (2004). Various forms for pharmaceutical applications include chitosan nanoparticles, microspheres prepared by spray drying, multiple emulsion/solvent evaporation and coacervation, hydrogels, films, fibers, and tablets. Various administration routes including oral, nasal, parenteral, transdermal, implants, ophthalmic preparations, and gene delivery have been proposed with chitin, chitosan and their derivatives carrier materials. Chitosan–PEO (polyethylene oxide) nanoparticles have shown protein entrapment efficiency up to 80%; O-carboxymethylate chitosan (o-CMC) nanoparticles could work for anticancer drugs and enzyme carriers; chitosan beads are applied as cancer chemotherapeutic carrier adriamycin; microbeads for oral delivery of nifedipine, ampicillin and various steroids; chitosan–calcium alginate beads for hemoglobin encapsulation containing a high hemoglobin concentration (up to 90%); encapsulation of dextran and bovine serum albumin in calcium alginate beads coated with chitosan are reported; chitosan/gelatin microspheres for controlled release of cimetidine (Ravi Kumar, 2000b). Chitosan and chitin-based materials were reviewed by Gupta and Ravi Kumar (2000) for the controlled release of several drugs including 5-fluorouracil, diclofenac sodium, cimetidine, hemoglobin, dextran, nifedipine, ampicillin, adriamycin, thiamine hydrochloride, chlorpheniramine maleate, cetylpyridinum chloride, propranolol hydrochloride, theophylline, chlorhexidine, silver sulfadiazine, vitamin A, vitamin E, riboflavin, levamisole, chloramphenicol, and ampicillin trihydrate. The materials discussed ranged from coated chitosan microspheres, chitosan/gelatin network polymer microspheres, chitosan–polyethylene oxide nanoparticles, chitosan/calcium alginate beads, multiporous chitosan beads, and cross-linked chitosan network beads to compressed tablets and hydrogels (chitosan–polyether interpenetrating polymer network hydrogel, semi-interpenetrating network hydrogel polymer networks of β-chitin and polyethylene glycol macromer, polyethylene glycol-co-poly(lactone) diacrylate macromers and β-chitin, polyethylene glycol macromer/β-chitosan, chitosan/gelatin hybrid polymer network, and chitosan-amine oxide gel).

6.2.5.3 Wound Healing Wound healing involves several stages. After hemostasis and clot formation, there are three overlapping phases: inflammation, proliferation, and scar maturation. It is important to heal a skin injury as soon as possible to avoid infections. The objectives of wound care are to prevent an infection, to maintain a humid environment, to protect the wound, and promote minimum scar formation (Muzzarelli et al., 2007; Muzzarelli, 2009). Chitosan was found to promote wound healing due to the biochemical activity for polymorphonuclear cell activation, fibroblast activation, cytokine production, giant cell migration, and stimulation of type IV collagen synthesis. Chitin accelerates macrophage migration and fibroblast proliferation, and promotes granulation and vascularization; and chitosan promotes granulation and organization, for that reason, they are beneficial in wound healing (Muzzarelli et al., 2005, 2007). Chitosan shows inhibition of the activation and expression of matrix metalloproteinases (MMP2)—this group of MMP can hydrolyze collagen IV since MMP2 cleave to collagen type IV and V. Moreover, chitin in the nanometer range (chitin nanofibrils) induced good tissue repair. Prepared as spray, gel, and gauze (freeze-dried chitin microfibrils into dibutyryl chitin) could be used as a first-aid for minor bleeding lesions (spray), for enhanced physiological repair in aesthetically important areas (gel) such as eyes area, and to restore the epidermis without scars. The incorporation of chitin microfibrils showed an improvement over plain dibutyryl chitin nonwoven tissues because the microfibrils helped preventing secondary infection when daily changes are necessary (Muzzarelli et al., 2007).

6.3 APPLICATIONS

6.3.1 Tissue Engineering

Polymer scaffolds with a high biocompatibility, appropriate biodegradability, and interaction with appropriate cells are preferred in engineering tissues and organs. Alternatively, modifying the physical characteristics of the surfaces of the biopolymers by enhancing the hydrophilic properties of surface; changing the pore sizes of the materials, and increasing the roughness of the surface can also enhance the growth of cells in biomaterials (Chung et al., 2006). Further enhancement can be achieved by incorporating bioactive factors and creating nanostructured surface features. For example, Chung et al. (2006) prepared composite films of chitosan over polycaprolactone (PCL) for fibroblasts cell culture. The chitosan surface was modified using a PCL mold previously etched with acetone to yield a nanostructured surface (nano-CTS/PCL). Nanostructured surface was analyzed using atomic force microscopy (AFM) to measure average roughness (Ra) values used to specify the nanostructure of the surface. Results indicate Ra value for the nano-CTS/PCL films is 106 nm, and greatly exceeds that of the CTS/PCL films. Figure 6.2 shows an AFM micrograph of the nano-CTS/PCL surface.

When culturing human dermal fibroblasts over this nanostructured chitosan film, it was observed that nanostructured chitosan promoted better growth of fibroblast cells

Ra : 106.0 ± 0.4 nm

FIGURE 6.2 Atomic force microscopy micrograph of nano-CTS/PCL prepared by Chung et al. (2006). The Ra value and vertical axis scale for nano-CS/PCL film are 106.0 ± 0.4 nm and 250 nm, respectively. The nanostructured surface of nano-CS/PCL films is observed. Reproduced with permission from Artificial Organs from Poly (ε-caprolactone) Grafted with Nanostructured Chitosan Enhances Growth of Human Dermal Fibroblasts. Chung, T.W., Wang, Y. Z., Huang, Y.Y., Pan, C.I. and S.S. Wang, 30, 2006.

than smooth chitosan/PCL. The growth of fibroblasts on nano-CTS/PCL and CTS/PCL films generally greatly exceeded those on PCL and nano-PCL films. Moreover, the growth of cells on the nano-CTS/PCL film was more pronounced than that on CTS/PCL film, but that on the nano-PCL film showed no effect compared with that on the PCL film. However, the growth rate of fibroblasts on the polystyrene culture plates is still better (Chung et al., 2006). This means that there is more to understand, if we find out the reason why growth rate of fibroblasts is better in PS (polystyrene) than in nano-CTS/PCL, we will be able to modify the biocompatible biopolymer to improve the cell growth rate and avoid the use of synthetic PS.

For repairing of the anterior cruciate ligament (ACL), the surgical intervention is necessary which may result in side effects including knee pain and morbidity at the donor site. Tissue engineering scaffold for a less invasive reconstruction is under investigation, but to develop a scaffold with sufficient tensile strength and attachment strength to bone is a big challenge. A possible approach includes the formation of soft tissues around the mid-substance of the scaffold and a direct attachment of the scaffold to the bone with the ingrowth of bony tissue to the scaffold. Hence, the strategy should be to enhance soft tissue formation and bone formation at the same time. Kawai et al. (2010) prepared chitin-coated nonwoven polyester fabric grafts for ACL reconstruction and performed animal experiments by utilizing a rabbit model. The chitin coating enhanced the formation of bone tissue in the femoral bone tunnel and soft tissue in the articular cavity, and increased the attachment strength of the graft to the bone. Thus, the efficacy of the chitin coating for the ACL reconstruction scaffolds was demonstrated. Figure 6.3 shows a scanning electron micrograph of the chitin-coated polyester nonwoven graft.

Fernandes et al. (2011) reported the physico-chemical properties and cytocompatibility (using a combined parameter assay) of collagen-chitosan porous scaffold for tissue repair. The scaffolds were prepared by freeze-drying. Figure 6.4 shows SEM micrographs of the open pore structure of chitosan scaffold with mean pore

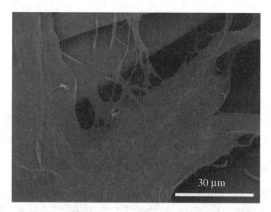

FIGURE 6.3 Scanning electron microscopy observation of chitin-coated polyester nonwoven graft prepared by Kawai et al. (2010). Reproduced with permission from Artificial Organs from Anterior Cruciate Ligament Reconstruction Using Chitin-coated Fabrics in a Rabbit Model. Kawai, T., Yamada, T., Yasukawa, A., Koyama, Y., Muneta, T. and K. Takakuda, 34, 2010.

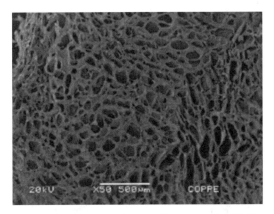

FIGURE 6.4 Scanning electron micrograph of chitosan scaffold at 50× (Fernandes et al., 2011). Reproduced with permission from Polímeros. Cytocompatibility of Chitosan and Collagen-Chitosan Scaffolds for Tissue Engineering. Fernandes, L.L., Resende, C.X., Tavares, D.S., Soares, G.A., Castro, L.O, and J.M. Granjeiro. Copyright 2011.

diameter of 16 μm and porosity of 50%. Figure 6.5 shows SEM micrographs of collagen–chitosan scaffold with mean pore size of 35 μm and porosity of 60%. Addition of collagen increases porosity and pore size and the collagen-chitosan blend become more cytocompatible.

A peripheral nerve is constituted by nervous fibers (axons) of different sizes whose function is to transmit impulses of the central nervous system. The spinal cord emits 36–37 spinal nerve pairs. The sciatic nerve emerges from the major sciatic notch and goes in direction to the limb's knee. Peripheral nerves are capable to regenerate by themselves; however, the functional recovery of a damaged peripheral nerve is still disappointing. Rosales-Cortés et al. (2008) proposed the use of chitosan prosthesis with pregnenolone neurosteroid to evaluate three different dog (French Poodle)

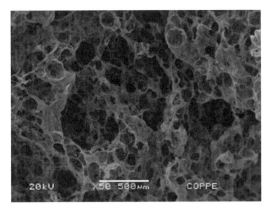

FIGURE 6.5 Scanning electron micrograph of collagen-chitosan scaffold at 50× (Fernandes et al., 2011). Reproduced with permission from Polímeros. Cytocompatibility of Chitosan and Collagen-Chitosan Scaffolds for Tissue Engineering. Fernandes, L.L., Resende, C.X., Tavares, D.S., Soares, G.A., Castro, L.O, and J.M. Granjeiro. Copyright 2011.

groups: (a) control group only axotomized, (b) axotomized and tubulized with chitosan prosthesis, and (c) axotomized and tubulized with chitosan prosthesis and 50 mg of pregnenolone preloaded into the matrix. After 60 days postlesion, chitosan prosthesis had completely disappeared and the nerves stumps were reconnected of 15 mm length. Hence, chitosan prosthesis is a therapeutic alternative for repairing defects of peripheral nerves.

6.3.2 Drug Delivery

Chitin, chitosan and derivatives are been widely studied for controlled drug release applications. Here are some examples for illustration. Cekic et al. (2007) prepared and evaluated microparticles with varying contents of calcium gelling ion and chitosan. These particles were loaded with phenytoin (an antiepileptic drug) in its acidic form, to improve drug release. All studied formulations had very high drug loading up to 90–96%. The formulation improved the irregular absorption that appears to occur in elderly patients. Chitosan particles were not spherical in shape (not shown). Liu et al. (2011) reported the preparation of chitosan microparticles using a novel freeze-drying technique. This technique uses a special microfluidic aerosol nozzle system, which is capable of atomizing the precursor solution into discrete droplets in a single stream. The droplets are then dispersed and dried in a microfluidic-jet spray drier. Figure 6.6 shows chitosan microparticles of great uniformity, the mean diameter varies according to chitosan concentration in the precursor solution. For chitosan solutions of 0.5, 1.0, and 1.5 wt%, mean diameters are 35.1 ± 2.22, 49.2 ± 2.42, and $56.2 \pm 4.59\,\mu m$, respectively. Therefore, microparticles of different sizes can be prepared for targeted drug release, by adjusting chitosan concentration in the precursor solution.

A hydrogel patch containing curcuminoids (a powerful antioxidant, anti-inflammatory, and antityrosinase agent) for application in cosmetics was developed by Boriwanwattanarak et al. (2008). The patch acted as a matrix, and chitosan from squid pen gave the patch with highest strength and flexibility. The release pattern of

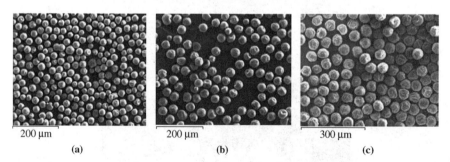

200 μm	200 μm	300 μm
(a)	**(b)**	**(c)**

FIGURE 6.6 SEM photographs and size distribution of chitosan microparticles prepared with different chitosan concentrations: (a) 0.5 wt%; (b) 1.0 wt%; (c) 1.5 wt%. Particles developed by Liu et al. (2011). Reproduced with permission from International Journal of Chemical Engineering. Uniform Chitosan Microparticles Prepared by a Novel Spray-Drying Technique. Liu, W., Wu, W.D., Selomulya, C. and X.D. Chen. Copyright 2011.

curcumin controlled by the patch prepared from the squid fitted well to the Higuchi's model. This implies that the diffusion through the pore is the major mechanism of curcumin release from the hydrogel patch. A porous structure of such patches resulted in a high rate and amount of release of curcumin. It was found that 96% of the initial curcumin content remained in the patch after 6 weeks at 4 °C and 75% RH. No signs of skin irritation were observed after patch application in under-eye area for 30 min.

6.3.3 Bone and Cartilage

For repairing bone and cartilage damages, a biocompatible material is needed that supports the growth and phenotypic expression of osteoblasts and chondrocytes. Chitosan was tested with human osteoblasts and articular chondrocytes. Both appeared spherical and refractile after 7 days of culture, and more than 90% of human osteoblasts and chondrocytes propagated on chitosan remained viable. Figure 6.7 shows a comparison between osteoblasts and chondrocytes on the surface of uncoated coverslips (a, c) and on the surface of chitosan-coated coverslips (b, d). It is evident that chitosan promotes growth and expression. Cells have greater affinity

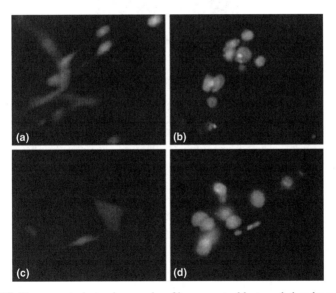

FIGURE 6.7 Fluorescent photomicrographs of human osteoblasts and chondrocytes labeled with live-dead assay (original magnification × 250). Cells analyzed at day 7 of culture: (a) osteoblasts on the surface of uncoated coverslips, (b) osteoblasts on the surface of chitosan-coated coverslips, (c) chondrocytes on the surface of uncoated coverslips, and (d) chondrocytes on the surface of chitosan-coated coverslips from Lahiji et al. (2000). Reprinted from Journal of Biomedical Materials Research. Lahiji, A., Sohrabi, A., Hungerford, D.S. and C.G. Frondoza. Chitosan supports the expression of extracellular matrix proteins in human osteoblasts and chondrocytes. 51:586–595. Copyright (2000) with permission from Wiley.

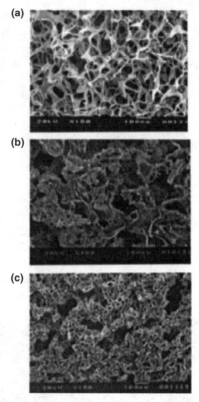

FIGURE 6.8 Scanning electron micrograph of surface of porous matrices. (a) Porous chitosan matrix, (b) PLLA–chitosan composite matrix, and (c) chitosan-coated PLLA matrix developed by Lee et al. (2002). Lee, J-Y., Nam, S-H., Im, S-Y., Park, Y-J., Lee, Y-M., Seol, Y-J., Chung, Ch-P. and S-J. Lee. Enhanced bone formation by controlled growth factor delivery from chitosan-based biomaterials. 187–197. Copyright (2002) with permission from Elsevier.

for films containing higher chitosan concentration, suggesting chitosan as a bio-compatible substrate for cell propagation (Lahiji et al., 2000).

Porous chitosan matrix and poly-L-lactic acid/chitosan (PLLA-CTS) composite porous matrix were developed as bone substitutes and tissue engineering scaffolds by Lee et al. (2002). Figure 6.8 shows differences in the surface micrographs of (upper) porous chitosan, (middle) PLLA-CTS composite matrix, and (lower) chitosan-coated PLLA matrix. Controlled release of platelet-derived growth factor-BB (PDGF-BB) from these matrices exerted significant osteoinductive effect in addition to the high osteoconducting capacity of the porous chitosan matrices. This study found that chitosan-based scaffolds significantly promoted bone healing and regeneration and can be used as a base material for scaffold devices (Lee et al., 2002).

Liji Sobhana et al. (2009) prepared hydroxyapatite nanoparticles onto gelatin–chitosan (GC) capped gold nanoparticles (GC-Au-HA). The novelty of this study lies in the usage of chitosan along with gelatin to act as a matrix for growing hydroxyapatite crystals. With this method, the yield of nano-HA is about 74%. Gold nanoparticles

FIGURE 6.9 SEM photomicrograph showing the porous structure of the hydroxyapatite–chitin matrix reported by Ge et al. (2004). Ge, Z., Baguenard, S., Lim, L.Y., Wee, A. and E. Khor. Hydroxyapatite-chitin materials as potential tissue engineered bone substitutes. 1049–1058. Copyright (2004) with permission of Elsevier.

promote hydroxyapatite growth. This novel composite can be applied in bone tissue repair and regeneration. Furthermore, hydroxyapatite-chitin (HA-CTN) materials were prepared by Ge et al. (2004) for tissue-engineered bone applications. Pore size of the HA-CTN matrix range from 200 to 400 μm with a porosity of 69%; this pore size range is optimum for osteoblast proliferation. Figure 6.9 shows a SEM photography of the porous HA-CTN matrix prepared by freeze-drying.

6.3.4 Scaffolds

Chitosan and chitosan/hydroxyapatite (CTS/HA) scaffolds were developed by freeze-drying for further usage on periodontal bone regeneration (Tigh et al., 2009). Hydroxyapatite beads were successfully incorporated into chitosan structure with pore size of about $\sim$100 μm. In their paper, Tigh et al. (2009) note that chitosan and CTS/HA porous structure and pore size are very similar. More hybrid scaffolds were prepared by Lee et al. (2004) using the freeze-drying and salt-leaching techniques in order to better control porosity and pore size along with mechanical properties. Collagen is added to chitin porous structure to increase cell attachment and proliferation. Figure 6.10 shows a scanning electron microphotography of the surface and cross-section morphologies of the chitin/collagen scaffold. They reported that fibroblasts proliferated and distributed well in the collagen-coated chitin scaffold after cell culture period of 14 days.

Phongying et al. (2007) report that chitosan nanoscaffold can be directly prepared from chitin whisker by reacting with NaOH. The preparation is simple and effective since it is a one-pot deacetylation and yields high amount of the chitosan nanoscaffold, and because no organic solvents or chemicals are involved, this is considered a safe scaffold. They reported a surface area of 55.75 m^2/g for chitosan nanoscaffolds, which is seven times greater than the surface area for chitin flakes (7.70 m^2/g). Average pore diameter is 15.75 nm and pore volume is 0.22 cm^3/g.

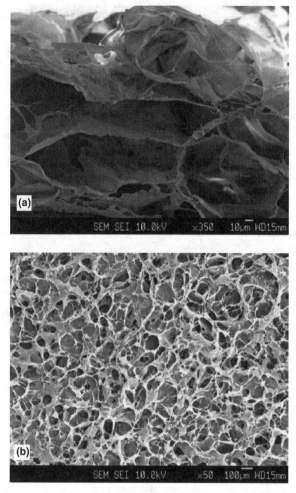

FIGURE 6.10 Morphologies of 0.5 wt% collagen-coated chitin scaffold: (a) cross-section and (b) surface as prepared by Lee et al. (2004). Lee, S.B., Kim, Y.H., Chong, M.S. and Y.M. Lee. Preparation and characteristics of hybrid scaffolds composed of β-chitin and collagen. 2309–2317. Copyright (2004) with permission of Elsevier.

Figure 6.11 shows TEM of chitosan scaffold nanostructure dispersed in water while Figure 6.12 shows SEM of lyophilized chitosan nanoscaffold (Phongying et al., 2007).

   Chitin nanofibers were successfully prepared by Louvier-Hernández et al. (2005) by supercritical carbon dioxide antisolvent method. Chitin was dissolved in hexa-fluoroisopropanol (HFIP) and sprayed into supercritical carbon dioxide (scCO$_2$) media. HFIP is miscible with scCO$_2$ and, because of the very rapid diffusion process, there is a fast precipitation of chitin in fiber form. The macrostructure of fibers with diameters about 50–80 μm is related to the nozzle diameter of 100 μm used for the solution injection into the scCO$_2$ media. Figure 6.13 shows the nanostructure of the fibers. Nanofibers with an average diameter of 84 nm forming a fiber web give structure to the microfiber. These materials have the potential for use in scaffolds for

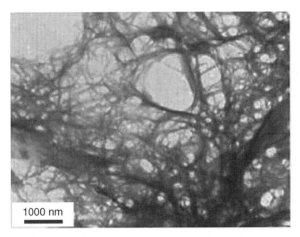

FIGURE 6.11 TEM micrograph of chitosan nanoscaffold in water as reported by Phongying et al. (2007).

tissue engineering or as wound dressing material. High surface area may promote wound healing and cell expression and differentiation. Moreover, the obtained chitin nanofibers are solvent-free and already sterilized due to processing in $scCO_2$.

Different kinds of scaffolds have been developed for different applications. Another example of these nanostructured fibers is the ultrafine fibrillar scaffold prepared by Skotak et al. (2008). L-Lactide oligomers were grafted into chitosan backbone to modify biodegradation rate and hydrophilicity. Using ethyl acetate and 2-butanone as solvents, electrospun fibers were prepared. Fiber diameters of about 3–5 μm were obtained. Figure 6.14 shows the morphology of the ultrafine

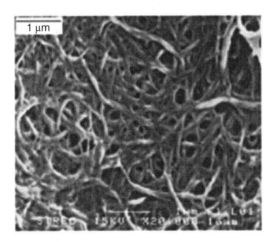

FIGURE 6.12 SEM micrograph of lyophilized chitosan nanoscaffold prepared by Phongying et al. (2007). Pore diameter of the nanostructure is ∼200 nm. Reprinted from Polymer. 48. Phongying, S., Aiba, S. and S. Chirachanchai. Direct chitosan nanoscaffold formation via chitin whiskers. 393–400. Copyright (2007) with permission from Elsevier.

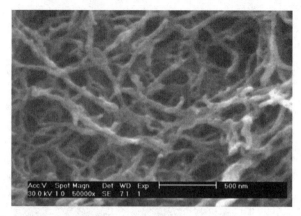

FIGURE 6.13 SEM micrographs of chitin fibers by supercritical antisolvent process at 50,000× (Louvier-Hernández, 2006). Adapted from Louvier-Hernández, J.F. 2006. Chitin nano-fibers processed by supercritical carbon dioxide and thermal properties of chitosan films. Doctoral Dissertation. Center for Research and Advanced Studies of the National Polytechnic Institute. Querétaro, México.

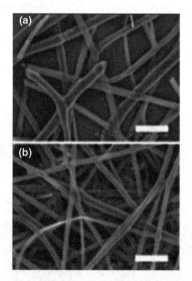

FIGURE 6.14 SEM images of electrospun fibers using solutions of 44% (w/w) high molecular weight chitosan-grafted poly-L-lactide in ethyl acetate. Samples with particular original chitosan to L-lactide ratios were used: (a) 1:6; (b) 1:12. Scale bar: 20 μm (Skotak et al., 2008). Reprinted from Biomacromolecules. Skotak, M., Leonov, A.P., Larsen, G., Noriega, S. and A. Subramanian. Biocompatible and Biodegradable Ultrafine Fibrillar Scaffold Materials for Tissue Engineering by Facile Grafting of L-Lactide onto Chitosan. 9:1902–1908. Copyright 2008. American Chemical Socitey.

fibers chitosan-poly-L-Lactide (CTS-PLA). The specimen with the highest molar ratio of L-lactide showed the most number of surviving fibroblasts cells.

Different poly-L-lactide side chain could fine-tune the hydrophobicity and bioabsorbability of the material.

6.4 COMMERCIAL PRODUCTS

From the scientific work done so far it is evident that chitin and chitosan-based materials are fine biopolymers that can be used for a wide variety of biomedical applications. Ten years ago, very few chitin-based commercial wound dressings were available. Nowadays, there are at least 15 products derived from chitin and chitosan with available literature in the World Wide Web. To mention some of them, Beschitin® from Unitika (Japan) is the wound dressing made from highly purified chitin extracted from the crab. Chitopack C® is a chitosan wound dressing and Chitipack® is an orthopedic product for veterinary use; both from Eisai Co. Ltd. (Japan). Chitopoly® is a registered trademark used for woven fabrics, loop knit fabrics, and nonwoven cloth, all used in the further manufacture of clothing and owned by Fuji Spinning Co., Ltd. (Japan). SyvekExcel® Patch: a 3 cm × 4 cm lyophilized pad of poly-N-acetylgluco-samine with foam backing; SyvekNT® is a soft, white, sterile nonwoven pad of a cellulosic polymer isolated from microalgae (poly-N-acetylglucosamine), which is attached to medical grade foam backing. mRDH Bandage™ is a trauma dressing intended for the temporary control of severely bleeding wounds such as surgical wounds (operative, postoperative, donor sites, dermatological, etc.) and traumatic injuries. Syvek and mRDH Bandage are from Marine Polymer Technologies Inc. (USA). Protasan™ ultrapure chitosan salts and bases are manufactured in the specially designed NovaMatrix production facility in Sandvika, Norway. Protasan chitosans are ideal for a wide variety of pharmaceutical, biomedical, biotechnology, and tissue engineering applications and are sold in different deacetylation degrees and molecular weights. Protasan is from NovaMatrix Ultrapure Polymer Systems a business unit of FMC Biopolymer (Norway). Kitomer™ chitosan from Marinard Biotech Inc. (Canada) used for cosmetic, agriculture, and food applications. Crabyon™ is a composite fiber of chitin and cellulose produced from a chitin/chitosan solution coextruded with cellulose viscose into a spin bath. This textile is antibacterial and excellent for sensitive skin, and is produced by Tec Service (Italy). The hydrocolloid dressing, Tegasorb from 3M Healthcare, is a polyurethane film coated with a layer of an acrylic adhesive and chitosan particles dispersed in polyisobutylene (USA). BioSyn-tech Inc. develops biotherapeutic themogels for tissue repair; it manufactures ultra pure chitosans under the Ultrasan® brand (Canada). HemCon Medical Technologies Inc. (USA) offer an antibacterial barrier against a wide range of gram positive and gram negative organisms with ChitoFlex hemostatic dressing, ChitoGauze and Guardacare highly flexible chitosan-coated hemostatic gauze dressings.

Behind all these commercially available products, there is much research done. Almost all of these products derive from scientific research papers (Khor, 2010). It is our belief that in the near future more chitin and chitosan-based biomaterials will be marketed, mainly because of the biocompatibility and nontoxicity of these polysaccharides.

6.5 SUMMARY

Chitin and chitosan are important biomaterials for a wide range of applications due to their unique properties including biodegradability, biocompatibility, hemostasis, antimicrobial activity, and so on. Chitin is produced from the processing waste of shellfish, krill, clams, oysters, and fungi and is the second most abundant polysaccharide in nature after cellulose. The physicochemical properties can vary based on the source material. Deacetylation of chitin results in chitosan that can be processed much more easily due to enhanced solubility in common organic acids. The peculiar biochemical properties of chitins and chitosans remain unmatched by other polysaccharides. Chitosan-based materials have been tested for various applications including tissue engineering, drug delivery, wound healing, bone and cartilage reconstruction, and in scaffolds. There are a number of commercial products in the market that utilize composites of chitosan. However, there is much work needed for these materials for a wider utilization.

REFERENCES

Abdou, E.S., K.S.A. Nagy, and M.Z. Elsabee. (2008). Extraction and characterization of chitin and chitosan from local sources. *Bioresource Technol.* **99**:1359–1367.

Bama, P., M. Vijayalakshimi, R. Jayasimman, P.T. Kalaichelvan, M. Deccaraman, and S. Sankaranarayanan. (2010). Extraction of collagen from cat fish (*Tachysurus maculatus*) by pepsin digestion and preparation and characterization of collagen chitosan sheet. *Int. J. Pharm. Pharm. Sci.* **2**:133–137.

Boriwanwattanarak, P., K. Ingkaninan, N. Khorana, and J. Viyoch, (2008). Development of curcuminoids hydrogel patch using chitosan from various sources as controlled-release matrix. *Int. J. Cosmetic Sci.* **30**:205–218.

Cauchie, H.-M. (2002). Chitin production by arthropods in the hydrosphere. *Hydrobiologia* **470**:63–96.

Cekic, N.D., S.D. Savic, J. Milic, M.M. Savic, Z. Jovic, and M. Malesevic. (2007). Preparation and characterisation of phenytoin-loaded alginate and alginate-chitosan microparticles. *Drug. Deliv.* **14**:483–490.

Chung, T.W., Y.Z. Wang, Y.Y. Huang, C.I. Pan, and S.S. Wang. (2006). Poly (ε-caprolactone) grafted with nano-structured chitosan enhances growth of human dermal fibroblasts. *Artif. Organs* **30**:35–41.

Dutta, J., and P.K. Dutta. (2010). Antimicrobial activity of chitin, chitosan, and their oligosaccharides. In: S-K. Kim, editor, *Chitin, Chitosan, Oligosaccharides and Their Derivatives. Biological Activities and Applications*, Boca Raton: CRC Press, pp. 195–214.

Fernandes, L.L., C.X. Resende, D.S. Tavares, G.A. Soares, L.O. Castro, and J.M. Granjeiro. 2011. Cytocompatibility of chitosan and collagen-chitosan scaffolds for tissue engineering. *Polímeros* **21**:1-6.

Francis Suh, J.-K., and H.W.T. Matthew. (2000). Application of chitosan-based polysaccharide biomaterials in cartilage tissue engineering: a review. *Biomaterials* **21**:2589–2598.

Ge, Z., S. Baguenard, L.Y. Lim, A. Wee, and E. Khor. (2004). Hydroxyapatite-chitin materials as potential tissue engineered bone substitutes. *Biomaterials* **25**:1049–1058.

Gupta, K.C., and M.N.V. Ravi Kumar. (2000). An overview on chitin and chitosan applications with an emphasis on controlled drug release formulations. *J. Macromol. Sci. R. M.* **C40**:273–308.

Jeuniaux, C., and M. F. Voss-Foucart. (1991). Chitin biomass and production in the marine environment. *Biochem. Syst. Ecol.* **19**:347–356.

Kawai, T., T. Yamada, A. Yasukawa, Y. Koyama, T. Muneta, and K. Takakuda. (2010). Anterior cruciate ligament reconstruction using chitin-coated fabrics in a rabbit model. *Artif. Organs* **34**:55–64.

Khor, E. (2001). *Chitin: fulfilling a biomaterials promise*. Amsterdam: Elsevier.

Khor, E., and L.Y. Lim. (2003). Implantable applications of chitin and chitosan. *Biomaterials* **24**:2339–2349.

Khor, E. (2010). Medical applications of chitin and chitosan: going forward. In: S-K. Kim, editor, *Chitin, Chitosan, Oligosaccharides and Their Derivatives. Biological Activities and Applications*, Boca Raton: CRC Press. pp. 405–413.

Khoushab, F. and M. Yamabhai. (2010). Chitin research revisited. *Mar. Drugs* **8**:1988–2012.

Kim, M.S., H.J. You, M.K. You, N.S. Kim, B.S. Shim, and H.M. Kim. (2004). Inhibitory effect of water-soluble chitosan on TNF-α and IL-8 secretion from HMC-1 cells. *Immunopharm. Immunot.* **26**:401–409.

Kim, M.M., and S.K. Kim. (2010). Anti-inflammatory activity of chitin, chitosan, and their derivatives. In: S-K. Kim, editor, *Chitin, Chitosan, Oligosaccharides and Their Derivatives. Biological Activities and Applications*, Boca Raton: CRC Press. pp. 215–221.

Kim, S.K., and W.K. Jung. (2010). Effects of chitin, chitosan, and their derivatives on human hemostasis. In: S.-K. Kim, editor, *Chitin, Chitosan, Oligosaccharides and Their Derivatives. Biological Activities and Applications*, Boca Raton: CRC Press pp. 251–262.

Kjartansson, G.T., S. Zivanovic, K. Kristbergsson, and J. Weiss. (2006). Sonication-assisted extraction of chitin from North Atlantic shrimps (*Pandalus borealis*). *J. Agric. Food Chem.* **54**:5894–5902.

Kurita, K., S. Hashimoto, H. Yoshino, S. Ishii, and S.I. Nishimura. (1996). Preparation of chitin/polystyrene hybrid materials by efficient graft copolymerization based on mercaptochitin. *Macromolecules* **29**:1939–1942.

Kurita, K. (2001). Controlled functionalization of the polysaccharide chitin. *Prog. Polym. Sci.* **26**:1921–1971.

Lahiji, A., A. Sohrabi, D.S. Hungerford, and C.G. Frondoza. (2000). Chitosan supports the expression of extracellular matrix proteins in human osteoblasts and chondrocytes. *J. Biomed. Mater. Res.* **51**:586–595.

Lee, J.-Y., S.-H. Nam, S.-Y. Im, Y.-J. Park, Y.-M. Lee, Y.-J. Seol, Ch.-P. Chung, and S.-J. Lee. (2002). Enhanced bone formation by controlled growth factor delivery from chitosan-based biomaterials. *J. Control. Release* **78**:187–197.

Lee, S.B., Y.H. Kim, M.S. Chong, and Y.M. Lee. (2004). Preparation and characteristics of hybrid scaffolds composed of β-chitin and collagen. *Biomaterials* **25**:2309–2317.

Liji Sobhana, S.S., J. Sundaraseelan, S. Sekar, T.P. Sastry, and A.B. Mandal. (2009). Gelatin-chitosan composite capped gold nanoparticles: a matrix for the growth of hydroxyapatite. *J. Nanopart. Res.* **11**:333–340.

Liu, W., W.D. Wu, C. Selomulya, and X.D. Chen. (2011). Uniform chitosan microparticles prepared by a novel spray-drying technique. *Int. J. Chem. Eng.* **2011**:267218.

Louvier-Hernández, J.F., G. Luna-Bárcenas, R. Thakur, and R.B. Gupta. (2005). Formation of chitin nanofibers by supercritical antisolvent. *J. Biomed. Nanotechnol.* **1**:109–114.

Louvier-Hernández, J.F. (2006). Chitin nano-fibers processed by supercritical carbon dioxide and thermal properties of chitosan films. Doctoral Dissertation, Center for Research and Advanced Studies of the National Polytechnic Institute, Querétaro, México.

Mano, J.F., G.A. Silva, H.S. Azevedo, P.B. Malafaya, R.A. Sousa, S.S. Silva, L.F. Boesel, J.M. Oliveira, T.C. Santos, A.P. Marques, N.M. Neves, and R.L. Reis. (2007). Natural origin biodegradable systems in tissue engineering and regenerative medicine: present status and some moving trends. *J. R. Soc. Interface* **4**:999–1030.

Muzzarelli, C., and R.A.A. Muzzarelli. (2002). Natural and artificial chitosan-inorganic composites. *J. Inorg. Biochem.* **92**:89–94.

Muzzarelli, R.A.A., M. Guerrieri, G. Goteri, C. Muzzarelli, T. Armeni, R. Ghiselli, and M. Cornelissen. (2005). The compatibility of dibutyryl chitin in the contexts of wound dressings. *Biomaterials* **26**:5844–5854.

Muzzarelli, R.A.A., P. Morganti, G. Morganti, P. Palombo, M. Palombo, G. Biagini, M.M. Belmonte, F. Giantomassi, F. Orlandi, and C. Muzzarelli. (2007). Chitin nanofibrils/chitosan glycolate composites as wound medicaments. *Carbohydr. Polym.* **70**:274–284.

Muzzarelli, R.A.A. (2009). Chitins and chitosans for the repair of wounded skin, nerve, cartilage and bone. *Carbohydr. Polym.* **76**:167–182.

Onishi, H., H. Takahashi, M. Yoshiyasu, and Y. Machida. (2001). Preparation and *in vitro* properties of N-succinylchitosan—or carboxymethylchitin—Mitomicin C conjugate microparticles with specified size. *Drug Dev. Ind. Pharm.* **27**:659–667.

Pavinatto, F.J., L. Caseli, and O.N. Oliveira, Jr., (2010). Chitosan in nanostructured thin films. *Biomacromolecules* **11**:1897–1908.

Peniche, C., Y. Solís, N. Davidenko, and R. García. (2010). Chitosan/hydroxyapatite based composites. *Biotecnología Aplicada* **27**:202–210.

Phongying, S., S. Aiba, and S. Chirachanchai. (2007). Direct chitosan nanoscaffold formation via chitin whiskers. *Polymer* **48**:393–400.

Prajapati, B. (2009). Chitosan a marine medical polymer and its lipid lowering capacity. *Internet J. Health* **9**.

Ravi Kumar, M.N.V. (2000a). A review of chitin and chitosan applications. *React. Funct. Polym.* **46**:1–27.

Ravi Kumar, M.N.V. (2000b). Nano and microparticles as controlled drug delivery devices. *J. Pharm. Pharmaceut. Sci.* **3**:234–258.

Ravi Kumar, M.N.V., R.A.A. Muzzarelli, C. Muzzarelli, H. Sashiwa, and A. J. Domb. (2004). Chitosan chemistry and pharmaceutical perspectives. *Chem. Rev.* **104**:6017–6084.

Rinaudo, M. (2008). Main properties and current applications of some polysaccharides as biomaterials. *Polym. Int.* **57**:397–430.

Roberts, G.A.F. (1992). *Chitin Chemistry*. Indianapolis, USA: MacMillan

Rosales-Cortés M., J. Peregrina-Sandoval, J. Hernández-Mercado, G. Nolasco-Rodríguez, M.E. Chávez-Delgado, U. Gómez-Pinedo, and E. Albarrán-Rodríguez. (2008). Regeneration of the dog's axotomized sciatic nerve with pregnenolone-saturated chitosan prosthesis implanted through the tubulization technique. *Vet. Méx.* **39**:55–66.

Saboktakin, M.R., R.M. Tabatabaie, A. Maharramov, and M.A. Ramazanov. (2010). Synthesis and characterization of chitosan-carboxymethyl starch hydrogels as nano carriers for colon-specific drug delivery. *J. Pharm. Educ. Res.* **1**:37–47.

Saburo, M., M. Minoru, O. Yoshiharu, S. Hiroyuki, and S. Yoshihiro. (2006). Biomedical materials from chitin and chitosan. In *Material Science of Chitin and Chitosan*. editors, T. Uragami, and S. Tokura. Japan: Springer pp. 191–218.

Sagheer, F.A., M.A. Al-Sughayer, S. Muslim, and M.Z. Elsabee. (2009). Extraction and characterization of chitin and chitosan from marine sources in Arabian Gulf. *Carbohydr. Polym.* **77**:410–419.

Sandford, P.A. (2003). Commercial sources of chitin and chitosan and their utilization. In: *Advances in Chitin Sciences*, **Vol. 6** editors, K. M. Vårum, A. Domard, and O. Smidsrød, Trondheim, Norway: Norwegian University of Science and Technology.

Shaidi, F., J.K.V. Arachchi, and Y.J. Jeon. (1999). Food application of chitin and chitosans. *Trends Food Sci. Technol.* **10**:37–51.

Skotak, M., A.P. Leonov, G. Larsen, S. Noriega, and A. Subramanian. (2008). Biocompatible and biodegradable ultrafine fibrillar scaffold materials for tissue engineering by facile grafting of L-lactide onto chitosan. *Biomacromolecules* **9**:1902–1908.

Tokura, S. and H. Tamura. (2001). O-Carboxymethyl-chitin concentration in granulocytes during bone repair. *Biomacromolecules* **2**:417–421.

Tigh, R.S., A.C. Akman, M. Gümüsderelioglu, and R.M. Nohutcu. (2009). *In vitro* release of dexamethasone or bFGF from chitosan/hydroxyapatite scaffolds. *J. Biomater. Sci.* **20**:1899–1914.

Vacanti, J. P. and R. Langer. (1999). Tissue engineering: the design and fabrication of living replacement devices for surgical reconstruction and transplantation. *Lancet* **354(SI)**:32–34.

Vikhoreva, G.A., L.S. Gal'braikh, I.N. Gorbacheva, and A.O. Chernyshenko. (2005). Advances in the Department of Chemical Fibre Technology in modification of chitin and chitosan. *Fibre Chem.* **37**:431–436.

Watthanaphanit, A., P. Supaphol, H. Tamura, S. Tokura, and R. Rujiravanit. (2010). Wet-spun alginate/chitosan whiskers nanocomposites fibers: preparation, characterization and release characteristic of the whiskers. *Carbohydr. Polym.* **79**:738–746.

Zhao, Y., W.T. Ju, and R. D. Park. (2010). Enzymatic modifications of chitin and chitosan. In: S-K. Kim, editor, *Chitin, Chitosan, Oligosaccharides and Their Derivatives. Biological Activities and Applications*, Boca Raton: CRC Press pp. 185–192.

Zhou, H.Y., X.G. Chen, C.S. Liu, X.H. Meng. C.G Liu, J. He, and L.J. Yu. (2006). Cellulose acetate/chitosan multimicrospheres preparation and ranitidine hydrochloride release *in vitro*. *Drug Deliv.* **13**:261–267.

WEB SITES

Unitika Ltd. web site (2011) Available at http://www.unitika.co.jp/e/products/hlth-bis/medical.html

Eisai Co. Ltd. web site (2011) Available at http://www.eisai.com/index.html

Syvek web site. Available at http://www.syvek.com/index.html

Novamatrix web site (2010) Available at http://www.novamatrix.biz/From FMC Corporation www.fmc.com

Marinard Biotech web site. Available at http://www.marinard.com/biotech/company.html

HemCon web site. Available at http://www.hemcon.com/Home.aspx

3M Skin and wound care web site (2011) Available at http://solutions.3m.com/wps/portal/3M/ en_US/3MSWC/Skin-Wound-Care/

7

CHITOSAN DERIVATIVES FOR BIOADHESIVE/HEMOSTATIC APPLICATIONS: CHEMICAL AND BIOLOGICAL ASPECTS

Mai Yamazaki and Samuel M. Hudson

7.1 INTRODUCTION

Recent advances in tissue engineering have been directed toward solving problems of patients who have suffered tissue/organ loss or skeletal defects (Barrera et al., 1993; Hu et al., 2003; Muzzarelli, 2009). Many natural, synthetic, or their hybrid matrices have been developed to cover wound sites to replace lost tissue functions and support cell growth. For example, aliphatic polyesters such as poly(lactic acid) and poly (glycolic acid) are versatile biomaterials due to their biodegradability and biocompatibility (Barrera et al., 1993; Hu et al., 2003; Moon et al., 2000, 2001; Shinoda et al., 2003). Polyurethane is also a well-known biomaterial with biocompatibility, and the mechanical and physical properties necessary for a blood-contacting material (Hsu and Chen, 2000; Lin et al., 1992; Takahashi et al., 2002).

The synthetic matrices have many advantages because their molecular designs, and mechanical or physical properties can be controlled and they can be manufactured on any scale. However, the usage of these synthetic scaffolds is still limited because of poor cell attachment/growth, adsorption of untargeted proteins, and induction of thrombogenesis on the surface. Furthermore, some of the synthetic polymers, including polyesters, are difficult to modify due to the lack of enough reactive functional groups. On the other hand, natural substrates such as adhesive

Polysaccharide Building Blocks: A Sustainable Approach to the Development of Renewable Biomaterials, First Edition. Edited by Youssef Habibi and Lucian A. Lucia.
© 2012 John Wiley & Sons, Inc. Published 2012 by John Wiley & Sons, Inc.

proteins have been extensively used because of excellent biocompatibility and bioactivity; however, they have batch-to-batch variations, and difficulties establishing large-scale processes. Many attempts for establishing novel biomedical applications have been studied by modification or combination of natural polymers including proteins (e.g., collagen (Wang et al., 2003a, 2006), silk fibroin (Hirano et al., 1999)) and polysaccharides (e.g., cellulose (Ishihara et al., 1992), hyaluronan (Yamane et al., 2005; Kukolikova et al., 2006), alginate (Knill et al., 2004; Remuñán-López and Bodmeier, 1997), chitin, and chitosan (Mori et al., 1997)).

As regard to polysaccharides, chitin and chitosan have been widely studied for use as biodegradable and biocompatible materials. Chitin is a poly-β(1 → 4)-N-acetyl-D-glucosamine which is known as a cell wall component of fungi and as a skeletal component of crustacea. The commercial source of chitin is mostly from crab, shrimp, and krill shells and fungi, of which large amounts are wasted in the food industry (Kumar and Hudson, 2004; Meanwell and Shama, 2006). Several groups have studied how to extract chitin and produce chitosan from these resources (Wang et al., 2006; Teng et al., 2001). A flow chart of chitin production is shown in Figure 7.1. Chitosan is a cationic poly-β(1 → 4)-2-amino-2-deoxy-D-glucose, obtained by deacetylation of chitin. When the degree of deacetylation of chitin is

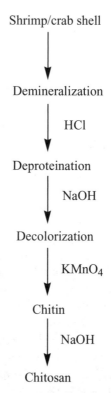

FIGURE 7.1 A flow chart of chitin and chitosan production.

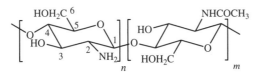

FIGURE 7.2 The chemical structure of chitosan. Chitosan is the copolymer with $n > m$.

more than 50%, it is generally considered as chitosan. The chemical structure of chitosan is described in Figure 7.2. Although chitin and chitosan have potential as resources for commercial use, both chitin and chitosan are inherently insoluble in water, which makes these materials difficult to process. Only very strong acidic conditions, such as with formic acid, di- or trichloroacetic acids, methansulfonic acid, and lithium chloride/amide, can solubilize chitin (Lin and Lin, 2003). Chitosan is soluble in milder acids such as aqueous acetic acid. Their insolubility is usually attributed to intermolecular hydrogen bonding in the solid state (Whang et al., 2005).

There are many examples of chemical modification and salt formation to make chitosan soluble in water or organic solvent systems (Muzzarelli and Ilari, 1994; Signini and Campana Filho, 1999; Kubota et al., 2000; Chen and Park, 2003; Ramos et al., 2003; Baumann and Faust, 2001; Kurita et al., 2002). *N*- or *O*- substitutions by various moieties such as the carboxymethyl group, are typical chemical modifications. However, studies for unmodified chitosan dissolution, almost all have reported molecular weight lowering of chitosan (Muzzarelli and Ilari, 1994; Signini and Campana Filho, 1999; Kubota et al., 2000). As a result, chitosan has been developed for a wide range of applications such as water clarification, flocculants, cosmetic and pharmaceutical uses, and biomedical devices (Table 7.1).

The biocompatibility, antigenicity, and bioadhesiveness of chitosan has led to extensive studies of biomedical applications including antibacterial activities and hemostatic agents for wound dressings. It has been reported that chitosan has an antibacterial nature and greater advantages over the other commercially available hemostatics (Whang et al., 2005; Neuffer, 2004; Alam et al., 2005). However, the biological effect of chitosan itself is still not fully understood in detail, because the chemical and physical properties of chitosan always depend on the molecular weight, degree of acetylation, degree of crystallinity, extent of ionization/free amino group, and so on (Whang et al., 2005). The optimum conditions for chitosan processing are wide-ranging. Although various types of hemostatic agents of chitosan and its derivatives have been suggested, the evaluation method for hemostatic properties is varied with each research group and has not completely been standardized among researchers. Although various preparation conditions of chitosan (e.g., molecular weight, pH, degree of deacetylation, viscosity, and solubility) have been examined chemically and physically, their results are not always reflected on the hemostatic applications and few reports have been published about the relationship between the chitosan condition information and their effectiveness as biological agents.

In this review, some properties of chitosan will first be described such as biocompatibility and cytotoxicity that effect applications such as scaffolds and

TABLE 7.1   Various Applications of Chitosan

Field	Application
Healthcare	Dentistry
	Contact lens/eye bandages
	Wound healing dressings
	Orthopedics
	Anticholestrol and fat binding
	Surgical sutures
	Hemostatics/bioadhesive
	Drug delivery
	Ophthalmology
	Transportation of cells
Food and beverages	Food stabilizer
	Flavor and tastes
	Food packaging
	Nutritional additives
	Fruit preservation
Agriculture	Seed treatments (coatings)
	Feed ingredients (animal feed)
	Nematocides and insecticides
Cosmetics and toiletries (personal care)	Hair treatment
	Skin care
	Oral care
Waste and water treatment (clarification)	Sewage effluents
	Drinking water
	Metal recovery
	Pools and spas
	Treating food waste (food processor waste water for protein recovery)
Product separation and recovery (bioapplications)	Membrane separations
	Chromatographic matrix
	Immobilization of enzymes and cells
	Recovery of bioproducts

drug delivery systems. Then, the use of chitosan and its derivatives for hemostatic applications and blood contacting test methods for clinical use will be discussed. This will demonstrate the interactions of these properties of chitosan for these applications.

7.2 BIOCOMPATIBILITY AND CYTOTOXICITY OF CHITOSAN-BASED MATRICES

One of the important aspects for a biocompatible material is the interaction between cells and material surfaces. The material surface is required to act as an artificial extracellular matrix (ECM) with a three-dimensional structure, which is vital for cell cycles (Seo et al., 2006; Peng et al., 2006). Most cell types cannot be viable without attachment to the scaffold otherwise apoptosis (programmed cell death) might be induced. Since these materials exist as extracellular matrices such as proteoglycans *in vivo*, many attempts for preparing artificial ECMs composed of these natural materials, have been reported to compensate for the shortage of tissue scaffolds, which could prompt cell proliferation, differentiation, and regeneration of the tissue functions. For instance, a highly porous scaffold made of chitosan derivative has been developed for a bioartificial liver device (Seo et al., 2006). It is well known that polysaccharides called glycosaminoglycans, control cell functions as well as bioadhesive proteins including collagen, fibronectin, laminine, and so on. On the other hand, natural hydrophilic polymers have also been identified as drug delivery carriers because of the potentially safe, sustainable, and controllable release of drugs. To characterize and evaluate these man-made matrices from the viewpoint of biocompatibility and cytotoxicity would be the primary task necessary to apply these matrices for tissue engineering. Such chitosan-based scaffolds for tissue engineering and capsules for drug targeting applications in different cell types will be described in Section 7.2.1.

7.2.1 Scaffolds for Tissue Engineering

Any matrices used for tissue engineering are required to support the cells and restore and improve the cell function, including induction of the cell-specific cytokine and gene expressions. It is reported that a chitosan supplement to the culture media accelerates the production of cytokines called IL-8, which are the marker cytokines of angiogenesis even though cell proliferation of mouse fibroblast (L929) was not significant (Mori et al., 1997). Other authors suggested that chitosan mediates the rat peritoneal exudates macrophage (PEM) activation for immune stimulation, which is confirmed by nitric oxide (NO) secretion (Peluso et al., 1994). Chitosan oligomers promote cell migration and proliferation of fibroblast (mouse fetal fibroblast; 3T6 cells) and vascular endothelium (HUVEC) even more (Figure 7.3, (Okamoto et al., 2002)).

However, chitosan is a highly thrombogenic material, so it is suggested for use as wound dressings rather than supplements, because of the formation of granulation on the tissue (Wang et al., 2003a, 2003b; Okamoto et al., 2002). Similarly, collagen, one of the structural proteins, is recognized as an appropriate tissue-culturing scaffold but induces thrombogenic features such as platelet aggregation and blood coagulation on the surface (Wang et al., 2003b). They concluded that a cross-linked collagen/chitosan multilayer matrix with few free carboxyl and amino groups contributed to decrease platelet adhesion but promote hepatocyte adhesion due to the remaining free amino groups of chitosan. However, there is no quantitative information about the remaining amino groups and the relationship between free carboxylic or amino group and the anticoagulant effect is still unclear.

Although chitosan itself has an affinity to many types of cells, it is anticipated that cell proliferation, growth, and differentiation can be further improved by chemically modifying the amino and hydroxyl groups of chitosan. Several peptide

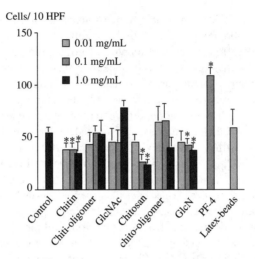

FIGURE 7.3 Direct affects of chitin, chitosan, and their derivatives on migration activity of 3T6 cells (Peng et al., 2006). Control: culture medium, GluNAc: N-acetyl-D-glucosamine, GlcN: D-glucosamine, chiti-oligomer: a mixture of GlcNAc1 and GlcNAc6, chito-oligomer: a mixture of GlcN1 and GlcN6, PF-4: human recombinant platelet factor-4. Data displays means $\pm$ SD for at least three replicates. $*p < 0.05$ from control.

sequences that play important roles for cell behaviors have been identified and RGD (Arg-Gly-Asp) and its analogs are particularly the most well-known adhesive peptide sequences (Lin et al., 1994; Li et al., 2006). To enhance the rat osteosarcoma cell (ROC) adhesion, C-terminal of RGDS peptide was covalently bonded to the amino group of chitosan by a water-soluble carbodiimide (Ho et al., 2005). There was no information how the amino group on the N-terminal peptide was protected when the immobilization was carried out. Chitosan-modified film and porous scaffolds were prepared by casting (for film) and lyophilization (for sponge). The morphology of the sponge was observed by SEM (Figure 7.4). They concluded that the peptide density (1×10^{-12} mol/cm^2) on the scaffolds would be sufficient to support the cell adhesion process and RGDS-modified chitosan improved the cell attachment and mineralization, which is a typical differentiation phenomenon for osteoblastic-like cells.

The peptide GRGD was photochemically grafted to chitosan film using a photoreactive spacer called SANPAH with an azide group (Chung et al., 2002, 2003). The film was further cross-linked with tripolyphosphate to enhance the adhesion and viability of human endothelial cells. The authors concluded that the peptide-grafted chitosan and its cross-linked chitosan improved the endothelial cell growth. A photochemical reaction may have also occurred between the azide group of the peptide-SANPAH fragment and hydroxyl group of chitosan as described in Figure 7.5. However, in general, the azide group was converted to a nitrene group. The nitrene reacts with the amino group, especially if the substrate contains a primary amine group. Furthermore, chitosan has two hydroxyl groups at the 3- and 6-positions

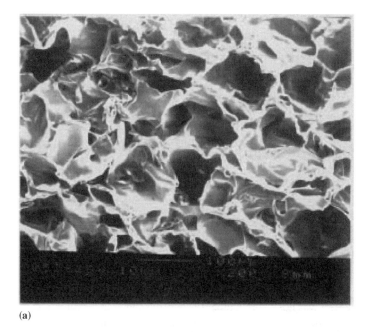

(a)

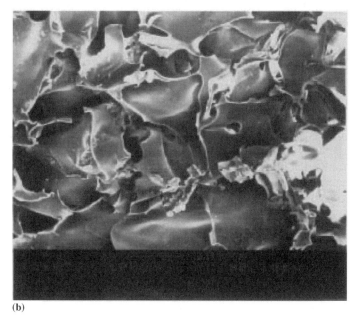

(b)

FIGURE 7.4 Morphology of the chitosan scaffold prepared by lyophilization (SEM, ×200) (Okamoto et al., 2002). (a) Before RGDS immobilization, (b) after RGDS immobilization.

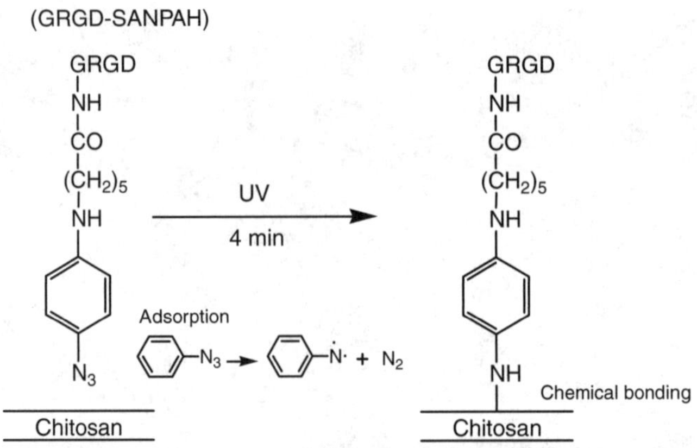

FIGURE 7.5 A synthetic scheme of photochemical immobilization of GRGD to chitosan.

in the structure (Figure 7.2) but no description was provided as to which hydroxyl group was most likely reacted with the cross-linker and to what extent the hydroxyl group was reacted. Generally, the primary alcohol at C-6 is considered more reactive in many cases. It was not clear whether the peptide and amino group of chitosan were indeed photochemically immobilized. Therefore, their interpretations of the data potentially include inappropriate conclusions. Further work is needed to improve the scaffolds for tissue engineering because polystyrene culture plates, widely used for *in vitro* studies, showed better cell growth than these scaffolds.

Bioadhesive hydrogel also provides a suitable environment for tissue adhesion due to its softness, because tissue has a rough surface and migrates onto such matrices (Zhao et al., 2001). In this work, deacetylated and *O*-carboxymethylated chitin, which resulted in molecular weight lowering after the modifications were conducted. The polyampholytic hydrogel was formed by mixing these solutions in the presence of glutaraldehyde as a cross-linking agent. An adhesion test was conducted to evaluate the interaction between the hydrogel and porcine dermis tissue *ex vivo*. They revealed that higher concentration (4 wt%, deacetylated chitin:carboxymethylated chitin = 95:5) of the polymer in the hydrogel possessed a lower water content, which induces higher adhesion strength between the tissue and hydrogel. The formation of ionic physical interactions between each polymer chain, contributed to the decreasing water content of the hydrogel. The hydrogel toughness improved with increasing polymer volume fraction, which enhanced the tissue adhesion because of the high binding energy of the elastic hydrogel. In addition, the increased polymer volume fraction provided a higher surface density of the bioactive polymer segment to interact and adhere to the tissue. It is anticipated that hydrogels made of chitosan and its derivatives would be another potential form of bioadhesive or hemostatic agents.

7.2.2 Microspheres and Capsules as a Drug Carrier

Chitosan is widely used in pharmaceutical applications due to its versatility, biocompatibility, digestibility, and low-cost. Chitosan is anticipated to be a promising drug delivery vehicle. Various applications including oral, nasal, mucosal, and transdermal delivery for drug targeting systems have been suggested by many researchers (Paul and Sharma, 2000). Chitosan is also used as a dietary supplement for controlling obesity because of its ability to bind with fat. Such sustained drug release vehicles include microspheres, beads, compressed tablets, nanoparticles, gels, and films. Various kinds of drugs, proteins, and enzymes encapsuled in microspheres have been introduced as chitosan-based microparticular delivery systems (Paul and Sharma, 2000; Lameiro et al., 2006).

Microsphere drug delivery is one of the models for oral delivery of drugs (Paul and Sharma, 2000; Lameiro et al., 2006; Cerchiara et al., 2003; Carreno-Gomez and Duncan, 1997; Jameela et al., 1998; Chen et al., 1996). It is important for a drug delivery system to localize the drug at a particular part of the body for an effective clinical treatment. Oral administration of such a site-specific drug delivery system has become of recent interest. Controlling the release of peptides and low molecular weight drugs is one of the key issues to overcome problems including exposure of drugs to acid environments and prevention of degradation by enzymes in the gastrointestinal tract. Chitosan is a candidate drug control-release carrier due to its nontoxic and bioabsorbable nature.

Since chitosan salt is a cationic polysaccharide, the sustained-release of drugs might be controlled in the presence of various fatty acids such as stearic, palmitic, myristic, and lauric acid under different pH conditions (Cerchiara et al., 2003). In this study, vancomycin hydrochloride, an antibiotic, was used as the drug. Chitosan salts were prepared by mixing chitosan with an aqueous solution of asparaginic acid and glutamic acid, or hydrochloric acid. The freeze-dried chitosan salts were physically mixed with vancomycin hydrochloride. Finally, each chitosan salt–drug mixture was added to various fatty acid solutions containing nonionic surfactant Span60. Drug containing microspheres with 1–5 µm diameters were obtained by a spray-drying method. The ability for sustained-release of drug was evaluated by the dialysis method. Solutions at pH 2.0 and 7.4, which assumed gastric and intestinal conditions, were used. At pH 2.0, the microspheres with longer alkyl chains (stearic and palmitic acid) suppressed the release of the drug significantly. On the other hand, at pH 7.4, overall drug release increased even in the drug coated with long alkyl chains. This might be caused by the increased solubility of the fatty acids at pH 7.4 compared to pH 2.0 due to the ionization of the carboxylic acid, which promoted the release of free drug. As a consequence, it was concluded that fatty acids retarded the release of drug in acidic conditions.

Several chitosan derivatives of different molecular weight and deacetylation, such as O-hydroxyethyl chitosan, chitosan hydrochloride, chitosan lactate, chitosan glutamate, and cross-linked chitosan were evaluated in terms of cytotoxicity, blood cell lysis, and horseradish peroxidase (HRP) release from the chitosan microspheres in vitro (Carreno-Gomez and Duncan, 1997). It is known that cationic polymers, such as poly-L-lysine, generally exhibit cytotoxicity toward cells in a concentration-dependent manner and cause blood cell lysis. The model drug, HRP, could be entrapped and retained in chitosan microspheres cross-linked by glutaral-dehyde although some active HRP was detected. All the soluble chitosan salts and

O-hydroxyethyl chitosan exhibited cytotoxicity toward murine melanoma cells (B16F10), depending on their concentrations, even though they were less toxic than the positive control, poly-L-lysine. The counterion of chitosan affects the interaction between the protonated amine group and negatively charged cells, leading to the observed differences. SEM observation indicated that the plasma membrane was damaged by interaction with the microspheres. Polymer molecular weight was also an important factor for cytotoxicity as higher molecular weight chitosan was more toxic. Although glutaraldehyde is frequently used as a model cross-linking agent due to its low toxicity, the glutaraldehyde cross-linked chitosan microspheres were even more toxic. Some studies demonstrated that the immune response by glutaraldehyde cross-linked microspheres is due to the residual aldehyde, which could be removed by a bisulfite wash (Jameela et al., 1998). *In vitro* release of progesterone from the cross-linked chitosan microspheres is shown in Figure 7.6 (Jameela et al., 1998). Red blood cell lysis accompanied with hemoglobin release, occurred for all chitosan derivatives in a time and molecular weight dependent manner even at low chitosan concentration (1–100 µg/mL) (Carreno-Gomez and Duncan, 1997).

The chain flexibility of chitosan also plays an important role in determining the capsule characteristics including shape, break force, thermal properties, and drug-

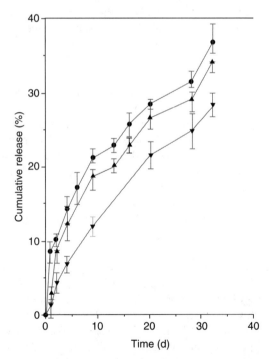

FIGURE 7.6 *In vitro* release of progesterone from chitosan microsphere cross-linked 10 mL of glutaraldehyde saturated toluene into phosphate buffer at 37 °C as a function of particle size (Zhao et al., 2001); 45–90 µm (●), 90–150 µm (▲), 150–300 µm (▼).

release ability of chitosan microspheres (Chen et al., 1996). Microspheres were prepared with different deacetylation (DD = 67.9–92.2%), molecular weight (M_w=1.8 × 10^5–31.8 × 10^5) with or without NaCl at various acidic conditions (pH = 2–4). Highly deacetylated chitosan resulted in more molecular weight loss. The more deacetylated, the more flexible the chitosan chain becomes, thus the chain tends to form a random coil, which has more intramolecular hydrogen bonds within the chain. This results in the chitosan chains being less entangled and more ellipsoid in shape. The enthalpy measured by DSC was higher due to the hydrogen bond formation in the capsules and therefore, their mechanical properties were generally weaker than those of less deacetylated microspheres. In contrast, the less deacetylated chitosan chain was more extended and had stronger intermolecular interactions, which made the chains more entangled. It contributed to reinforce the capsule structure with a more spherical shape. Molecular weight also influenced the number of the hydrogen bonds and entanglements and to the capacity for capsule formation. However, the molecular weight of the microspheres was also varied with the degree of deacetylation. Thus, it seems to be difficult to conclude what parameters, either molecular weight or degree of acetylation, or both and to what extent both actually affected the chain flexibility and extent of entanglement (Chen et al., 1996).

7.3 ANTIBACTERIAL ACTIVITY OF CHITOSAN AND ITS DERIVATIVES

It is important to clarify the antibacterial actions of scaffolds or wound dressings as well as biocompatibility and blood compatibility for hemostatic applications. When tissue is burned and damaged, the wound healing process is disrupted because of the surrounding normal tissue barrier system is trying to prevent microbial infections (Loke et al., 2000). Keeping the wound tissue moist but not debilitated by microbes is the primary requirement for wound dressing materials. Microbes also prefer to range in the damaged tissue area where moisture is, which leads to sepsis, and perhaps morbidity. Antimicrobial agents have been incorporated in the dressings but it is difficult to control the sustained-release of these materials, which are usually toxic. Wound dressings should also absorb the exudate or body fluid produced from the wound area but not adhere to the wound surface too strongly, to avoid damaging newly formed tissues. Consequently, preventing wound invasion by microorganisms and removing excess exudate from the damaged tissue accelerates wound closure. Because of the antimicrobial and hemostatic effects of chitosan, wound dressings made of chitosan derivatives have been extensively studied to overcome those limitations (Yamane et al., 2005; Loke et al., 2000; Mi et al., 2001; Ono et al., 2000; Ishihara et al., 2002).

7.3.1 Antibacterial Actions

Chitosan and its derivatives are known as antimicrobial agents against a wide variety of bacteria and fungi, as introduced in many papers (Rabea et al., 2003;

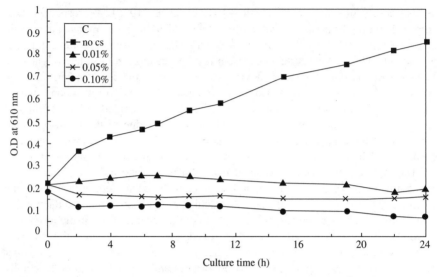

FIGURE 7.7 OD versus culture time for the chitosan ($M_v = 5.11 \times 10^4$) whose C in the medium at 0.01, 0.05, and 0.10% against *E. coli* (Lameiro et al., 2006).

Sarasam et al., 2006). Figure 7.7 shows culture media containing chitosan-suppressed bacterial growth (Liu et al., 2001). One mechanism suggested that the polycation of chitosan interacts with negatively charged bacterial surfaces to inhibit the bacterial growth (Liu et al., 2001). Chitosan solutions at lower concentrations induce permeability changes, which led to leakage of intracellular components of bacteria. At higher concentration, chitosan accumulated on the bacterial surface to disturb mass transportation (Rabea et al., 2003; Liu et al., 2001). However, the antibacterial activity of chitosan itself is exhibited only in acidic solutions due to the poor solubility above pH 6.5 and lack of cationic charges. Therefore, water-soluble chitosan derivatives including carboxymethylated chitosan and quaternary ammonium salts of chitosan are good candidates for antibacterial applications (Liu et al., 2001; Zhao et al., 2003; Jia et al., 2001). For this reason, *N,O*-carboxymethylated and *O*-carboxymethylated chitosan with different molecular weights and deacetylations were prepared and their antimicrobial activity against *Escherichia coli* was demonstrated in different culture conditions (Liu et al., 2001). Although the antibacterial activity increased with increasing molecular weight of chitosan on some level, too high a molecular weight or too high a concentration affected the antimicrobial action adversely. Furthermore, only *N*-carboxymethylated chitosan did not prevent bacterial growth, which was attributed to the chemical modification of the amino group. They also postulated that too many amino groups on a single chain, might form a pseudo-cross-linked structure through intramolecular hydrogen bondings and the chitosan could not then interact with bacterial surfaces. The optimum molecular weight of chitosan was around $M_v = 9.16 \times 10^4$. In similar experiments, however, the antibacterial activity increased with increasing deacetylation $(M_v = 12.7–27.4 \times 10^4$, $DDA = 74–96\%)$ and concentration $(M_w = 5000$, $DDA = 73\%$, $C = 0–0.50$ w/v%) of chitosan. The highest deacetylated sample and the highest concentration of chitosan exhibited the best antibacterial activity. Unfortunately, there was no information about the theoretical amount of ammonium salt forming at the pH utilized, depending on the molecular weight of the chitosan. Further comparisons between ionized amino groups chitosan with high molecular weight and with higher concentration in the culture system would be needed.

In another attempt to increase water solubility, the N-alkyl chitosan derivatives: N,N,N-trimethyl chitosan, N-N-propyl-N,N-dimethyl chitosan, N-furfuryl-N,N-dimethyl chitosan quaternary salts, were prepared through Schiff base intermediates with different molecular weights of chitosan (Jia et al., 2001). The N-furfuryl-N,N-dimethyl chitosan with the lowest molecular weight $(M_v=7.80 \times 10^3)$ showed the highest water solubility. It was attributed to the quaternary salt formed and low molecular weight. The minimum inhibitory concentration (MIC) and minimum bacterial concentration (MBC) effects of quaternized chitosan against $E.\ coli$ depended on the molecular weight. The antibacterial activity increased while increasing the alkyl chain length and also increased in the presence of acetic acid. They concluded that quaternary chitosan ammonium salts exhibited a higher antibacterial effect than chitosan itself and that an acidic medium also contributed to enhance the antibacterial activity.

Some studies about the biospecific fraction of chitosan have also been reported (Sasaki et al., 2003; Strand et al., 2003). According to the article, a low fraction of N-acetylated units (F_A) specifically bind to lysozymes and chitinases without cleavage of glycosidic linkages, whereas fully deacetylated chitosan did not (Sasaki et al., 2003). The acetyl group was essential for binding with lysozymes and the affinity was strongly dependent on pH and ionic strength. The effect of chemical composition of chitosan (F_A) against $E.\ coli$ was also examined in terms of various conditions such as molecular weight, pH, and ionic strength (Strand et al., 2003). The chitosan adsorbed to $E.\ coli$ strongly increased with pH. However, it decreased with increasing molecular weight, which was not consistent with other authors. Interestingly, chitosan with a highly acetylated fraction $(F_A = 0.49)$ flocculated $E.\ coli$ most effectively, although the details of this mechanism are still in question. Nevertheless, an acetylated fraction of chitosan might interact with particular enzymes and cells biospecifically, whereas the amino group of chitosan might physically interact with them.

7.3.2 Bioadhesives for Wound Dressings

At this time, several kinds of wound dressings are commercially available to support wound healing processes. Sponges, hydrogels, woven and nonwoven dressings derived from natural, and abundant polymers have been developed for practical use. Since chitosan is acknowledged as a biodegradable, biocompatible, and bacteriostatic polysaccharide, many chitosan matrices for wound dressings with bioadhesiveness have been studied extensively (Venter et al., 2006; Wittaya-Areekul and Prahsarn, 2006; Mo et al., 2010).

A dual layer chitosan-based wound dressing was fabricated by combining carboxymethylated chitin hydrogel as an upper layer with chitosan acetate foam as a lower layer (Loke et al., 2000). Although not specified, we assume the foam is the acetate salt of chitosan. The matrix was designed that the upper hydrogel layer was able to absorb wound exudates and block microbial invasion, and the lower foam layer could serve as an antibacterial material. Indeed, the upper hydrogel was swollen and absorbed four times its own weight of water and its 50% vapor permeability was sufficient to prevent accumulation of exudates. The chitosan acetate foam incorporated chlorhexidine gluconate, whose optimal loading concentration was 1% (w/v), and released it to inhibit bacterial growth of *Pseudomonas aeruginosa* and *Staphlycoccus aureus* around the foam discs. Further *in vitro* and *in vivo* experiments would be expected to ensure their results.

An asymmetric chitosan membrane consisting of a top layer containing an interconnected microporous skin surface and sponge-like macroporous sublayer, was prepared by an immersion-precipitation phase inversion (IPPI) method (Mi et al., 2001). The advantage of this structure was to achieve both the prevention of bacterial invasion and regulation of evaporative water loss. The thickness and density of the membrane was also controllable, depending on the perevaporation conditions. The membrane exhibited moderate water evaporation, oxygen permeability, and fluid drainage ability due to the dense skin surface, and the thick and porous sublayer of chitosan membrane. Furthermore, it inhibited bacterial invasion and penetration into the membrane because of the dense surface and antibacterial nature of chitosan. A rat skin wound area covered with chitosan membrane stopped bleeding since chitosan is a hemostatic agent and rapid epithelialization was facilitated.

Hydrogels are considered a substitute for fibrin glue. Fibrin glue has been widely applied for medical purposes such as sealing wounded tissue. It contains fibrinogen, thrombin, factor III, and protease inhibitor, which are vital for hemostasis and blood coagulation. However, due to the difficulty of its mass production and contaminant suppression, a novel biological adhesive with better properties has been desired (Ono et al., 2000). Although cross-linked gelatin and cyanoacrylate polymers have been developed, they are not suitable for biomedical applications because of their high toxicity. In contrast, polysaccharides and their derivatives including chitin, chitosan, and hyaluronan are known as biocompatible materials and have been used for healing processes and supporting tissue defects. Since chitosan has antibacterial and hemostatic effects, a chitosan derivative with both azide and lactose moieties were photochemically cross-linked to prepare a hydrogel adhesive (Ono et al., 2000; Ishihara et al., 2002). Lactobionic acid and 4-azidebenzoic acid were introduced to chitosan by stepwise condensation reactions with TEMED (*N,N,N',N'*-tetramethylethylenediamine) and EDC (1-ethyl-3-(3-dimethylaminopropyl) carbodiimide). The reactive azide group was converted to nitrene by UV irradiation, which reacted with the remaining amino groups of chitosan, causing hydrogel formation. There was no information about the degree of modified functional groups and the remaining amino groups in the chitosan. Water solubility (around neutral pH) of modified chitosan increased with increasing lactose concentration. Small amounts of azide did not affect the water solubility. The time required for UV irradiation to form hydrogel was 30 s and it is faster than that of fibrin glue. Binding and sealing strength of Az-CH-LA increased with increasing concentration. Even though cultured cells (human skin fibroblast, coronary smooth muscle cells, and endothelial cells) did not adhere onto the chitosan surface very well, the cell viability was retained without toxicity for the cell culturing. The authors also demonstrated with *in vivo* experiments, that advanced granulation tissue formed and epithelialization occurred, when treated with chitosan hydrogel, which facilitated rapid wound occlusion (Ishihara et al., 2002).

7.4 HEMOSTATIC POTENTIAL OF CHITOSAN AND ITS DERIVATIVES

Hemostatic materials are required to have different surface properties from ordinary blood contacting materials, which seek to avoid the induction of thrombosis. When developing artificial biomaterials that contact blood, antithrombogenic material surfaces are a primary consideration. Thrombosis on the material surface has always been a serious problem even though the material itself has good blood compatibility. As an example of thrombosis, a small caliber vascular graft results in thrombus formation in the short term and intimal hyperplasia occurs in the long term (Takahashi et al., 2002). It is caused by plasma protein adhesion, and then followed by platelet adhesion and activation. On the other hand, polyurethane (PU) is a well-known biomaterial with good biocompatibility and mechanical properties for biomedical applications. However, the usage of PU is still limited because of the poor cell attachment/growth or the induction of thrombinogenesis on the surface. Many attempts have been made, such as chemically or photochemically modifying the PU surface to overcome these problems (Takahashi et al., 2002; Hsu and Chen, 2000). Bioactive molecules such as prostaglandin E_1, albumin, and heparin or its derivatives with sulfate groups have also been used to modify the material surface in order to decrease the thrombogenicity (Chandy and Sharma, 1992; Kottke-Marchant et al., 1989; Beena et al., 1995; Ko et al., 1993). They have an anti-coagulation property, however, which also induces hemorrhage of the tissue. When chitosan is applied as a blood contacting material, such anticoagulant modifications are usually applied (Chandy and Sharma, 1992; Beena et al., 1995). The effect of chemical structure modifications and physical form of the chitosan upon hemostasis was recently reviewed (Whang et al., 2005; Recinos et al., 2008). In addition, several commercially available hemostatic agents including a chitosan derivative approved by Food and Drug Administration have been used in recent combat operations and their effectiveness has been reported (Wedmore et al., 2006; Pusateri et al., 2006; Brown et al., 2009). Chitosan, itself, is not suitable for a blood contacting material in spite of its biocompatibility.

Due to this blood coagulating nature, chitosan has been a desirable material as a hemostatic agent in addition to its biocompatible and antibacterial characteristics. This hemostatic activity of chitosan is important for understanding the mechanism of action as a coagulant and developing medical bandages which control bleeding during surgery. Several studies have reported that the hemostatic mechanism induced from chitosan was independent of the classical blood coagulation cascade (Whang et al., 2005; Rao and Sharma, 1997; Klokkevold et al., 1999; Chou et al., 2003). However, because of the complexity of the blood coagulation mechanism itself, it is still difficult to clarify how chitosan affects this mechanism. A wide variety of evaluation methods have been adopted to examine the hemostatic properties, and will be introduced in a later section. Nevertheless, it is important to understand what factors in blood influence the hemostatic mechanism, especially for the potential cascade induced by chitosan.

7.4.1 Factors Involving the Blood Coagulation Mechanism

The blood coagulation mechanism is still under investigation because of the complexity of the hemostatic–thrombotic system. Hemker et al. (2004) introduced the first law of hemostasis and thrombosis: increasing thrombin formation causes more thrombosis but less bleeding, and decreasing thrombin formation causes more bleeding but less thrombosis. Recent studies allowed monitoring the thrombin generation in platelet-poor plasma (PPP) and platelet-rich plasma (PRP) by using fluorogenic thrombin substrates (Hemker et al., 2004). However, it is still challenging to evaluate the blood coagulation as close as possible *in vivo*, because the actual blood circulation time is shorter than that time required for whole blood clotting.

Currently, two major pathways are believed to trigger blood coagulation (Figure 7.8, (Mao et al., 2004; Hanson, 2004)). One is the adsorption of plasma

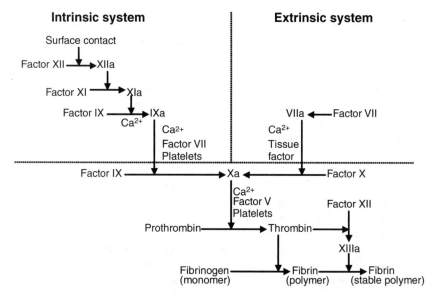

FIGURE 7.8 Blood coagulation pathways. Clotting factors (proenzymes), identified by Roman numerals, interact in a sequential series of enzymatic activation reactions (coagulation cascade) leading to the amplified production of the enzyme thrombin, which in turn activates fibrinogen to form a fibrin polymer that stabilizes a clot or thrombus.

proteins such as albumin, γ-globulin, fibrinogen, and prothrombin onto the material surfaces. Factor XII, an intermediate of the intrinsic coagulation pathways is activated following the protein adsorption to initiate the clotting cascade. The other pathway is cell-bound thrombin generation involving tissue factor cells called monocytes and perivascular cells and platelets. It is pointed out that the activated platelets induce the interaction between platelet membrane glycoprotein (e.g., GPIIb/IIIa complex, von Willebrand factor) and subendothelials (Mao et al., 2004). Thrombin and collagen also play an important role for platelet activation to produce procoagulant phospholipids (Hemker et al., 2004). Several biomaterials such as polyethylene and polyurethane that induce thrombosis on surfaces have been used in conjunction with several plasma factors including prothrombin, fibrinopeptide, blood coagulation inhibitor called protein kinase C, and thrombin–antithrombin III complex (Hanson, 2004; Cenni et al., 1996; Bordenave et al., 1993). A platelet reaction at artificial surfaces is considered as a trigger of thrombosis (Figure 7.9).

In the case of chitosan, several hemostasis evaluations, which are usually used in coagulation-deficient diagnosis, have been reported. However, many test methods have been applied to different types of tissue in different articles. Chitosan's hemostasis might be caused by the nonclassic coagulation pathway, whose mechanism and detailed phenomena are still unclear. Nevertheless, these studies mainly focus on the blood coagulation time and platelet adhesion and aggregation induced by chitosan. The effect of chitosan and its derivatives on hemostasis potential and the hemostasis tests suggested in several studies, will be introduced in the following section.

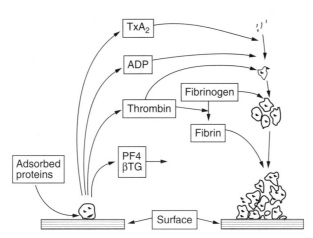

FIGURE 7.9 Platelet reaction at material surfaces (Kottke-Marchant et al., 1989). Following protein adsorption, platelets adhere and release their α-granule contents, including ADP. Thrombin production is catalyzed locally by platelet membrane phospholipids. Thranbaxane A_2 (TxA_2) is synthesized. ADP, TxA_2, and thrombin then recruit additional circulating platelets into an enlarging platelet mass that is stabilized by fibrin.

7.4.2 Hemostasis Evaluations for Chitosan and its Derivatives

When examining baseline blood conditions in the normal state *in vitro*, the parameters such as hematocrit, hemoglobin concentration, platelet count, prothrombin time, activated partial thromboplastin time, and plasma fibrinogen concentration are evaluated (Pusateri et al., 2003). So many hemostasis tests have been carried out in different studies for chitosan hemostatic evaluations, that standardized methods have not been recognized. As mentioned earlier, the chitosan hemostasis pathway seems to be different from the classic coagulation pathway. Since it is already known that platelet aggregation was observed on chitosan materials microscopically, the morphology of platelets was considered to affect the blood coagulation system. Thus, platelet-related cytokines and proteins were also examined in addition to general coagulation tests.

In vitro examination of blood clotting ability is widely used for diagnosing clotting disorders in the medical field and, is assessed in regard to several coagulation factors. For example, coagulation tests are carried out by measuring prothrombin time (PT), partial thromboplastin time (PTT), thrombin time (TT), activated partial thromboplastin time (APTT), clot retraction time (CRT), plasma recalcification time (PRT) [71], and whole blood clotting time (WBCT) (Lin and Lin, 2003; Rao and Sharma, 1997; Queiroz et al., 2003). Similar coagulation tests, such as blood coagulation time (BCT) (Okamoto et al., 2003), fibrin clot formation time (Fischer et al., 2004), R_{APTT}, R_{TT}, R_{PT} values (ratio of APTT, TT, or PT to those of the control assays) (Ronghua et al., 2003) were examined. These test methods were varied depending on the blood specimen and conditions, which reflected the effect of heparin and its concentration. Platelet aggregation (PA) is one of the key phenomena in blood coagulation induced by chitosan derivatives. PA was monitored by a light penetration (Okamoto et al., 2003) or turbidimetric device (Lin and Lin, 2003; Chou et al., 2003), and scanning electron microscopy (SEM) (Rao and Sharma, 1997; Queiroz et al., 2003; Okamoto et al., 2003; Ronghua et al., 2003; Zhu et al., 2002). The platelet activation was assessed by counting the number of Coomassie brilliant blue-stained platelets (Amiji, 1995). Some studies also focused on red blood cell morphology by SEM (Figure 7.10, (Klokkevold et al., 1999)) and cell aggregation was measured spectrophotometrically (OD_{528}) (Fischer et al., 2004). Some cytokines released from platelets have been considered as important factors to enhance early wound healing processes. Thus, platelet-derived growth factor (PDGF-AB) and transforming growth factor $\beta1$ (TGF-$\beta1$) release was measured by enzyme immunoassay (Okamoto et al., 2003). Related to platelet activation, intracellular calcium level in platelets and glycoprotein IIb/IIIa complex on platelet surface were also examined to elucidate details of the chitosan-induced blood coagulation mechanism (Chou et al., 2003).

Various animals with different lacerations were treated with chitosan derivatives as another approach to *in vivo* testing. Sutures coated with chitosan, *N*-acyl-chitosans, chitosan-tropocollagen fibers and their *N*-modified fibers were introduced into the lumen of dog's jugular and femoral veins (Hirano and Noishiki, 1985; Hirano et al., 2000). The blood clot formed around the fibers was observed macroscopically.

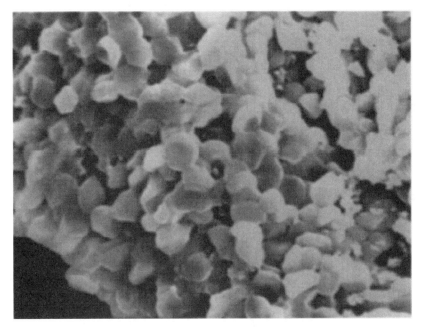

FIGURE 7.10 SEM of a blood clot formation from chitosan-treated lingual incision (Klokkevold et al., 1999). Red blood cells have lost their typical biconcave morphology and they appear to have an unusual affinity toward one another (original magnification ×2000).

In other studies, lingual bleeding time, systematic bleeding time, and systematic coagulation time was measured when chitosan solution was introduced onto lacerated rabbit tongue (Klokkevold et al., 1999). Severe swine liver injury due to the lacerated major vessels was treated with a chitosan acetate salt dressing and the effect of the dressing was studied by monitoring blood loss, fluid use, hemostasis time, and survival (Table 7.2, (Pusateri et al., 2003)). Furthermore, hemostasis performance of poly-*N*-acetyl glucosamine (p-GlcNac) fiber was observed until bleeding was stopped by continuous compression at the wound area in swine spleen (Fischer et al., 2004).

7.4.3 Effects of Chitosan and its Derivatives on Blood Coagulation

Various kinds of chitosan-based hemostatic agents have been developed and tested for their potential usability *in vitro* or *in vivo* (Whang et al., 2005; Yang et al., 2008; Kranokpiraksa et al., 2010). One of the early studies was the development of a silk polyfilament coated with chitosan and *N*-acetyl and *N*-hexanoyl chitosans (Hirano and Noishiki, 1985). *In vivo* study demonstrated that a thick coagulum was formed on the chitosan suture inserted to the lumen of a dog's peripheral veins, due to the rough fibril surface but less on *N*-modified chitosan because of the smooth surfaces. Blood components tend to adsorb physically onto rough surfaces (Hirano et al., 2000).

TABLE 7.2 Effect of a Chitosan-Based Hemostatic Dressing on Blood Loss and Survival in Swine

Variable	Gauge Sponge Control Group	Chitosan Group	p Value of Difference
Posttreatment blood loss (mL)	2,879 (95% CI, 788–10,513)	264 (95% CI, 82–852)	<0.01
Posttreatment blood loss (mL/kg body weight)	102.4 (95% CI, 28.2–371.8)	9.4 (95% CI, 2.9–30.3)	<0.01
Fluid use (mL)	6614 (95% CI, 2519–17363)	1793 (95% CI, 749–4291)	0.03
Survival (%)	2/7 (28.6)	7/8 (87.5)	0.04
Survival time (min: nonsurvivors only)	38.4 ± 5.8	10.0	N/A
	(n = 5)	(n = 1)	
Hemostasis at 1 mn (%)	0	50	0.08
Hemostasis at 2 mn (%)	0	50	0.08
Hemostasis at 3 mn (%)	0	62	0.03
Hemostasis at 4 mn (%)	0	62	0.03

Posttreatment blood loss, fluid use, time for hemostasis, and survival are tabulated CI, confidence interval.

Chitosan-tropocollagen, *N*-acetyl chitosan, and *N*-acetyl chitosan-tropocollagen fibers were inserted into the lumen of a dog's jugular and femoral veins in a similar manner (Hirano et al., 2000). Chitosan-tropocollagen and *N*-acetylated chitosan fibers caused weak blood coagula to form but almost no blood coagula formed on *N*-acetyl chitosan-tropocollagen and tropocollagen fibers. Not only chitosan but also chitin improves the blood coagulation to some extent. Introducing carboxy groups into the *N*- and *O*-positions increased ATPP and TT more than that just due to the amino groups (Ronghua et al., 2003). Thus, carboxylation of *N*- and *O*-positions tended to prolong the blood coagulation time.

The effect of chitosan on hemostasis has been investigated for several chitosan physical forms. It was found that chitosan solution in 2% aqueous acetic acid caused hemagglutination even at low concentration and the whole blood clotting time was reduced by 40% of normal clotting time (Rao and Sharma, 1997). In addition, chitosan did not induce red blood cell lysis. Chitosan interacted with the red blood cell membrane and led to the erythrocyte aggregation as seen by SEM observation (Figure 7.10). The hemostatic property was attributed to electrostatic interaction by the cationic nature of chitosan at pH's below 6.4. The authors concluded that chitosan was independent of the normal blood clotting cascade but dependent on red blood cell agglutination. Similar conclusions were derived from other studies. For example, topical administration of chitosan dissolved in 0.2% glacial acetic acid solution shortened the lingual bleeding time of rabbit to 43% of the control (Klokkevold et al., 1999). SEM observation demonstrated that the red blood cells treated with chitosan solution lost their typical biconcave morphology and coalesced into a clot. It has been hypothesized that chitosan promotes hemostasis by linking erythrocytes together to form a lattice to entrap the cells.

Platelet activation has been another line of study for understanding hemostasis. Chitosan-coated microtiter plates enhanced platelet adhesion and aggregation concentration and time dependency (Chou et al., 2003). Significantly, chitosan influenced the intracellular calcium level known as the second messenger in platelets and GPIIb/IIIa expression on platelets. Since platelets do not adhere to endothelial cells under normal conditions, this feature would indicate platelet activation for initiating hemostasis or the thrombosis step. It is known that fibrinogen-GPIIb/IIIa binding could be observed in the final common pathway of blood coagulation. Therefore, chitosan might account for the interaction between activated platelets and damaged tissue to promote the wound healing process.

Chitin also seems to have hemostatic potential. Chitin and chitosan suspension in phosphate-buffered saline (PBS) reduced blood coagulation time in a dose-dependent manner (Okamoto et al., 2003). They shortened the BCT to 30–40% of the control, even at a low concentration (0.1 mg/mL). Chitin and chitosan have a hemostatic effect due to a physical binding effect and the amino group in their chemical structure. Chitosan was more effective than chitin for hemostasis, whereas chitin could induce more platelet aggregation than chitosan. Therefore, they concluded that blood coagulation might be regulated by not only platelet aggregation but also erythrocytes morphology. Cytokine (PDGF-AB and TGF-β1) release, which is related to the wound healing process, was greater for a chitosan suspension than chitin. It was much less for latex and other controls. As a result, chitosan formed stronger platelet aggregates by interacting with platelets on membranes. Porous sponges of chitosan acetate salt also significantly reduced blood and other fluid loss for a swine liver injury, compared to the swine treated with control gauze sponge (Pusateri et al., 2003). The authors indicated that a fully N-acetylated chitin sheet (p-GlcNAc) reduced hemostasis time compared to the control. Fully N-acetylated chitin is found in some algal sources and is known as chitan by some authors (McLachlan et al., 1965; Smucker, 1991).

The dependency of the hemostatic effect on molecular weight was demonstrated with water-soluble chito-oligosaccharide (COS) and highly deacetylated chito-oligosaccharide (HDCOS) (Lin and Lin, 2003). Both 10% of COS and HDCOS prolonged the WBCT compared to the PBS control. There was no significant difference between COS and HDCOS in WBCT. Furthermore, these polymers did not have a significant effect on the platelet aggregation compared to the control. These results indicated that low molecular weight chitosans did not have a hemostatic effect and that only chitosan polymers of a minimum critical molecular weight have hemostatic potential.

In contrast, chitosan-based hemostatic agents did not always exhibit a better result than Syvek p-GlcNAc (Vournakis et al., 1997). The p-GlcNAc is a fiber with a crystalline beta-form three-dimensional structure (Fischer et al., 2004; Vournakis et al., 2004). It is a purified N-acetylglucosamine produced by microalgae (Vournakis et al., 1997). It is distinguished from other N-acetyl glucosamine-based polymers such as chitin, chitosan, and hyaluronan. Chitin and chitosan are the copolymers of N-acetylglucosamine and N-glucosamine, and which are not defined exactly, in terms of the degree of deacetylation and the polymer chain structure (Vournakis et al., 1997).

A commercially available chitosan-based hemostatic agent, Clo-Sur® PAD is entirely composed of chitosan. Chitoseal® consists of a thin layer of chitosan coated onto PET filament. These were compared to examine the hemostatic potential of chitosan derivatives (Fischer et al., 2004). Red blood cell aggregation was observed with p-GlcNAc in a concentration-dependent manner but not for the other two chitosan-based materials. Platelets absorbed onto p-GlcNAc were inferred to be in an activated state, which was similar to fibrin-platelet interactions, whereas chitosan materials were unable to do so. Furthermore, p-GlcNAc was more effective in stopping bleeding in swine splenic trauma compared to the other two. These different behaviors between p-GlcNAc and other chitosan materials may result from their structural differences. Although the aligned beta three-dimensional structure of p-GlcNAc is rare in nature, its fibril strand size was similar to that of fibrin. It was concluded that the unique, large surface area structure of p-GlcNAc might be one of the reasons why it is effective in its interaction with blood proteins and cells for promoting hemostasis. However, it is still a question why the chitosan-based hemostatic agents were not effective in this study. Overwhelming though, the effectiveness of chitosan derivatives as hemostatic agents has been demonstrated, however, the mechanism regulated by chitosan derivatives is still unknown.

7.5 CONCLUSIONS

Chitosan and its derivatives have been found to have hemostatic potential as well as antimicrobial activity and biocompatibility. Chitin is a natural and abundant polymer. Chitosan is easily obtained from it by a relatively simple chemical reaction. Therefore, chitin and chitosan derivatives have been prepared as relatively inexpensive hemostatic agents because currently available fibrinogen-based hemostatic agents are potentially infectious and expensive. In most studies, chitosan and its derivatives enhanced platelet and erythrocyte aggregation, which is necessary to initiate blood coagulation. It has been clarified that not only the physical form of chitosans, along with molecular weight, degree of deacetylation, surface characteristics but also their chemical structures affect the hemostatic and antibacterial actions. However, various chitosan compositions with different molecular weight, deacetylation, counter ions, and solvents have been used in different antibacterial or blood coagulation tests, which makes it difficult to compare all results. More details of the blood coagulation cascade and its mechanism induced by chitosans are also expected to clarify and distinguish the action of chitosan on the classic cascade. In conclusion, fully chemically and physically characterized chitosan and its derivatives are needed in order to elucidate the effects on antibacterial and hemostatic actions. Standardized test methods are needed to compare the results in a more meaningful way.

REFERENCES

Alam, H. B., D. Burris, J. A. DaCorta, and P. Rhee (2005). Hemorrhage control in the battlefield: role of new hemostatic agents. *Military Med.* **170**:63–69.

Amiji, M. M. (1995). Permeability and blood compatibility properties of chitosan-poly (ethylene oxide) blend membranes for haemodialysis. *Biomaterials.* **16**:593–599.

Barrera, D. A., E. Zylstra, P. T. Lansbury, and R. Langer (1993). Synthesis and RGD peptide modification of a new biodegradable copolymer: poly(lactic acid-*co*-lysin). *J. Am. Chem. Soc.* **115**:11010–11011.

Baumann, H., and V. Faust (2001). Concepts for improved regioselective placement of *O*-sulfo, *N*-sulfo, *N*-acetyl, and *N*-carboxymethyl groups in chitosan derivatives. *Carbohydr. Res.* **331**:43–57.

Beena, M. S., T. Chandy, and C. P. Sharma (1995). Heparin immobilized chitosan-poly ethylene glycol interpenetrating network: antithrombogenicity. *Art. Cells Blood Subs. Immob. Biotech.* **23**:175–192.

Bordenave, L., C. Lbaquey, R. Bareille, F. Lefebvre, C. Lauroua, V. Guerin, F. Rouais, N. More, C. Vergnes, and J. M. Anderson (1993). Anderson endothelial-cell compatibility testing of 3 different pelletanes. *J. Biomed. Mater. Res.* **27**:1367–1381.

Brown, M., M. Daya, and J. Worley (2009). Experience with chitosan dressings in a civilian EMS system. *J. Emerg. Med.* **37**(1): 1–7.

Carreno-Gomez, B., and R. Duncan (1997). Evaluation of the properties of soluble chitosan and chitosan microspheres. *Int. J. Pharmac.* **148**:231–240.

Cenni, E., G. Ciapetti, M. Cervellati, D. Cavedagna, G. Falsone, S. Gamberini, and A. Pizzoferrato (1996). Activation of the plasma coagulation system induced by some biomaterials. *J. Biomed. Mater. Res.* **31**:145–148.

Cerchiara, T., B. Luppi, F. Bigucci, M. Petrachi, I. Orienti, and V. Zecchi (2003). Controlled release of vancomycin from freeze-dried chitosan salts coated with different fatty acids by spray-drying. *J. Microencapsul.* **20**:473–478.

Chandy, T, and C. P. Sharma (1992). Prostaglandin E1-immobilized poly(vinyl alcohol)-blended chitosan membranes: blood compatibility and permeability properties. *J. Appl. Polym. Sci.* **44**:2145–2156.

Chen, R. H., M L. Tsaih, and W. C. Lin (1996). Effects of chain flexibility of chitosan molecules on the preparation, physical, and release characteristics of the prepared capsule. *Carbohydr. Polym.* **31**:141–148.

Chen, X. G., and H. J. Park (2003). Chemical characteristics of *O*-carboxymethyl chitosans related to the preparation conditions. *Carbohydr. Polym.* **53**:355–359.

Chou, T. C., E. Fu, C. J. Wu, and J. H. Yeh (2003). Chitosan enhances platelet adhesion and aggregation. *Biochem. Biophys. Res. Commun.* **302**:480–483.

Chung, T. W., Y. F. Lu, S. S. Wang, Y. S. Lin, and S. H. Chu (2002). Growth of human endothelial cells on photochemically grafted Gly-Arg-Gly-Asp (GRGD) chitosans. *Biomaterials* **23**:4803–4809.

Chung, T. W., Y. F. Lu, H. Y. Wang, W. P. Chen, S. S. Wang, Y. S. Lin, and S. H. Chu (2003). Growth of human endothelial cells on different concentration of Gly-Arg-Gly-Asp (GRGD) grafted chitosan surface. *Artif. Organs.* **27**:155–161.

Fischer, T. H., R. Connolly, H. S. Thatte, and S. S. Schwaitzberg (2004). Comparison of structural and hemostatic properties of the poly-*N*-acetyl glucosamine Syvek Patch with products containing chitosan. *Microsc. Res. Tech.* **63**:168–174.

Hanson, S. R. (2004). Encyclopedia of Biomaterials and Biomedical Engineering, New York: Dekker Publications, pp. 144–154.

Hemker, H. C., R. A. Dieri, and S. Béguin (2004). Thrombin generation assays: accruing clinical relevance. *Curr. Opin. Hematol.* **11**:170–175.

Hirano, S., and Y. Noishiki (1985). The blood compatibility of chitosan and *N*-acylchitosans. *J. Biomed. Mater. Res.* **19**:413–417.

Hirano, S., T. Nakahira, M. Nakagawa, and S. K. Kim (1999). The preparation and applications of functional fibres from crab shell chitin. *J. Biotech.* **70**:373–377.

Hirano, S., M. Zhang, M. Nakagawa, and T. Miyata (2000). Wet spun chitosan-collagen fibers, their chemical *N*-modifications, and blood compatibility. *Biomaterials* **21**:997–1003.

Ho, M. H., D. M. Wang, H. J. Hsieh, H. C. Liu, T. Y. Hsien, J. Y. Lai, and L. T. Hou (2005). Preparation and characterization of RGD-immobilized chitosan scaffolds. *Biomaterials* **26**:3197–3206.

Hsu, S. H., and W. C. Chen (2000). Improved cell adhesion by plasma-induced grafting of L-lactide onto polyurethane surface. *Biomaterials* **21**:359–367.

Hu, Y., S. R. Winn, I. Krajbich, and J. O. Hollinger (2003). Porous polymer scaffolds surface-modified with arginine-glycine-aspartic acid enhance bone cell attachment and differentiation *in vitro. J. Biomed. Mater. Res.* **64A**:583–590.

Ishihara, K., R. Takayama, N. Nakabayashi, K. Fukumoto, and J. Aoki (1992). Improvement of blood compatibility on cellulose dialysis membrane. *Biomaterials* **13**:235–239.

Ishihara, M., K. Nakanishi, K. Ono, M. Sato, M. Kikuchi, H. Yura, T. Matsui, H. Hattori, M. Uenoyama, and A. Kurita (2002). Photocrosslinkable chitosan as a dressing for wound occlusion and accelerator in healing process. *Biomaterials* **23**:833–840.

Jameela, S. R., T. V. Kumary, A. V. Lal, and A. Jayakrishnan (1998). Progesterone-loaded chitosan microspheres: a long acting biodegradable controlled delivery system. *J. Control. Release* **52**:17–24.

Jia, Z., D. Shen, and W. Xu (2001). Synthesis and antibacterial activities of quaternary ammonium salt of chitosan. *Carbohydr. Res.* **333**:1–6.

Klokkevold, P. R., H. Fukuyama, E. C. Sung, and C. N. Bertolami (1999). The effect of chitosan (poly-*N*-acetyl glucosamine) on lingual hemostasis in heparinized rabbits. *J. Oral. Maxillofac. Surg.* **57**:49–52.

Knill, C. J., J. F. Kennedy, J. Mistry, G. Smart, M. R. Groocock, and H. J. Williams (2004). Alginate fibres modified with unhydrolyzed and hydrolyzed chitosans for wound dressings. *Carbohydr. Polym.* **55**:65–76.

Ko, T. M., J. C. Lin, and S. L. Cooper (1993). Surface characterization and platelet adhesion studies of plasma-sulphonated polyethylene. *Biomaterials* **14**:657–664.

Kottke-Marchant, K., J. M. Anderson, Y. Uemura, and R. E. Marchant (1989). Effect of albumin coating on the *in vitro* blood compatibility of Dacron® arterial prostheses. *Biomaterials* **10**:147–155.

Kranokpiraksa, P., D. Pavcnik, H. Kakizawa, B. Uchida, M. Jeromel, F. Keller, and J. Rosch (2010). Hemostatic efficacy of chitosan based bandage for closure of percutaneous arterial access sites. *Radiol. Oncol.* **44**(2): 86–91.

Kubota, N., N. Tastumoto, T. Sano, and K. Toya (2000). A simple preparation of half *N*-acetylated chitosan highly soluble in water and aqueous organic solvents. *Carbohydr. Res.* **324**:268–274.

Kukolikova, L., D. Bakos, P. Alexy, S. Hanzelova, and W. Zhong (2006). Optimization of the properties of chitosan lactate/hyaluronan film. *J. Appl. Polym. Sci.* **100**:1413–1419.

Kumar, M. N. V. R., and S. M. Hudson (2004). *Encyclopedia of Biomaterials and Biomedical Engineering* New York: Dekker Publication, pp. 310–323.

Kurita, K., H. Ikeda, Y. Yoshida, M. Shimojoh, and M. Harata (2002). Chemoselective protection of the amino groups of chitosan by controlled phthaloylation: facile preparation of a precursor useful for chemical modifications. *Biomacromolecules* **3**:1–4.

Lameiro, M. H., A. Lopes, L. O. Martins, P. M. Alves, and E. Melo (2006). Incorporation of a model protein into chitosan-bile salt microparticles. *Intl. J. Pharm.* **312**:119–130.

Li, J., H. Yun, Y. Gong, N. Zhao, and X. Zhang (2006). Investigation of MC3T3-E1 cell behavior on the surface of GRGDS-coupled chitosan. *Biomacromolecules* **7**:1112–1123.

Lin, C. W., and J. C. Lin (2003). Characterization and blood coagulation evaluation of the water-soluble chitooligosaccharides prepared by a facile fractionation method. *Biomacro-molecules* **4**:1691–1697.

Lin, H. B., C. Garcia-Echeverria, S. Asakura, W. Sun, D. F. Mosher, and S. L. Cooper (1992). Endothelial cell adhesion on polyurethanes containing covalently attached RGD-peptides. *Biomaterials* **13**:905–914.

Lin, H. B., W. Sun, D. F. Mosher, C. Carcia-Echeverria, K. Schaufelberger, P. I. Lelkes, and S. L. Cooper (1994). Synthesis, surface, and cell-adhesion properties of polyurethanes containing covalently grafted RGD-peptides. *J. Biomed. Mater. Res.* **28**:329–342.

Liu, X. F., Y. L. Guan, D. Z. Yang, Z. Li, and K. D. Yao (2001). Antibacterial action of chitosan and carboxymethylated chitosan. *J. Appl. Polym. Sci.* **79**:1324–1335.

Loke, W. K., S. K. Lau, L. L. Yong, E. Khor, and C. K. Sum (2000). Wound dressing with sustained anti-microbial capability. *J. Biomed. Mater. Res.* **53**:8–17.

Mao, C., Y. Qiu, H. Sang, H. Mei, A. Zhu, J. Shen, and S. Lin (2004). Various approaches to modify biomaterial surfaces for improving hemocompatibility. *Adv. Colloid Interface Sci.* **110**:5–17.

Mclachlan, J., A. G. Mcinnes, and M. Falk (1965). Studies on chitan (chitin-poly-*N*-acetylglucosamine) fibers of diatom thalassiosira fluviatilis hustedt. 1. Production and isolation of chitan fibers. *Can. J. Bot.* **43**:707.

Meanwell, J. L. R., and G. Shama (2006). Chitin in a dual role as substrate for *Streptomyces griseus* and as adsorbent for streptomycin produced during fermentation. *Enzyme Microb. Tech.* **38**:657–664.

Mi, F. L., S. S. Shyu, Y. B. Wu, S. T. Lee, J. Y. Shyoug, and R. N. Huang (2001). Fabrication and characterization of a sponge-like asymmetric chitosan membrane as a wound dressing. *Biomaterials* **22**:165–173.

Mo, X., H. Iwata, and Y. Ikada (2010). A tissue adhesives evaluated *in vitro* and *in vivo* analysis. *J. Biomed. Mater. Res. Part A* **94a**(1): 326–332.

Moon, S. I., C. W. Lee, M. Miyamoto, and Y. Kimura (2000). Melt polycondensation of L-lactic acid with Sn(II) catalysts activated by various proton acids: a direct manufacturing route to high molecular weight poly(L-lactic acid). *J. Polym. Sci. Part A* **38**:1673–1679.

Moon, S. I., C. W. Lee, M. Miyamoto, and Y. Kimura (2001). Melt/solid polycondensation of L-lactic acid: an alternative route to poly(L-lactic acid) with high molecular weight. *Polymer* **42**:5059–5062.

Mori, T. M. Okumura, M. Matsuura, K. Ueno, S. Tokura, Y. Okamoto, S. Minami, and T. Fujinaga (1997). Effects of chitin and its derivatives on the proliferation and cytokine production of fibroblasts *in vitro*. *Biomaterials* **18**:947–951.

Muzzarelli, R. A. A., and P. Ilari (1994). Solubility and structure of *N*-carboxymethlchitosan. *Int. J. Biol. Macromol.* **16**:177–180.

Muzzarelli, R. A. A. (2009). Chitins and chitosans for the repair of wounded skin, nerve, cartilage and bone. *Carbohydr. Polym.* **76**(2): 167–182.

Neuffer, M. C. (2004). Hemostatic dressings for the first responder: a review. *Military Med.* **169**:716–720.

Okamoto, Y., M. Watanabe, K. Miyatake, M. Morimoto, Y. Shigemasa, and S. Minami (2002). Effects of chitin/chitosan and their oligomers/monomers on migrations of fibroblasts and vascular endothelium. *Biomaterials* **23**:1975–1979.

Okamoto, Y., R. Yano, K. Miyatake, I. Tomohiro, Y. Shigemasa, and S. Minami (2003). Effects of chitin and chitosan on blood coagulation. *Carbohydr. Polym.* **53**:337–342.

Ono, K., Y. Saito, H. Yura, K. Ishikawa, A. Kurita, T. Akaike, and M. Ishihara (2000). Photocrosslinkable chitosan as a biological adhesive. *J. Biomed. Mater. Res.* **49**:289–295.

Paul, W., and C. P. Sharma (2000). Chitosan, a drug carrier for the 21st century: a review. *S.T.P. Pharma. Sci.* **10**:5–22.

Peluso, G., O. Petillo, M. Ranieri, M. Santin, L. Ambrosio, D. Calabró, B. Avallone, and G. Balsamo (1994). Chitosan-mediated stimulation of macrophage function. *Biomaterials* **15**:1215–1220.

Peng, C. K., S. H. Yu, F. L. Mi, and S. S. Shyu (2006). Polysaccharide-based artificial extracellular matrix: preparation and characterization of three-dimensional, macroporous chitosan and chondroitin sulfate composite scaffolds. *J. Appl. Polym. Sci.* **99**: 2091–2100.

Pusateri, A. E., S. J. McCarthy, K. W. Gregory, R. A. Harris, L. Cardenas, A. T. McManus, and C. W. Goodwin (2003). Effect of a chitosan-based hemostatic dressing on blood loss and survival in a model of severe venous hemorrhage and hepatic injury in swine. *J. Trauma.* **54**:177–182.

Pusateri, A. E., J. B. Holcomb, B. S. Kheirabadi, H. B. Alam, C. E. Wade, and K. L. Ryan (2006) Making sense of the preclinical literature on advanced hemostatic products. *J. Trauma.* **60**:674–682.

Queiroz, A. A. A., H. G. Ferraz, G. A. Abraham, M. Fernandez, A. L. Bravo, and J. S. Roman (2003). Development of new hydroactive dressings based on chitosan membranes: characterization and *in vivo* behavior. *J. Biomed. Mater. Res.* **64A**:147–154.

Rabea, E. I., M. E. T. Badawy, C. V. Stevens, G. Smagghe, and Steurbaut (2003). Chitosan as antimicrobial agent: applications and mode of action. *Biomacromolecules* **4**: 1457–1465.

Ramos, V. M., N. M. Rodríguez, M. F. Díaz, M. S. Rodríguez, A. Heras, and E. Agulló (2003). *N*-Methylene phosphonic chitosan. Effect of preparation methods on its properties. *Carbohydr. Res.* **52**:39–46.

Rao, S. B., and C. P. Sharma (1997). Use of chitosan as a biomaterial: studies on its safety and hemostatic potential. *J. Biomed. Mater. Res.* **34**:21–28.

Recinos, G., K. Inaba, J. Dubose D. Demetriades, and P. Rhee, (2008). Local and systematic hemostatics in trauma. *Turk. J. Trauma Emerg. Surg.* **14**(3): 175–181.

Remuñán-López, C., and R. Bodmeier (1997). Mechanical, water uptake and permeability properties of crosslinked chitosan glutamate and alginate films. *J Control. Release.* **44**:215–225.

Ronghua, H., D. Yumin, and Y. Jianhong (2003). Preparation and anticoagulant activity of carboxybutyrylated hydroxyethyl chitosan sulfates. *Carbohydr. Polym.* **51**:431–438.

Sarasam, A., R. K. Krishnaswamy, and S. V. Madihally (2006). Blending chitosan with polycaprolactone: effects on physicochemical and antibacterial properties. *Biomacromolecules* **7**:1131–1138.

Sasaki, C., A. Kristiansen, T. Fukamizo, and K. M. Vårum (2003). Biospecific fractionation of chitosan. *Biomacromolecules* **4**:1686–1690.

Seo, S. J., Y. J. Choi, T. Akaike, A. Higuchi, and C. S. Cho (2006). Alginate/galactosylated chitosan/heparin scaffold as a new synthetic extracellular matrix for hepatocytes. *Tissue Eng.* **12**:33–44.

Shinoda, H., Y. Asou, A. Suetsugu, and K. Tanaka (2003). Synthesis and characterization of amphiphilic biodegradable copolymer, poly(aspartic acid-*co*-lactic acid). *Macromol. Biosci.* **3**:34–43.

Signini, R., and S. P. Campana Filho (1999). On the preparation and characterization of chitosan hydrochloride. *Polym. Bull.* **42**:159–166.

Smucker, R. A. (1991). Chitin primary production. *Biochem. Syst. Ecol.* **19**:357–369.

Strand, S. P., K. M. Vårum, and K. Østgaard (2003). Interactions between chitosans and bacterial suspensions: adsorption and flocculation. *Colloids Surf. B* **27**:71–81.

Takahashi, A., R. Kita, and M. Kaibara (2002). Effects of thermal annealing of segmented-polyurethane on surface properties, structure and antithrombogenicity. *J. Mater. Sci. Mater. Med.* **13**:259–264.

Teng, W. L., E. Khor, T. K. Tan, L. Y. Lim, and S. C. Tan (2001). Concurrent production of chitin from shrimp shells and fungi. *Carbohydr. Res.* **332**:305–316.

Venter, J. P., A. F. Kotzé, R. Auzély-Velty, and M. Rinaudo (2006). Synthesis and evaluation of the mucoadhesivity of a CD-chitosan derivative. *Intl. J. Pharm.* **313**:36–42.

Vournakis, J., E. R. Pariser, S. Finkielszlein, M. Helton, Poly-N-acetyl glucosamine. US patent 5,623,064. April 22, 1997.

Vournakis, J. N., M. Demcheva, A. Whitson, R. Guirca, and E. R. Pariser (2004). Isolation, purification, and characterization of poly-N-acetyl glucosamine use as a hemostatic agent. *J. Trauma.* **57**:S2–S6.

Wang, X. H., D. P. Li, W. J. Wang, Q. L. Feng, F. Z. Cui, Y. X. Xu, X. H. Song, and M. Werf (2003a). Crosslinked collagen/chitosan matrix for artificial livers. *Biomaterials* **24**:3213–3220.

Wang, X. H., D. P. Li, W. J. Wang, Q. L. Feng, F. Z. Cui, Y. X. Xu, and X. H. Song (2003b). Covalent immobilization of chitosan and heparin on PLGA surface. *Int. J. Biol. Macromol.* **33**:95–100.

Wang, X., Y. Yan, and R. Zhang (2006). A comparison of chitosan and collagen sponges as hemostatic dressings. *J. Bioact. Compat. Polym.* **21**:39–53.

Wedmore, I., J. G. McManus, A. E. Pusateri, and J. B. Holcomb (2006). A special report on the chitosan-based hemostatic dressing: experience in current combat operations. *J. Trauma.* **60**:655–658.

Whang, H. S., W. Kirsch, and S. M. Hudson (2005). Hemostatic agents from chitin and chitosan. *J. Macromol. Sci. Polym. Rev.* **45**:309–323.

Wittaya-Areekul, S., and C. Prahsarn (2006). Development and *in vitro* evaluation of chitosan-polysaccharides composite wound dressings. *Intl. J. Pharm.* **313**:123–128.

Yamane, S., N. Iwasaki, T. Majima, T. Funakoshi, T. Masuko, K. Harada, A. Minami, K. Monde, and S. Nishimura (2005). Feasibility of chitosan-based hyaluronic acid hybrid biomaterial for a novel scaffold in cartilage tissue engineering. *Biomaterials* **26**:611–619.

Yang, J., F. Tian, Z. Wang, Q. Wang, Y. Zeng, and S. Chen (2008). Effect of chitosan molecular weight and deacetylation degree on hemostasis. *J. Biomed. Mater. Res. Part B* **84B**(1): 131–137.

Zhao, L., H. Mitomo, M. Zhai, F. Yoshii, N. Nagasawa, and T. Kume (2003). Synthesis of antibacterial PVA/CM-chitosan blend hydrogels with electron beam irradiation. *Carbohydr. Polym.* **53**:439–446.

Zhao, X., K. Kato, Y. Fukumoto, and K. Nakamae (2001). Synthesis of bioadhesive hydrogels from chitin derivatives. *Int. J. Adhes. Adhes.* **21**:227–232.

Zhu, A., M. Zhang, J. Wu, and J. Shen (2002). Covalent immobilization of chitosan/heparin complex with a photosensitive hetero-bifunctional crosslinking reagent on PLA surface. *Biomaterials* **23**:4657–4665.

8

CHITIN NANOFIBERS AS BUILDING BLOCKS FOR ADVANCED MATERIALS

Youssef Habibi and Lucian A. Lucia

8.1 INTRODUCTION

Chitin (*N*-acetyl-D-glucosamine) is one of the most abundant biopolymers naturally produced in the biosphere in addition to cellulose. The word "chitin" is derived from the French word "chitine" (1830–1840), which comes from the Greek "chitōn" that translates as "mollusk." It was first isolated and identified in 1884 and since then its intrinsic features (biosynthesis, organization, chemical functions, structure, etc.) as well as its excellent properties continue to attract attention in the research community. It is the main constituent in a number of naturally occurring biomaterials, including the exoskeletons of arthropods, crustaceans, fungal cell walls, mollusks, and insects. Structurally, it resembles cellulose because it possesses the same 180° inverted (corkscrewed) β-1,4-D-glucopyranoside monomer backbone, while functionally it mimics the utility of the protein keratin.

From a chemical perspective, it is conjectured that the pendant (acetyl)-amino groups on the polysaccharide backbone play a crucial role in modulating various unique properties. In nature, chitin occurs fully acetylated with a very dense hydrogen-bonded network, which makes its completely insoluble in all usual organic solvents and thereby severely compromises its processing and final end uses. However, a more manageable and useful by-product, chitosan, can be obtained by partial deacetylation of chitin under alkaline conditions or by enzymatic hydrolysis in the presence of a chitin deacetylase. Chitosan is the most important chitin

Polysaccharide Building Blocks: A Sustainable Approach to the Development of Renewable Biomaterials, First Edition. Edited by Youssef Habibi and Lucian A. Lucia.

derivative in terms of applications. It is defined as the by-product of chitin upon achieving a degree of acetylation below 50%. Chitin and its derivatives, mainly chitosan, have displayed numerous applications in agriculture, medicine, paper making, materials, and food industries, tissue engineering, wound dressing, drug delivery, and so on. Chitin and its deacetylated form, chitosan, are attracting increasingly more attention recently as its inherent appealing biological and physicochemical characteristics are being better elaborated and understood. Several reviews have appeared recently that have highlighted advances in chitin/chitosan chemistry and engineering acquired over the past several decades.

From an engineering perspective, one of the most prominent features of this biopolymer is its ability to be processed in a variety of nanostructural shapes ranging from nanorods to more elongated particles known as nanofibrils. These nanosized building blocks have garnered tremendous attention over the past several decades toward the development of promising engineered nanomaterials such as tissue scaffolds, wound healing adjuvants, composites nanomaterials, and so on.

The purpose of this chapter is to summarize several key advances in the preparation and applications of nanochitin-based materials. As opposed to the flourishing literature abounding on the studies, modification, and applications of chitosan, published data focusing exclusively on chitin-based materials are quite rare. Therefore, this chapter will focus on chitin nanofibers as the raw material for further processed applications. We expect that the work related here will provide fundamental insight into extended uses of chitin-based nanobiomaterials in the burgeoning arena of nanobiotechnology.

8.2 CHITIN IN THE CELL WALL

Crustaceans typically protect themselves from environmental stressors including predators and pathogens by secreting an acellular exoskeleton (outer shell layer) that is composed of calcite crystals within a fibrillar organic network made up of chitin. These organic–inorganic hybrid composite tissues reveal a high degree of hierarchical organization (Figure 8.1), where the most characteristic feature is the presence of mesoscale arch structures commonly referred to as "plywood" patterns. This hybrid composite exoskeleton resembles to a great degree the structure and functionality of nacre, another organic–inorganic composite made of aragonite (calcium carbonate crystallites) arranged in a continuous parallel lamina separated by an organic matrix that can include chitin (but also includes proteins), but is generally deposited within the inner portion of the shell layer. This composite is renowned for its intrinsically high Young's modulus of 70 GPa as a direct result of the brickwork patterning of the organic–inorganic matrix. Likewise, within the exoskeleton of crustaceans, about 15–28 molecular chains of chitin are assembled in the form of narrow and long crystalline units, which are wrapped by proteins, forming nanofibrils of about 2–5 nm diameter and about 300 nm length. These nanofibrils are clustered into long chitin–protein fibers of about 50–300 nm diameter that are in turn cross-linked to form a woven-like network. The spacing between these strands is filled by a

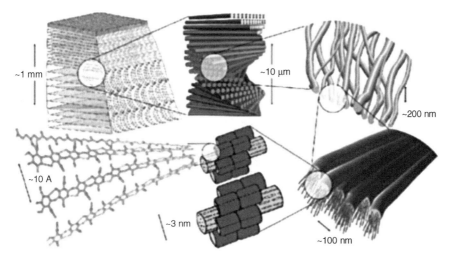

FIGURE 8.1 Structural elements of the exoskeleton material (Raabe et al., 2005). Reproduced with permission from Elsevier, © 2005.

variety of proteins and minerals. Several fibrous chitin–protein layers are stacked in helicoidal planes that are gradually rotated about their normal axis.

8.3 CHITIN EXTRACTION FROM ITS NATURAL SOURCES

Chitin, a natural polymer from marine resources, is found almost exclusively in the shells of crustaceans such as crabs, mollusks, and shrimps, the cuticles of insects, and the cell walls of fungi. It is one of the most abundant biopolymers with an annual production of approximately 10^{10}–10^{12} tons.

Zooplankton cuticles (in particular small shrimps constituting krill), a class of heterotrophic plankton, are perhaps the most important source of chitin. However, exploiting these microorganisms (can span in size from small protozoa to large metazoan or up to a few millimeters in length) is too difficult to consider for any wide scale and economical industrial use. Despite the ubiquitous occurrence of chitin, the shellfish canning waste industry (shrimp or crab shells), in which the chitin content ranges only between 8% and 33%, constitutes the main industrial source of this biopolymer.

In the cell wall, chitin is intimately associated with proteins, minerals, lipids, and pigments. Therefore, to quantitatively remove it from its associated components and achieve a satisfactory level of purity with concomitant limited degradation, the extraction process of chitin involves different steps aiming mainly to dissolve minerals, for example, calcium carbonate, and to remove proteins (see Figure 8.2).

Demineralization is generally performed through the use of acids including hydrochloric, nitric, sulfurous, acetic, and formic acids. Nevertheless, hydrochloric acid seems to be the preferred reagent and it is used with a concentration between

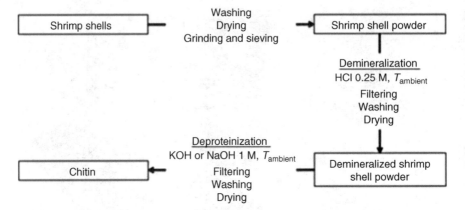

FIGURE 8.2 General extraction procedure of chitin from shrimp shells.

0.275 and 2 M for 1–48 h at temperatures varying from room temperature to 100 °C. Deproteinization is usually achieved by the application of alkaline treatments such as 1 M aqueous solutions of sodium hydroxide or potassium hydroxide as the most employed treatments from 1 to 72 h at temperatures ranging from room temperature to 100 °C. Preferentially, the treatment is conducted for 24 h at room temperature. Optionally and in order to obtain a colorless product, a decolorization (bleaching) step is often carried out to remove residual chromophores. These treatments must be adapted to each chitin source, owing to differences in the ultrastructure of the starting materials.

8.4 STRUCTURAL FEATURES

Chitin is a linear polysaccharide of $\beta\,(1 \rightarrow 4)$-2-acetamido-2-deoxy-D-glucopyranose, where all residues are comprised entirely of N-acetyl-D-glucosamine residues (Figure 8.3).

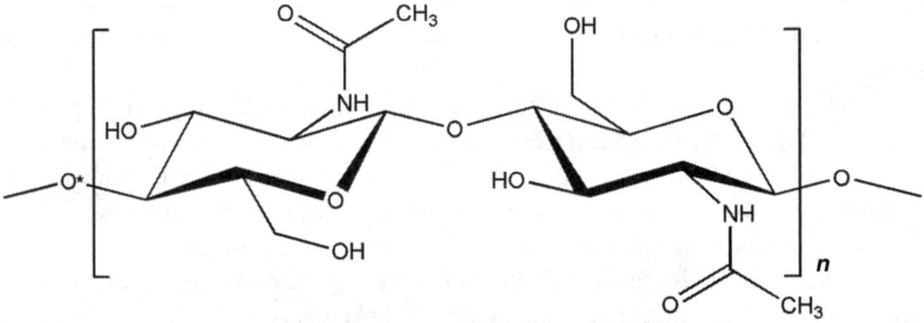

FIGURE 8.3 Chemical structure of chitin.

From a structural point of view, chitin is very similar to cellulose, in which an acetylamino group at the 2-position substitutes for a hydroxyl group on each monomeric residue. This structural feature enhances the hydrogen bonding between adjacent polymer chains, therefore strengthening the material.

Native chitin is highly crystalline and exists in three polymorphic solid-state forms identified as α-, β-, and γ-chitin that are readily differentiated by infrared and solid-state NMR spectroscopy together with X-ray diffraction. In both α- and β-forms, the chitin chains are organized in sheets, where they are tightly held by a number of intrasheet hydrogen bonds. In α-chitin, all chains are arranged in an antiparallel fashion whereas the β-form consists of a parallel arrangement (Figure 8.4) (Atkins, 1985). α-Chitin is the most abundant and stable form owing to its dominant presence within arthropod cuticles and mushroom cellular walls. It also occurs in fungal and yeast cell walls, krill, lobster, and crab tendons and shells, shrimp shells, and insect cuticles. In addition to the native chitin, α-form systematically results from recrystallization from solution (Persson et al., 1992; Helbert and Sugiyama, 1998), *in vitro* biosynthesis (Bartnicki-Garcia et al., 1994), or enzymatic polymerization (Sakamoto et al., 2000). The rarest β-chitin is found in association with proteins in squid pens (Rudall and Kenchington, 1973), tubes synthesized by pogonophoran and vestimentiferan worms (Blackwell et al., 1965; Gaill et al., 1992), aphrodite *chaetae* (Lotmar and Picken, 1950b), and lorica built by some seaweeds or protozoa (Herth et al., 1977). From a detailed analysis, it seems that the γ-form is just a variant of the α-form (Salmon and Hudson, 1997).

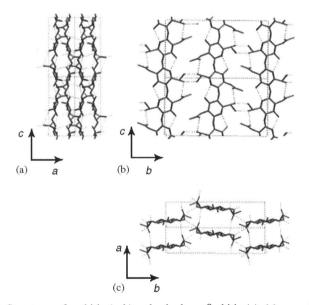

FIGURE 8.4 Structures of α-chitin (a, b) and anhydrous β-chitin (c): (a) *ac* projection; (b) *bc* projection; (c) *ab* projection (Rinaudo, 2006). Reproduced with permission from Elsevier, © 2006.

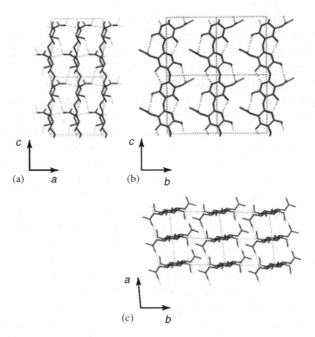

FIGURE 8.4 (*Continued*)

Similarly to the transformation of cellulose I to cellulose II by crystal swelling in alkali solution, well known as "mercerization," β-chitin is known to convert to α-chitin. Different swelling agents for chitin including HNO_3 (Lotmar and Picken, 1950a), 6 N HCl (Saito et al., 1997), and 50% NaOH (Li et al., 1999) have been studied. Although the action of HCl was elucidated for highly crystalline α-chitin of tube worm (Rudall, 1963; Saito et al., 2000), that of NaOH has been studied for a poorly crystalline sample from diatom spine for NaOH concentration between 20% and 30% (Noishiki et al., 2003).

8.5 ARCHITECTURE OF CHITIN NANOFIBERS

Chitin is known to form fibrillar arrangements embedded in a protein matrix, and the elementary nanofibrils have diameters ranging from 2.5 to 2.8 nm depending on the origin (Brine and Austin, 1975; Revol and Marchessault, 1993; Muzzarelli, 2011). Each individual nanofibril is, in turn, composed of a succession of crystalline and amorphous domains. This hierarchical structure can be broken down by different means into their individual building blocks. We distinguish between elongated nanosized fibrils, having 3–4 nm thickness and several tens of microns in length, called hereafter nanofibrils, and shorter rod-like nanocrystals or nanowhiskers. As their morphologies differ, their preparation methods are also different.

8.5.1 Preparation of Chitin Nanofibrils

After purification of chitin fibers and the removal of their associated components, chitin macrofibers are tightly held by a strong hydrogen bonding network making them highly crystalline. Thus, chitin is well known for its intractability and insolubility in water or common organic solvents. Consequently, production of chitin nanofibers remains very challenging. Most of the production methods rely upon top-down destruction approach to prepare chitin nanofibrils. The disintegration and individualization of chitin fibers along their axis to their substructural units can be realized under mechanically induced shear. Indeed, individual chitin nanofibrils 3–4 nm in cross-sectional width and at least a few microns in length have been successfully prepared from squid pen β-chitin by simple mechanical grinding in water under acidic conditions (pH 3–4, see Figure 8.5a) (Fan et al., 2008b). In this case, cationic charges were formed at the glucosamine units on the fibril surfaces, which then brought about interfibrillar electrostatic repulsion in water. The occurrence of such electrostatic repulsion promotes the individualization of the nanofibrils. However, this method of individualization of fibrils was not applicable to α-chitins.

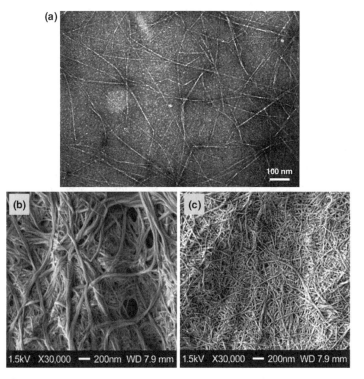

FIGURE 8.5 Nanofibers obtained after mechanical grinding without any chemical modification of β-chitin under acidic conditions (a) (Fan et al., 2008b) and of α-chitin under neutral (b) and acidic (c) conditions (Ifuku et al., 2009). Reprinted with permission from the American Chemical Society, © 2008, 2009.

Only squid pen β-chitin can be converted to long and individual fibrils by this simple method, probably because of its low crystallinity and some structural specificity such as lower content of acetyl groups. Grinding of native α-chitin in water under acidic conditions provided only fibril bundles having 10–20 nm width (Figure 8.5b and c) (Ifuku et al., 2009, 2010b). The same bundles were also obtained even by using ultrasonication to disintegrate the fibers (Zhao et al., 2007). Attempts to further individualize these bundles by the reduction of the degree of acetylation of α-chitin, before the mechanical treatment, have led to individual, but short nanoparticles (Fan et al., 2010). After partial desacetylation with 33% NaOH to reach a degree of acetylation of about 0.74–0.70, mechanical disintegration of α-chitin, under conditions promoting its cationization (acidic conditions), yields nanofibers having an overall length of 250 nm and width of 6 nm with only small numbers of long nanofibrils (length greater than 500 nm).

TEMPO (2,2,6,6-tetramethylpiperidine-1-oxyl radical)-mediated oxidation was successfully applied to β-chitins, extracted from tubeworm to introduce ionic charges and therefore enhance their defibrillation. The mechanical disintegration of the β-chitin water-insoluble fraction led to viscous and translucent gels. The gels consisted of mostly nanofibrils having 20–50 nm width and at least several microns in length. However, no nanofibrils were obtained from the controlled TEMPO-mediated oxidation of squid pen β-chitins; indeed, when an excess of the primary oxidizing agent was employed, the fibers became completely soluble polyuronic acids.

Although the bottom-up approach is widely used to produce nanofibers from chitosan because of practical solubility considerations, there have been only few attempts to produce nanofibers from pristine chitin because of its insolubility in common solvents. There are only a few solvents that can dissolve chitin, including dimethyl acetamide containing LiCl (DMAc/LiCl), methanol saturated with calcium chloride dehydrate, and highly polar fluorinated solvents such as hexafluoroisopropanol (HFIP) and hexafluoroacetone sesquihydrate. By solubilizing chitin fibers in such solvents, the dissolved molecules can be assembled into regenerated nanofibrils using different techniques. Various spinning technologies have been used, but most of the spun fiber diameters were in the micron range; these findings have already received a significant discussion by Pillai et al. in a recent review (Pillai et al., 2009). Processing chitin fibers with diameters in the nanoscale range is very challenging and has therefore been the reason for the lack of any meaningful published studies. A promising processing technique to achieve nanoscale dimensions is electrospinning, which has been performed on pure chitin with HFIP as a spinning solvent (Min et al., 2004; Noh et al., 2006). However, the spinning produced chitin nanofibers displaying a broad fiber diameter distribution; yet most of the fiber diameters were less than 100 nm. Chitin nanofibrils were also successfully produced by using a supercritical antisolvent process. Chitin nanofibrils have also been generated by precipitation from HFIP-based solution in supercritical carbon dioxide (Louvier-Hernández et al., 2005). The latter process is a fast precipitation process that appeared to have prevented strong hydrogen bonds and therefore the occurrence of large agglomerations of chitin fibers and consequently providing for a reduction of their diameters. Nevertheless, the nanofibrous chitin obtained from the latter process was very porous and tacky, possessed a very low bulk density, and the average diameter of the nanofibers was 84 nm.

One of the chief goals in the field of nanochemistry is to build materials in a controlled manner from the bottom-up. Ultrathin cytocompatible self-assembled biogenic chitin nanofibrils have been recently produced under mild conditions with a facile and gentle bottom-up strategy (Zhong et al., 2010). This approach allowed for the production of ultrafine (3 nm) nanofibers after drying a chitin/HFIP solution, and larger (10 nm) nanofibers were precipitated from DMAc/LiCl upon addition of ample amount of water.

Chitin derivatives, especially alkyl chitin, were also converted to nanofibers while chitin was regenerated to its original crystalline by posttreatments. Chitin/polymer blends especially poly(glycolic acid) were also converted to nanofibers (Pillai et al., 2009).

8.5.2 Chitin Nanocrystals

Nanocrystals result from transversal cleavage of chitin fibers at their amorphous regions that are more sensitive as compared to their crystalline sections. Although other procedures have been studied, acid-catalyzed hydrolysis is still the most common method used to prepare chitin nanocrystals. The low crystalline region of purified chitin can be hydrolyzed by acid treatment, whereas the water-insoluble, highly crystalline sections remain as nanorods. In a typical procedure, dry chitin powder is treated with 3 M HCI at 104 °C for 1–2 h. The sample is then washed with distilled water by successive low-speed centrifugation–dilution cycles until the supernatant reached a pH of about 2. At this pH, the coarse dispersion from the residue of the chitin fragments begins to convert spontaneously into a colloidal suspension. The washing is further continued by dialysis against distilled water until neutrality. Due to acid hydrolysis of chitin fibers, typically a 30–40% mass loss occurs after the acid treatment. To promote the dispersion, suspensions of the resulting crystallites have to be submitted, for few minutes, to mechanical shearing, usually sonication. Upon observation of the dried suspensions, the needle-shaped crystals were revealed. The diameter and length of these nanorods depend on the origin of chitin fibers. Examples of transmission electron microscopy images of these nanocrystals, obtained from different sources, are depicted in Figure 8.6, whereas Table 8.1 lists the corresponding dimensions.

Chitin nanocrystals have been successfully prepared by controlled TEMPO-mediated oxidation of chitin fibers in water at pH 10 followed by ultrasonic oxidation (Habibi and Vignon, 2004; Vignon et al., 2004; Fan et al., 2008a). When sufficient levels of the primary oxidizing agent were added, individualized chitin nanocrystals were obtained, and the average nanocrystal length and width were 340 and 8 nm, respectively (Figure 8.7a–c). The oxidation occurred only at the surface of the nanocrystals because the native crystalline structure was not impaired (Figure 8.7d and e). In addition, no N-deacetylation occurred on the TEMPO-oxidized chitins.

Nanocrystals, with average width and length of 6 and 250 nm, respectively (Figure 8.8), resulted also from the partial desacetylation of α-chitin by 33%

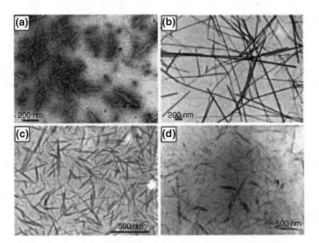

FIGURE 8.6 Transmission electron micrographs of dilute suspension of chitin nanocrystals from (a) squid pen (Paillet and Dufresne, 2001), (b) Riftia tubes (Morin and Dufresne, 2002), (c) crab shell (Nair and Dufresne, 2003a), and (d) shrimps (Sriupayo et al., 2005b). All images reproduced with permission from Elsevier, © 2001, 2002, 2005.

NaOH treatment at 90 °C for 2–4 h followed by mechanical disintegration at pH 3–4. The resulting nanocrystals retained their integrity and their native crystalline structure demonstrating that desacetylation occurred only at their surface and not at their core.

8.6 NANOCHITIN-BASED MATERIALS

Many reviews have been written on functionalizing chitin that is mainly derived from crab and shrimp shells and studying its applications in different arenas such as pharmaceutical and biomedical, paper production, textile finishes, photographic products, cements, heavy metal chelating agents, membranes, hollow fibers,

TABLE 8.1   Geometrical Characteristics of Chitin Nanocrystals from Different Sources

Source	Length (nm)	Diameter (nm)	References
Crab shell	80–600	8–50	Nair and Dufresne (2003a), Nge et al. (2003), Lu et al. (2004)
	80–350	8–12	Nge et al. (2003)
	200–500	5–20	Yamamoto et al. (2010)
Riftia tubes	500–10,000	18	Morin and Dufresne (2002)
Shrimp	50–300	5–70	Sriupayo et al. (2005a, 2005b)
	200–500	10–15	Goodrich and Winter (2007)
	231–969	12–65	Junkasem et al. (2006)
Squid pen	150–800	10	Paillet and Dufresne (2001)

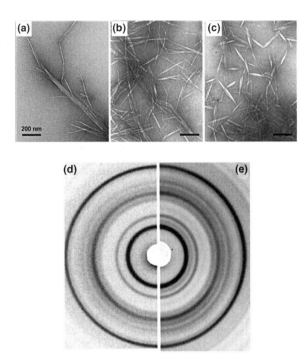

FIGURE 8.7 Transmission electron micrographs of dilute suspension of chitin nanocrystals from TEMPO-mediated oxidation of chitin fibers with various amounts of NaOCl added: (a) 2.5, (b) 5.0, and (c) 10.0 mmol/g (Fan et al., 2008a). X-ray diffraction patterns of chitin nanocrystals before (d) after (e) oxidation (Habibi and Vignon, 2004). (a), (b) and (c) reprinted with permission from the American Chemical Society, © 2008.

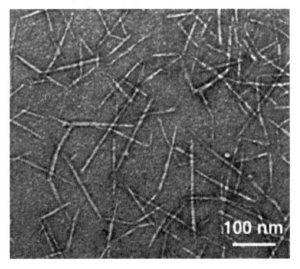

FIGURE 8.8 TEM images of partially deacetylated chitin with DNAc 0.73 prepared from α-chitin by 33% NaOH treatment at 90°C for 3 h followed by disintegration in water at pH 3–4 (Fan et al. 2010). Reprinted with permission from Elsevier, © 2010.

waste removal, affinity chromatography, biosensing, and industrial pollutant remediation (Skjåk-Braek et al., 1989; Peter, 1995; Dutta et al., 2004; Kurita, 2006;
Rinaudo, 2006; Harish Prashanth and Tharanathan, 2007; Jayakumar et al., 2010a).

In the biochemical arena, chitin is widely used to immobilize enzymes and whole
cells, allowing applications in the food industry, such as clarification of fruit juices
and processing of milk (Krajewska, 2004). Chitin can also be processed in the form of
films and fibers. Nonallergic, deodorizing, antibacterial, and moisture controlling
chitin fibers can be obtained by wet spinning of chitin dissolved in a 14 N NaOH
solution (Pillai et al., 2009). The use of regenerated chitin derivative fibers as binders
in the paper making process improves the strength of paper. However, the main
development of chitin films and fibers is in medical and pharmaceutical applications
as a wound-dressing material. In fact, chitin and its derivative, mainly chitosan, are
biocompatible, biodegradable, nontoxic, bioabsorbable, antimicrobial, and hydrat-
ing. The presence of amino groups on the polysaccharide backbone no doubt plays a
fundamental role in assisting to perform these unique properties. Due to these latter
attributes, chitin-based materials show good biocompatibility and positive effects on
wound healing, cosmetic, and pharmaceutical applications. Wound dressing, absorbable sutures, and scaffolds for tissue engineering are probably the largest groups of
biomaterials available based on nanofibers of chitin and its derivatives (Morganti and
Morganti, 2008; Jayakumar et al., 2010a, 2010b, 2011). Chitin nanofibrils exhibit an
enormous surface development that allows them to interact with enzymes, platelets,
and other cell compounds present in living tissues. Thus, the ability for faster
adequate granulation and tissue formation is accompanied by angiogenesis and
regular deposition of collagen fibers, with the consequent enhanced and correct repair
of dermoepidermal lesions (Figure 8.9). These nanofibrils have in fact relevant
biologic significance, activating the polymorphonuclear cells and fibroblasts,
increasing cytokine production, favoring giant cells migration, and stimulating
type IV collagen synthesis. Chitin nanofibrils are also able to capture enzymes
and proteins, such as drugs or other active principles, and may support hemostatic

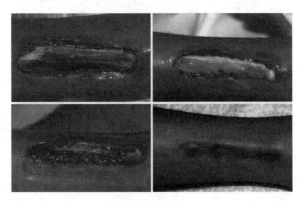

FIGURE 8.9 Repairing activity of an innovative medical device based on chitin nanofibrils
(Morganti and Morganti, 2008). Reprinted with permission from Elsevier, © 2008.

biocompatibility in human cells. Chitin nanofibers are also used as blends with other natural or synthetic polymers particularly poly(glycolic acid) (PGA) largely used in sutures and implants.

Chitin has low toxicity, is inert in the gastrointestinal tract of mammals, and is biodegradable owing to the presence of chitinases widely distributed in nature and found in bacteria, fungi, plants, and digestive systems of many animals. For these reasons, chitin fibers are used in controlled drug release and as an excipient and drug carrier in film, gel, or powder form for applications involving mucoadhesivity. They also present ability to regulate the secretion of the inflammatory mediators such as interleukin 8, prostaglandin E, interleukin 1β, and others (Kato et al., 2003). Chitin nanofibers have also been reported to aid in the production of stable, coalescent oil-in-water emulsions for a period up to 1 month (Tzoumaki et al., 2011). This property allows their inclusion in cosmetic and pharmaceutical solutions and emulsions used in cosmetic dermatology. In fact, chitin nanofibrils, protecting both corneocytes and intracorneal lamellae, help to maintain cutaneous homeostasis, neutralize the activity of free radicals and trap them in their structure, and regulate correct cell turnover (Morganti et al., 2007; Morganti and Morganti, 2008). They can also help in the formation of hygroscopic molecular films for water retention and contribute to skin hydration and therefore protect the molecules of the natural moisturizing factor present at the level of the corneocytes' membrane (Morganti et al., 2007).

A chitosan nanoscaffold has been directly obtained by deacetylation of chitin whiskers under alkaline conditions (Phongying et al., 2007). This procedure was improved by the use the microwave technique (Lertwattanaseri et al., 2009) or supercritical carbon dioxide as a reaction medium (Rinki et al., 2009).

Owing to their impressive mechanical properties, chitin nanocrystals have been incorporated into polymeric composites to enhance their mechanical properties. Although it has never been specifically measured, the stiffness of chitin nanocrystals is at least 150 GPa, based on the observation that the stiffness of cellulose is about 130 GPa and the extra bonding in the chitin crystallite will stiffen it even further (Vincent and Wegst, 2004). Solid films were obtained by either freeze drying and hot pressing or solution casting with different polymers such as soy proteins (Lu et al., 2004), poly(styrene-*co*-butyl acrylate) (Paillet and Dufresne, 2001), natural rubber (Nair and Dufresne, 2003a, 2003b; Nair et al., 2003), polycaprolactone (Morin and Dufresne, 2002; Wu et al., 2007), and poly(vinyl alcohol) (PVA) (Sriupayo et al., 2005a, 2005b). Highly transparent composites were made with acrylic resin and pristine chitin nanofibers (Shams et al., 2011) or their acetylated counterparts with a fiber content of approximately 25% (Ifuku et al., 2010a). Layer-by-layer deposition was also employed to construct chitosan–chitin whisker three-dimensional nanocomposites (Wang et al., 2010).

In the field of composites, 50:50 (w/w) blends of hyaluronan–gelatin nanocomposite scaffolds were reinforced with chitin whiskers. The resulting porous scaffolds showed improved mechanical properties as well as better thermal stability and resistance to biodegradation. Composites with 10% nanowhisker loading were good candidates for the proliferation of cultured human osteosarcoma cells (Hariraksapitak and Supaphol, 2010). Cytocompatible, highly porous, spongy silk fibroin scaffolds were filled by chitin whiskers to improve their dimensional stability and limit their shrinkage upon environmental changes. The presence of chitin whiskers embedded into a silk fibroin sponge not only improved its dimensional stability, but also enhanced its compression strength. Moreover, when compared to the neat silk fibroin sponge, the incorporation of chitin whiskers into the silk fibroin matrix was found to promote cell spreading (Wongpanit et al., 2007). Successful fabrication of α-chitin whisker-reinforced PVA nanocomposite nanofibers by electrospinning has also been accomplished (Junkasem et al., 2006, 2010). The incorporation of chitin whiskers within the as-spun nanocomposite fiber mats increased the Young's modulus by about four to eight times over that of the neat as-spun PVA fiber mat.

Another important application that has witnessed significant promise is the ability of chitin/chitosan to act as a sorbent. Recently, it has been reported that commercial dyes can be efficiently scavenged from aqueous-based solutions by the use of chitin-based hydrogels (Copello et al., 2011). Not surprisingly, it was shown that the absorption phenomenon was pH dependent in which it was purported that the mechanism was based on a spontaneous charge-associated interaction of the dye with the chitin materials. Interestingly, the chitin-based hydrogels behaved as well if not better than their analogous chitosan-based materials.

Finally, in a very elegant application variety, colloidal suspensions of chitin nanocrystals have shown ability to self-organize in liquid crystalline fashion at a given concentration. When the suspension in its chiral nematic order is dried to a solid film, the chiral nematic order is maintained mimicking the helicoid organization characteristic of the chitin microfibrils in the cuticle of arthropods (Revol and Marchessault, 1993). Related to this interesting property, a bioinspired mineralization route was chosen to prepare $CaCO_3$/chitin whisker hybrids using the liquid crystalline suspension of the chitin whisker as a template (Yamamoto et al., 2010). Suspensions of chitin nanowhiskers, exhibiting lyotropic liquid crystalline behavior, were converted to a gel form when they were exposed to ammonium carbonate vapor. $CaCO_3$ crystals formed after 30 days in chitin gels as templates. The nucleation of $CaCO_3$ occurred in the liquid crystalline chitin matrix and the $CaCO_3$ crystals deposited in the chitin gels to form hybrids that possessed an interpenetrated three-dimensional structure.

8.7 CONCLUSIONS

This chapter has attempted to provide a general overview of the salient nanotechnological features, functional considerations, and applications of chitin-based nanomaterials. A review of chitin and its derived materials is extremely relevant and timely especially given the current impetus toward developing a bio-based application platform for commercial end uses when considering their abundance, chemical tunability, biocompatibility, biodegradability, renewability, and functionality. To achieve a broad understanding of its relevance within such a platform, we especially focused on the origin of chitin, its structural features, preparation, development of nanofibrils and crystals, their characterization, and finally touched upon selected important chemical modifications that can lead to advanced materials to serve the pharmaceutical, biomedical, paper, cosmetic, food, and industrial remediation industries.

ACKNOWLEDGMENTS

We are especially grateful for receiving support from a number of U.S. federal agencies that allowed us to contribute to the development of this chapter, including the Department of Energy (Cooperative Agreement DE-FC36-04GO14308) and the Department of Agriculture (Cooperative Agreement 2006-38411-17035). We are also thankful to the members of the Laboratory of Soft Materials and Green Chemistry for their active contributions.

REFERENCES

Atkins, E. (1985). Conformations in polysaccharides and complex carbohydrates. *J. Biosci.* **8**:375–387.

Bartnicki-Garcia, S., J. Persson, and H. Chanzy (1994). An electron microscope and electron diffraction study of the effect of Calcofluor and Congo Red on the biosynthesis of chitin *in vitro*. *Arch. Biochem. Biophys.* **310**:6–15.

Blackwell, J., K. D. Parker, and K. M. Rudall (1965). Chitin in pogonophore tubes. *J. Mar. Biol.* **45**:659–661.

Brine, C. J., and P. R. Austin (1975). Renatured chitin fibrils, films and filaments. In: T. D. Church,editor, *Marine Chemistry in the Coastal Environment*, ACS Symposium Series, Vol. 18, Washington, DC: American Chemical Society, pp. 505–518.

Copello, G. J., A. M. Mebert, M. Raineri, M. P. Pesenti, and L. E. Diaz (2011). Removal of dyes from water using chitosan hydrogel/SiO_2 and chitin hydrogel/SiO_2 hybrid materials obtained by the sol-gel method. *J. Hazard. Mater.* **186**:932–939.

Dutta, P. K., J. Dutta, and V. Tripathi (2004). Chitin and chitosan: chemistry, properties and applications. *J. Sci. Ind. Res.* **63**:20–31.

Fan, Y., T. Saito, and A. Isogai (2008a). Chitin nanocrystals prepared by TEMPO-mediated oxidation of α-chitin. *Biomacromolecules* **9**:192–198.

Fan, Y., T. Saito, and A. Isogai (2008b). Preparation of chitin nanofibers from squid pen β-chitin by simple mechanical treatment under acid conditions. *Biomacromolecules* **9**:1919–1923.

Fan, Y., T. Saito, and A. Isogai (2010). Individual chitin nano-whiskers prepared from partially deacetylated α-chitin by fibril surface cationization. *Carbohydr. Polym.* **79**:1046–1051.

Gaill, F., J. Persson, J. Sugiyama, R. Vuong, and H. Chanzy (1992). The chitin system in the tubes of deep sea hydrothermal vent worms. *J. Struct. Biol.* **109**:116–128.

Goodrich, J. D., and W. T. Winter (2007). α-Chitin nanocrystals prepared from shrimp shells and their specific surface area measurement. *Biomacromolecules* **8**:252–257.

Habibi, Y., and M. R. Vignon (2004). TEMPO-mediated oxidation of chitin nanocrystals. Unpublished results.

Hariraksapitak, P., and P. Supaphol (2010). Preparation and properties of α-chitin-whisker-reinforced hyaluronan–gelatin nanocomposite scaffolds. *J. Appl. Polym. Sci.* **117**:3406–3418.

Harish Prashanth, K. V., and R. N. Tharanathan (2007). Chitin/chitosan: modifications and their unlimited application potential—an overview. *Trends Food Sci. Technol.* **18**:117–131.

Helbert, W., and J. Sugiyama (1998). High-resolution electron microscopy on cellulose II and α-chitin single crystals. *Cellulose* **5**:113–122.

Herth, W., A. Kuppel, and E. Schnepf (1977). Chitinous fibrils in the lorica of the flagellate chrysophyte *Poteriochromonas stipitata* (syn *Ochromonas malhamensis*). *J. Cell Biol.* **73**:311–321.

Ifuku, S., M. Nogi, K. Abe, M. Yoshioka, M. Morimoto, H. Saimoto, and H. Yano (2009). Preparation of chitin nanofibers with a uniform width as α-chitin from crab shells. *Biomacromolecules* **10**:1584–1588.

Ifuku, S., S. Morooka, M. Morimoto, and H. Saimoto (2010a). Acetylation of chitin nanofibers and their transparent nanocomposite films. *Biomacromolecules* **11**:1326–1330.

Ifuku, S., M. Nogi, M. Yoshioka, M. Morimoto, H. Yano, and H. Saimoto (2010b). Fibrillation of dried chitin into 10–20 nm nanofibers by a simple grinding method under acidic conditions. *Carbohydr. Polym.* **81**:134–139.

Jayakumar, R., D. Menon, K. Manzoor, S. V. Nair, and H. Tamura (2010a). Biomedical applications of chitin and chitosan based nanomaterials: a short review. *Carbohydr. Polym.* **82**:227–232.

Jayakumar, R., M. Prabaharan, S. V. Nair, and H. Tamura (2010b). Novel chitin and chitosan nanofibers in biomedical applications. *Biotechnol. Adv.* **28**:142–150.

Jayakumar, R., M. Prabaharan, P. T. Sudheesh Kumar, S. V. Nair, and H. Tamura (2011). Biomaterials based on chitin and chitosan in wound dressing applications. *Biotechnol. Adv.* **29**:322–337.

Junkasem, J., R. Rujiravanit, and P. Supaphol (2006). Fabrication of α-chitin whisker-reinforced poly(vinyl alcohol) nanocomposite nanofibres by electrospinning. *Nanotechnology* **17**:4519–4528.

Junkasem, J., R. Rujiravanit, B. P. Grady, and P. Supaphol (2010). X-ray diffraction and dynamic mechanical analyses of α-chitin whisker-reinforced poly(vinyl alcohol) nanocomposite nanofibers. *Polym. Int.* **59**:85–91.

Kato, Y., H. Onishi, and Y. Machida (2003). Application of chitin and chitosan derivatives in the pharmaceutical field. *Curr. Pharm. Biotechnol.* **4**:303–309.

Krajewska, B. (2004). Application of chitin- and chitosan-based materials for enzyme immobilizations: a review. *Enzyme Microb. Technol.* **35**:126–139.

Kurita, K. (2006). Chitin and chitosan: functional biopolymers from marine crustaceans. *Mar. Biotechnol.* **8**:203–226.

Lertwattanaseri, T., N. Ichikawa, T. Mizoguchi, Y. Tanaka, and S. Chirachanchai (2009). Microwave technique for efficient deacetylation of chitin nanowhiskers to a chitosan nanoscaffold. *Carbohydr. Res.* **344**:331–335.

Li, J., J. F. Revol, and R. H. Marchessault (1999). Alkali induced polymorphic changes of chitin. In: S. H. Imam, R. V. Greene, and B. R. Zaidi, editors, *Biopolymers: Utilizing Nature's Advanced Materials*, Vol. **723**, Washington, DC: American Chemical Society, pp. 88–96.

Lotmar, W. and L. Picken (1950a). A new crystallographic modification of chitin and its distribution. *Cell. Mol. Life Sci.* **6**:58–59.

Lotmar, W., and L. E. R. Picken (1950b). A new crystallographic modification of chitin and its distribution. *Experientia* **6**:58–59.

Louvier-Hernández, J. F., G. Luna-Bárcenas, R. Thakur, and R. B. Gupta (2005). Formation of chitin nanofibers by supercritical antisolvent. *J. Biomed. Nanotechnol.* **1**:109–114.

Lu, Y., L. Weng, and L. Zhang (2004). Morphology and properties of soy protein isolate thermoplastics reinforced with chitin whiskers. *Biomacromolecules* **5**:1046–1051.

Min, B.-M., S. W. Lee, J. N. Lim, Y. You, T. S. Lee, P. H. Kang, and W. H. Park (2004). Chitin and chitosan nanofibers: electrospinning of chitin and deacetylation of chitin nanofibers. *Polymer* **45**:7137–7142.

Morganti, P., and G. Morganti (2008). Chitin nanofibrils for advanced cosmeceuticals. *Clin. Dermatol.* **26**:334–340.

Morganti, P., L. Yuanhong, and G. Morganti (2007). Nano-structured products: technology and future. *J. Appl. Cosmetol.* **25**:161.

Morin, A., and A. Dufresne (2002). Nanocomposites of chitin whiskers from Riftia tubes and poly(caprolactone). *Macromolecules* **35**:2190–2199.

Muzzarelli, R. A. A. (2011). Chitin nanostructures in living organisms. In: N. S. Gupta, editor, *Chitin*, Vol. 34, The Netherlands: Springer, pp. 1–34.

Nair, K. G., and A. Dufresne (2003a). Crab shell chitin whisker reinforced natural rubber nanocomposites. 1. Processing and swelling behavior. *Biomacromolecules* **4**:657–665.

Nair, K. G., and A. Dufresne (2003b). Crab shell chitin whisker reinforced natural rubber nanocomposites. 2. Mechanical behavior. *Biomacromolecules* **4**:666–674.

Nair, K. G., A. Dufresne, A. Gandini, and M. N. Belgacem (2003). Crab shell chitin whiskers reinforced natural rubber nanocomposites. 3. Effect of chemical modification of chitin whiskers. *Biomacromolecules* **4**:1835–1842.

Nge, T. T., N. Hori, A. Takemura, H. Ono, and T. Kimura (2003). Phase behavior of liquid crystalline chitin/acrylic acid liquid mixture. *Langmuir* **19**:1390–1395.

Noh, H. K., S. W. Lee, J.-M. Kim, J.-E. Oh, K.-H. Kim, C.-P. Chung, S.-C. Choi, W. H. Park, and B.-M. Min (2006). Electrospinning of chitin nanofibers: degradation behavior and cellular response to normal human keratinocytes and fibroblasts. *Biomaterials* **27**:3934–3944.

Noishiki, Y., H. Takami, Y. Nishiyama, M. Wada, S. Okada, and S. Kuga (2003). Alkali-induced conversion of β-chitin to α-chitin. *Biomacromolecules* **4**:896–899.

Paillet, M., and A. Dufresne (2001). Chitin whisker reinforced thermoplastic nanocomposites. *Macromolecules* **34**:6527–6530.

Persson, J. E., A. Domard, and H. Chanzy (1992). Single crystals of α-chitin. *Int. J. Biol. Macromol.* **14**:221–224.

Peter, M. G. (1995). Applications and environmental aspects of chitin and chitosan. *J. Macromol. Sci. A* **32**:629–640.

Phongying, S., S.-I. Aiba, and S. Chirachanchai (2007). Direct chitosan nanoscaffold formation via chitin whiskers. *Polymer* **48**:393–400.

Pillai, C. K. S., W. Paul, and C. P. Sharma (2009). Chitin and chitosan polymers: chemistry, solubility and fiber formation. *Prog. Polym. Sci.* **34**:641–678.

Raabe, D., P. Romano, C. Sachs, A. Al-Sawalmih, H. G. Brokmeier, S. B. Yi, G. Servos, and H. G. Hartwig (2005). Discovery of a honeycomb structure in the twisted plywood patterns of fibrous biological nanocomposite tissue. *J. Cryst. Growth* **283**:1–7.

Revol J. F., and R. H. Marchessault (1993). *In vitro* chiral nematic ordering of chitin crystallites. *Int. J. Biol. Macromol.* **15**:329–335.

Rinaudo, M. (2006). Chitin and chitosan: properties and applications. *Prog. Polym. Sci.* **31**:603–632.

Rinki, K., S. Tripathi, P. K. Dutta, J. Dutta, A. J. Hunt, D. J. Macquarrie, and J. H. Clark (2009). Direct chitosan scaffold formation via chitin whiskers by a supercritical carbon dioxide method: a green approach. *J. Mater. Chem.* **19**:8651–8655.

Rudall, K. M. (1963). The chitin/protein complexes of insect cuticles. In: J. W. L. Beament, J. E. Treherne, and V. B. Wigglesworth,editors, *Advances in Insect Physiology*, Vol. 1, New York: Academic Press, pp. 257–313.

Rudall, K. M., and W. Kenchington (1973). The chitin system. *Biol. Rev.* **48**:597–633.

Saito, Y., J. L. Putaux, T. Okano, F. Gaill, and H. Chanzy (1997). Structural aspects of the swelling of β-chitin in HCl and its conversion into α-chitin. *Macromolecules* **30**:3867–3873.

Saito, Y., T. Okano, F. Gaill, H. Chanzy, and J. L. Putaux (2000). Structural data on the intra-crystalline swelling of β-chitin. *Int. J. Biol. Macromol.* **28**:81–88.

Sakamoto, J., J. Sugiyama, S. Kimura, T. Imai, T. Itoh, T. Watanabe, and S. Kobayashi (2000). Artificial chitin spherulites composed of single crystalline ribbons of α-chitin via enzymatic polymerization. *Macromolecules* **33**:4155–4160.

Salmon, S., and S. M. Hudson (1997). Crystal morphology, biosynthesis, and physical assembly of cellulose, chitin, and chitosan. *Polym. Rev.* **37**:199–276.

Shams, M., S. Ifuku, M. Nogi, T. Oku, and H. Yano (2011). Fabrication of optically transparent chitin nanocomposites. *Appl. Phys. A* **102**:325–331.

Skjåk-Braek, G., T. Anthossen, and P. A. Sandford, editors (1989). *Chitin and Chitosan: Sources, Chemistry, Biochemistry, Physical Properties and Applications*, London: Elsevier Applied Science.

Sriupayo, J., P. Supaphol, J. Blackwell, and R. Rujiravanit (2005a). Preparation and characterization of α-chitin whisker-reinforced chitosan nanocomposite films with or without heat treatment. *Carbohydr. Polym.* **62**:130–136.

Sriupayo, J., P. Supaphol, J. Blackwell, and R. Rujiravanit (2005b). Preparation and characterization of α-chitin whisker-reinforced poly(vinyl alcohol) nanocomposite films with or without heat treatment. *Polymer* **46**:5637–5644.

Tzoumaki, M. V., T. Moschakis, V. Kiosseoglou, and C. G. Biliaderis (2011). Oil-in-water emulsions stabilized by chitin nanocrystal particles. *Food Hydrocolloids*

Vignon, M., S. Montanari, and Y. Habibi, (2004). Crystalline polysaccharide derivatives in the form of water-insoluble aggregates of microcrystals, for use e.g. as viscosity modifiers or super-absorbers, manufactured by controlled oxidation of primary alcohol groups. FR2854161.

Vincent, J. F. V., and U. G. K. Wegst (2004). Design and mechanical properties of insect cuticle. *Arthropod Struct. Dev.* **33**:187–199.

Wang, Z., Q. Hu, and L. Cai (2010). Chitin fiber and chitosan 3D composite rods. *Int. J. Polym. Sci.* doi:10.1155/2010/369759.

Wongpanit, P., N. Sanchavanakit, P. Pavasant, T. Bunaprasert, Y. Tabata, and R. Rujiravanit (2007). Preparation and characterization of chitin whisker-reinforced silk fibroin nanocomposite sponges. *Eur. Polym. J.* **43**:4123–4135.

Wu, X., F. G. Torres, F. Vilaseca, and T. Peijs (2007). Influence of the processing conditions on the mechanical properties of chitin whisker reinforced poly(caprolactone) nanocomposites. *J. Biobased Mater. Bioenergy* **1**:341–350.

Yamamoto, Y., T. Nishimura, T. Saito, and T. Kato (2010). $CaCO_3$/chitin-whisker hybrids: formation of $CaCO_3$ crystals in chitin-based liquid-crystalline suspension. *Polym. J.* **42**:583–586.

Zhao, H. P., X. Q. Feng, and H. J. Gao (2007). Ultrasonic technique for extracting nanofibers from nature materials. *Appl. Phys. Lett.* 90.

Zhong, C., A. Cooper, A. Kapetanovic, Z. Fang, M. Zhang, and M. Rolandi (2010). A facile bottom-up route to self-assembled biogenic chitin nanofibers. *Soft Matter* **6**:5298–5301.

9

ELECTRICAL CONDUCTIVITY AND POLYSACCHARIDES

Axel Rußler and Thomas Rosenau

9.1 INTRODUCTION

Conductivity and polymers are two terms that have seldom used in a common context until the mid 1970s when Alan J. Heeger, Alan G. MacDiarmid, and Hideki Shirakawa discovered electrically conductive polymers and were honored for this achievement with the Nobel Prize in 2000 (Shirakawa et al., 1977).

Since that time, intrinsically conductive polymers (ICPs) have made constant progress and a growing diversity of different derivatives with specialized and improved characteristics has been developed. But still there is not much common ground between this interesting new class of materials and polysaccharides. Polysaccharides by themselves do not show electrical conductivity and in fact are normally known as good insulators. Nevertheless, there exists a multiplicity of possibilities for the creation and use of conductive polysaccharides and composites based on this class of natural polymers (Pei et al., 1992; Kuhn et al., 1995; MacDiarmid, 1997; Planès et al., 1999; Schultze et al., 1999; Rehahn, 2003; Lu et al., 2010a and 2010b).

In this chapter, electrical conductivity and semiconductivity of polysaccharidic materials is in focus. One has to separate this phenomenon that is based on the delocalization of π-electrons in conjugated systems from ion conductivity or electrolytic conductivity that can be found in solutions of charged polymers, and is well known also from solutions of charged polysaccharides (Finkenstadt, 2005).

Polysaccharide Building Blocks: A Sustainable Approach to the Development of Renewable Biomaterials, First Edition. Edited by Youssef Habibi and Lucian A. Lucia.
© 2012 John Wiley & Sons, Inc. Published 2012 by John Wiley & Sons, Inc.

9.1.1 Conductive Components in Composites with Polysaccharides

There are different possibilities and approaches to combine conductivity with polysaccharides. Besides the proper functionalization of polysaccharides, a challenge hardly met until now, a more feasible way seems to be the creation of composite materials involving conductive components. These conductive substances can again be carbon species, metal particles, and intrinsic conductive polymers.

Besides conventional conducting forms of carbon, such as conducting carbon black and graphite, nanosized carbon species such as single- or multiwalled carbon nanotubes (SWCNTs, MWCNTs) are used for the production of conductive composites. CNTs exhibit an extreme aspect ratio with diameters of about 1–50 nm and a length of up to several millimeters. They are not only conductive or semiconductive with a very high ampacity and thermal conductivity, depending on their intrinsic structure, but also offer an extremely high tensile strength of up to 63 GPa for MWCNTs. Steel exhibits only a strength of up to 2 GPa but has fivefold higher density and a somewhat equal ratio for the Young's modulus. Thus, together with the implementation of conductivity, the utilization of CNTs in polysaccharide matrices typically entails additionally significant improvements for the mechanical characteristics of the composite (Yu et al., 2000a, 2000b; Bellucci, 2005). However, a problem in the proper application of carbon nanoparticles, such as CNTs, is their tendency to agglomerate. Powerful dispersing agents are therefore needed. Charged polysaccharide species can help to overcome this problem (Hu and Hu, 2009; Sarrazin et al., 2009).

In contrast, conducting polymers usually exhibit not very good mechanical properties, a reason for the use in composites with components that supply sufficient strength. Their conductivity does not reach that of CNTs, of which some species exhibit even superconductivity at low temperatures, but is sufficient for most applications. On the other hand, the manufacturing of ICTs is known to be environment friendly and the products are classified as safe for the human health, whereas there are some severe concerns about health risks of carbon nanoparticles (Genaidy et al., 2009; Savolainen et al., 2010).

The first polymer discovered to exhibit conductivity was polyacetylene (PA). However, polyaniline (PAni) and polypyrrole (PPy) play a major role in the literature and applications today due to their good environmental stability, polythiophene (PThio) and polyfuran (PF) as well as their derivatives being equally well in focus. For the use in industry meanwhile the conducting polymer complex poly(3,4-ethylenedioxythiophene)/poly(styrene sulfonate) (PEDOT:PSS) is quite established, because of its good processability, stability, and optical characteristics. There are already a lot more such substances known and research is very intense on this topic. Figure 9.1 shows the structures of some intrinsic conductive polymers (Kuhn et al., 1995; MacDiarmid, 1997; Groenendaal et al., 2000; Rehahn, 2003; Lu et al., 2010b).

Conductive polymers gain their properties by the conjugation of π-electrons due to the overlapping of carbon p-orbitals along their polymeric chain. Doping with either electron-rich (n-dopant) or electron-poor (p-dopant) agents can further increase the conductivity and is in some cases obligatory. Doping with ionic substances or oxidative

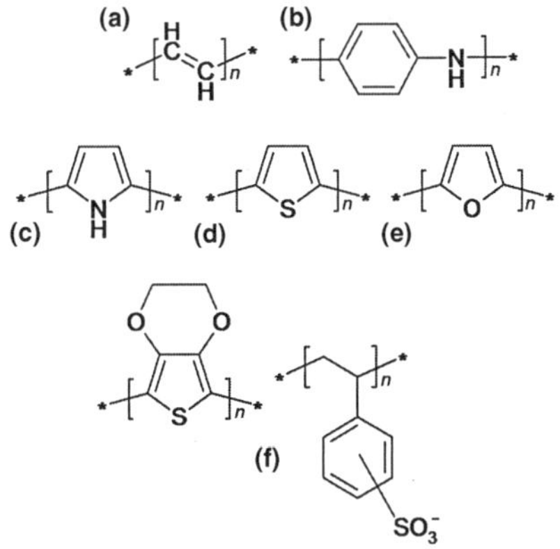

FIGURE 9.1 Structures (monomeric units) of some ICPs: (a) polyacetylene, (b) polyaniline, (c) polypyrrole, (d) polythiophene, (e) polyfuran, and (f) poly(3,4-ethylenedioxythiophene)/poly(styrene sulfonate).

doping helps to narrow the energy band gap between the HOMO and LUMO of the ICP by improving the mobility of charge carriers. Typical dopants are Cl^-, ClO_4^-, $Fe[CN]_6^{3-}$, NO_3^-, PF_6^-, BF_4^-, alkyl sulfonates, alkylbenzene sulfonates, alkyl sulfates, and poly (styrene sulfonate) (PSS). Investigations on the influence of different dopants used in the PPy–cellulose system showed significant differences in the resulting conductivity as well as in the morphology of deposited PPy on cellulose surfaces. Dopants with larger molecular structures were thereby found to show better results according to conductivity and achievable dopant level, which was accounted to a stronger interaction of these molecules with cellulose as well as with PPy. These investigations underline the importance of a careful choice of the dopant (Ding et al., 2010).

Ionic polysaccharides can also act as such dopant agents. Carboxymethyl cellulose (CMC) was used very effectively together with PPy (Sasso et al., 2008) as well as heparin and hyaluronic acid (Zhou et al., 1999; Collier et al., 2000; Finkenstadt, 2005).

Besides these unique features, conducting polymers typically lack good processability due to their insolubility in all or nearly all solvents; in addition, they exhibit only poor mechanical properties. Some soluble derivatives of conducting polymers are meanwhile known, but they are still very expensive and show no significantly improved mechanical properties (Arslan et al., 2007; Esfandiari, 2008; Karlsson et al., 2009).

The synthesis of conjugated polymers can follow in general two routes: the electrochemical polymerization of the monomers on electrode surfaces and the chemical polymerization with the help of a catalyst, usually an oxidizing reagent, often iron(III) chloride ($FeCl_3$) or ammonium persulfate (($NH_4)_2S_2O_8$). This polymerization occurs normally in the presence of dopants, but doping can also be applied afterward. PAni exists in different oxidative forms; a greenish form called emeraldine is the most commonly used type with the highest stability and best conductivity. A unique feature of PAni is its sensitivity to acidic doping. Only the protonated form exhibits considerable conductivity. This effect can be used in detector devices (Ansari, 2006; Bhadra et al., 2009).

PPy is oxidized very easily. It forms a black substance and needs no additional dopant if synthesized chemically. In contrast to PAni, which is soluble in the base form at least in *N*-methylpyrrolidone, PPy is insoluble in any solvent and therefore has to be synthesized *in situ* or handled in solid form as it is.

A big advantage of PAni as well as PPy is their reasonable environmental stability and the ease of polymerization that can be done even in aqueous media, which is of great benefit for the work with polysaccharide matrices, as they often show hygroscopic properties (Feast et al., 1996; Malinauskas, 2001; Bhadra et al., 2009; Cerqueira et al., 2009; Lu et al., 2010a).

A large variety of possible high-tech applications such as field-effect transistors (FETs), polymer light-emitting diodes (PLEDs), batteries, actuators, solar cells, and chemical and biochemical sensors, but also common applications such as antistatic coatings, render the usage of conductive polymers a promising research field in general. The merger of specific properties of polysaccharides and their nearly endless number of possible derivatives on the one hand and conducting structures on the other hand promises a great variety of useful and reasonable applications (Kuhn et al., 1995; Kathirgamanathan et al., 2000; Rehahn, 2003; Ryu et al., 2004; Hu et al., 2005; Guimard et al., 2007; Bouvree et al., 2009; Nyström et al., 2009; Kim et al., 2010).

9.1.2 General Ways to Prepare Conductive Polysaccharide Materials

The combination of conductive materials with polysaccharides can follow very different approaches to produce composite materials suitable for the aimed purpose and allowing for convenient handling and sufficient processability of the materials used.

Besides ordered mechanical compounds such as textiles that contain polysaccharide as well as conducting filaments (see (a) in Figure 9.2), there is the possibility of a more random organization of the components as a network of fibers or particles. These mixtures can be produced by physical mixing of the objects, such as in mixed powders or suspensions, by *in situ* synthesis, or by precipitation of one of the components within the network of the other component (see (b) and (c) in Figure 9.2).

The modification of shaped polysaccharide objects such as fibers and films itself can follow several different routes. In that case, the polysaccharide can always be understood as the substrate. The incorporation of conducting particles is one way for this modification (see (d) in Figure 9.2). Here the conducting substance is distributed more or less uniformly within the polysaccharide matrix. In contrast, only the surface can be modified. This is done by depositing a conductive substance on the polysaccharide surface, which can be done either by applying a continuous conductive layer (see (e) in Figure 9.2) or via the attachment of individual particles (see (f) in Figure 9.2), mostly nanoparticles, which on their part build up a continuous conducting phase. The fourth possibility is that of a real polymer blend, where both materials are intimately merged to give a macroscopically homogeneous, solid mixture (see (g) in Figure 9.2).

A layer-by-layer deposition with phases of conducting and nonconducting polymers or particles is a possible composite that is, however, until now hardly used for

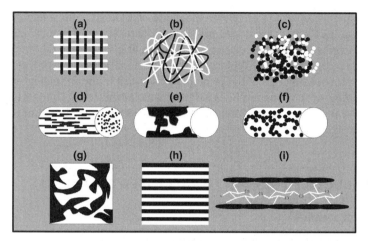

FIGURE 9.2 Different general principles for introducing conductivity into polysaccharides: (a) textile produced from polysaccharide and conducting filaments, (b) random network of polysaccharide and conducting fibers, (c) mixture of polysaccharide and conducting particles, (d) conducting particles incorporated into a polysaccharide, (e) conducting material coated on the surface of a polysaccharide, (f) conducting nanoparticles coated onto a polysaccharide matrix, (g) blend of polysaccharide and conducting material, (h) layer-by-layer deposition of polysaccharide and conducting material, and (i) chemical derivatization of a polysaccharide (cellulose) with substituents forming an electrically conducting phase.

conducting polysaccharide composites (see (h) in Figure 9.2). One rare example is the spray layer-by-layer deposition of chitosan and carbon nanoparticles in one phase as investigated by Bouvree et al. (2009) to design a sensor for polar vapors.

In any case, evidently it is necessary to form a continuous conducting phase, either on a surface or within a polysaccharide matrix, to find the typical characteristics of the conducting phase also in the finished composite. The better this continuity, the better the resulting conductivity. As this is normally influenced not only by the morphology but also by the ratio between the conducting part and the polysaccharide matrix, in composites there will be always an interplay between conductivity performance and mechanical properties of the product.

Only very few examples exist also for the final possibility, a chemical derivatization that forms a covalent bond between the polysaccharide and the conductive component (see (i) in Figure 9.2) (Russler et al., 2010).

The given possibilities for the production of conducting polysaccharide composites are illustrated in Figure 9.2.

9.2 TEXTILES

The oldest combination of conductive materials and polysaccharide is found in textiles produced by weaving metallic filaments into the fabric. This was usually used for the electrical heating of these textiles, for example, for blankets, to equip it with antistatic properties, for electromagnetic shielding, or to integrate simple sensor systems. Such materials are still used for antistatic professional clothing in electrical engineering. However, nowadays this fabrication route does not play an important role anymore. The replacement of the metallic filaments by conducting filaments produced by one of the principles discussed in the following is increasingly opening up new possibilities (Kathirgamanathan et al., 2000; Hakansson et al., 2004; Bhat et al., 2006; Micusík et al., 2007; Knittel and Schollmeyer, 2009; Jourand et al., 2010).

9.3 CONDUCTIVE POLYSACCHARIDE-BASED COMPOSITES

9.3.1 Incorporation

A more advanced approach than imparting different functions on a textile by combining different types of filaments within the fabric is the incorporation of particles into the filament matrix, which carry the specific functionality and transfer it into the product composite. Products commercially available today combining conductivity with polysaccharides usually are based on the incorporation of carbon species into the polysaccharide matrix, mostly cellulose, but also ICP particles are known as functional fillers (Evans et al., 2006; Lu et al., 2010a).

Regenerated cellulose fibers can be doped with conductive black during fabrication according to the NMMO process. Fibers produced this way still comprise enough strength to be normally processed if they contain 50% of conductive black and comprise a specific electrical resistance of about 0.4–0.5 Ω cm (Table 9.1) (Taeger et al., 1997; Meister et al., 2003).

A more advanced approach was made by Chen et al. (2009) who incorporated multiwalled carbon nanotubes into a wet-spun fiber made of regenerated bacterial cellulose. By this, not only a low conductivity was introduced into the material by the addition of only 1 wt% of the MWCNTs, but simultaneously the mechanical properties were also significantly enhanced.

Evans et al. (2006) presented a patent that claims a direct introduction of carbonaceous particles into bacterial cellulose by adding the carbon to the cultivation medium. They gained a conductive material that was meant to act as membrane electrode in fuel cells.

To produce conductive cellulose films, containing MWCNTs as the conducting component, Yoon et al. (2006) chose a path that stands somehow between coating and incorporation. They incorporated MWCNTs into a water-swollen network of bacterial cellulose from *Gluconacetobacter xylinus* by simple dipping into a surfactant-stabilized dispersion of the nanotubes and subsequent drying of the so produced composite. With an incorporation of 9.6 wt% of MWCNTs, an electrical conductivity of up to 1.4×10^{-1} S/cm was achievable (Yoon et al., 2006).

Wei et al. (2010) used cellulose xanthate, a well-known cellulose derivative for the production of regenerated cellulose structures (viscose/rayon fibers), as an intermediate to mix the polymer with carbon nanotubes and subsequently

TABLE 9.1 Overview of the Approaches Presented for Section 9.3.1: Incorporation

Conductive Component	Polysaccharide	Method Used	References
Conductive black	Cellulose	NMMO process	Taeger et al. (1997), Meister et al. (2003)
MWCNT	Cellulose	DMAc/LiCl spinning	Chen et al. (2009)
Carbonaceous particles	Cellulose	Addition to bacterial cellulose cultivation medium	Evans et al. (2006)
MWCNT	Cellulose	Dipping of bacterial cellulose gel and subsequent drying	Yoon et al. (2006)
Carbon nanotubes	Cellulose	Viscose process	Wei et al. (2010)
Carbon nanoparticles	Chitosan	Drying of solution with suspended particles	Bouvree et al. (2009)
PAni nanofibers	Chitosan	Drying of solution with suspended particles	Du et al. (2009)
PPy	Cellulose	*In situ* polymerization	Dall'Acqua et al. (2004), Beneventi et al. (2006)
PPy, PAni	Cellulose	*In situ* polymerization and incorporation into IL solutions and gels with subsequent curing	Rußler et al. (2011)

by regeneration of the cellulose came to a cellulose–CNT composite (Wei et al., 2010).

Also on the basis of carbon nanoparticles, this time round in shape and with chitosan instead of cellulose as the polysaccharide component, Bouvree et al. (2009) produced a nanobiocomposite that they found to be a good candidate for a chemical sensor to determine polar solvent vapors, for example, for the sensing of water vapor.

Recently, Du et al. (2009) also used chitosan as a matrix to produce a biosensor for the determination of hydrogen peroxide. They used PAni nanofibers as the conductive component, produced at the interface of carbon tetrachloride and aqueous solution. The composite material was easy to obtain and readily processable. The detector was completed by immobilizing horseradish peroxidase within the PAni/chitosan matrix.

A different approach lying between incorporation and blending was followed by Dall'Acqua et al. in 2004. A composite of cellulose and PPy was made by an *in situ* polymerization within fibers of regenerated cellulose. For that the fibers are soaked in an aqueous solution containing the catalyst (FeCl$_3$) and dopant (antraquinone-2,6-disulfonic acid disodium salt) before the monomer is added to start polymerization. In this way, a homogeneous PPy network is formed within the fibers. The conductivity in this case is directly related to the fiber structure as well as obviously to the content of PPy and the oxidant-to-dopant ratio. Viscose fibers show a much higher PPy uptake and a more uniform distribution than lyocell fibers. This seems to be a consequence of the higher crystallinity and the bigger crystallites of lyocell fibers. Hence, the PPy network was supposed to be formed especially in the amorphous parts of the polysaccharide matrix. More recently, we made a similar approach by incorporation and formation of PPy and PAni particles within cellulose solutions and cellulose gels with subsequent regeneration of the polysaccharide (Dall'Acqua et al., 2004; Rußler et al., 2011).

In 2006, Beneventi et al. focused on the necessary concentration of iron ions within the cellulose matrix and the formation of PPy on the fiber surface (Beneventi et al., 2006).

Incorporation of conducting particles proved to be a good possibility that is easy to handle if the particles are either formed *in situ* or are small enough to be well dispersed in the polysaccharide matrix and stable enough to survive the processing conditions for the polysaccharide, which can be quite harsh, as for instance in the NMMO process. The result is a material that offers by its intrinsic structure a protection to the conducting particles by enclosure within the polysaccharide component. On the other hand, complete shielding or encapsulation of the conducting particles by the surrounding polysaccharide phase is also unwanted as it would decrease or even prevent conductivity. So incorporation of conducting material is more attractive if a general conducting or semiconducting effect of the composite product is aimed at. For sensor applications, where big surfaces with electrochemical activity or interface interaction are needed, these kinds of materials appear less suitable.

9.3.2 Coating

As already mentioned, polysaccharides can be utilized as a supporting material for ICTs and help to open up many perspectives for this kind of composites.

Especially in the modification of fibrous structures, coating is an elegant way to combine the advantageous features of polysaccharides with electrical conductivity. As most of the common conductive polymers known are insoluble in any solvent, an *in situ* polymerization on the polysaccharide surface is necessary if a more or less complete encapsulation of the polysaccharide objects is required.

Already in 1993 and 1996, Bhadani et al. performed preliminary studies on the electrochemical coating of natural fibers from cotton, silk, and wool with PAni and PPy (Bhadani et al., 1993, 1996), and more recently, Hosseini and Pairovi (2005) as well as Esfandiari (2008) compared the chemical polymerization of pyrrole on cotton and silk in vapor and liquid phases (Table 9.2).

Nyström et al. (2010) showed very recently the production of a composite material based on nanocellulose and PPy by coating of microfibrillated cellulose.

A very old approach was renewed by Johnston et al. (2009), presenting a study for the coating of paper with PPy and PAni under different conditions. He reached full coverage of the cellulose particles within the paper and conducted SEM studies and cyclic voltammetry on the prepared samples.

TABLE 9.2   Overview of the Approaches Presented for Section 9.3.2: Coating

Conductive Component	Polysaccharide	Method Used	References
PAni, PPy	Cotton	Electrochemical coating	Bhadani et al. (1993, 1996)
PPy	Cotton	Chemical coating in vapor and liquid phases	Hosseini and Pairovi (2005), Esfandiari (2008)
PPy	Microfibrillated cellulose, algae cellulose	Chemical coating	Nyström et al. (2009, 2010)
PAni, PPy	Cellulose paper	Chemical coating	Johnston et al. (2009)
PEDOT:PSS	Cellulose	Chemical coating	Knittel and Schollmeyer (2009), Montibon (2009), Montibon et al. (2009)
PPy	Cellulose	Chemical coating	Huang et al. (2005)
PPy	Cellulose	Chemical and electrochemical coating	Babu et al. (2009)
PPy	Cellulose	Chemical coating with preliminary surface modification	Micusík et al. (2007)
PAni, PPy	Cellulose paper	Electrochemical coating	Kim et al. (2006, 2007)
Gold	Cellulose and chitosan films	Electrochemical coating	Wang et al. (2007), Cai and Kim (2008)
MWCNT	Cellulose	Film coating	Yun and Kim (2007, 2008)
PPy, ionic liquid	Cellulose	Film coating	Mahadeva and Kim (2009, 2010)

The coating of cellulose with the more advanced, commercially available, and industrially used conducting polymer complex PEDOT:PSS was the focus of studies by Montibon and Knittel, who demonstrated the interaction of both polymers (Knittel and Schollmeyer, 2009; Montibon, 2009; Montibon et al., 2009).

The abrasive wear of ICP coatings still can be a problem in the use of conductively modified polysaccharides. Damage can also be due to oxidative changes by the influence of light and atmosphere. This problem is tackled by applying a secondary coating for the protection of the conductive layer. Anyway, the adhesion of ICPs, for example, PPy and PAni, on cellulose is reasonable (Dall'Acqua et al., 2004; Knittel and Schollmeyer, 2009).

Kelly et al. (2007) and Johnston et al. (2005, 2009) concluded that PAni as well as PPy is bound to cellulose by interactions between their nitrogen functions and the surface hydroxyl groups of the polysaccharide. These bonds may also involve hydrogen bond interactions between the nitrogen-bound hydrogen atoms of the ICP and the lone electron pairs of the cellulose hydroxyl groups, as well as between

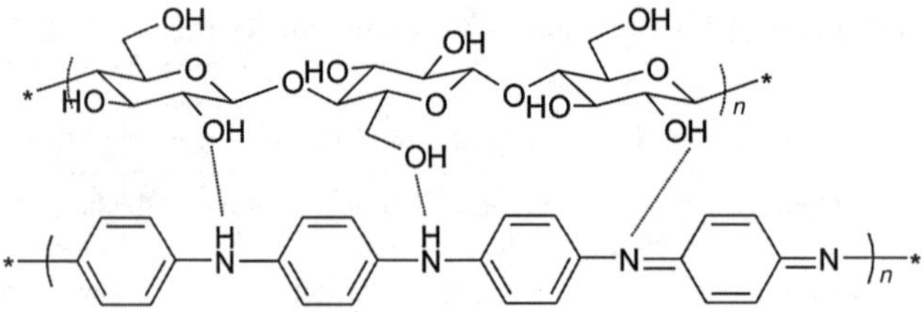

FIGURE 9.3 Binding of PAni (as emeraldine base) to cellulose according to Kelly et al. (2007).

the hydrogen atoms of the hydroxyl groups of cellulose and the free electron pair of the nitrogen atoms of PAni and PPy. A formula to explain the bonding is given in Figure 9.3.

Huang et al. (2005) showed that nanocoating of natural cellulose fibers with PPy is also a way to coat morphologically complex cellulose substances, without disrupting the hierarchical network, while Babu et al. (2009) compared chemical and electro-chemical microstructured deposition of PPy on cotton fabrics.

In the case of a PPy coating on cellulose, a preliminary surface modification with pyrrole bearing silane groups was investigated to improve the binding of the coating (Micusík et al., 2007).

For the production of so-called paper actuators, based on a cellulose matrix, different strategies are available in the literature. Paper actuators are low-cost electromechanical building elements that provide a large deformation, low actuation voltage, and low power consumption, as well as biodegradability. Constructions with coatings of ionic liquids that act through an ion migration effect are known, as well as with PPy, carbon nanotube, or gold coating, where a piezoelectric effect is predom-inant (Kim et al., 2006, 2007, 2010; Wang et al., 2007; Yun and Kim, 2007, 2008; Cai and Kim, 2008; Mahadeva and Kim, 2009, 2010).

The big surface of algae cellulose allowed the construction of an ultrafast battery by a coating of this structure with PPy. Nyström et al. (2009) could exceed conventional setups with this bio-based approach.

A lot of applications of ICP-coated polysaccharides have been introduced over the past several years. Such devices have opened up an extremely interesting field for high-tech application of polysaccharides, which can bring in all their inherent advantages. This begins with the mechanical strength that forms a stable base for all kinds of devices, combined with a high degree of flexibility at the same time, necessary especially for actuators. Similarly, it is not very difficult to generate structures with very high specific surfaces from some polysaccharide materials, such as aerogels (Liebner et al., 2007, 2008). These surfaces can be used as active reaction surfaces in sensors or batteries. A typical advantage of polysaccharides is also the possibility of a variety of subsequent derivatizations and surface mod-ifications originating from the hydroxyl functionalities present, so that the characteristics can be tuned in a wide range according to the envisaged applications.

TABLE 9.3   Overview of the Approaches Presented for Section 9.3.3: Nanoparticles for Coating

Nanoparticle	Polysaccharide	Method Used	References
PThio	Cellulose	Deposition after application of CMC to the surface	Sarrazin et al. (2009)
Bacteria	Chitosan/CNT	Deposition of living cells onto conducting composite	Odaci et al. (2008)

9.3.3 Nanoparticles for Coating

The basic principle of a coating with nanoparticles is the strong adhesion between the matrix and the coating particle, which is mediated by electrostatic forces. Cellulose fibers, if treated in alkaline or moderately acidic medium, show a slightly negative charge on their surface due to the ionization of surface acidic groups derived from hemicelluloses, lignin, or cellulose carboxylic groups introduced by oxidative side reactions during processing. Thus, cellulose surfaces get attractive for cationic particles. To enforce this principle, it is also possible to modify the polysaccharide surface by chemical derivatization with low-degree substitution or by a primary coating with a charged species. Carboxymethyl cellulose is a very promising candidate for such a modification as it combines the introduction of the required negative charges with a good affinity and compatibility to cellulose.

Sarrazin et al. (2009) used the mentioned strategy of an intensified charge density by the adsorption of CMC onto cellulose fibers for the subsequent deposition of poly (3-octylthiophene) nanoparticles (Table 9.3).

Odaci et al. (2008) showed an advanced approach where nanoparticles are not directly used for coating, but a composite bearing nanoparticles enabled the adsorption of cells. They used a chitosan–CNT composite to prepare an electrode surface suitable for coating with living cells of *Pseudomonas fluorescens* bacteria. This assembly was then used as a "living" biosensor for sensing of multiple substances.

The formation or deposition of conducting nanoparticles onto the surface of polysaccharides will gain more attention in the future, since more nanaoparticles with a wider spectrum of properties will become available, enabling a specific morphological structuring of the surface of the polysaccharide matrix. Especially, the interaction of such composites with other substances or biological tissues is a field of fast growing interest.

9.3.4 Polysaccharides Helping the Production of Nanoparticles

Polysaccharides can act not only as a component of a conductive polysaccharide-based composite, but also as auxiliary to influence the morphology and shape of ICP nanoparticles or as helping agents to control the growth or dispersion of those

TABLE 9.4   Overview of the Approaches Presented for Section 9.3.4: Polysaccharides Helping the Production of Nanoparticles

Nanoparticle	Polysaccharide	Method Used	Reference
PPy nanotubes and fibers	Heparin	Morphology directing during chemical and electrochemical polymerization	Shi et al. (2006), Serra Moreno et al. (2008)
PPy particles	CMC	Dispersion during polymerization	Sasso et al. (2008)
PAni particles	Sodium alginate	Complex formation with monomers	Yu et al. (2006)
PAni nanofibers	Schizophyllan	Complex formation during polymerization	Numata et al. (2004)
Carbon nanotubes	Chitosan	Dispersion stabilization	Hu and Hu (2009)
Ag particles	Cellulose	Growth of Ag particles on PAni-coated surface	Stejskal et al. (2008)

nanoparticles. The influence is thereby exerted by the structure of the matrix of the polysaccharide (template effects) or via charges from functional groups along the polysaccharide chain. The concentration, structure, or molecular geometry of the polysaccharides can thereby be used to control dimensions and morphologies of the nanoparticles formed by polymerization in a reaction medium.

Heparin was used by two groups as an effective morphology-directing agent for the growth of PPy nanotubes or nanowires with chemical as well as electrochemical polymerization of the conducting polymer. Thus, the polysaccharide can act as an anion dopant at the same time due to its substituents (Table 9.4) (Shi et al., 2006; Serra Moreno et al., 2008).

Carboxymethyl cellulose acted as an anionic group-bearing polymer, as a dopant, and as an effective dispersing agent during the formation of PPy particles (Sasso et al., 2008).

Similar effects can be used during the synthesis of PAni. Yu et al. (2006) established a method for the production of large amounts of uniform PAni nanofibers of which the diameters were easily controllable by the concentration of sodium alginate in the solution during the synthesis of the conducting polymer. They demonstrated that the nanofiber synthesis was dominated by a complex formation between the biopolymer and the aniline monomers.

Numata et al. (2004) succeeded with a similar approach for the formation of PAni nanofibers with the help of schizophyllan, a fungal polysaccharide with a β-1,3-linked glucose backbone and β-1,6-glycosidic side chains.

Chitosan as another polysaccharide was used for the stabilization of dispersions of carbon nanotubes, which enabled some new applications of this nanomaterial in biomedical sensors (Hu and Hu, 2009).

Stejskal et al. (2008) developed a process that leads to the formation of nano-particles at particular positions. They prepared a composite by coating PAni onto cellulose fibers and subsequently decorating the surface with silver nanoparticles that were formed by an electrochemical reduction of silver nitrate on the conductive surface of the composite.

Despite some growing concerns about the implications of nanoparticles on human life and health, nanoscaled new materials will shape future technologies to a large extent. Conducting nanoparticles will be an important part of these innovations. Polysaccharides were shown to possess ideal prerequisites to play an important role in the morphological design of such materials as their variety of structures and electrochemically active groups can be used for tailor-making such nanoscaled products.

9.3.5 Blends

The blending of intrinsically conductive polymers with polysaccharides is a highly desirable task, as it provides a macroscopically homogeneous material that combines the intrinsic characteristics of the two base substances and could provide a product that is easy to process further. At the same time, this task of blending is not so easy to achieve. The classical blending of polymers is blending of the melts of two components, a task hard to accomplish in the present case as ICPs as well as unaltered polysaccharides are normally not meltable. Another problem lies in the very different material characteristics of the two components. As polysaccharides mostly appear in their natural form as rather hydrophilic substances, ICPs as conjugated polymers are normally strongly hydrophobic by nature.

As already mentioned, cellulose acetate is a favorable and often used polysac-charide derivative for the production of conductive composites due to its hydrophobic character. It was used in combination with PPy (Table 9.5) (Dubitsky and Zhubanov, 1993) and with PAni (De Paoli et al., 1991; Pron et al., 1997; Niziol and Laskam, 1999; Marques et al., 2002; Al-Ahmed et al., 2004; Cerqueira et al., 2009). However, cellulose acetate is strongly determined in its properties by the degree of substitution (DS). Cerqueira et al. (2009) observed a strong influence of the conductivity of composites on the DS of different cellulose acetate samples used. They found a higher conductive performance at a DS of 2 (CDA) than with a DS of 3 (CTA) and attributed this behavior to the fact that PAni in CDA is present in the well conducting, partly protonated form of emeraldine, whereas in the case of CTA it is found in the leucoemeraldine form that is fully protonated and less conductive.

The use of unmodified cellulose is preferable in some cases, since cellulose derivatization evidently drastically alters the properties of the polysaccharide. Nonderivatized cellulose shows a response to humidity, swelling, and shrinking and the possibility for a subsequent surface modification that might be used for sensor devices and related applications.

Rußler et al. (2011) investigated the different ways for the production of composites based on pure cellulose and conducting polymers. Figure 9.4 gives an overview on different principles for the production of such composites. The approaches start either

TABLE 9.5 Overview of the Approaches Presented for Section 9.3.5: Blending

Conductive Component	Polysaccharide	Method Used	References
PPy	Cellulose acetate	Solution blending	Dubitsky and Zhubanov (1993)
PAni	Cellulose acetate	Solution blending	De Paoli et al. (1991), Pron et al. (1997), Niziol and Laskam (1993), Marques et al. (2002), Cerqueira et al. (2009)
PPy, PAni	Cellulose gel, cellulose solution	Solution blending and *in situ* polymerization	Rußler et al. (2011)
PPy	Cellulose	Diffusion-driven polymerization	Dubitsky and Zhubanov (1993)
PAni	Cellulose	Heterogeneous blending	Mo et al. (2009)
PAni	Cross-linked cellulose	*In situ* polymerization	Yin et al. (1997)
PAni	Cellulose	Solution blending in ionic liquid	Holbrey et al. (2005)
PAni	CA, CP, CAB, CAHP, CMC	Solution blending	Yin et al. (1997), Marques et al. (2002)
CNT	Chitosan	Solution blending	Lau et al. (2008), Liu et al. (2009)
CNT	Chitosan	Solution blending and coating with enzymes	Ghica et al. (2009), Kaushik et al. (2010)
PPy	Chitosan	Solution blending with polycaprolactone	Lu et al. (2010a)

from solid or from dissolved materials or use *in situ* synthesis. The variations of these systems proved to be limited especially by the dissolution of the components in different solvent systems, the reaction medium, and the processed form of the solids used. Different compositions resulted in very different conductivity performances of the obtained composites (Rußler et al. (2011)). The highest resulting conductivities were found if the two solids, PPy or PAni, were introduced as powders and the cellulose in the form of a cellulose gel, microcrystalline cellulose (MCC), activated by mechanical shearing as a suspension in water. All other approaches resulted in lower conductivities. Also, the mixing of the two dissolved components, which is possible in ionic liquids in the case of PAni as the conducting component and cellulose, provides a composite that is less conductive after regeneration and redoping than its counterpart obtained by mixing the components in suspension only.

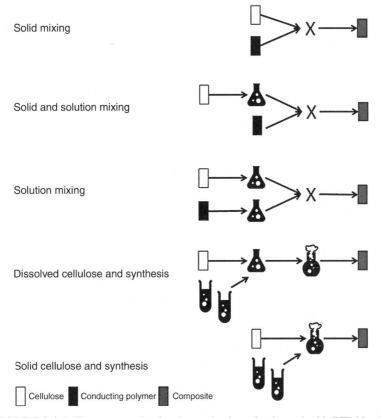

FIGURE 9.4 Different strategies for the production of polysaccharide/ICP blends.

If only cellulose was dissolved or the conducting polymer was synthesized within the cellulose solution, no or only insufficient conductivity was found. It was concluded that the conductivity was reduced due to the fact that the ICP particles are covered by a layer of cellulose that isolates them from each other. In addition, the synthesis of ICPs in the cellulose solution in fact seems to be hindered by the high viscosity of the solution, thus being an inadequate reaction medium.

The covering of cellulose structures by synthesis of an ICP in suspension, that is, not in solutions of cellulose, on the other hand is possible without problems, but leads again to particles separated from each other, while the gel-like structure and film-forming ability are lost (Rußler et al. (2011)). As a solution, the strategy of Dubitsky and Zhubanov (1993) to produce cellulose-based conductive films, a diffusion-driven process in which the ICP is introduced into a ready-made film made of the polysaccharide, is more promising.

Altogether in the pertinent literature, blends of polysaccharide with conducting substances are mainly limited to cellulose, cellulose acetate, and chitosan as the polysaccharide component and PPy, PAni, and CNT as the conducting component.

Mo et al. (2009), for instance, prepared a composite of cellulose and PAni heterogeneously, and established that a doping with dibasic acids exhibited more favorable conductivity than the use of monobasic acids.

In another approach to a cellulose–PAni interpenetrating network, tetraethyl orthosilicate cross-linked cellulose was used and showed an ability to increase the conductivity in comparison to not cross-linked cellulose–PAni composites (Yin et al., 1997).

Holbrey et al. (2005) present in their patent application a method that uses ionic liquids to blend cellulose with PAni.

As already mentioned, the hydrophobic character of cellulose esters promotes their application in combination with conducting polymers. Cellulose acetate, therefore, is frequently used for the production of PAni-containing composites as well as for PPy-containing ones (De Paoli et al., 1991; Dubitsky and Zhubanov, 1993; Pron et al., 1997; Niziol and Laskam, 1999; Al-Ahmed et al., 2004; Cerqueira et al., 2009).

Some other cellulose esters were utilized by Marques et al. (2002), whereas Yin et al. (1997) used cross-linked CMC for the fabrication of PAni composites.

Whereas these composites are designed for a more general application as a conducting polymer composite due to good mechanical as well as conductive properties, composites containing chitosan as the polysaccharide component normally are developed with regard to the outstanding biocompatibility of chitosan and with the aim of producing materials suitable for biosensors and other applications in the context of living cells and tissues.

Lau et al. (2008) used for this purpose carbon nanotubes embedded into a macroporous chitosan structure, whereas Liu et al. (2009) used poly(styrene sulfonic acid)-modified CNTs. Other chitosan-based composites are additionally coated with enzymes such as glucose oxidase or immobilized proteins such as rabbit immunoglobulin and bovine serum albumin for immunosensors (Ghica et al., 2009; Kaushik et al., 2010; Yavuz et al., 2010). A composite with polycaprolactone as a third polymer besides chitosan and PPy exhibits a possible application for nerve repairs (Lu et al., 2010a).

The given examples for the research on blends composed of polysaccharides and conducting substances demonstrate not only the difficulties that are implied in this special fusion, but also the strategies to overcome these obstacles. Blends that allow an easy further procession are highly desirable as they could be formed and structured after their production and therefore during production of different kinds of devices no subsequent processing steps have to be passed with a specially formed mold, as in the case of some coated materials.

The introduction of mediating substances (compatibilizers) to help with this merging is not in focus until today, but could help in the future to make more powerful composites and mixtures of a bigger variety of components.

9.4 DERIVATIVES

The most elegant way to introduce conductivity into polysaccharides is an adequate derivatization. This modification can be made in the form of grafting, where known

TABLE 9.6 Overview of the Approaches Presented for Section 9.4: Derivatives

Conductive Component	Polysaccharide	Method Used	References
PAni	Chitosan	Grafting	Varghese et al. (2010)
PAni	Gum arabic	Grafting	Tiwari (2007), Tiwari and Singh (2008)
PPy	Cellulose	Grafting	Dall'Acqua et al. (2004)
PPy	Alginate	Grafting	Abu-Rabeah and Marks (2009)
PAni	Chitosan	Grafting	Ramaprasad et al. (2009)
CNT	Chitosan	Covalent binding	Carson et al. (2009)
Porphyrin	Cellulose	Covalent binding and additional doping	Sakakibara et al. (2006, 2007a, 2007b), Sakakibara (2008), Sakakibara and Nakatsubo (2008)
Ferrocene, gold	Chitosan	Covalent binding of ferrocene and doping with gold nanoparticles	Qiu et al. (2009)

inherently conducting polymers are covalently bond to polysaccharides. Grafting of PAni and PPy seems to be a quite conventional possibility. Examples for this approach can be found for chitosan as well as for gum arabic for PAni as the conducting component and for alginate and cellulose in the case of PPy as ICP (Table 9.6) (Dall'Acqua et al., 2004; Tiwari, 2007; Tiwari and Singh, 2008; Abu-Rabeah and Marks, 2009; Ramaprasad et al., 2009; Varghese et al., 2010).

Carson et al. (2009) could covalently bind carbon nanotubes to a chitosan matrix, but did not report on the conductivity of the synthesized derivative.

Sakakibara and coworkers showed in their work that approaches on derivatizing cellulose can not only lead to just conductive bio-based substances, sensors, and the like, but also open doors to different electronic functionalities (Sakakibara et al., 2006, 2007a, 2007b; Sakakibara, 2008; Sakakibara and Nakatsubo, 2008). In their research, they succeeded in preparing films that bear photovoltaic properties based on a natural substance. For that purpose, cellulose was substituted with porphyrin moieties on the C6 position, whereas the two other OH groups were derivatized with fatty acid chains to get to the possibility to produce Langmuir–Blodgett films from the material. In this way, the regioselective substitution leads to a uniform arrangement of the photoelectroactive moieties, exhibiting one "side" of the derivatized cellulose chain as the light harvesting one. The dotation of the porphyrin ring systems with some metal ions further improved the performance, as did the introduction of C_{60} fullerenes that were positioned between the porphyrin rings and improved effective charge separation.

However, for the production of composites of higher orders with more than two components in some cases also the combination of different principles of production might be adequate, for example, for a decoration of the surface of a composite. Qiu et al. (2009) give a good example for such a composite material where material for an immunosensor is made by first derivatizing chitosan with branches of ferrocene and then doping the surface with gold nanoparticles.

The derivatization of polysaccharides with groups and side chains that provide potential electronic properties is just at the beginning, but the possibility for various (also regioselective) modifications of this renewable base material is very promising. Increasing future research activities on this field can certainly be expected.

9.5 SUMMARY AND OUTLOOK

The combination of the properties of polysaccharides with conducting features already bears the possibility to create numerous applications and constructions in very different fields of electrical and electronic devices, circuitry, and building elements, reaching from down-to-earth approaches to high-tech applications.

Nevertheless, this research area is still just an emerging one; compared to other fields, only rather few investigations have been carried out until now, and the degree of complexity of the approaches is still limited. This means on the other hand that still a lot of possibilities are undiscovered and that there are still great chances to open up new fields of applications for these very special composites. All possible approaches for the formation of such composites can be valuable to solve specific problems, for which reason they have been described in this chapter.

The very different chemical nature of polysaccharides on one side and conducting substances on the other side requires creative solutions for their combination within one material or even within one molecule. However, these different characteristics also offer the possibility for tailor-made material solutions, as especially the polysaccharide components offer a nearly unlimited potential in tuning their mechanical and chemical properties. Research in this field is today very much focused on ICPs and conducting nanoparticles. As these substance classes are hot topics for themselves, lots of material innovations can be expected to be derived from the interesting combination of polysaccharide and (semi)conductive substances.

REFERENCES

Abu-Rabeah, K., and R. S. Marks (2009). Impedance study of the hybrid molecule alginate–pyrrole: demonstration as host matrix for the construction of a highly sensitive amperometric glucose biosensor. *Sens. Actuators B* **136**(2):516–522.

Al-Ahmed, A., F. Mohammad, and M. Zaki Ab. Rahman (2004). Composites of polyaniline and cellulose acetate: preparation, characterization, thermo-oxidative degradation and stability in terms of DC electrical conductivity retention. *Synth. Met.* **144**(1):29–49.

Ansari, R. (2006). Polypyrrole conducting electroactive polymers: synthesis and stability studies. *E-J. Chem.* **3**(4):15.

Arslan, A., Ö. Türkarslan, C. Tanyeli, I. M. Akhmedov, and L. Toppare (2007). Electrochromic properties of a soluble conducting polymer: poly(1-(4-fluorophenyl)-2,5-di(thiophen-2-yl)-1*H*-pyrrole). *Mater. Chem. Phys.* **104**(2–3):410–416.

Babu, K. F., R. Senthilkumar, M. Noel, and M. A. Kulandainathan (2009). Polypyrrole microstructure deposited by chemical and electrochemical methods on cotton fabrics. *Synth. Met.* **159**(13):1353–1358.

Bellucci, S. (2005). Carbon nanotubes: physics and applications. *Phys. Status Solidi c* **2**(1):34–47.

Beneventi, D., S. Alila, S. Boufi, D. Chaussy, and P. Nortier (2006). Polymerization of pyrrole on cellulose fibres using a FeCl$_3$ impregnation–pyrrole polymerization sequence. *Cellulose* **13**(6):725–734.

Bhadani, S. N., S. K. Sen Gupta, and M. K. Gupta (1993). Electrically conducting natural fibers. *Indian J. Fibre Text.* **18**(1):2.

Bhadani, S. N., S. K. Sen Gupta, G. C. Sahu, and M. Kumari (1996). Electrochemical formation of some conducting fibers. *J. Appl. Polym. Sci.* **61**:207–212.

Bhadra, S., D. Khastgir, N. K. Singha, and J. H. Lee (2009). Progress in preparation, processing and applications of polyaniline. *Prog. Polym. Sci.* **34**(8):783–810.

Bhat, N. V., D. T. Seshadri, M. M. Nate, and A. V. Gore (2006). Development of conductive cotton fabrics for heating devices. *J. Appl. Polym. Sci.* **102**:4690–4695.

Bouvree, A., J.-F. Feller, M. Castro, Y. Grohens, and M. Rinaudo (2009). Conductive polymer nano-biocomposites (CPC): chitosan–carbon nanoparticle a good candidate to design polar vapour sensors. *Sens. Actuators B* **138**(1):138–147.

Cai, Z., and J. Kim (2008). Characterization and electromechanical performance of cellulose–chitosan blend electro-active paper. *Smart Mater. Struct.* **17**(3):8.

Carson, L., C. Kelly-Brown, M. Stewart, A. Oki, G. Regisford, Z. Luo, and V. I. Bakhmutov (2009). Synthesis and characterization of chitosan–carbon nanotube composites. *Mater. Lett.* **63**(6–7):617–620.

Cerqueira, D. A., A. J. M. Valente, G. R. Filho, and H. D. Burrows (2009). Synthesis and properties of polyaniline–cellulose acetate blends: the use of sugarcane bagasse waste and the effect of the substitution degree. *Carbohydr. Polym.* **78**(3):402–408.

Chen, P., H.-S. Kim, S.-M. Kwon, Y. S. Yun, and H.-J. Jin (2009). Regenerated bacterial cellulose/multi-walled carbon nanotubes composite fibers prepared by wet-spinning. *Curr. Appl. Phys.* **9** (2, Suppl. 1): e96–e99.

Collier, J. H., J. P. Camp, T. W. Hudson, and C. E. Schmidt (2000). Synthesis and characterization of polypyrrole–hyaluronic acid composite biomaterials for tissue engineering applications. *J. Biomed. Mater. Res.* **50**:574–584.

Dall'Acqua, L., C. Tonin, R. Peila, F. Ferrero, and M. Catellani (2004). Performances and properties of intrinsic conductive cellulose–polypyrrole textiles. *Synth. Met.* **146**(2):213–221.

De Paoli, M.-A., E. R. Duek, and M. A. Rodrigues (1991). Poly(aniline)/cellulose acetate composites: conductivity and electrochromic properties. *Synth. Met.* **41**(3):973–978.

Ding, C., X. Qian, G. Yu, and X. An (2010). Dopant effect and characterization of polypyrrole–cellulose composites prepared by *in situ* polymerization process. *Cellulose* **17**(6):1067–1077.

Du, Z., C. Li, L. Li, M. Zhang, S. Xu, and T. Wang (2009). Simple fabrication of a sensitive hydrogen peroxide biosensor using enzymes immobilized in processable polyaniline nanofibers/chitosan film. *Mater. Sci. Eng.* **29**(6):1794–1797.

Dubitsky, Y. A., and B. A. Zhubanov (1993). Polypyrrole–poly(vinyl chloride) and poly-pyrrole–cellulose acetate conducting composite films by opposite-diffusion polymerization. *Synth. Met.* **53**(3):303–307.

Esfandiari, A. (2008). PPy covered cellulosic and protein fibres using novel covering methods to improve the electrical property. *World Appl. Sci. J.* **3**(3):6.

Evans, B. R., H. M. ÓNeill, and J. Woodward (2006). Electrically conductive cellulose composite. U.S. Patent No. 11,153,146.

Feast, W. J., J. Tsibouklis, K. L. Pouwer, L. Groenendaal, and E. W. Meijer (1996). Synthesis, processing and material properties of conjugated polymers. *Polymer* **37**(22):5017–5047.

Finkenstadt, V. L. (2005). Natural polysaccharides as electroactive polymers. *Appl. Microbiol. Biotechnol.* **67**(6):735–745.

Genaidy, A., T. Tolaymat, R. Sequeira, M. Rinder, and D. Dionysiou (2009). Health effects of exposure to carbon nanofibers: systematic review, critical appraisal, meta analysis and research to practice perspectives. *Sci. Total Environ.* **407**(12):3686–3701.

Ghica, M. E., R. Pauliukaite, O. Fatibello-Filho, and C. M. A. Brett (2009). Application of functionalised carbon nanotubes immobilised into chitosan films in amperometric enzyme biosensors. *Sens. Actuators B* **142**(1):308–315.

Groenendaal, L., F. Jonas, D. Freitag, H. Pielartzik, and J. R. Reynolds (2000). *Poly(3,4-ethylenedioxythiophene) and Its Derivatives: Past, Present, and Future*, Weinheim: Wiley-VCH Verlag GmbH.

Guimard, N. K., N. Gomez, and C. E. Schmidt (2007). Conducting polymers in biomedical engineering. *Prog. Polym. Sci.* **32**(8–9):876–921.

Hakansson, E., A. Kaynak, T. Lin, S. Nahavandi, T. Jones, and E. Hu (2004). Characterization of conducting polymer coated synthetic fabrics for heat generation. *Synth. Met.* **144**(1):21–28.

Holbrey, J., R. P. Swatloski, J. Chen, D. Daly, and R. D. Rogers (2005). Polymer dissolution and blend formation in ionic liquids. U.S. Patent No. 11,087,496.

Hosseini, S. H., and A. Pairovi (2005). Preparation of conducting fibres from cellulose and silk by polypyrrole coating. *Iran. Polym. J.* **14**(11):7.

Hu, C., and S. Hu (2009). Carbon nanotube-based electrochemical sensors: principles and applications in biomedical systems. *J. Sens.* 1–40.

Hu, E., A. Kaynak, and Y. Li (2005). Development of a cooling fabric from conducting polymer coated fibres: proof of concept. *Synth. Met.* **150**(2):139–143.

Huang, J., I. Ichinose, and T. Kunitake (2005). Nanocoating of natural cellulose fibers with conjugated polymer: hierarchical polypyrrole composite materials. *Chem. Commun.* (13):1717–1719.

Johnston, J. H., J. Moraes, and T. Borrmann (2005). Conducting polymers on paper fibres. *Synth. Met.* **153**(1–3):65–68.

Johnston, J. H., F. M. Kelly, K. A. Burridge, and T. Borrmann (2009). Hybrid materials of conducting polymers with natural fibres and silicates. *Int. J. Nanotechnol.* **6**(3–4):312–328.

Jourand, P., H. De Clercq, and R. Puers (2010). Robust monitoring of vital signs integrated in textile. *Sens. Actuators A* **161**(1–2):288–296.

Karlsson, R. H., A. Herland, M. Hamedi, J. A. Wigenius, A. Åslund, X. Liu, M. Fahlman, O. Inganäs, and P. Konradsson (2009). Iron-catalyzed polymerization of alkoxysulfonate-functionalized 3,4-ethylenedioxythiophene gives water-soluble poly(3,4-ethylenedioxythiophene) of high conductivity. *Chem. Mater.* **21**(9):1815–1821.

Kathirgamanathan, P., M. J. Toohey, J. Haase, P. Holdstock, J. Laperre, and G. Schmeer-Lioe (2000). Measurements of incendivity of electrostatic discharges from textiles used in personal protective clothing. *J. Electrostat.* **49**(1–2):51–70.

Kaushik, A., P. R. Solanki, M. K. Pandey, K. Kaneto, S. Ahmad, and B. D. Malhotra (2010). Carbon nanotubes—chitosan nanobiocomposite for immunosensor. *Thin Solid Films* **519**(3):1160–1166.

Kelly, F. M., J. H. Johnston, T. Borrmann, and M. J. Richardson (2007). Functionalised hybrid materials of conducting polymers with individual fibres of cellulose. *Eur. J. Inorg. Chem.* **2007**(35):5571–5577.

Kim, J., S. D. Deshpande, S. Yun, and Q. Li (2006). A comparative study of conductive polypyrrole and polyaniline coatings on electro-active papers. *Polym. J.* **38**(7):659–668.

Kim, J., S.-R. Yun, and S. D. Deshpande (2007). Synthesis, characterization and actuation behavior of polyaniline-coated electroactive paper actuators. *Polym. Int.* **56**:1530–1536.

Kim, J., S. Yun, S. K. Mahadeva, K. Yun, S. Y. Yang, and M. Maniruzzaman (2010). Paper actuators made with cellulose and hybrid materials. *Sensors* **10**(3):1473–1485.

Knittel, D., and E. Schollmeyer (2009). Electrically high-conductive textiles. *Synth. Met.* **159**(14):1433–1437.

Kuhn, H. H., A. D. Child, and W. C. Kimbrell (1995). Toward real applications of conductive polymers. *Synth. Met.* **71**(1–3):2139–2142.

Lau, C., M. J. Cooney, and P. Atanassov (2008). Conductive macroporous composite chitosan–carbon nanotube scaffolds. *Langmuir* **24**(13):7004–7010.

Liebner, F., A. Potthast, T. Rosenau, E. Haimer, and M. Wendland (2007). Ultralight-weight cellulose aerogels from NBnMO-stabilized lyocell dopes. *Res. Lett. Mater. Sci.* **2007**:4.

Liebner, F., A. Potthast, T. Rosenau, E. Haimer, and M. Wendland (2008). Cellulose aerogels: highly porous, ultra-lightweight materials. *Holzforschung* **62**(2):129–135.

Liu, Y.-L., W.-H. Chen, and Y.-H. Chang (2009). Preparation and properties of chitosan/carbon nanotube nanocomposites using poly(styrene sulfonic acid)-modified CNTs. *Carbohydr. Polym.* **76**(2):232–238.

Lu, X., Z. Qiu, Y. Wan, Z. Hu, and Y. Zhao (2010a). Preparation and characterization of conducting polycaprolactone/chitosan/polypyrrole composites. *Compos. Part A* **41**(10):1516–1523.

Lu, X., W. Zhang, C. Wang, T.-C. Wen, and Y. Wei (2010b). One-dimensional conducting polymer nanocomposites: synthesis, properties and applications. *Prog. Polym. Sci.* **36**(5):671–712.

MacDiarmid, A. G. (1997). Polyaniline and polypyrrole: where are we headed? *Synth. Met.* **84**(1–3):27–34.

Mahadeva, S. K., and J. Kim (2009). Electromechanical behavior of room temperature ionic liquid dispersed cellulose. *J. Phys. Chem. C* **113**(28):12523–12529.

Mahadeva, S. K., and J. Kim (2010). Nanocoating of ionic liquid and polypyrrole for durable electro-active paper actuators working under ambient conditions. *J. Phys. D* **43**(20):205502.

Malinauskas, A. (2001). Chemical deposition of conducting polymers. *Polymer* **42**(9):3957–3972.

Marques, A. P., C. M. A. Brett, H. D. Burrows, A. P. Monkman, and B. Retimal (2002). Spectral and electrochemical studies on blends of polyaniline and cellulose esters. *J. Appl. Polym. Sci.* **86**(9):2182–2188.

Meister, F., D. Vorbach, F. Niemz, and T. Schulze (2003). High-Tech-Cellulose-Funktionspolymere nach dem ALCERU®-Verfahren. *Materialwiss. Werkstofftech.* **34**(3):262–266.

Micusík, M., T. Nedelcev, M. Omastová, I. Krupa, K. Olejníková, P. Fedorko, and M. M. Chehimi (2007). Conductive polymer-coated textiles: the role of fabric treatment by pyrrole-functionalized triethoxysilane. *Synth. Met.* **157**(22–23):914–923.

Mo, Z.-L., Z.-L. Zhao, H. Chen, G.-P. Niu, and H.-F. Shi (2009). Heterogeneous preparation of cellulose–polyaniline conductive composites with cellulose activated by acids and its electrical properties. *Carbohydr. Polym.* **75**(4):660–664.

Montibon, E. (2009). Preparation of electroconductive paper by deposition of conducting polymer. Faculty of Technology and Science, Karlstad University.

Montibon, E., L. Järnström, and M. Lestelius (2009). Characterization of poly(3,4-ethylene-dioxythiophene)/poly(styrene sulfonate) (PEDOT:PSS) adsorption on cellulosic materials. *Cellulose* **16**(5):807–815.

Niziol, J., and J. Laskam (1999). Conductivity of blends of polyaniline with PMMA and cellulose acetate: aging studies. *Synth. Met.* **101**(1–3):720–721.

Numata, M., T. Hasegawa, T. Fujisawa, K. Sakurai, and S. Shinkai (2004). β-1,3-Glucan (schizophyllan) can act as a one-dimensional host for creation of novel poly(aniline) nanofiber structures. *Org. Lett.* **6**(24):4447–4450.

Nyström, G., A. Razaq, M. Strømme, L. Nyholm, and A. Mihranyan (2009). Ultrafast all-polymer paper-based batteries. *Nano Lett.* **9**(10):3635–3639.

Nyström, G., A. Mihranyan, A. Razaq, T. Lindström, L. Nyholm, and M. Strømme (2010). A nanocellulose polypyrrole composite based on microfibrillated cellulose from wood. *J. Phys. Chem. B* **114**(12):4178–4182.

Odaci, D., S. Timur, and A. Telefoncu (2008). Bacterial sensors based on chitosan matrices. *Sens. Actuators B* **134**(1):89–94.

Pei, Q., H. Jarvinen, J. E. Osterholm, O. Inganaes, and J. Laakso (1992). Poly[3-(4-octylphenyl) thiophene], a new processable conducting polymer. *Macromolecules* **25**(17):4297–4301.

Planès, J., Y. Cheguettine, and Y. Samson (1999). Nanostructure of polyaniline blends. *Synth. Met.* **101**(1–3):789–790.

Pron, A., M. Zagorska, Y. Nicolau, F. Genoud, and M. Nechtschein (1997). Highly conductive composites of polyaniline with plasticized cellulose acetate. *Synth. Met.* **84**(1–3):89–90.

Qiu, J.-D., R.-P. Liang, R. Wang, L.-X. Fan, Y.-W. Chen, and X.-H. Xia (2009). A label-free amperometric immunosensor based on biocompatible conductive redox chitosan–ferrocene/ gold nanoparticles matrix. *Biosens. Bioelectron.* **25**(4):852–857.

Ramaprasad, A. T., V. Rao, G. Sanjeev, S. P. Ramanani, and S. Sabharwal (2009). Grafting of polyaniline onto the radiation crosslinked chitosan. *Synth. Met.* **159**(19–20):1983–1990.

Rehahn, M. (2003). Elektrisch leitfähige Kunststoffe: Der Weg zu einer neuen Materialklasse. *Chem. Unserer Zeit.* **37**(1):18–30.

Rußler, A., K. Sakakibara, T. Rosenau (2011). Cellulose as matrix component of conducting films, *Cellulose*, "http://pisces.boku.ac.at/han/2487/www.springerlink.com/content/0969-0239/18/4/" "Verweis auf Issue dieser Article" **18**(4):937–944.

Ryu, K. S., Y. Lee, K.-S. Han, and M. G. Kim (2004). The electrochemical performance of polythiophene synthesized by chemical method as the polymer battery electrode. *Mater. Chem. Phys.* **84**(2–3):380–384.

Sakakibara, K. (2008). Fabrication of cellulose supramolecular architectures towards photo-current generation systems by the Langmuir–Blodgett technique. Ph.D. thesis, University of Kyoto, Kyoto, p. 117.

Sakakibara, K., and F. Nakatsubo (2008). Effect of fullerene on photocurrent performance of 6-*O*-porphyrin-2,3-di-*O*-stearoylcellulose Langmuir–Blodgett films. *Macromol. Chem. Phys.* **209**(12):1274–1281.

Sakakibara, K., S. Ifuku, Y. Tsujii, H. Kamitakahara, T. Takano, and F. Nakatsubo (2006). Langmuir–Blodgett films of a novel cellulose derivative with dihydrophytyl group: the ability to anchor beta-carotene molecules. *Biomacromolecules* **7**(6):1960–1967.

Sakakibara, K., H. Kamitakahara, T. Takano, and F. Nakatsubo (2007a). Redox-active cellulose Langmuir–Blodgett films containing beta-carotene as a molecular wire. *Bioma-cromolecules* **8**(5):1657–1664.

Sakakibara, K., Y. Ogawa, and F. Nakatsubo (2007b). First cellulose Langmuir–Blodgett films towards photocurrent generation systems. *Macromol. Rapid Commun.* **28**(11):1270–1275.

Sarrazin, P., D. Chaussy, O. Stephan, L. Vurth, and D. Beneventi (2009). Adsorption of poly(3-octylthiophene) nanoparticles on cellulose fibres: effect of dispersion stability and fibre pre-treatment with carboxymethyl cellulose. *Colloids Surf. A* **349**(1–3):83–89.

Sasso, C., M. Fenoll, O. Stephan, and D. Beneventi (2008). Use of wood derivatives as doping/dispersing agents in the preparation of polypyrrole aqueous dispersions. *Bioresources* **3**(4):1187–1195.

Savolainen, K., H. Alenius, H. Norppa, L. Pylkkänen, T. Tuomi, and G. Kasper (2010). Risk assessment of engineered nanomaterials and nanotechnologies: a review. *Toxicology* **269**(2–3):92–104.

Schultze, J. W., T. Morgenstern, D. Schattka, and S. Winkels (1999). Microstructuring of conducting polymers. *Electrochim. Acta* **44**(12):1847–1864.

Serra Moreno, J., S. Panero, and B. Scrosati (2008). Electrochemical polymerization of polypyrrole–heparin nanotubes: kinetics and morphological properties. *Electrochim. Acta* **53**(5):2154–2160.

Shi, W., D. Ge, J. Wang, Z. Jiang, L. Ren, and Q. Zhang (2006). Heparin-controlled growth of polypyrrole nanowires. *Macromol. Rapid Commun.* **27**(12):926–930.

Shirakawa, H., E. J. Louis, A. G. MacDiarmid, C. K. Chiang, and A. J. Heeger (1977). Synthesis of electrically conducting organic polymers: halogen derivatives of polyacetylene, (CH)$_x$. *J. Chem. Soc., Chem. Commun.* (16):578–580.

Stejskal, J., M. Trchová, J. Kovářová, J. Prokeš, and M. Omastová (2008). Polyaniline-coated cellulose fibers decorated with silver nanoparticles. *Chem. Pap.* **62**(2):181–186.

Taeger, E., K. Berghof, R. Maron, F. Meister, C. Michels, and D. Vorbach (1997). Eigenschaftsänderungen im Alceru-Faden durch Zweitpolymere. *Lenzinger Berichte* **76**:126–131.

Tiwari, A. (2007). Gum arabic-*graft*-polyaniline: an electrically active redox biomaterial for sensor applications. *J. Macromol. Sci. A* **44**(7):735–745.

Tiwari, A., and V. Singh (2008). Microwave-induced synthesis of electrical conducting gum acacia-*graft*-polyaniline. *Carbohydr. Polym.* **74**(3):427–434.

Varghese, J. G., A. A. Kittur, P. S. Rachipudi, and M. Y. Kariduraganavar (2010). Synthesis, characterization and pervaporation performance of chitosan-*g*-polyaniline membranes for the dehydration of isopropanol. *J. Membr. Sci.* **364**(1–2):10.

Wang, N., Y. Chen, and J. Kim (2007). Electroactive paper actuator made with chitosan–cellulose films: effect of acetic acid. *Macromol. Mater. Eng.* **292**:748–753.

Wei, B., P. Guan, L. Zhang, and G. Chen (2010). Solubilization of carbon nanotubes by cellulose xanthate toward the fabrication of enhanced amperometric detectors. *Carbon* **48**(5):1380–1387.

Yavuz, A. G., A. Uygun, and V. R. Bhethanabotla (2010). Preparation of substituted polyaniline/chitosan composites by *in situ* electropolymerization and their application to glucose sensing. *Carbohydr. Polym.* **81**(3):712–719.

Yin, W., J. Li, Y. Li, Y. Wu, T. Gu, and C. Liu (1997). Conducting IPN based on polyaniline and crosslinked cellulose. *Polym. Int.* **42**(3):276–280.

Yoon, S. H., H. J. Jin, M. C. Kook, and Y. R. Pyun (2006). Electrically conductive bacterial cellulose by incorporation of carbon nanotubes. *Biomacromolecules* **7**(4):4.

Yu, M.-F., B. S. Files, S. Arepalli, and R. S. Ruoff (2000a). Tensile loading of ropes of single wall carbon nanotubes and their mechanical properties. *Phys. Rev. Lett.* **84**(24):5552.

Yu, M.-F., O. Lourie, M. J. Dyer, K. Moloni, T. F. Kelly, and R. S. Ruoff (2000b). Strength and breaking mechanism of multiwalled carbon nanotubes under tensile load. *Science* **287**(5453):637–640.

Yu, Y., S. Zhihuai, S. Chen, C. Bian, W. Chen, and G. Xue (2006). Facile synthesis of polyaniline–sodium alginate nanofibers. *Langmuir* **22**(8):3899–3905.

Yun, S., and J. Kim (2007). A bending electro-active paper actuator made by mixing multi-walled carbon nanotubes and cellulose. *Smart Mater. Struct.* **16**(4):6.

Yun, S., and J. Kim (2008). Characteristics and performance of functionalized MWNT blended cellulose electro-active paper actuator. *Synth. Met.* **158**(13):521–526.

Zhou, D., C. O. Too, and G. G. Wallace (1999). Synthesis and characterisation of polypyrrole/heparin composites. *React. Funct. Polym.* **39**(1):19–26.

10

POLYSACCHARIDE-BASED POROUS MATERIALS

Peter S. Shuttleworth, Avtar Matharu, and James H. Clark

10.1 INTRODUCTION

The annual worldwide production of biomass is estimated at 170×10^9 tons, of which carbohydrates account for 75%. Only 3.5% (6×10^9 tons) of this total biomass is being utilized by humans (Röper, 2002; Demirbas et al., 2009; Bain, 2007). Apart from the use of biomass as a source for housing and energy (33%), the main use by mankind is for food (62%), with only a fraction (5%) being used for nonfood applications such as in the chemical industry (Stevens, 2004). If the 3.5% of biomass used by man was utilized more efficiently, and only a fraction of the 96.5% unused biomass was exploited, then there would be an ample supply for future food, material, and chemical demands.

Starch is one of the most abundant polysaccharides. The United Kingdom annually uses 880,000 tons, of which 75% is for both human and animal consumption, while the remainder is for nonfood uses such as in the paper, detergents, and plastics industry (Entwistle et al., 1998). Within Europe as a whole, the majority of the maize starch (82%) consumption is in the form of animal feed with the remainder divided almost equally between industrial manufacture and human food use (defra, 2003). It can therefore be seen that any development using starch in future chemical industry will not adversely affect potential food utilization of this material with correct land management.

Historically, starch has seen widespread use as an adhesive for bonding various paper, wood, and cardboard products (Kennedy, 1989). The majority of starch-based adhesives are used within the paper and textile industry as binders and sizing agents

Polysaccharide Building Blocks: A Sustainable Approach to the Development of Renewable Biomaterials, First Edition. Edited by Youssef Habibi and Lucian A. Lucia.
© 2012 John Wiley & Sons, Inc. Published 2012 by John Wiley & Sons, Inc.

(Baumann and Conner, 1994). Over the past 15–20 years, interest in this material has grown due to assured supply and its renewable nature. Starch is a material with inherently high levels of hydroxyl functionality, and therefore has significant opportunity to modify its structure. However, this functionality is greatly hindered due to the dense packing of polysaccharide chains within the starch granule. Chemical modification of native starch surface hydroxyl groups is restricted, generally resulting in low degrees of substitution unless harsh pretreatments are adopted. However, a simple process of gelatinization in water followed by a period of retrogradation converts the once dense structure into an easily accessible porous structure, which subsequently leads to more effective chemical modifications and thus higher degrees of substitution (Doi et al., 2002).

The aim of this chapter is to provide a comprehensive overview of polysaccharide-based porous materials and their potential applications, with special emphasis on Starbon® mesoporous materials.

10.2 POROUS POLYSACCHARIDES

Starch-based microcellular foams (expanded starch) were first developed by Glenn and coworkers from retrograded aqueous starch gels (aquagels) (Glenn and Irving, 1995; Glenn and Stern, 1999). These foams were formed after displacing water within an aquagel matrix with ethanol and then drying to form a low-density mesoporous solid (a type of microcellular foam). Continued research within this area has found that this technology can be applied to other polysaccharide and non-polysaccharide sources such as alginic acid (White et al., 2010a), pectin (White et al., 2010b), and poly(vinyl alcohol) (PVA) (Hunt et al., 2009). The understanding of why some polymers and not others can form these types of porous solids has not yet been definitively concluded, but the general rationale is that if the linear chain fragment of polymers can form a helical type arrangement, then porous solids may result. This has created an increased interest within this area, as each porous material tested has a unique textural property (surface area and pore volume).

10.2.1 Preparation

The general method as outlined in Figure 10.1 to prepare these types of materials involves gelation to open up the helical structure, retrogradation to create a high surface area aqueous gel network, and then a solvent displacement stage. Exchanging the water with a low surface tension solvent, followed by drying, yields a highly porous powder. Direct air drying of aquagels leads to formation of a dense nonporous material because as water is removed its high surface tension causes stresses on the pore system, thus leading to collapse.

10.2.1.1 Gelation Starch gelatinization in the presence of a solvent can be described as a phase transition where the starch granules simultaneously swell while going from an ordered to a disordered state with the loss of birefringence

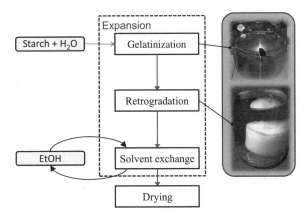

FIGURE 10.1 Preparation of porous polysaccharide materials.

(Hermansson and Svegmark, 1996). We will only consider preparation of PPDMs in water. However, it can be achieved at room temperature in nonaqueous solvents such as liquid ammonia and dimethyl sulfoxide, or mechanically by milling (solvent-free). The critical temperature required to achieve gelatinization depends upon parameters such as chain length, crystal structure, granule size, and the ratio of amylopectin to amylose (the polysaccharides that make up starch).

10.2.1.2 Retrogradation Retrogradation refers to the association and crystallization of starch in water, causing effects such as precipitation, gelation, and changes in consistency and opacity (Hermansson and Svegmark, 1996). Irrespective of the native starch structure, retrogradation produces a starch with a β-type crystalline configuration (Fredriksson et al., 1998). In gelatinized starch, structural transformations such as chain aggregation or recrystallization change the gel from an amorphous state to a more ordered crystalline state. This change can be seen on both a molecular level with changes in the mobility of the starch and water and a macroscopic level with changes in the textural properties such as gel thickening and water loss (syneresis) that can leave as little as 0.4% of starch in solution (Hermansson and Svegmark, 1996; Zobel, 1984).

10.2.1.3 Solvent Exchange and Drying Once the gelatinized starch has been retrograded for a given time period, water contained within the retrograded starch matrix needs to be gradually exchanged for ethanol, which has a lower surface tension than water. Evaporation of ethanol followed by drying in a vacuum oven set at 50 °C overnight affords a high surface area material (determined by N_2 porosimetry). However, if the starch is dried without exchanging the water for ethanol, the pore apertures collapse because strong hydrogen bonds are formed between adjacent molecular chains of the starch granule as water is removed.

10.2.2 Applications

10.2.2.1 Switchable Adhesives One area that is looking to adopt this technology is the carpet tile manufacturing sector. With the shift toward more office-based work, this sector is now seriously looking at its environmental credentials and the nonrenewable and resource-intensive components that are used to make up carpet tiles. This is further pushed with increasing consumer and industry awareness, and predicted tougher future legislation.

A carpet tile consists of a bitumen backing layer onto which a nylon 6,6 fabric layer is adhered using SBR latex/PVC adhesive. These are ideal as they are relatively easy to apply with very strong fixative/adhesive properties. However, the obstacle comes at the end of the tile's life as the adhesives continue to bind the layers making separation problematic. Mechanical separation is possible, but adhesive contamination of the expensive nylon fabric prevents its recyclability, leaving only the option of downcycling. Therefore, virgin nylon has to be used in the fabric layer of all new carpet tiles produced. The environmental burden of this is enormous, as virgin nylon is energy intensive to make and releases the high global warming potential (GWP) nitrous oxide while making adipic acid, an intermediate in the process. Through simply being able to reuse the nylon, this burden could be avoided.

Recent work has shown that by using expanded (porous) starch to facilitate high levels of modification, a renewable switchable adhesive can be formed (Shuttleworth et al., 2010). This adhesive as well as conventional adhesives not only performed well under standard testing conditions, but also imparted excellent flame retardancy to the tile. In addition, it was possible to "switch" adhesive properties at the end of life, enabling component separation and nylon reuse. This emphasizes that these new generation renewable polymers have the criteria to compete with synthetic counterparts.

10.2.2.2 Other Applications   Utilization of such porous, high surface area starches has also been recently described in the separation of organic compounds in nonaqueous solutions (Budarin et al., 2005) and in odor control. Apart from applications based on adsorbency and adhesives, there is a growing potential for these high surface area and pore volume characteristics in a vast amount of technological areas. The remainder of this chapter will be devoted to one such area: the use of porous polysaccharides to form mesoporous carbons (MCs) with interesting properties and diverse applications.

10.3 STARBON® MESOPOROUS CARBONS

10.3.1 Background

A mesoporous material is defined by the presence of pores with diameters between 2 and 50 nm. The outstanding potential of mesoporous carbons is in part due to their extended pore diameter network that may be utilized in a wide range of technologically important applications that are critically dependent on the mass transport of chemicals to the carbon surface (Sayari, 1996; Joo et al., 2001; Poole, 2003).

Template-assisted synthesis is the most common method for achieving control over pore size distribution enabling the creation of large pore diameter networks. A typical procedure involves filling mesoporous silica with a carbon precursor (e.g., sucrose) that is subsequently carbonized through a series of high-temperature processes (Ryoo et al., 1999). The template is then removed by using hydrofluoric acid or caustic soda. The resultant carbons possess well-ordered mesoporous structures with a large specific volume (Lee et al., 1999; Kim et al., 2003; Gadiou et al., 2005; Hartmann et al., 2005).

However, templating synthesis is limited to the production of graphitic carbons as unlike higher oxygen-containing forms, they are stable to the highly aggressive chemicals required to remove the templates. When unstable, the material will simultaneously solubilize with the silica templates during their removal (Jeong and Werth, 2005; Eusterhues et al., 2003).

Another disadvantage of material produced by this method is the low adsorption potential that results from wide mesopore radii and the subsequent lack of functional groups. Adsorption capacity and selectivity of the MCs are significantly reduced, thus restricting the applications in important areas such as catalysis, water purification, and chromatography. One way to combat these deficiencies and fulfill the goals of a "greener" nanotechnology is to introduce high concentrations of specific functional groups on the surface of mesoporous carbon (e.g., hydrophilic, acidic, aromatic, and stereoselective) to enhance adsorption. In order to functionalize these carbon surfaces and open up new chemistries, further difficult chemical modifications are required, which come at the cost of reducing available mesopores (Li and Dai, 2005; Li et al., 2005).

Nature can help with this; for example, starch and other helix-forming carbohydrates naturally have a nanochannelled biopolymer structure, which can be utilized as a template. These materials, treated first to form an expanded mesoporous solid as described in Section 10.2.1, can template carbonization reactions to produce mesoporous carbonaceous materials, through doping with acid and heating. This approach has generated a completely novel family of porous carbonaceous materials that have the trademark Starbon® and can have a range of carbon surfaces depending on the temperature used for preparation as shown in Figure 10.2 (Budarin et al., 2006).

10.3.2 Starbon® Synthesis

Starbon® synthesis is based on using expanded polysaccharides as a precursor for mesoporous carbon material without the need for a templating agent (White et al., 2008; Milkowski et al., 2004; Budarin et al., 2005). This method utilizes, for example, the natural ability of the polysaccharide polymer chains within starch (amylose and amylopectin) to assemble into organized, largely mesoporous structures. The method comprises of two key stages (Figure 10.3). First, the prepared PPDMs are doped with a catalytic amount of an organic acid (e.g., p-toluenesulfonic acid, p-TSA), and then heated under vacuum (Budarin et al., 2006). Alternatively, this method can be used for the polysaccharides pectin and alginic acid, which do not require a catalyst and can be converted into carbonaceous materials purely on heating.

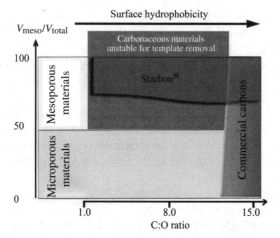

FIGURE 10.2 Starbon® relative to other types of carbon.

A variety of mesoporous carbonaceous materials of controllable surface and bulk characteristics have been produced by heating to different temperatures ranging from 100 to 1400 °C. These types of mesoporous carbonaceous solids can be in powder form or in the form of stable monolith as shown in Figure 10.3.

In conclusion, Starbon® technology is very advantageous because it is

(i) *green*: process avoids the use of harmful chemicals;

(ii) *sustainable*: polysaccharides are renewable resources that are widely available in many countries;

(iii) *simple*: methodology comprises of just three stages;

(iv) *environmentally benign*: nonpersistent, nonbioaccumulative, and nontoxic.

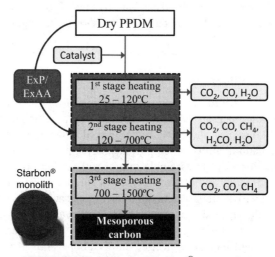

FIGURE 10.3 Method of Starbon® preparation.

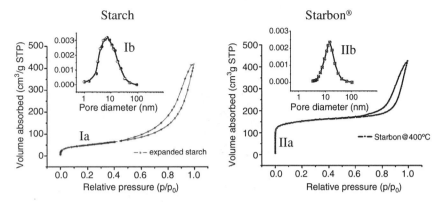

FIGURE 10.4 (I) Textural properties of mesoporous expanded starch: (a) isotherm of adsorption; (b) BJH desorption isotherm. (II) Textural properties of Starbon® prepared at 400 °C: (a) isotherm of adsorption; (b) BJH desorption isotherm.

All of these features make Starbon® technology promising for large-scale production so that it is not limited to academic curiosity alone.

10.3.3 Starbon® Properties

10.3.3.1 Textural Properties The porous structure of the Starbon® comes from the expanded polysaccharide, which removes the need for synthetic templates such as silica used to define structure (see Figure 10.4). It also negates the problem of micelle collapse associated with micelle-templated polymer methods for MC synthesis.

The total pore volume and the average pore diameter in the mesoporous region remain essentially constant throughout the carbonization process. The average pore diameter in the mesoporous region is around 10 nm indicating a predominance of mesopores in the structure of Starbon®. In addition, Figure 10.5 illustrates how the morphology of Starbon® is controlled by the inherent nature of the polysaccharide carbonized (morphology of the sample particles is largely preserved during pyrolysis).

It can be seen that the starch-derived Starbon® has a spherical shape with particle sizes in the range of 5–15 μm. This is consistent with the size and shape of the original starch granules. However, the pectin- and alginic acid-derived Starbon® materials demonstrate a continuous phase network, with the alginic acid-derived materials organized in a rod-like morphology into mesoscale-sized domains, generating large mesopore volumes.

Temperature is another factor that can alter Starbon® materials' textural properties, as can be seen from the data shown in Table 10.1.

Although there is a substantial increase in the contribution of the microporous region to the total surface area with increased temperature, the actual volume that this corresponds to is small in comparison to the total mesoporous volume. Moreover, the level of mesoporosity (up to 98%) of Starbon® dramatically increases when it is made from pectin and alginic acid due to pore diameters changing from 7 to 25 nm. This

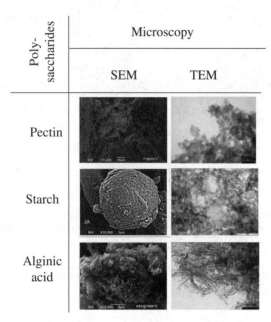

FIGURE 10.5 TEM and SEM images of Starbon® obtained from different polysaccharides.

emphasizes our control over the textural properties, first from temperature and second from the original polysaccharide source.

These structural properties make Starbon® materials particularly suitable for applications such as catalysis, chromatography, and adsorption of large organic molecules.

10.3.3.2 Surface Functionality The surface functionality of the acid-doped expanded starch does not significantly differ on heating until above 150 °C when

TABLE 10.1 Surface Properties of Starbon®

Surface Properties	Starch			Alginic Acid			Pectin		
	ExS[a]	300 °C[b]	800 °C[b]	ExA[c]	300 °C[b]	800 °C[b]	ExP[d]	300 °C[b]	700 °C[b]
S_{BET} (m²/g)	184	293	600	209	216	349	200	174	298
Pore volume (cm³/g)									
Micropore	0.01	0.16	0.2	0.01	0.01	0.05	0.08	0.08	0.13
Mesopore	0.61	0.37	0.43	2.65	0.9	1.0	0.38	0.79	0.97
Pore diameter (nm)	7.6	17.2	7.0	25	18	15.8	7.3	28.8	25.0
C:O ratio	1.2	3.43	8.6	1.17	3.86	8.59	1.36	5.0	9.56

[a]Expanded starch.
[b]Starbon® preparation temperature.
[c]Expanded alginic acid.
[d]Expanded pectin.

TABLE 10.2 Change in C:O Ratio and Surface Energy with Starbon® Preparation Temperature

	Ex-st[a]	100 °C[b]	150 °C[b]	300 °C[b]	350 °C[b]	450 °C[b]	600 °C[b]	700 °C[b]	800 °C[b]
C:O (EA)	1.2	1.26	1.55	3.43	5	6.01	7.53	8.54	8.6
C:O (XPS)	1.1	1.34	1.99	3.79	5.1	6.04	7.55	8.5	8.6
E_{DR}	7.4	6.9	6.5	17.7	18.2	20.6	24.4	26.6	25.8

[a]Ex-st: expanded starch.
[b]Starbon® preparation temperature.

an almost linear change in the C:O ratio (tested via elemental analysis) is observed with oxygen diminishing up to 800 °C as seen in Table 10.2 (Budarin et al., 2004). This change to predominately carbon is verified further with the change in surface energy (E_{DR}) as measured by the Dubinin–Astakhov method.

The results of XPS not only concur with the general C:O ratio trend found by elemental analysis but also indicate how the carbonization may be taking place. Although XPS suggests a consistently greater C:O ratio compared to those obtained from elemental analysis, up to 700 °C, the results differ less between techniques as Starbon® preparation temperature increases.

This indicates as expected that carbonization proceeds from the outside of the material, and that thermal diffusion increases as it becomes more predominately carbon.

The properties of Starbon® have been further characterized by a number of techniques such as TGA, ^{13}C MAS NMR, and DRIFT with the overall properties and possible applications at each temperature range of preparation summarized in Figure 10.6.

It can be seen that from starch there is a progressive increase in the hydrophobicity of the functional groups present until a graphite-like structure is formed above 700 °C. This can be approximately broken down into three temperature ranges. In the 150–200 °C range, some of the hydroxyl functionality condenses to form ether groups. From 200 to 300 °C, carbonyl groups can be seen that conjugate with olefinic groups to form aliphatic and alkene/aromatic functions. As the material is heated to 300 °C and to progressively higher temperatures, the aliphatic functionality converts to aromatic π-systems. Materials could also be prepared up to > 1000 °C (showing extended graphitic character as confirmed by XPS analysis) with no decrease in the quality of the textural properties or alteration in the structural morphology.

For further information on this area, the reader is referred to an excellent review on this topic (White et al., 2009).

10.3.4 Starbon® Applications

Although the potential for Starbon® mesoporous carbons is vast due to their easily adaptable surface properties and array of pore structures, there are two main application areas: chromatography and heterogeneous catalytic support, which have shown the most promise.

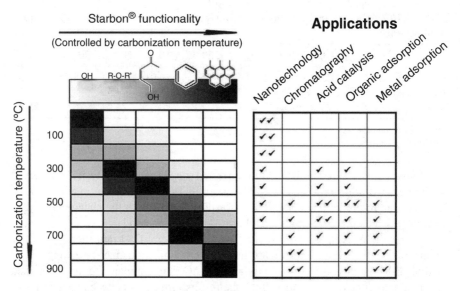

FIGURE 10.6 Changing of Starbon® surface functionality with preparation temperature and the materials' respective application.

10.3.4.1 *Chromatography*

Porous graphitic carbon (PGC) HPLC columns have proven to be useful in the separation of polar analytes, for example, sugars, among others (Knox and Ross, 1997). However, separation of sugars, such as glucose, still poses a substantial challenge regardless of the continued advancement in this technology.

Columns packed with second-generation Starbon® materials derived from alginic acid are particularly attractive as chromatographic stationary phase materials, as they present minimal micropore and good mesopore content. The lack of micropore content aids the efficiency of separation by reducing the amount of irreversible, high-energy analyte adsorption that typically occurs in sub-2 nm pores. These types of stationary phases were found to be particularly efficient at separating the sugars, for example, glucose (monosaccharide), sucrose (disaccharide), and raffinose (trisaccharide). The resultant ion chromatograms had good peak shape and near-baseline resolution (White et al., 2010a).

The performance of the Starbon® stationary phase proved to be comparable to commercially available columns, with potential improvements in the separation of glucose. With further refinement of particle size, and an easily modifiable surface, significant improvements in separation efficiency are expected.

10.3.4.2 *Heterogeneous Acid Catalysis*

Solid acid catalysts have been used in a wide range of important chemical transformations. Carbon materials were found to be good catalytic supports that after sulfonation can act as strong Brønsted acid catalysts.

Esterification in Aqueous Media   Starbon® is an excellent support for heterogeneous catalysis where its unique and tunable surface characteristics are appropriate for many reactions including unexpected esterification reactions conducted in aqueous media. This is particularly important in biomass fermentation reactions, which produce a range of organic acids that can be utilized as platform molecules in applications such as the production of polymers and higher value intermediates. Esterification is one of the key upgrading steps for these acids. The fermentation process is carried out in aqueous media and the resulting aqueous broths require resource-intensive separation steps before the acids can be upgraded. The new Starbon® catalysts can overcome this problem.

Dicarboxylic acids were chosen to explore the esterification reaction due in part to their inclusion in the list of top biomass-derived platform molecules for future large-scale applicability. Succinic, fumaric, and itaconic acids were esterified in the aqueous ethanol, which showed selectivity toward Starbon® acids produced at 400, 450, and 550 °C, respectively. These results further emphasize how important a tunable surface functionality is in acid-catalyzed reactions. For further information on these types of reactions, the reader can refer to the work of Budarin and coworkers (Budarin et al., 2007a, 2007b, 2007c; Clark et al., 2008).

Glycerol Esterification and Etherification Reactions Glycerol, a by-product from biodiesel manufacture and a starting point for many other higher value chemicals, is ideally suited for Starbon® acid catalysis.

Esterification of glycerol and acetic acid with Starbon® acid, prepared at 400 °C under microwave conditions, resulted in a quantitative conversion with approximately 75% monoester. Adjustment of the ratio of glycerol to acetic acid, 1:3 and 1:6, respectively, yielded conditions that were more favorable for di- and triester formation (70–80% selectivity). Overall, Starbon® acid was found to be the most active and selective catalyst compared to a number of other commercially available catalysts (Luque et al., 2008).

Etherification of glycerol with various aryl and alkyl alcohols with Starbon® acid under microwave conditions gave good conversions. The Starbon® acid exhibited a similar activity to that of the silica sulfonated material and an improved performance compared to some other acid catalysts including beta zeolite and *p*-TSA). A slight improvement in the selectivity of monoether in the 1-position was also obtained with respect to the other solid acids tested in the etherification reaction (Luque et al., 2008).

Preparation of Amides *N*-Acylation of a range of aliphatic and aromatic amines with acetic acid using Starbon® acid under microwave conditions was found to work better than zeolites, Al-MCM-41, and acidic clays. The reaction was complete within 10 min with >90% selectivity to acylated product. In addition, *N*-acylation of aniline, under similar reaction conditions, was found to work 4–10 times better than commercial catalysts, and twice as fast as sulfonated microporous carbonaceous materials using Starbon® acid (Luque et al., 2009a).

10.3.4.3 Heterogeneous Catalysis: Nanoparticles

Hydrogenation Metal nanoparticle (Pt, Pd, Rh, and Ru, 5 wt% loading)-supported Starbon® prepared at 300 °C was used to hydrogenate succinic acid. Good conversions to 1,4-butanediol and γ-butyrolactone were found, with the greatest turnover numbers (TON) and turnover frequencies (TOF) with the Pt- and Ru-supported Starbon® materials. This was attributed to these metal nanoparticles being smaller and more evenly distributed throughout the Starbon® matrix. For reactions involving Ru-supported Starbon® materials, THF was also reported as a product obtained in significant quantities. On repeated tests (five reaction cycles), all catalysts maintained > 95% activity with minimal leaching < 0.5 ppm (Luque et al., 2009b). The Ru-supported Starbon® materials have also been found to be effective for the organic acids such as fumaric, itaconic, levulinic, and pyruvic acids under similar reaction conditions (Luque and Clark, 2010).

Glycerol Oxidation Reactions   Pd (9.2 wt% loading)-supported Starbon® prepared at 400 °C and hydrogen peroxide were used to oxidize glycerol under microwave heating conditions. It was found that the overall conversion was 60–95 mol%. Selectivity to glycolic acid and oxalic acid was 35–90 and 5–60 mol%, respectively. Selectivities were not significantly different on catalyst reuse (Luque et al., 2008).

10.4 SUMMARY AND FUTURE PERSPECTIVES

Polysaccharides are an abundant renewable supply of valuable chemicals that are utilized in a wide range of industries. Continued development and research into this fascinating class of materials is needed as current industrial economies that are largely dependent on oil (90% by weight) transcend to one that is more based on local renewable feedstock. Expansion of the polysaccharide structure allowing their rich functionality to be more accessible is one method that is helping add value to this material, and help it compete with traditional oil-based polymers, such as in the case of "switchable" adhesive technology.

Expanded polysaccharides and their natural ability to form nanochannelled structures without the need of a templating agent have opened up a completely new chapter on the formation of mesoporous carbons and carbonaceous materials that are trademarked under the name Starbon®. The ability to diversify its surface functionality by simply altering the temperature of preparation has increased its potential to a diverse range of applications in comparison to conventional mesoporous carbons. This, simultaneously with high surface area in the mesoporous region, mechanical stability and it being a readily scalable technology makes the Starbon® family particularly suitable for applications such as catalysis, chromatography, and water purification.

ACKNOWLEDGMENTS

We thank colleagues at the York Green Chemistry Centre, past and present, for their intellectual input. We also thank Dr. Vtaly Budarin and Miss Jo Parker for their assistance with diagrams and corrections.

REFERENCES

Bain, R. (2007). World Biofuels Assessment. Worldwide Biomass Potential: Technology Characterizations, National Renewable Energy Laboratory.

Baumann, M., and A. Conner (1994). Carbohydrate polymers as adhesives. In: A. Pizzi and K. Mittal,editors, *Handbook of Adhesive Technology*, New York: Marcel Dekker, Inc.

Budarin, V., J. Clark, and S. Tavener (2004). Surface energy and surface area measurements by ^{19}F MAS NMR of adsorbed trifluoroacetic acid. *Chem. Commun.* 524–525.

Budarin, V., J. H. Clark, F. E. I. Deswarte, J. J. E. Hardy, A. J. Hunt, and F. M. Kerton (2005). Delicious not siliceous: expanded carbohydrates as renewable separation media for column chromatography. *Chem. Commun.* 2903–2905.

Budarin, V., J. H. Clark, J. J. E. Hardy, R. Luque, K. Milkowski, S. J. Tavener, and A. J. Wilson (2006). Starbons: new starch-derived mesoporous carbonaceous materials with tunable properties. *Angew. Chem., Int. Ed.* **45**:3782–3786.

Budarin, V., R. Luque, D. J. Macquarrie, and J. H. Clark (2007a). Towards a bio-based industry: benign catalytic esterifications of succinic acid in the presence of water. *Chem. Eur. J.* **13**:6914–6919.

Budarin, V. L., J. H. Clark, R. Luque, and D. J. Macquarrie (2007b). Versatile mesoporous carbonaceous materials for acid catalysis. *Chem. Commun.* 634–636.

Budarin, V. L., J. H. Clark, R. Luque, D. J. Macquarrie, A. Koutinas, and C. Webb (2007c). Tunable mesoporous materials optimised for aqueous phase esterifications. *Green Chem.* **9**:992–995.

Clark, J. H., V. Budarin, T. Dugmore, R. Luque, D. J. Macquarrie, and V. Strelko (2008). Catalytic performance of carbonaceous materials in the esterification of succinic acid. *Catal. Commun.* **9**:1709–1714.

defra (2003). Supply chain impacts of further regulation of products consisting of, containing, or derived from, genetically modified organisms [online]. LMC International Ltd. Available at http://www.defra.gov.uk/environment/gm/research/pdf/epg_1-5-212.pdf (accessed June 3, 2008).

Demirbas, M. F., M. Balat, and H. Balat (2009). Potential contribution of biomass to the sustainable energy development. *Energy Convers. Manage.* **50**:1746–1760.

Doi, S., J. Clark, D. Macquarrie, and K. Milkowski (2002). New materials based on renewable resources: chemically modified expanded corn starches as catalysts for liquid phase organic reactions. *Chem. Commun.* 2632–2633.

Entwistle, G., S. Bachelor, E. Booth, and K. Walker (1998). Economics of starch production in the UK. *Ind. Crops Products* **7**:175–186.

Eusterhues, K., C. Rumpel, M. Kleber, and I. Kögel-Knabner (2003). Stabilization of soil organic matter by interactions with minerals as revealed by mineral dissolution and oxidative degradation. *Org. Geochem.* **34**:1591–1600.

Fredriksson, H., J. Silverio, R. Andersson, A. Eliasson, and P. Åman (1998). The influence of amylose and amylopectin characteristics on gelatinization and retrogradation properties of different starches. *Carbohydr. Polym.* **35**:119–134.

Gadiou, R., S.-E. Saadallah, T. Piquero, P. David, J. Parmentier, and C. Vix-Guterl (2005). The influence of textural properties on the adsorption of hydrogen on ordered nanostructured carbons. *Microporous Mesoporous Mater.* **79**:121–128.

Glenn, G., and D. Irving (1995). Starch-based microcellular foams. *Cereal Chem.* **72**:155–161.

Glenn, G., and D. Stern (1999). Starch-based microcellular foams. U.S. Patent No. 5,958,589.

Hartmann, M., A. Vinu, and G. Chandrasekar (2005). Adsorption of vitamin E on mesoporous carbon molecular sieves. *Chem. Mater.* **17**:829–833.

Hermansson, A., and K. Svegmark (1996). Developments in the understanding of starch functionality. *Trends Food Sci. Technol.* **7**:345–353.

Hunt, A. J., V. L. Budarin, S. W. Breeden, A. S. Matharu, and J. H. Clark (2009). Expanding the potential for waste polyvinyl-alcohol. *Green Chem.* **11**:1332–1336.

Jeong, S., and C. Werth (2005). Evaluation of methods to obtain geosorbent fractions enriched in carbonaceous materials that affect hydrophobic organic chemical sorption. *Environ. Sci. Technol.* **39**:3279–3288.

Joo, S., S. Choi, I. Oh, J. Kwak, Z. Liu, O. Terasaki, and R. Ryoo (2001). Ordered nanoporous arrays of carbon supporting high dispersions of platinum nanoparticles. *Nature* **412**:169–172.

Kennedy, H. (1989). Starch- and dextrin-based adhesives. In: R. Hemingway, A. Conner, and S. Branham,editors, *Adhesives from Renewable Resources*, Washington, DC: ACS Publications.

Kim, T.-W., I.-S. Park, and R. Ryoo (2003). A synthetic route to ordered mesoporous carbon materials with graphitic pore walls. *Angew. Chem., Int. Ed.* **42**:4375–4379.

Knox, J. H., and P. Ross (1997). *Carbon-Based Packing Materials for Liquid Chromatography—Structure, Performance, and Retention Mechanisms*, Advances in Chromatography, Vol. 37, New York: Marcel Dekker.

Lee, J., S. Yoon, T. Hyeon, S. Oh, and K. Kim (1999). Synthesis of a new mesoporous carbon and its application to electrochemical double-layer capacitors. *Chem. Commun.* 2177–2178.

Li, Z., and S. Dai (2005). Surface functionalization and pore size manipulation for carbons of ordered structure. *Chem. Mater.* **17**:1717–1721.

Li, Z., W. Yan, and S. Dai (2005). Surface functionalization of ordered mesoporous carbons: a comparative study. *Langmuir* **21**:11999–12006.

Luque, R., and J. H. Clark (2010). Water-tolerant Ru-Starbon® materials for the hydrogenation of organic acids in aqueous ethanol. *Catal. Commun.* **11**:928–931.

Luque, R., V. Budarin, J. H. Clark, and D. J. Macquarrie (2008). Glycerol transformations on polysaccharide derived mesoporous materials. *Appl. Catal. B* **82**:157–162.

Luque, R., V. Budarin, J. H. Clark, and D. J. Macquarrie (2009a). Microwave-assisted preparation of amides using a stable and reusable mesoporous carbonaceous solid acid. *Green Chem.* **11**:459–461.

Luque, R., J. H. Clark, K. Yoshida, and P. L. Gai (2009b). Efficient aqueous hydrogenation of biomass platform molecules using supported metal nanoparticles on Starbons®. *Chem. Commun.* 5305–5307.

Milkowski, K., J. Clark, and S. Doi (2004). New materials based on renewable resources: chemically modified highly porous starches and their composites with synthetic monomers. *Green Chem.* **6**:189–190.

Poole, C. (2003). *The Essence of Chromatography*, Elsevier Science.

Röper, H. (2002). Renewable raw materials in Europe: industrial utilisation of starch and sugar. *Starch* **54**:89–99.

Ryoo, R., S. Joo, and S. Jun (1999). Synthesis of highly ordered carbon molecular sieves via template-mediated structural transformation. *J. Phys. Chem. B* **103**:7743–7746.

Sayari, A. (1996). Catalysis by crystalline mesoporous molecular sieves. *Chem. Mater.* **8**:1840–1852.

Shuttleworth, P. S., J. H. Clark, R. Mantle, and N. Stansfield (2010). Switchable adhesives for carpet tiles: a major breakthrough in sustainable flooring. *Green Chem.* **12**:798–803.

Stevens, C. (2004). Industrial products from carbohydrates, wood and fibres. In: C. Stevens and R. Verhé,editors, *Renewable Bioresources, Scope and Modification for Non-Food Applications*, Chichester, UK: Wiley.

White, R. J., V. Budarin, R. Luque, J. H. Clark, and D. J. Macquarrie (2009). *Tuneable porous carbonaceous materials from renewable resources*. Chem. Soc. Rev. **38**:3401–3418.

White, R. J., C. Antonio, V. L. Budarin, E. Bergstrom, J. Thomas-Oates, and J. H. Clark (2010a). Polysaccharide-derived carbons for polar analyte separations. *Adv. Funct. Mater.* **20**:1834–1841.

White, R. J., V. L. Budarin, and J. H. Clark (2010b). Pectin-derived porous materials. *Chem. Eur. J.* **16**:1326–1335.

White, R. J., V. L. Budarin, and J. H. Clark (2008). Tuneable mesoporous materials from alpha-D-polysaccharides. CHEMSUSCHEM. **1**(5):408–411

Zobel, H. (1984). Gelatinization of starch and mechanical properties of starch pastes. In: R. Whistler, J. Berniller, and E. Paschell,editors, *Starch: Chemistry and Technology*, 2nd ed., New York: Academic Press.

11

STARCH-BASED BIONANOCOMPOSITES: PROCESSING AND PROPERTIES

Visakh P. M., Aji P. Mathew, Kristiina Oksman, and Sabu Thomas

11.1 INTRODUCTION

Starch is a well-known polymer, naturally produced by plants in the form of granules by all green plants as an energy store and is a major food source for humans. Starch or amylum is a polysaccharide carbohydrate consisting of a large number of glucose units joined together by glycosidic bonds (Buléon et al., 1998). Pure starch is a white, tasteless, and odorless powder that is insoluble in cold water or alcohol. Chemically, it consists of two types of molecules (a) linear and helical amylose and (b) the branched amylopectin. Depending on the source, starch generally contains 20–25% amylose and 75–80% amylopectin (Vandeputte and Delcour, 2004; Oates, 1997). Amylose is a semicrystalline biopolymer and is soluble in hot water, while amylopectin is highly crystalline and is insoluble in hot water (Vandeputte and Delcour, 2004). The chemical structure of starch is shown in Figure 11.1.

Traditionally, starch that is obtained from a great variety of crops, have been used as a thickening, stiffening, or gluing agent after dissolving in warm water (Angellier et al., 2006). More recently, starch, is considered as the most promising material for the production of biodegradable plastic because it is a versatile biopolymer with immense potential and low price for use in the nonfood industries. Starch has been investigated widely for the potential manufacture of products such as water-soluble pouches for detergents and insecticides, flushable liners and bags, and medical delivery systems and devices (Fishman et al., 2004). In these applications, native

Polysaccharide Building Blocks: A Sustainable Approach to the Development of Renewable Biomaterials, First Edition. Edited by Youssef Habibi and Lucian A. Lucia.
© 2012 John Wiley & Sons, Inc. Published 2012 by John Wiley & Sons, Inc.

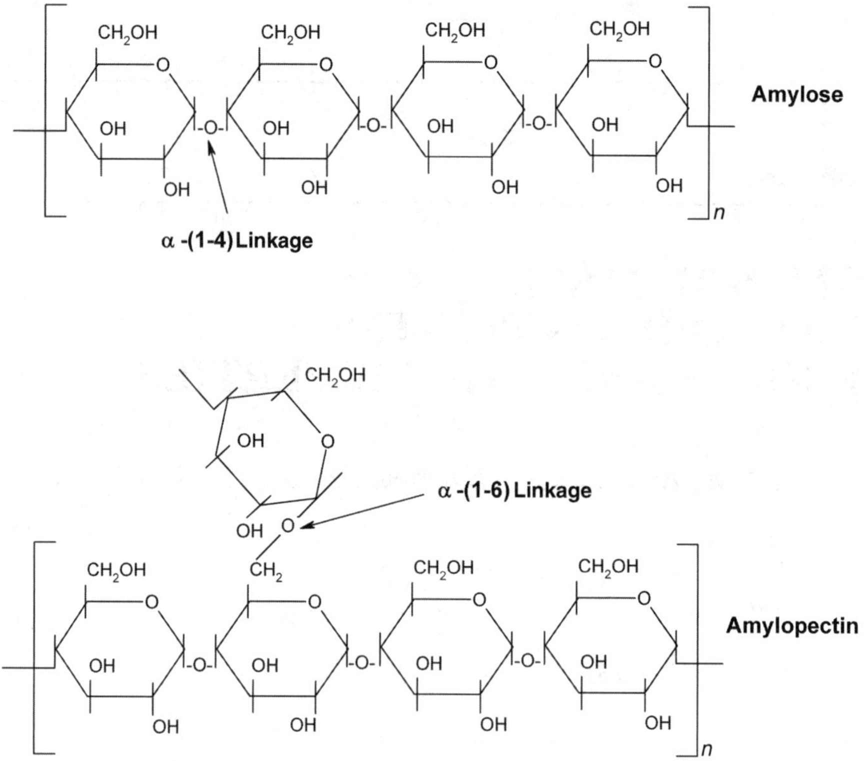

FIGURE 11.1 Chemical structure of starch with amylose and amylopectin units.

starch that exists in a granular structure is usually processed into continuous phase, i.e., thermoplastic starch (TPS) by a process called gelatinization in the presence of heat, moisture, and other plasticizers (Mathew and Dufresne, 2002a; Ma et al., 2007). However, plasticized starch still exhibits problems such as high water absorption, brittleness (in absence of plasticizer), and the mechanical properties are highly affected by the relative humidity (Kumar and Singh, 2008).

To maintain the biodegradability and water resistance of thermoplastic starch without compromising on mechanical properties, starch is usually blended with other synthetic polymers (Huang et al., 2006) or reinforcing fillers (Lu et al., 2005). Compostable multilayers (Lu et al., 2004; Santayanon and Wootthikanokkhan, 2003) or blends (Ray and Okamoto, 2003; Carvalho et al., 2003) have been developed by association between thermoplastic starch and different biodegradable polymers. Funke et al. (1998) have reported a significant improvement in water resistance of starch by adding commercial fibers up to 15%. Another approach is the addition of a nano filler as reinforcement for TPS. Many types of reinforcements have been utilized into the plasticized-starch matrix, such as layer silicates, carbon nanotubes, carbon black, metal (platinum, palladium, and silver), and metal oxide nanoparticles (Chivrac et al., 2008; Avella et al., 2005; Huang et al., 2006; Manno et al., 2008; Lucía et al., 2010; Virendra et al., 2010). Recently, much attention has been paid to preparing totally biodegradable composites with natural polymers and their derivatives (Averous, 2004; Mohanty et al., 2002; Kuciel and Liber-Knec, 2009). In the recent years, nanoreinforcements in the form of biobased nanofibrils or nanocrystals, and so on are being utilized to process starch-based bionanocomposites (Angellier et al., 2005; Anglés and Dufresne, 2000; Mathew and Dufresne, 2002b; Dufresne et al., 2000; Lopez-Rubio et al., 2007).

Cellulosic nanoreinforcements namely, nanofibrils and nanowhiskers have generated much attention recently in thermoplastic starch and has shown great potential in improving the mechanical properties, thermal properties, and barrier properties. In addition nanochitin have also been used to reinforce thermoplastic starch in few studies (Chang et al., 2010). Apart from these, starch itself in nanocrystal or nanoplatelet form are being used as reinforcing phase on bionanocompsites.

Therefore, starch-based bionanocomposites can be divided broadly into two groups having

(i) Thermoplastic starch as the matrix phase where nanocelluloses act as the reinforcing phase and
(ii) Nanocomposites where starch nanocrystals act as the reinforcing phase in a biopolymer matrix.

These two types of starch-based nanocomposites will be focused on in this chapter.

The most commonly used and easy method to prepare reinforced starch nanocomposites is by solution casting where a mixture of starch, plasticizer, and the reinforcement are in water medium gelatinized by heating, casted, and solvent evaporated. Alternatively the mixture of starch, plasticizer and nanoreinforcement in water medium can be melt compounded to form thermoplastic starch-based nanocomposites. In this chapter, different starch nanocomposite studies will be discussed based on the type of plasticizer, type of nanoreinforcement, and processing method. Finally the challenges and future possibilities in this field will be discussed.

11.2 TYPES OF NANOREINFORCEMENTS

The different types of nanoreinforcements used in starch nanocomposites include cellulose nanofibers, nanocrystals, and starch nanocrystals. These nanoreinforcements will be discussed and the microscopy images showing the structures of these nanoreinforcements are shown in Figure 11.2.

11.2.1 Cellulose Microfibrils and Nanofibrils

The general structure of cellulose has been reviewed (O'Sullivan, 1997) and more recent studies based on neutron diffraction provide details of the crystal structure (Nishiyama et al., 2003; Müller et al., 2000). Cellulose nanofibrils are the 3–10 nm thick fibrils formed during cellulose biosynthesis in higher plants (Brett and Waldron, 1996) whereas the microfibrillated cellulose nanofibers consist of

(a) (b) (c) (d)

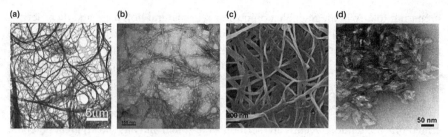

FIGURE 11.2 Morphology of (a) cellulose fibrils (Mathew et al., 2006), (b) cellulose nano-crystals (Petersson et al., 2007), (c) bacterial cellulose (Lee et al., 2009), and (d) starch nanocrystals (Angellier et al., 2005).

aggregates of elementary fibrils. Microfibrillated or nanofibrillated cellulose could be isolated from the cell wall of lignocellulosic materials using very high mechanical shearing. The cellulose nanofibers are obtained by disintegration of the plant fiber cell wall and they typically have nanometric width and length of several micrometers and posses high specific surface area.

The cellulose nanofiber is a highly attractive organic biodegradable reinforcement in polymer nanocomposites, due to its high aspect ratio, good inherent mechanical properties and its ability to form a network (Chang et al., 2010; Petersson et al., 2007). Microfibrillated cellulose is a promising natural nanomaterial that could be isolated from the cell wall of lignocellulosic materials using very high mechanical shearing. Different technologies for fibrillation of plant fibers have employed mechanical treatments such as high-pressure homogenizing (Seydibeyoğlu and Oksman, 2008), grinding (Zimmermann et al., 2005), cryo-crushing (Pääkkö et al., 2007; Hubbe et al., 2008), or ultrasonication (Wegner and Jones, 2006). Very often pretreatment methods such as enzymatic methods (Samir et al., 2005; Taniguchi and Okamura, 1998) and chemo-mechanical methods (Alemdar and Sain, 2007) are used to facilitate the fibrillation process.

Svagan et al. (2007) prepared the starch-based cellulose nanocomposites and they reported the diffusion of moisture in pure and glycerol plasticized high-amylopectin starch, reinforced with different amounts of microfibrillated cellulose nanofibers. Svagan and coworkers have explained the effects of microfibrillated cellulose nanofibers from wood on the moisture sorption kinetics (30% RH) of glycerol plasticized and pure high-amylopectin starch films were studied. The presence of a nanofiber network (70 wt% cellulose nanofibers) reduced the moisture uptake to half the value of the pure plasticized starch film. The moisture diffusivity decreased rapidly with increasing nanofiber content and the diffusivity of the neat cellulose network was, in relative terms, very low. It was possible to describe the strong decrease in zero-concentration diffusivity with increasing cellulose nanofiber/matrix ratio, simply by assuming only geometrical blocking using the model. Still, also constraining effects on swelling from the high modulus/hydrogen bonding cellulose network and reduced amylopectin molecular mobility due to strong starch–cellulose molecular interactions were suggested to contribute to the reductions in moisture diffusivity. This material has interesting mechanical properties and mimics biological-plant structures in several ways. Processing was possible at room temperature in a water medium. The matrix and reinforcement phases are both polysaccharides, where the cellulose is nanostructured as in plants. The matrix phase is very soft with nearly viscous behavior and the mechanical material integrity derives from the cellulose nanofiber network.

Another study of authors on glycerol plasticized starch materials, reinforced by cellulose nanofibers have also shown favorable interactions between cellulose and the polymer matrix (Svagan et al., 2009). In this study, the source of nanofibers was bleached sulfite softwood cellulose pulp that was disintegrated by combining mild enzymatic hydrolysis with mechanical shearing and high-pressure homogenization.

11.2.2 Cellulose Nanowhiskers

Cellulose nanowhiskers are rod-shaped crystals formed when native cellulose is subjected to strong acid hydrolysis (Bondeson et al., 2006; Araki et al., 1998). The dimensions of the resulting nanowhiskers vary according to the source, but can range from 10 to 20 nm in diameter and from 100 to 1000 nm in length (Rodriguez et al., 2006; Kvien et al., 2005; Grunert and Winter, 2002). The modulus of cellulose nanowhiskers has been experimentally and theoretically calculated to range from 138 GPa (experimentally determined) to 167 GPa (theoretically determined) (Tashiro and Kobayashi, 1991; Sakurada et al., 1962).

Any cellulose-based materials can be used as starting materials to produce nanowhiskers, as example saw dust, different straws, vegetables, or paper fibers can be used as well as microcrystalline cellulose (MCC) because of its purity. Acid hydrolysis (HCl/H_2SO_4), follows by neutralization techniques and concentration are carried out to produce nanowhiskers from commercially available MCC. Bondeson et al. (Hubbe et al., 2008) have optimized the method of preparation of cellulose nanowhiskers from the microcrystalline cellulose by the acid hydrolysis (see Figure 11.3 to understand the process flow).

In this process, nanocrystals were prepared by treating MCC suspension with 64% H_2SO_4 at 45 °C for 130 min. The nanocrystals isolated (shown in Figure 11.2b) had diameters in the range of 5–10 nm and lengths in the range of 150–300 nm (Taniguchi and Okamura, 1998). It was also seen that HCl hydrolyzed nanowhiskers are more

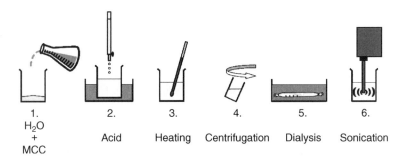

FIGURE 11.3 Isolation process of cellulose nanowhiskers from microcrystalline cellulose.

thermally stable than H_2SO_4 hydrolyzed nanowhiskers but the whiskers produced by HCl are not as well separated as whiskers from H_2SO_4 depending on the lack of negative charging on the nanowhiskers surfaces (Bondeson et al., 2007). Cellulose nanowhiskers isolated from MCC was used by Kvien et al. (2007) to reinforce thermoplastic starch.

Dufresne and coworkers have used cellulose whiskers from tunicate (a sea animal) as reinforcing phase in thermoplastic starch (Angellier et al., 2005; Anglés and Dufresne, 2000). To isolate the nanowhiskers, the mantles of tunicate were first cut into small fragments that were deproteinized by three successive bleaching treatments. The bleached mantle (the tunicin) was then disintegrated in water with a Waring blender. The resulting aqueous tunicin suspension was treated with H_2SO_4 at an acid/water concentration of 55 wt% at 60 °C for 20 min under strong stirring. The suspension was neutralized with water washing. The dispersion of cellulose whiskers was completed by two successive ultrasonic treatments during 3 min each. The suspension did not sediment or flocculate as a consequence of surface sulfate groups created during the sulfuric acid treatment (Angellier et al., 2005). It is constituted of individual cellulose fragments consisting of slender parallelepiped rods that have a broad distribution in size. These fragments have a length ranging from 500 nm up to 1–2 μm and they are almost 10 nm in width. The average aspect ratio of these whiskers was estimated to be close to 70 (Angellier et al., 2005).

11.2.3 Bacterial Cellulose

The nanofibers of cellulose is synthesized by polymerization of glucose molecules, in the interior of bacterial cells. The synthesized nanofibers from a hydrogel network on the surface of culture medium and can be successfully isolated and cleaned to obtain bacterial cellulose (BC) nanofibers (Cristian et al., 2008). Alternatively, the isolated bacterial cellulose can be acid hydrolyzed like any other cellulose to obtain nanocrystalline cellulose. Both nanofibers and nanocrystals from bacterial cellulose are being used as reinforcements in starch (Cristian et al., 2008). Starch was added to the culture medium that used to grow the bacterial cellulose, which resulted in a uniform layer of starch around the BC nanofibers. This processing method resulted in a homogeneous nanocomposites where the original structure of the BC fiber network was retained.

The composites were prepared in a single step with cornstarch by adding glycerol/ water as the plasticizer and bacterial cellulose (1% and 5% w/w) as the reinforcing agent by Ivo et al. (2009). Vegetable cellulose was also tested as reinforcement for comparison and found that bacterial cellulose displayed better mechanical properties than those with vegetable cellulose fibers.

Wan et al. (2009) in his work, used bacterial cellulose nanofibers as the biodegradable reinforcement. The BC nanofibers were incorporated in the starch plasticized with glycerol via a solution impregnation method. Tensile properties of the BC/starch biocomposites were tested and compared with those of the unreinforced starch. Moisture absorption of the biocomposites under 75% RH at 25 °C was analyzed. Additionally, the BC/starch biocomposite (15 wt% BC) and starch were submitted to biodegradation by soil burial experiments in perforated boxes. Tensile strength after exposure to moisture and microorganism attacks was measured. It is found that the moisture sorption mechanism in the BC/starch biocomposites follows a Fickian diffusion mode. The presence of BC nanofibers improves the tensile properties and the resistance to moisture and microorganism attacks.

11.2.4 Starch Nanocrystals

Angellier et al. have extracted nanostarch crystals from macro-sized maize granules (Angellier et al., 2006; Funke et al., 1998). According to the extraction procedure, 36.72 g of native waxy maize starch granules were mixed with 250 mL of 3.16 M H_2SO_4 for 5 days at 40 °C, with a stirring speed of 100 rpm. The resulting suspension was washed by successive centrifugations with distilled water until neutrality. The aqueous suspensions obtained had a weight concentration of about 3.4 wt%. Waxy maize starch nanocrystals consist of platelet-like particles with a thickness of 6–8 nm, a length of 40–60 nm, a width of 15–30 nm, and a density of 1.55 g/cm^3. Such nanocrystals are generally observed in the form of aggregates. Waxy maize starch nanocrystals obtained by hydrolysis of native granules are used as the reinforcing filler in glycerol plasticized starch (GPS) (Angellier et al., 2006; Funke et al., 1998). Another method for the preparation of starch nanoparticles is the delivery of ethanol into starch paste solutions and the synthesis of citrate starch nanoparticles by cross-linking starch nanoparticles with citric acid, which cannot be gelatinized in hot water (Ma et al., 2008).

In a recent study, starch nanocrystals were used in polymer matrices to produce hydrogels, which showed potential in drug delivery application (Zhang et al., 2010). The polysaccharide nanocrystals increased the stability of the hydrogel framework and showed prominent sustained release profiles. Additionally, the incorporation of polysaccharide nanocrystals did not show additional cytotoxicity compared to the native hydrogel where as the inherited shear-thinning property of the nanocomposite hydrogels increased their potential as injectable biomaterials.

11.3 PROCESSING OF STARCH NANOCOMPOSITES

11.3.1 Starch as Matrix Phase

11.3.1.1 Solution Casting The most common method for preparing starch-based nanocomposites is solution casting from aqueous medium. Solution casting is an easy method to process nanocomposites with homogeneous dispersion of nanoreinforcements in starch taking advantage of the solubility in aqueous medium. However, one drawback of this process is that this is a lab scale process.

Anglés and Dufresne (2000) used solution-casting method to prepare starch-based nanocomposites reinforced with tunicin whiskers, using glycerol as the plasticizer. Different percentage of starch and glycerol as the plasticizer were mixed in different ratios to obtain composite films with a homogeneous dispersion. The plasticizer content was fixed at 33 wt%(dry basis of starch matrix). The cellulose whiskers content was varied between 0 and 25 wt%. The mixture was gelatinized using an autoclave and degassed and casted to films. Mathew and Dufresne (2002a) also have successfully prepared the starch/tunicin whiskers nanocomposites using casting method but used sorbitol as plasticizer instead of glycerol.

Kaushik et al. have prepared nanocomposite films using solution casting method with thermoplastic starch and cellulose nanofibrils (CNFs) extracted from wheat straw (Kaushik and Mandeep, 2011; Kaushik et al., 2010). Cellulose nanofibers were dispersed in distilled water and sonicated for almost 3 h. Maize starch was added with 30% glycerol and shear mixed for 10 min by using homogenizer (10,000 rpm). Dispersion of cellulose nanofibers was added to the starch, glycerol mixer, and further shear mixed for 20 min. Starch, glycerol, and CNFs mixture was continuously stirred (at 80–100 rpm) using a mechanical stirrer and heated at $75 \pm 30\,°C$. After the solution became viscous, it was poured on to glass Petri dishes and kept at around $37\,°C$ for 2 days until it was completely dry. Solution cast films of TPS cellulose nanocomposites were made with 5%, 10%, and 15% nanofibers (as per dry weight of nanocomposites). The films of thickness approximately 80 μm were obtained and conditioned at 43% RH for 15 days before testing was done.

Lu and coworkers have prepared plasticized starch (PS)/cellulose nanocrystallites composites, which showed an increase in Young's modulus and tensile strength from 56 to 480 MPa and 2.8 to 6.9 MPa, respectively with increasing fillers contents (Lu et al., 2006). They have prepared cellulose crystals from ramie fibers by acid hydrolysis and used to reinforce plasticized starch matrix, by casting method.

11.3.1.2 Solution Impregnation Very few reports are available on starch-based nanocomposites produced by impregnation technique. Wan and coworkers have developed bacterial cellulose reinforced starch nanocomposites, plasticized with glycerol by solution impregnation method (Sakurada et al., 1962). The BC/starch biocomposites prepared by this method showed much higher tensile strength and modulus than the unreinforced starch, but show lower elongation at break.

11.3.1.3 Twin Screw Extrusion In our laboratory (LTU, Sweden) we have attempted the processing of thermoplastic starch reinforced with cellulose nanocrystals isolated from microcrystalline cellulose. A wet feeding method where slurry with all components premixed was pumped into the extruder was developed. It was expected that the gelatinization of starch to form TPS as well as compounding of starch with nanocrystals to form the nanocomposite will occur during the single step extrusion process.

The screw design and experimental parameters are given in Figure 11.4. A small amount of kneading elements was present for distributive and dispersive purpose and to increase the residence time since it is wanted that as much water as possible will evaporate. Two atmospheric venting channels are present to transfer moisture out of the barrel. Screw speed has a crucial role for residence time (Klemens and Werner, 2008). A screw speed of 150 rpm was used so as to provide the best possible residence time enabling the evaporation of water from the extruded material.

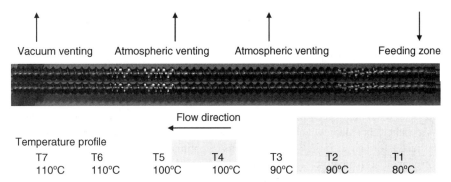

FIGURE 11.4 Screw set up and temperature profile for processing of starch and cellulose nanocrystals based nanocomposites produced by twin-screw extrusion.

11.3.2 Starch as the Reinforcing Phase

11.3.2.1 Solution Casting Unlike the starch-based nanocomposites where starch acts as the matrix phase processed by different methods such as solution casting, melt-compounding, or resin impregnation, solution casting is the only reported method for the processing of starch bionanocomposites with starch nanocrystals as the reinforcing phase.

Casting technique leading to the formation of films was used for the preparation of latex-based starch nanocomposites (Angellier et al., 2006; Funke et al., 1998). According to the procedure, the aqueous suspension of starch nanocrystals and the natural rubber latex were mixed in various proportions in order to obtain dry films between 200 μm and 1 mm thick depending on the weight fractions of dry starch nanocrystals within the NR matrix ranging from 0 to 50 wt%. After mixing, the mixtures were stored under vacuum and stirred on a rota-vapor during about 10 min in order to degas the mixture and thereby avoid the formation of irreversible bubbles during the water evaporation step. Then, the films were cast in Teflon moulds and evaporated at 40 °C in a ventilated oven for 6–8 h (depending on the water content) and then heated at 60 °C under vacuum for 2 h. Resulting dry films were conditioned at room temperature in desiccators containing P_2O_5 salt until being tested. Other matrices such as starch, PLA, PVA, polyurethanes, and so on where starch nanocrystals are used as nanoreinforcements are also reported (Deéborah et al., 2010; Chen et al., 2008; Zheng et al., 2009).

11.4 PROPERTIES OF STARCH NANOCOMPOSITES

11.4.1 Effect of Plasticizers on Starch-Based Nanocomposites

The performance of native starch is indicative of a vitreous material with a glass transition temperature (T_g) above ambient temperatures. The melting temperature of a starch granule is close to its degradation temperature. Hence, plasticizers are required to destroy the intermolecular hydrogen bonding in the crystalline regions of starch granules and decrease the melting temperature during thermoplastic processing. Plasticizers are generally added to convert starch into a so-called thermoplastic starch and to extrude or mold an object. Plasticizers can be defined as low molecular weight substances that are incorporated into the polymer matrix to increase the film flexibility and processibility. They increase the free volume or molecular mobility of polymers by reducing the H-bonding between the polymer chains. Small molecules such as glycerol and water that can form hydrogen bonds with starch can serve as plasticizers in starch-based materials (Mathew and Dufresne, 2002a). Water is the most common primary plasticizer for starch and glycerol is the most commonly used secondary plasticizer. However, apart from glycerol other plasticizers are also used and in this chapter use of different polyols and their impact on starch properties will be discussed. Incorporating plasticizing agent, such as water and/or poly-alcohols, starch can be made thermoplastic called thermoplastic starch or plasticized starch (PS) through destructurization by the introduction of mechanical and heat energy (Mathew and Dufresne, 2002a). Mathew and Dufresne (2002a) have also investigated in detail the possibilities of using different polyols as the plasticizer for starch and were reported earlier. The chemical structures of some of the common plasticizers for starch are shown in Figure 11.5.

It was noticed, however, that the plasticizers used to produce TPS can interact and interfere with the nanoreinforcements and greatly influence the performance of the nanocomposite materials. The reports on starch-based nanocomposites demonstrate the influence of chemistry, molecular weight, and concentration of plasticizer on the performance (Van Soest et al., 1996).

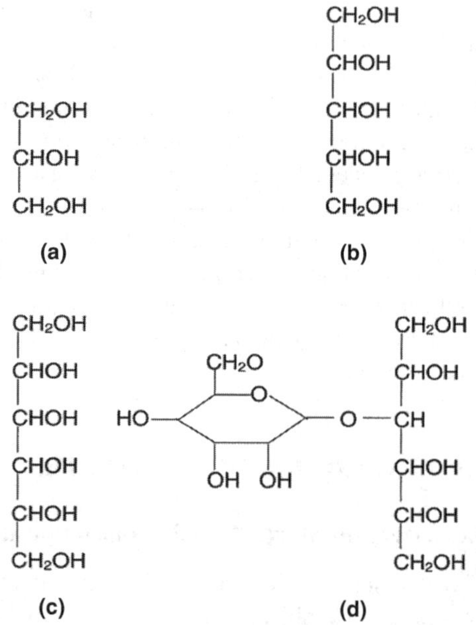

FIGURE 11.5 The chemical structure of some common plasticizers for starch (a) glycerol, (b) xylitol, (c) sorbitol, and (d) maltitol (Mathew and Dufresne, 2002a).

Russo et al. (2007) showed that the moisture diffusivity in high-amylose starch, blended with a small amount of a water-soluble polyol (1–10%), increased as an exponential function of the water content and moisture-dependent diffusivity was therefore evaluated. The objective was to assess the effects of cellulose nanofibers on the moisture transport properties of amylopectin films. A strong reduction in moisture diffusivity is observed and two different theoretical models are used to predict the reduction in zero-concentration moisture diffusivity with increasing cellulose content. The authors explained the effect of plasticizer on the properties of nanocomposites (Bondeson et al., 2006; Chen et al., 2008). The moisture uptake decreased with increasing content of cellulose. Cellulose nanofibers are less hygroscopic than starch due to the higher degree of molecular order. In cellulose, the disordered regions are likely to be preferred sorption sites, as discussed in the introduction. Neat cellulose nanofiber films sorbed only 40% and 56% of the amount of moisture sorbed in, respectively, the plasticized and glycerol-free amylopectin. The moisture/nanofiber film interactions are complex. Since the relative humidity was low, moisture is not believed to be accumulating in the voids. The authors also studied the glass transition temperature of amylopectin/ glycerol film, which is well below 33 °C (Bondeson et al., 2006). Thus, the "extra" plasticizing effect from sorbed water was small. With 40 and 70 wt% cellulose nanofibers, good fit to the water sorption curves was obtained only with a use of a moisture concentration dependent diffusivity. This indicated that, in this case, the plasticizing action of moisture played a more important role. The average and initial diffusion coefficients decreased with increasing cellulose nanofiber content and increased with increasing glycerol contents would increase with a decreasing chain mobility (polymer relaxation rate) and moisture diffusivity (Anglés and Dufresne, 2001). However, diffusion coefficients decreased with increasing glycerol content.

Dufresne and coworkers have reported a relatively low reinforcing effect upon the addition of tunicin whiskers in glycerol-plasticized starch (Anglés and Dufresne, 2000). This unexpected low mechanical performance was due to the accumulation of the main plasticizer toward the cellulose/amylopectin interfacial zone, which interferes with hydrogen bonding in the system. The coating of the cellulose whiskers by plasticizer hindered the stress transfer at filler/matrix interface, resulting in poor mechanical properties. Therefore, further studies are taken up with a different plasticizer to understand the possibilities of creating an efficient filler/ matrix interface, which will not interfere with the stress transfer mechanism. However, in sorbitol-plasticized tunicin whisker-reinforced starch nanocomposites the mechanical properties were much higher and indicated an absence of accumulation of plasticizer on the interface and a resultant efficient stress transfer between the matrix and the reinforcement (Mathew et al., 2008).

11.4.2 Effect of Nanoreinforcements Type on Starch Nanocomposites

Very few reports are available where a systematic study is presented about the effect of nanoreinforcement on the properties of starch nanocomposites.

Chen et al. have studied the impact on micro- and nanosized cellulose from pea hull in pea starch. The study showed that the nanoscaled cellulose composites showed significantly higher light transmission, mechanical properties, and higher glass transition temperature than the matrix and the micro composites (Chen et al., 2009).

Kvien et al. (2007) have compared the effect of cellulose nanowhiskers and nanoclay in starch matrix plasticized with sorbitol. The DMTA study showed that at room temperature the storage modulus of CNW and LS nanocomposites were improved by 74 MPa (17%) and 705 MPa (162%), respectively, compared to the pure starch film. At 60 °C the improvement was 102.7 MPa (160%) and 250 MPa (388%) for the CNW and the LS nanocomposite compared to the pure starch film.

Orts et al. have compared the effect of different types of cellulose nanofibrils in starch nanocomposites produced by melt blending. Significant changes in mechanical properties, especially maximum load and tensile strength, were obtained for fibrils derived from several cellulosic sources, including cotton, softwood, and bacterial cellulose. For extruded starch plastics, the addition of cotton-derived microfibrils at 10.3% (w/w) concentration increased Young's modulus by fivefold relative to a control sample with no cellulose reinforcement (Orts et al., 2005) (Figure 11.6).

11.4.3 Effect of Moisture Conditions

Starch has poor moisture resistance and the incorporation of nanoreinforcements is an effective way of decreasing its moisture sensitivity and to obtain better mechanical properties. In CNW-based nanocomposites the CNW do not have affinity to water and cellulose nanofibers are less hygroscopic than starch due to higher degree of molecular order making them effective in improving the barrier. In addition the restriction of molecular chain mobility of the matrix phase in a nanocomposite decreases the moisture uptake. Starch nanocrystals on the other hand acts as platelets

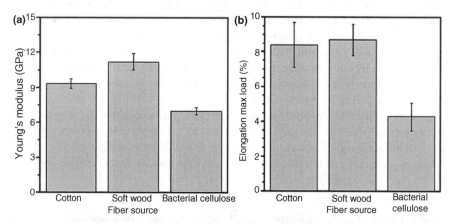

FIGURE 11.6 Maximum load (a) and elongation to break (b) of wheat starch thermoplastic monofilaments with the addition of 5% added cellulose microfibrils isolated from cotton, soft wood, and bacterial cellulose (Orts et al., 2005).

or sheets and provides moisture barrier via a mechanism similar to that of clay nanoparticles (Zhang et al., 2010; Kristo and Biliaderis, 2007).

In spite of this decreased moisture sensitivity in presence of nanoreinforcements, the properties of starch-based nanocomposites are highly varied depending on the relative humidity conditions used for storing as well as testing. In glycerol-plasticized and sorbitol-plasticized starch nanocomposite systems the T_g decreased with increased moisture content due to the plasticizing effect of water (Angellier et al., 2005; Anglés and Dufresne, 2000). The effect of relative humidity conditions on mechanical properties is demonstrated well by Dufresne and coworkers in their study on tunicin whiskers based starch nanocomposites (Deéborah et al., 2010; Chen et al., 2008). In glycerol-plasticized composites, the tensile failure showed brittle behavior for RH upto 58%. But at 75% and 98% RH conditions the same composites showed more tearing before failure (Deéborah et al., 2010).

In sorbitol-plasticized nanocomposites system the mechanical properties also showed a strong dependence on relative humidity conditions (Zheng et al., 2009). For example, the Young's modulus increases from 208 MPa (98% RH) to 838 MPa (31% RH) for 25 wt% whisker composites. The plasticization by water thus appears to have a very significant effect on the composite performance. Generally in these composites the tensile strength and Young's modulus are high at lower RH levels and elongation at break remained constant, irrespective of RH and filler content. Figure 11.7 shows the effect of RH conditions and whiskers concentration on the mechanical properties of sorbitol-plasticized starch/tunicin whisker nanocomposites.

11.4.4 Effect of Processing Method

No systematic study on the effect of processing technique on the properties of starch nanocomposites are available. However, in our laboratory sorbitol-plasticized starch nanocomposites reinforced with cellulose nanowhiskers were prepared by solution casting methods as well as twin-screw extrusion.

In Figure 11.8, the scanning electron microscopy images of pure starch and the two composites prepared by twin-screw extrusion showed that no starch granules indicating that the starch is completely gelatinized. Furthermore, no CNW agglomerates in macrosize can be found which means that either the nano reinforcement is in nanoscale or the adhesion is strong between the matrix and reinforcement. One can also see an increase in surface roughness with higher CNW content.

The mechanical properties of the starch film and the nanocomposites prepared by twin-screw extrusion are given in Table 11.1. The nanocomposites showed a reduction in E-modulus and strain at break while a slight improvement in tensile strength.

The developed materials had good optical clarity and are compared in Figure 11.9. Starch without reinforcement has highest transparency, but there is only a slight difference between the nonreinforced and 4% CNW. This indicates that there are no CNW agglomerates, since these would have reduced the transparency. When raising the CNW content to 8% one can see that the transmission is decreased further but still had good transparency.

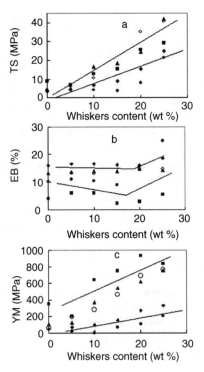

FIGURE 11.7 (a) Tensile strength, (b) elongation at break, and (c) Young's modulus as a function of cellulose content in sorbitol plasticized starch/tunicin whiskers composites at 31 (■), 43 (O), 58 (▲), 75 (♦), and 98 (○). Solid lines serve to guide the eye (Mathew et al., 2008).

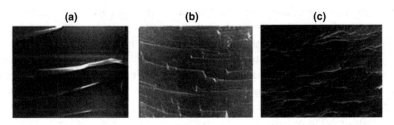

FIGURE 11.8 SEM pictures of starch for starch and cellulose nanocrystals based nanocomposites produced by twin-screw extrusion. (a) TS, (b) TS-CNW4, and (c) TS-CNW8.

TABLE 11.1 Mechanical Properties for Starch and Cellulose Nanocrystals Based Nanocomposites Produced by Twin-Screw Extrusion

Materials	E-Modulus (GPa)	Tensile Strength (MPa)	Strain at Break (%)
TS	1.36 ± 0.33	12.65 ± 2.97	11.4 ± 4.9
TS-CNW4	1.31 ± 0.35	14.98 ± 2.94	4.7 ± 2.3
TS-CNW8	1.13 ± 0.31	15.55 ± 1.18	3.1 ± 0.9

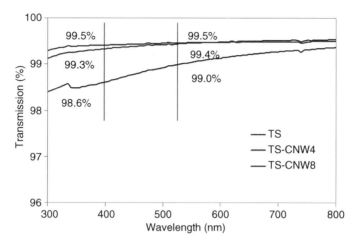

FIGURE 11.9 Transmission in different wavelength for starch and cellulose nanocrystals based nanocomposites produced by twin-screw extrusion.

In the case of solution-casted nanocomposites also starch was found to be gelatinized completely and the transparency was retained as on the case of twin-screw extruded ones (Kvien et al., 2007). Kvien et al. (2007) showed a significant improvement in E-modulus and in strain at break. However, the extruded composites showed a high E-modulus (1.3 GPa) for both pure starch and the nanocomposite samples compared to the ones reported by Kvien et al. (370 MPa). Earlier studies where solution casting has been used have shown the similar response for modulus, tensile strength, and strain at break when adding whiskers as reinforcement (Anglés and Dufresne, 2001; Mathew et al., 2008).

11.5 CONCLUSIONS

Starch-based bionanocomposites are a new class of biobased nanomaterials, which are biodegradable, optically transparent, and having higher mechanical properties, thermal stability, and barrier properties than thermoplastic starch. Starch-based bionanocomposites can be broadly divided into (1) those with starch as matrix and (2) those with starch nanocrystals as reinforcing phase. Cellulose nanofibers, cellulose nanocrystals, and bacterial cellulose are the most used nanoreinforcements in starch matrix. Although these types of starch nanocomposites were usually prepared by solution casting method, other processing methods such as twin-screw extrusion, polymer impregnation, and so on are being used to a limited extent. The studies on starch-based bionanocomposites have indicated that the properties of these materials vary widely based on the moisture conditions used, nature and concentrations of plasticizer as well as nature and concentration of the nanoreinforcement used. It is possible to obtain materials with wide range of properties, starting from brittle material to ductile materials based on the afore-mentioned factors.

The potential application areas of starch-based bionanocomposites are in packaging, coatings, and so on. Processing methods such as twin-screw extrusion should be developed further and optimized to produce starch-based nanocomposite films, foams, and so on for packing application. In the case of medical application, the biocompatibility and nontoxicity of starch nanocomposites will be an added advantage and will find application in drug-delivery, burn healing, hydrogels, and so on.

ACKNOWLEDGMENTS

The authors acknowledge the financial support for bilateral collaboration under Swedish research link (SIDA); Project No. 348-2008-6040. Martin Nilsson M.Sc. is acknowledged for the studies on twin-screw extrusion of starch nanocomposites.

REFERENCES

Alemdar, A., and M. Sain (2007). Isolation and characterization of nanofibers from agricultural residues—wheat straw and soy hulls. *Bioresource Technol.* **99**(6):1664–1671.

Angellier, H., S.M. Boisseau, L. Lebrun, and A. Dufresne (2005). Processing and structural properties of waxy maize starch nanocrystals reinforced natural rubber. *Macromolecules* **38**:3783–3792.

Angellier, H. S. Molina-Boisseau, P. Dole, and A. Dufresne (2006). Thermoplastic starch—waxy maize starch nanocrystals nanocomposites. *Biomacromolecules* **7**(2):531–539.

Anglés, M. N., and A. Dufresne (2000). Plasticized starch/tunicin whiskers nanocomposites. 1. Structural analysis. *Macromolecules* **33**:8344–8353.

Anglés, M. N., and A. Dufresne (2001). Plasticized starch/tunicin whiskers nanocomposites. 2. Mechanical behavior. *Macromolecules* **34**:2921–2931.

Araki, J., M. Wada, S. Kuga, and T. Okano (1998) *J. Colloids Surf. A Physiochem. Eng. Aspects* **142**:75.

Avella, M., J. J. D. Vlieger, M. E. Errico, S. Fischer, P. Vacca, and M. G. Volpe (2005). Biodegradable starch/clay nanocomposite films for food packaging applications. *J. Food Chem.* **93**:467–474.

Averous, L. (2004). Biodegradable multiphase system based on plasticized starch: a review. *Macromol. Sci. Polym. Rev.* **C44**:231–274.

Bondeson, D., A. P. Mathew, and K. Oksman, (2006). Optimization of the isolation of nanocrystals from microcrystalline cellulose by acid hydrolysis. *Cellulose* **13**:171–180.

Bondeson, D., P. Syre, and K. Oksman (2007). All cellulose nanocomposites produced by extrusion. *J. Biomater. Bioenerg.* **1**:367–371.

Brett, C. T., and K. W. Waldron *Physiology and Biochemistry of Plant Cell Walls*. 2nd ed. London: Chapman and Hall (1996).

Buléon, A., P. Colonna, V. Planchot, and S. Ball (1998). Starch granules: structure and biosynthesis. *Int. J. Biol. Macromol.* **23**:85–112.

Carvalho, A. J. F., A. E. Job, N. Alves, A. A. S. Curvelo, and A. Gandini (2003). Thermoplastic starch/natural rubber blends. *Carbohydr. Polym.* **53**:95–99.

Chang, P.R., R. Jian, J. Yu, and X. Ma (2010). Starch-based composites reinforced with novel chitin nanoparticles. *Carbohydr. Polym.* **80**(2):421–426.

Chen, Y., X. Cao, P.R. Chang, and M. A. Huneault (2008). Comparative study on the films of poly(vinyl alcohol)/pea starch nanocrystals and poly(vinyl alcohol)/native pea starch. *Carbohydr. Polym.* **73**(1):8–17.

Chen, Y., C. Liu, P. R. Chang, D. P. Anderson, and M. A. Huneault (2009). Pea starch-based composite films with pea hull fibers and pea hull fiber-derived nanowhiskers. *J. Polym. Eng. Sci.* **49**(2):369–378.

Chivrac, F., E. Pollet, M. Schmutz, and L. Averous (2008). New approach to elaborate exfoliated starch-based nanobiocomposites. *Biomacromolecules* **9**:896–900.

Cristian, J. G., G. T. Fernando, M. G. Clara, P. T. Omar, C. F. Joseph, and M. P. Juan, (2008). Morphological characterisation of bacterial cellulose-starch nanocomposites. *J. Polym. Polym. Compos.* **16**(3):181–185.

Deéborah, L. C., B. Julien, and A. Dufresne (2010). Starch nanoparticles: a review. *Biomacromolecules* **11**:1153–1162.

Dufresne, A., D. Dupeyre, and M. R. Vignon (2000). Cellulose microfibrils from potato tuber cells: processing and characterization of starch–cellulose microfibril composites. *J. Appl. Polym. Sci.* **76**(14):2080–2092.

Fishman, M. L., D. R. Coffin, C. I. Onwulata, and R. P. Konstance (2004). Extrusion of pectin and glycerol with various combinations of orange albedo and starch. *Carbohydr. Polym.* **57**:401–413.

Funke, U., W. Bergthaller, and M. G. Lindhauer (1998). Processing and characterization of biodegradable products based on starch. *Polym. Degrad. Stabil.* **59**(1–3):293–296.

Grunert, M., and W. T. Winter (2002). Nanocomposites of cellulose acetate butyrate reinforced with cellulose nanocrystals. *J. Polym. Environ.* **10**(1):27–30.

Huang, M., J. Yu, and X. Ma (2006). High mechanical performance MMT-urea and form-amide-plasticized thermoplastic cornstarch biodegradable nanocomposites. *Carbohydr. Polym*, **63**:393–399.

Hubbe, M. A., O. J. Rojas, L. A. Lucia, and M. Sain (2008). Cellulosic nanocomposites: a review. *BioResources* **3**(3):929–980.

Ivo, M. G. M., P. M. Sandra, O. Lúcia, S. R. Carmen, A. Freire, J. D. Silvestre, P. N. Carlos, and G. Alessandro (2009). New biocomposites based on thermoplastic starch and bacterial cellulose. *Compos. Sci. Technol.* **69**:2163–2168.

Kaushik, A., and S. Mandeep (2011). Isolation and characterization of cellulose nanofibrils from wheat straw using steam explosion coupled with high shear homogenization. *Carbohydr. Res.* **346**:76–85.

Kaushik, A., M. Singh, and G. Verma (2010). Green nanocomposites based on thermoplastic starch and steam exploded cellulose nanofibrils from wheat straw. *Carbohydr. Polym.* **82**:337–345.

Klemens, K., and W. Werner (2008). Co-rotating twin-screw extruders, fundamentals. *Technol. Appl.*

Kristo, E., and C. G. Biliaderis (2007). Physical properties of starch nanocrystal-reinforced pullulan films. *Carbohydr. Polym.* **68**(1):146–158.

Kuciel, S., and A. Liber-Knec (2009). Biocomposites on the base of thermoplastic starch filled by wood and kenaf fiber. *J. Biobased Mater. Bioenerg.* **3**(3):269–274.

Kumar, A. P. and R. P. Singh (2008). Biocomposites of cellulose reinforced starch: Improvement of properties by photo-induced crosslinking. *Bioresource Technol.* **99**(18):8803–8809.

Kvien, I., J. Sujiyama, M. Votrubec, and K. Oksman (2007). Characterization of starch based nanocomposites. *J. Mater. Sci.* **42**(19):8163–8171.

Kvien, I., B. S. Tanem, and K. Oksman (2005). Characterization of cellulose whiskers and their nanocomposites by atomic force and electron microscopy. *Biomacromolecules* **6**:3160–3165.

Lee, K.-Y., J. J. Blaker, and A. Bismarck (2009). Surface functionalisation of bacterial cellulose as the route to produce green polylactide nanocomposites with improved properties. *Compos. Sci. Technol.* **69**:2724–2733.

Lopez-Rubio, A., J. M. Lagaron, M. Ankerfors, T. Lindstrom, D. Nordqvist, and A. Mattozzi (2007). Enhanced film forming and film properties of amylopectin using microfibrillated cellulose. *Carbohydr. Polym.* **68**(4):718–27.

Lu, Y., L. Tighzert, F. Berzin, and S. Rondot (2005). Innovative plasticized starch films modified with waterborne polyurethane from renewable resources. *Carbohydr. Polym.* **61**:174–182.

Lu, Y., L. Weng, and L. Zhang (2004). Morphology and properties of soy protein isolate thermoplastics reinforced with chitin whiskers. *Biomacromolecules* **5**:1046–1051.

Lu, Y., L. Weng, and X. Cao (2006). Morphological thermal and mechanical properties of ramie crystallites-reinforced plasticized starch biocomposites. *Carbohydr. Polym.* **63**:198–204.

Lucía, M. F., P. Valeria, N. G. Silvia, and R. B. Celina (2010). Starch/multi-walled carbon nanotubes composites with improved mechanical properties. *Carbohydr. Polym.* **83**:1226–1231.

Ma, X., R. Jian, R. Peter, P. R. Chang, and J. Yu (2008). Fabrication and characterization of citric acid-modified starch nanoparticles/plasticized-starch composites. *Biomacromolecules* **9**(11):3314–3320.

Ma, X., J. Yu, and N. Wang (2007). Production of thermoplastic starch/MMT-sorbitol nanocomposites by dual-melt extrusion processing. *Macromol. Mater. Eng.* **292**:723–728.

Manno, D., E. M. Filippo, D. Giulio, and A. Serra (2008). Synthesis and characterization of starch-stabilized Ag nanostructures for sensors applications. *Non-Crystalline Solids* **345**:5515–5520.

Mathew, A. P. and A. Dufresne (2002a). Plasticized waxy maize starch: effect of polyols and relative humidity on material properties. *Biomacromolecules* **3**(5):1101–1108.

Mathew, A. P., and A. Dufresne (2002b). Morphological investigation of nanocomposites from sorbitol plasticized starch and tunicin whiskers. *Biomacromolecules* **3**:609–617.

Mathew, A.P., A. Chakraborty, K. Oksman, and M. Sain (2006). The structure and mechanical properties of cellulose nanocomposites prepared by twin screw extrusion. *ACS Symp. Series* **938**:114–131.

Mathew, A. P., W. Thielemans, and A. Dufresne (2008). Mechanical properties of nanocomposites from sorbitol plasticized starch and tunicin whiskers. *J. Appl. Polym. Sci.* **109**:4065–4074.

Mohanty, A. K., M. Misra, and L. T. Drzal (2002). Sustainable bio-composites from renewable resources: opportunities and challenges in the green materials world. *J. Polym. Environ.* **10**:19–26.

Müller, M., C. Czihak, H. Schober, Y. Nishiyama, and G. Vogl (2000). All disordered regions of native cellulose show common low-frequency dynamics. *Macromolecules* **33**:1834–40.

Nishiyama, Y., J. Sugiyama, H. Chanzy, and P. Langan (2003). Crystal structure and hydrogen bonding system in cellulose Ia from synchrotron X-ray and neutron fiber diffraction. *J. Am. Chem. Soc.* **47**:14300–14306.

O'Sullivan, A. C. (1997). Cellulose: the structure slowly unravels. *Cellulose* **4**:173–207.

Oates, C. G. (1997). Towards an understanding of starch granule structure and hydrolysis. *Trends Food Sci. Technol.* **8**(11):375–382.

Orts, W. J., J. Shey, S. H. Imam, G. M. Glenn, M. E. Guttman, and J.-F. Revol (2005). Application of cellulose microfibrils in polymer nanocomposites. *J. Polym. Environ.* **13**: (4):301–306.

Pääkkö, M., M. Ankerfors, H. Kosonen, A. Nykänen, S. Ahola, and M. Österberg (2007). Enzymatic hydrolysis combined with mechanical shearing and high-pressure homogenization for nanoscale cellulose fibrils and strong gels. *Biomacromolecules* **8**(6):1934–1941.

Petersson, L., I. Kvien, and K. Oksman (2007). Structure and thermal properties of poly(lactic acid)/cellulose whiskers nanocomposite materials. *Compos. Sci. Technol.* **67**:2535–2544.

Ray, S. S., and M. Okamoto (2003). Biodegradable polylactide and its nanocomposites: opening a new dimension for plastics and composites. *Macro. Rapid Commun.* **24**:815–840.

Rodriguez, N. L. G., W. Thielemans, and A. Dufresne (2006). Sisal cellulose whiskers reinforces polyvinyl acetate nanocomposites. *Cellulose* **12**:261–270.

Russo, M. A. L., E. Strounina, M. Waret, T. Nicholson, R. Truss, and P. J. Halley (2007). A study of water diffusion into a high-amylose starch blend: the effect of moisture content and temperature. *Biomacromolecules* **8**:296–301.

Sakurada, I., Y. Nikushina, and T. Ito (1962). Experimental determination of the elastic modulus of crystalline regions in oriented polymers. *J. Appl. Polym. Sci.* **57**:651–660.

Samir, M. A. S. A., F. Alloin, and A. Dufresne (2005). Review of recent research into cellulosic whiskers, their properties and their application in the nanocomposite field. *Biomacromolecules* **6**:612–626.

Santayanon, R., and J. Wootthikanokkhan (2003). Modification of cassava starch by using propionic anhydride and properties of the starch-blended polyester polyurethane. *Carbohydr. Polym.* **51**:17–24.

Seydibeyoğlu, M. Ö., and K. Oksman (2008). Novel nanocomposites based on polyurethane and micro fibrillated cellulose. *Compos. Sci. Technol.* **68**(3–4):908–914.

Svagan, A. J., S. M. A. S. Azizi, and L. A. Berglund (2007). Biomimetic polysaccharide nanocomposites of high cellulose content and high toughness. *Biomacromolecules* **8**(8):2556–63.

Svagan, A.J., M. S. Hedenqvist, and L. Berglund (2009). Reduced water vapour sorption in cellulose nanocomposites with starch matrix. *Compos. Sci. Technol.* **69**:500–506.

Taniguchi, T., and K. Okamura (1998). New films produced from microfibrillated natural fibers. *Inter. Polym. J.* **47**(3):291–294.

Tashiro, K., and M. Kobayashi (1991). Theoretical evaluation of three-dimensional elastic constants of native and regenerated celluloses: role of hydrogen bonds. *Polymer* **32**:1516–1526.

Van Soest, J. J. C., D. De Wit, and J. F. C. Vliegenthart (1996). Mechanical properties of thermoplastic waxy maize starch. *J. Appl. Polym. Sci.* **61**:1927–1937.

Vandeputte, G. E. and J. A. Delcour (2004). From sucrose to starch granule to starch physical behaviour: a focus on rice starch. *Carbohydr. Polym.* **58**(3):245–266.

Virendra, P., A. J. Shaikh, A. A. Kathe, D. K. Bisoyi, A. K. Verma, and N. Vigneshwaran (2010). Functional behaviour of paper coated with zinc oxide–soluble starch nanocomposites. *Mater. Proc. Technol.* **210**:1962–1967.

Wan, Y. Z., L. Honglin, F. He, H. Liang, Y. Huang, and X. L. Li (2009). Mechanical, moisture absorption, and biodegradation behaviours of bacterial cellulose fibre-reinforced starch biocomposites. *Compos. Sci. Technol.* **69**:1212–1217.

Wegner, T. H., and P. E. Jones (2006). Advancing cellulose-based nanotechnology. *Cellulose* **13**:115–118.

Zhang X., J. Huang, P. R. Chang, J. Li, Y. Chen, D. Wang, J. Yu, and J. Chen (2010). Structure and properties of polysaccharide nanocrystal-doped supramolecular hydrogels based on Cyclodextrin inclusion. *Polymer* **51**:4398–4407.

Zheng, H., F. Ai, P. R. Chang, J. Huang, and A. Dufresne (2009). Structure and properties of starch nanocrystal-reinforced soy protein plastics. *Polym. Compos.* **30**(4):474–480.

Zimmermann, T., E. Pöhler, and P. Schwaller (2005). Mechanical and morphological properties of cellulose fibril reinforced nanocomposites. *Adv. Eng. Mater.* **7**:1156–1161.

12

STARCH-BASED SUSTAINABLE MATERIALS

Luc Avérous

12.1 INTRODUCTION

The potential of biodegradable polymers and more particularly that of polymers obtained from agro-resources (agro-polymers) such as the polysaccharides (e.g., starch) has long been recognized. For instance, agro-polymers could preserve petrol resources replacing some polymers based on fossil resources for some applications, in agreement with the concept of sustainability. However, nowadays, these polymers, which are largely used in product packaging applications (e.g., food industry), have not found extensive applications in, for example, widespread packaging applications or agriculture, to replace conventional plastic materials. In some special cases, agro-polymers can also find biomedical applications (e.g., tissue scaffolds, drug delivery) linked to with their intrinsic properties (e.g., biocompatibility, degradation). In fact, agro-polymers could be an interesting way to overcome the limitations and increasing prices of petrochemical resources and could contribute to carbon footprint reduction in the future.

Agro-polymers are mainly extracted from plants, are compostable and renewable polymers and show some common characteristics such as a hydrophilic character. Most of them can be processed directly, as fillers, plasticized or chemically modified. Specifically there are different families of agro-polymers such as polysaccharides, proteins (e.g., gluten or zein) and lignins. The most abundant are the polysaccharides, with different products and structures such as cellulose, chitin, and starch. In this family, starch is an important material as it is the main storage supply in botanical

Polysaccharide Building Blocks: A Sustainable Approach to the Development of Renewable Biomaterials, First Edition. Edited by Youssef Habibi and Lucian A. Lucia.
© 2012 John Wiley & Sons, Inc. Published 2012 by John Wiley & Sons, Inc.

(a)   (b)

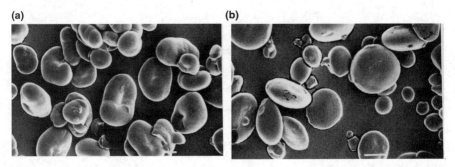

FIGURE 12.1 SEM observations of native starches wheat (a) and pea starch (b). White scale = 10 μm.

resources (cereals, legumes, and tubers), a widely available raw material and after processing, can be useful for many different industrial applications such as food manufacturing, paper, textile, or adhesives.

12.2 NATIVE STARCH

Starch granules can be easily isolated from plants sources. Main sources are wheat, potato, maize, rice, cassava, and pea. Native starch granules typically have dimensions ranging from one to two hundreds micrometers and appear in a variety of shapes. Figure 12.1 shows for instance some SEM micrographs with different starch granules (pea and wheat) granules. Besides, Table 12.1 gives starch properties and granule diameters for different starch types.

Starch is a polysaccharide consisting in D-glucose units, referred to as homoglucan or glucopyranose. Starch is composed of two main macromolecules, amylose and amylopectin. Amylose is a sparsely branched carbohydrate mainly based on $\alpha(1-4)$ bonds with a molecular weight of 10^5-10^6. The number of macromolecular configurations based on $\alpha(1-6)$ links are directly proportional to the amylose molecular weight. The chains show spiral-shaped single or double helixes with a rotation on the $\alpha(1-4)$ link and with six glucoses per rotation. Amylopectin is a highly multiple-branched biopolymer with a very high molecular weight (10^7-10^9). It is based on $\alpha(1-4)$ (around 95%) and $\alpha(1-6)$ links (around 5%), constituting branching points are localized every 22–70 glucose units (Averous, 2004), generating a kind of grape branched-like structure with pending chains.

Table 12.1 shows that starch composition and macromolecular structure is dependent upon botanical origin. Indeed some mutant plant species present some singular compositions. Typically one can find rich-amylose starch as the amylomaize (up to 80% amylose) and some rich-amylopectine starch, like the waxy maize (>99% amylopectin). Also after different industrial stages of isolation and refining, starch usually shows some traces of lipids, gluten, or phosphate, which can interfere with the starch properties, for example, by the formation of lipid complexes or by Maillard reactions (with the residual proteins).

TABLE 12.1 Composition and Characteristics of Different Starches

Starch	Amylose Content[a] (%)	Amylopectin Content[a] (%)	Lipid Content[a] (%)	Protein Content[a] (%)	Phosphorus Content[a] (%)	Moisture Content[b] (%)	Granule Diameter (μm)	Crystallinity (%)
Wheat	26–27	72–73	0.63	0.30	0.06	13	25	36
Maize	26–28	71–73	0.63	0.30	0.02	12–13	15	39
Waxy starch	<1	99	0.23	0.10	0.01	n.d.	15	39
Amylomaize	50–80	20–50	1.11	0.50	0.03	n.d.	10	19
Potato	20–25	79–74	0.03	0.05	0.08	18–19	40–100	25

n.d. = not determined.

Source: Averous and Halley (2009).

[a]Determined on a dry basis.

[b]Determined after equilibrium at 65% RH, 20°C.

Native starch is a crystalline product (Table 12.1). It shows a special granular organization with a high degree of radial organization from hilum (Noel et al., 1992). Macromolecules are mainly oriented according to the radial axis. This ultrastructure is obtained by inter-macromolecular hydrogen links, between hydroxyls groups, with the participation of water molecules. Amylose and the branching regions of amylopectin form the amorphous zone in the granule. Amylopectin is the dominating crystalline component in native starch with double helix organizations from the pending chains. We can also find cocrystallization with amylose and single helical crystallization between amylose and free fatty acids or lipids. Several types of crystallinity are observed and four starch allomorphic structures exist, such as the following:

- *A-Type.* This structure is synthesized in cereals in dry and warm conditions. The chains are organized in a double helix conformation with six glucose units per turn. These helices are organized into a monoclinic structure containing eight water molecules per unit cell.
- *B-Type.* This structure is synthesized in tubers and in starch with high amylose contents. It is also organized in a double left helix with six glucose units per turn into a hexagonal system containing 36 water molecules per unit cell.
- *C-Type.* This structure is a mix of the A and B structures and is mainly synthesized in vegetables.
- *V-Type.* This structure is synthesized in the presence of small molecules such as iodine or fatty acids. This crystalline form is characterized by a simple left helix with six glucose units per turn.

The starch granule organization consists of alternating crystalline and amorphous zones leading to a concentric domain structure. The crystalline regions (15–45%) are arranged as thin lamellar domains, perpendicular to the radial axis in a kind of decentralized onion structure.

12.3 PLASTICIZED STARCH

12.3.1 General

Except for applications as fillers to produce reinforced plastics (Shah et al., 1995), native starch is typically chemically or/and physically modified and used as destructured starch for food or nonfood applications. Starch is gelatinized with heat combined with high water content, which is the destructuring agent. Gelatinization causes the disruption of the highly granule organization and the starch swells, forming a viscous paste with destruction of most of the inter-macromolecule hydrogen links and a reduction of the melting and the glass transition temperature (T_g).

Depending on the destructurization level (thermomechanical input associated with the water content), one can obtain different products and applications.

For instance, we can achieve expanded structures at a medium water content and medium destructurizing input. Such closed cells structures (foams) have been developed to obtain shock absorbable (e.g., loose fill) and isothermal packaging. At high destructuring level and low water content, we obtain plasticized starch (PLS), or so-called "thermoplastic starch" (Shogren, 1992). The first patents and articles on PLS were published at the end of the 1980s (Averous, 2004).

Most starch applications require water addition and partial or complete gelatinization. By decreasing the moisture content (to less than 20 wt%), the melting temperature tends to be close to the degradation temperature. For instance, the melting temperature of pure and dry starch is 220–240°C compared to the temperature of the beginning of starch decomposition, 220°C (Russell, 1987). To overcome this issue, a nonvolatile (at the process temperature) plasticizer such as glycerol or others polyols (sorbitol, polyethylene glycol, and so on) is added to decrease the melting and process temperature (Gaudin et al., 1999a; Mathew and Dufresne, 2002a, 2002b). A mixture of different polyols can also be utilized, for example, glycerol and sorbitol (Chivrac et al., 2010a, 2010b). Other compounds such as those containing nitrogen (urea, ammonium derived, amines) can also be used (Shogren et al., 1992; Ma et al., 2004, 2005, 2006). Thus, PLS in practice often combines starch with a nonvolatile and high-boiling plasticizer, mainly polyols (Averous, 2004). Water is often added as a destructuring agent and a volatile plasticizer. PLS is then usually transformed under thermomechanical treatment as a thermoplastic, using conventional machines for plastic processing (e.g., extruder, internal mixer, injection moulder).

12.3.2 Process

The disruption of granular starch is the transformation of the semicrystalline granule into a homogeneous, mainly amorphous material with a partial destruction of intermacromolecular hydrogen bonds. Disruption can be accomplished by casting (e.g., with dry drums) or with thermomechanical energy in a continuous process, using destructuring agents, where the combination of thermal and mechanical inputs can be obtained, for example, by extrusion. Processing can be in one or two stages. In a one-stage process, usually a twin-screw extruder is fed with native starch. Along the barrel, water and plasticizer are then successively introduced. The water in excess is often eliminated before the extruder exit by vacuum venting to avoid the formation of a foam structure at the die exit. The starch is fully destructured and melted. In a two-stage process (Figure 12.2), the first stage is a dry blend preparation (Averous et al., 2000a, 2000b). For example, in a turbo-mixer, under high speed, the plasticizer is slowly added into the native starch. Then, the mixture is placed in a vented oven allowing the diffusion of the plasticizer into the granule, where the plasticizer swells the starch. After cooling, the right amount of water is added to the mixture using a turbo-mixer. This dry blend is then introduced into an extruder. During the extrusion, the starch granules are initially fragmented, and then under temperature and shearing, starch is destructured, plasticized, and melted but also under the thermomechanical input, partially depolymerized. At the end, a homogeneous molten phase is obtained.

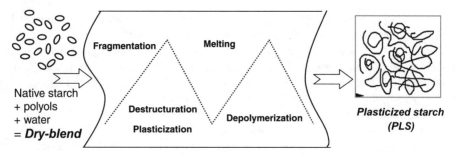

FIGURE 12.2 Schematic of starch process by extrusion (two-step process).

12.3.3 Plasticized Starch Properties

12.3.3.1 Crystallinity Compared to native starch, plasticized starch shows a reduced crystallinity with a strong modification of the types and the organization of the crystallinity. According to Van Soest et al. (1996), two kinds of crystallinity are obtained after PLS processing, (i) residual crystallinity from native starch (A, B, and C types) and (ii) processing-induced crystallinity (V and E types). Crystallinity induced by processing is influenced by parameters such as the extrusion residence time, the screw speed, and the process temperature. It is mainly caused by the fast recrystallization of amylose into single-helical structures. After postprocessing aging, Van Soest and Knooren (1997) proposed a complex model for crystallization of plasticized starch with amorphous and crystalline amylose and amylopectin and probable cocrystallization between amylose and amylopectin.

12.3.3.2 Plasticizer and Water Interactions The evolution of the moisture impacts the properties and transitions temperatures of PLS, such as the T_g, because water also acts as a plasticizer (Hulleman et al., 1998). Water is a volatile plasticizer that is equilibrated in mechanisms of sorption–desorption with the environment. Lourdin et al. (1997) and Mathew and Dufresne (2002a, 2002b) have determined the moisture content after equilibrium on starch as a function of the relative humidity and the nature and content of the plasticizer. Godbillot et al. (2006) have recently proposed a phase diagram for a water/glycerol/starch system, as a function of the plasticizer content. These last authors highlight the different interactions taking place in these multiphase systems (Figure 12.3). Depending on the glycerol content and the relative humidity, the plasticizer can be more or less linked with the polysaccharide chains. The plasticizer could occupy specific sorption sites when the water and plasticizer content are low, or become free when the sorption sites are fulfilled, at high relative humidity and glycerol content. Between these two limiting cases, complex interactions between the different components could be established. Thus, in certain conditions a phase separation occurs, generating multiphase structures with rich and poor plasticizer domains (Lourdin et al., 1997, 1998).

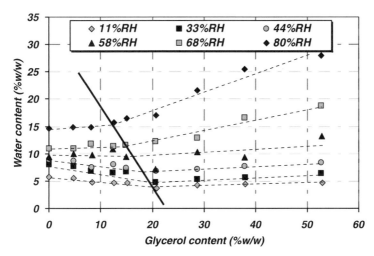

FIGURE 12.3 Water content at equilibrium as a function of glycerol content and relative humidity (%RH). *Source*: Godbillot et al. (2006).

12.3.3.3 Postprocessing Aging Different authors have shown that after processing (Shogren, 1992; Van Soest and Knooren, 1997; Averous et al., 2000a, 2000b; Averous et al., 2001; Doungjai and Sanguansri, 2007; Thirathumthavorn and Charoenrein, 2007), plasticized starch exhibits aging with for instance a strong effect on the mechanical properties such as the increase of the Young's modulus over several weeks aging. PLS present two kinds of aging behavior depending on T_g value. In the sub-T_g domain, PLS shows a physical aging versus time, with a material densification. At a temperature above T_g, PLS shows retrogradation phenomena with the evolution of the crystallinity and rearrangements of plasticizer molecules into the material as a function of storage time (Van Soest and Knooren, 1997). The retrogradation kinetics depends on the macromolecules mobility, the plasticizer type and content (Smits et al., 1999a, 1999b).

12.3.3.4 Physical Properties Compared to native starch, plasticized starch shows some particular properties. Depending on the plasticizer/starch ratio, PLS presents a wide range of properties. Table 12.2 shows that increasing glycerol/starch ratios modify different properties, including physical and mechanical behavior.

In certain formulations, the C_p variations at glass transitions are very low and difficult to determine via DSC, thus DMTA determinations are often more desirable to obtain PLS glass transitions. DMTA profiles are shown in Figure 12.4 for the different formulations, given in Table 12.2. The evolution of tan delta versus temperature shows two transitions. The broad, main relaxation (α transition) can be linked to the PLS glass transition (Averous et al., 2000a, 2000b). The secondary relaxation (between -50 and $-60°C$) could be connected to the glass transition of glycerol (Averous et al., 2000a, 2000b). According to Lourdin et al. (1997), this latter relaxation could be an indicator of the level of interactions between the plasticizer

TABLE 12.2 Plasticized Starch: Formulations and Physical Properties

	Glycerol/Dry Starch Ratio[a] (w/w)	Water[a] (wt%)	Density[a]	T_g (by DSC) (°C)	α-Transition (by DMTA) (C°)	Modulus[a] (MPa)	Max. Tensile[a] Strength (MPa)	Elongation[a] at Break[a] (%)
S74G10W10[b]	0.14	9	1.39	43	63	1144 (42)	21.4 (5.2)	3 (0)
S75G18W12[b]	0.25	9	1.37	8	31	116 (11)	4.0 (1.7)	104 (5)
S67G24W9[b]	0.35	12	1.35	−7	17	45 (5)	3.3 (0.1)	98 (5)
S65G35W0[b]	0.50	13	1.34	−20	1	11 (1)	1.4 (0.1)	60 (5)

Standard deviations are in brackets.

Source: Averous and Halley (2009).

[a] Contents and properties after equilibrium at 23°C and 50% RH, 6 weeks.

[b] Initial formulation (SxGyWz): S = starch (x wt%), G = glycerol(y wt%), W = water (z wt%).

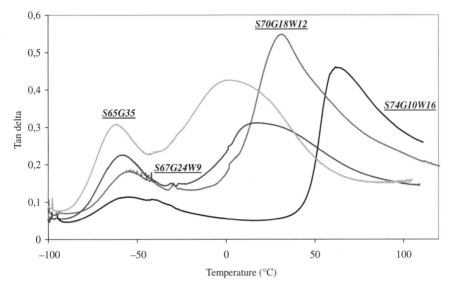

FIGURE 12.4 DMTA evolutions with different glycerol/starch ratios (formulations shown in Table 12.2). *Source*: Averous et al. (2000a, 2000b).

and the polysaccharides. Figure 12.4 shows that this relaxation temperature decreases when glycerol content increases along with the phase segregation.

Table 12.2 shows that we can achieve a large range of mechanical properties and permeability (moisture and oxygen) by variation of the plasticizer content (Lourdin et al., 1997; Gaudin et al., 1999b, 1999c). Although in PLS the water permeability is high due to its polar character, note the oxygen permeability is found lower than most polyesters (Van Tuil et al., 2000). Thus, there appears to be potential for use of PLS as an oxygen barrier material in multilayer structures (Dole et al., 2005), to try to replace, for example, EVOH. With increasing plasticizer content, the permeability increases drastically at the glass transition and continues to rise through the rubber plateau.

To understand the behavior of PLS during processing, the rheological behavior of this material has been studied in the molten state. A previous publication has shown the high dependency of the specific mechanical energy (SME) on the PLS viscosity (Martin et al., 2003).

12.3.3.5 PLS Issues and Strategies As a material, PLS shows strong end-use attributes such as total compostability without toxic residues, and the renewability of the resource. Besides, compared to synthetic thermoplastics, it is a rather cheap material. Compared to fossil resources, the price of starch resources remains currently rather stable. In addition, PLS can be easily processed with plastic processing machines and does not need the development of special equipment. Also, depending on the plasticizer level and the starch botanical source, a wide properties range may be obtained. However, unfortunately, PLS shows poor moisture

sensitivity and rather weak mechanical properties compared to conventional synthetic polymer, and these latter are important properties for more widespread polymer
applications.

To overcome these weaknesses, different strategies have been examined such as
chemical modification of starch. This approach has been carried out since the first
half of the 1970s, in with focus on modified starches and cellulose. For example,
starch esterification (e.g., by acetylation) improves the water resistance of this
material (Fringant et al., 1996, 1998). In addition, by controlling the degree of
substitution (between 0 and 3), an accurate hydrophobic character can be obtained.
Additionally these modified starches can be plasticized (e.g., with ester citrate).
However, the strategy of chemical modification is strongly limited as far as
potential toxicity and diversity of by-products obtained during the chemical
reactions are concerned. Another limitation lies in the cost of the additional process
stages for modification and product purification (to eliminate the by-products).
Also, the chemical reactions can lead to the decrease in polysaccharide molecular
weight due to chain scission. Consequently, the final properties are altered.
However, another more promising strategy has developed which is the association
of PLS with others compounds to obtain compostable multiphase materials.
Additionally depending on the processing conditions, we can obtain different
structures for compostable multiphase materials, such as blends with for example,
biodegradable polyesters (PCL, PLA, PHA, and so on) or biocomposites (biodegradable composites) (Averous et al., 2000a, 2000b, 2001; Averous and Fringant
2001; Averous, 2004; Averous and Boquillon, 2004; Averous and Le Digabel, 2006;
Averous and Halley, 2009). This approach, mainly based on materials formulation,
induces some issues linked with the final nano/microstructure, the quality of the
interfaces/interphases, the continuity and the compatibility between the phases,
which need to be controlled.

12.4 STARCH-BASED BIOCOMPOSITES

Biocomposites are a special class of composite materials. They are obtained by
blending biodegradable polymers with fillers (e.g., lignocellulose fibers). Tailoring
new composites within a perspective of eco-design or sustainable development is a
philosophy that is applied to more and more materials. Depending on the filler size,
we can obtain macrobiocomposite or nanobiocomposites. For this latter form, the
filler is submicrometer in size and controls the matrix nanostructure.

12.4.1 Starch-Based Macrobiocomposites

12.4.1.1 Cellulose Fiber Reinforcement Various types of (ligno-)cellulose fibers
or microfibrils were tested in association with plasticized starch such as microfibrils
from potato pulp (Dufresne et al., 2000), bleached leafwood fibers (Funke
et al., 1998; Averous et al., 2001; Averous and Boquillon, 2004; Averous, 2007),
fibers from bleached eucalyptus pulp (Curvelo et al., 2001), flax, and jute fibers

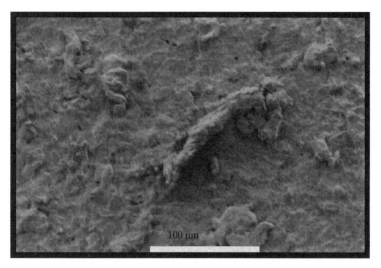

FIGURE 12.5 SEM observation. Cryogenic fracture of composites PLS-leafwood cellulose fibers (white scale = 100 mµ). *Source*: Averous et al. (2001).

(Wollerdorfer and Bader, 1998). These different authors have shown high compatibility between starch and these fibers. For instance, Averous et al. (2001) have found a strong T_g increase by addition of cellulose fibers into a PLS matrix. This behavior is linked to the fibers–matrix interactions, which decreases starch chains mobility. For instance, Figure 12.5 shows a SEM image of a cryogenic fracture. The cellulose fibers are fully embedded into the starchy matrix. Similar results have also been found by Curvelo et al. (2001).

After mixing, authors have found spectacular properties with large improvement in the materials performances. Some of these enhancements are linked to typical matrix reinforcement (Averous, 2007), and some others are linked to the fiber–matrix interface interactions. For instance, one can highlight the following unexpected improvements:

- higher mechanical properties (Averous, 2004). Compared to biopolyesters-based biocomposites, starch-based biocomposites present superior properties linked to higher interactions between the matrix and filler,
- higher thermal resistance (Dufresne and Vignon, 1998), due to the transition shift of T_g and an increase in the rubber plateau,
- reduced water sensitivity due to fiber–matrix interactions and to the higher hydrophobic character of the cellulose, which is linked to its high crystallinity (Dufresne and Vignon, 1998; Funke et al., 1998; Dufresne et al., 2000; Averous et al., 2001; Curvelo et al., 2001),
- reduced postprocessing aging, due to the formation of a 3D network between the different matrix–filler carbohydrates based on hydrogen bonds (Averous and Boquillon, 2004).

12.4.1.2 Lignin and Mineral Fillers   Different types of fractioned or modified lignins (i.e., Kraft lignins, Acell® lignins) have been tested in association with PLS. Biocomposites were obtained by film casting preparation and by extrusion. According to Baumberger et al. (1998a, 1998b), the lignins act as fillers or as extenders for the PLS matrix, with the soluble lignin fractions interacting with the PLS matrix.

Mineral microfillers were tested into a PLS matrix (De Carvalho et al., 2001). Kaolin particles, with micrometer size, were incorporated by extrusion. Due to a significant compatibility between matrix and filler, subsequent behavior such as a glass transition increase, a reduction of water uptake, and an increase of the stiffness can be observed.

12.4.2 From Macro to Nanobiocomposites

Nanobiocomposites are organic/inorganic hybrid biomaterials composed of nano-sized fillers (nanofillers) incorporated into a biopolymer matrix (Bordes et al., 2009; Chivrac et al., 2009). Depending on the nanofiller chosen, the nanocomposite materials could exhibit drastic modifications in their properties, such as improved mechanical properties, barrier properties, or change in their thermal and electrical conductivity (Alexandre and Dubois, 2000). Such properties enhancements rely both on the nanofiller geometry, on the nanofiller surface area (e.g., $700 \, m^2/g$ for the montmorillonite when the nanofiller is fully exfoliated) and nanofiller surface chemistry (Sinha Ray and Okamoto, 2003). According to the literature, three main types of nanoparticles have been incorporated into PLS nanocomposites, (i) whiskers obtained from cellulose, (ii) nanocrystals from starch, and (iii) natural or organo-modified nanoclays (Chivrac et al., 2009).

12.4.2.1 Whisker-Based Nanobiocomposites   Through a long pretreatment, whiskers can be isolated from their original biomass through acid hydrolysis with concentrated mineral acids under strictly controlled conditions of time and temperature (Azizi Samir et al., 2004, 2005). Acid action results in a decrease of the amorphous parts by removing polysaccharide material closely bonded to the crystallite surface and breaks down portions of glucose chains in most accessible, noncrystalline regions and by acid hydrolysis of cellulosic materials. Although chitin can be used, whiskers are typically cellulose monocrystals. Some authors used tunicin (seafood cellulose) whiskers (Angles and Dufresne, 2000, 2001; Mathew and Dufresne, 2002a, 2002b; Mathew et al., 2008), which are slender parallelepiped rods of 500 nm to 1–2 μm length and 10 nm width, into PLS matrices. The whiskers–matrix interactions are important and the high shape ratio of the nanoparticles (50–200) and the high specific area ($\approx 170 \, m^2/g$) increase the interfacial phenomena. Compared to the common PLS-cellulose macrocomposites, the global behavior of nanowhisker-based material is primarily driven by the matrix/nanofiller interface, which in turn controls the subsequent performance properties (mechanical properties and permeability). For instance, tunicin whiskers favor starch crystallization due to the nucleating effect of the nanofiller (Mathew and Dufresne, 2002a, 2002b).

TABLE 12.3 Smectites Classification

Di-Octahedral Phyllosilicate	Tri-Octahedral Phyllosilicate
Montmorillonite, $(Si_8)(Al_{4-y}Mg_y)$ $O_{20}(OH)_4$, M_y^+	Hectorite, $(Si_8)(Al_{6-y}Li_y)O_{20}(OH)_4$, M_y^+
Beidellite, $(Si_{8-x}Al_x)Al_4O_{20}(OH)_4$, M_x^+	Saponite, $(Si_{8-x}Al_x)(Mg_6)O_{20}(OH)_4$, M_x^+
Illites, $(Si_{8-x}Al_x)(Al_{4-y}M_y^{2+}O_{20}(OH)_4$, K_{x+y}^+	Vermiculite, $(Si_{8-x}Al_x)(Mg_{6-y}M_y^{3+})O_{20}(OH)_4$, K_{x-y}^+

12.4.2.2 Starch Nanocrystals-Based Nanobiocomposites Waxy maize starch nanocrystals were obtained by acid hydrolysis of native granules by strictly controlling the temperature of the process, the acid and starch concentrations, the hydrolysis duration, and the stirring speed. Waxy maize starch nanocrystals consist of 5–7 nm thick platelet-like particles with a length ranging from 20 to 40 nm and a width in the range 15–30 nm. They were used as a reinforcing agent in a waxy maize starch matrix plasticized with glycerol, where Angellier et al. have shown that the reinforcing effect of starch nanocrystals can be attributed to strong filler/filler and filler/matrix interactions due to the establishment of increased hydrogen bonding. The presence of starch nanocrystals leads to a slowing down of the recrystallization of the matrix during aging in high humidity atmospheres (Angellier et al., 2006).

12.4.2.3 Clay-Based Nanobiocomposites The most intensive research on nano-biocomposites are focused on layered particles (Raquez et al., 2007; Chivrac et al., 2009), and especially on nanoclays such as montmorillonite (MMT) due to their availability, versatility, and low environment and health concerns.

Nanoclays Phyllosilicates are a wide family in which smectites with different structure, texture, or morphology can be found. Table 12.3 presents for instance montmorillonites, which are anisotropic flexible particles with a high aspect ratio, a width of hundreds nanometers, and a thickness of around 1 nm.

The distance observed between two platelets of the primary particle, termed as the inter-layer spacing or *d*-spacing (d_{001}), depends on the silicate type. This value not only depend on the layer crystal structure but also on the type of the counter cation and on the hydration state of the silicate. To increase the *d*-spacing and then enhance the nanostructure effects, a chemical modification of the clay surface, with the aim to match the polymer matrix polarity, is often carried out (Alexandre and Dubois, 2000). Cationic exchange is the most common technique for chemical surface modification, but other original techniques such as the organosilane grafting (Dai and Huang, 1999; Ke et al., 2000) and the use of ionomers (Lagaly, 1999; Shen et al., 2002) or block copolymers adsorption (Fischer et al., 1999) are also used.

The cationic exchange consists of substitution of the original inorganic cations by organic ones. These surfactants are often quaternary alkylammonium cations (Table 12.4). The ionic substitution is performed in water because of its ability to

320

TABLE 12.4 Nanoclays Types and the Counter-Ion Chemical Structures

Code	Name	Counter-Cation
MMT-Na	Natural sodium montmorillonite	Na +
OMMT-Alk1	Cloisite® 15A—Southern Clay	$H_3C-\overset{\overset{CH_3}{\vert}}{\underset{\underset{HT}{\vert}}{N^+}}-HT$ Dimethyl-dihydrogenated tallow ammonium
OMMT-Alk2 OMMT-Bz	Cloisite® 6A—Southern Clay Cloisite® 10A—Southern Clay	$H_3C-\overset{\overset{CH_3}{\vert}}{\underset{\underset{HT}{\vert}}{N^+}}-CH_2$ Dimethyl-benzyl-hydrogenated tallow ammonium
OMMT-OH	Cloisite® 30B—Southern Clay	$H_3C-\overset{\overset{CH_2CH_2OH}{\vert}}{\underset{\underset{CH_2CH_2OH}{\vert}}{N^+}}-T$ Methyl-tallow-bis-2-hydroxyethyl ammonium
OMMT-CS	—	Cationic starch

T = Tallow (~65% C18; ~30% C16; ~5% C14).
HT = Hydrogenated tallow.

swell clay, washed to remove the salt formed during the organo-modifier adsorption and the surfactant excess and then, lyophilized to obtain organo-modified clays. In addition to the modification of the clay surface polarity, organo-modification increases the interlayer spacing, which will also further facilitate the polymer chains intercalation by reputation (Lagaly, 1986).

Structures of MMT-Based Nanobiocomposites Depending on the process conditions and on the matrix/nanoclay affinity, different nanocomposite structures can be obtained, such as the following (Vaia and Giannelis, 1997; Alexandre and Dubois, 2000):

- *Microcomposites.* where the polymer chains have not penetrated into the clay interlayer spacing and the clay particles are mainly stacked and aggregated. The aggregates size is micrometer-sized, and the corresponding behavior is close to that predicted by composite theory. Strictly in this case, the nanocomposite designation is incorrect.
- *Intercalated Nanocomposites.* where this structure shows regularly alternating layered silicates and polymer chains. This polymer diffusion into the clay interlayer spacing leads to an increase in the d_{001}.
- *Exfoliated Nanocomposites.* where the nanofillers are individually delaminated and homogeneously and fully dispersed into the matrix.

To reach exfoliation, different nanofillers and dispersion protocols have been tested to produce plasticized starch-based nanobiocomposites (Chivrac et al., 2009). Rather hydrophobic nanofillers, were incorporated into plasticized starch based on wheat (Chiou et al., 2005), potato (Park et al., 2002, 2003; Chen and Evans, 2005), or corn (Zhang et al., 2007), with MMT content varying from 0 to 9 wt%. It was clearly demonstrated that the incorporation of OMMT-Alk1, OMMT-Alk2, or OMMT-Bz (see Table 12.4 for the OMMT designations) led to the formation of microbiocomposites (Park et al., 2002; Chen and Evans, 2005; Chiou et al., 2005; Zhang et al., 2007), evidenced by the unchanged values of the d_{001}. Higher spacing results were obtained with OMMT-OH, which presents a more hydrophilic character, with a slight d_{001} shift and a strong decrease in the diffraction peak intensity (Park et al., 2003; Chiou et al., 2005, 2006), corresponding to a higher dispersion. This state was likely achieved due to the hydrogen bounds established between the clay surfactant and the starch chains (Park et al., 2003).

Nanobiocomposites were also elaborated with natural sodium MMT (MMT-Na) due to the starch hydrophilic characters and the Na-based nanofiller (Park et al., 2002, 2003; Huang et al., 2004; Avella et al., 2005; Chen and Evans, 2005; Chen et al., 2005; Chiou et al., 2005, 2006, 2007; Pandey and Singh, 2005; Zhang et al., 2007; Cyras et al., 2008). These materials were prepared with corn starch (Huang et al., 2004; Pandey and Singh, 2005; Zhang et al., 2007), wheat (Chiou et al., 2005, 2007), or potato (Park et al., 2002, 2003; Avella et al., 2005; Chen and Evans, 2005; Chen et al., 2005; Cyras et al., 2008). It was highlighted that for glycerol content higher than 10 wt%,

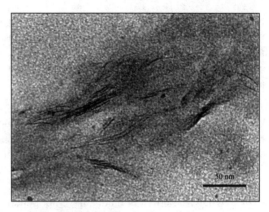

FIGURE 12.6 TEM micrograph of nanobiocomposites based on plasticized starch/OMMT-CS 3 wt% (black scale = 50 nm). *Source*: Chivrac et al. (2008).

such systems led to the formation of an intercalated structure with a d_{001} increasing to 18 Å, a value which is commonly reported in the literature as generally attributed to glycerol intercalation (Wilhelm et al., 2003; Chiou et al., 2006). The influence of the plasticizer on the MMT dispersion and on the exfoliation state was also highlighted by Dean et al. (2007). Corresponding results showed a homogeneous dispersion with an exfoliated structure in agreement with results of Cyras et al. (2008), which have demonstrated that for glycerol content lower than 10 wt%, exfoliation was achieved. In certain conditions, MMT-Na seemed suitable to achieve exfoliation. These results were in agreement with some studies that have highlighted the formation of hydrogen bonds and deep interactions between glycerol and MMT platelets (Wilhelm et al., 2003; Huang et al., 2004; Pandey and Singh, 2005). Huang et al. (2005a, 2005b, 2005c, 2006) have demonstrated that by changing the plasticizer nature, with for example, urea or urea/formamide, exfoliation can also be reached. Nevertheless, these compounds generate eco-toxic residues after biodegradation or composting and cannot be used for safe biodegradable materials. Kampeerapappun et al. (2007) have focused their attention on the use of a new eco-friendly compatibilizer, chitosan, to promote the MMT platelets exfoliation. But, only a small increase in the d_{001} was achieved. Nevertheless, this approach was successfully applied (Chivrac et al., 2008), with cationic starch (CS) as MMT organo-modifier. According to these authors, no diffraction peak was observed by X-ray diffraction, suggesting an exfoliated morphology, which was confirmed by TEM analysis (Figure 12.6).

Properties of Nanoclay-Based Nanobiocomposites Large improvements in the material performance were found with nanoclay-based nanobiocomposites, some of them are linked to traditional matrix reinforcement, some others are brought by the high interface area between nanoclay and the matrix, and by the corresponding dispersion state.

As usual, these nanocomposites displayed substantial improvement in mechanical properties such as Young's modulus, which is correlated with the clay loading for

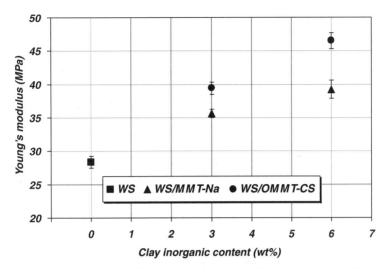

FIGURE 12.7 Variations of the Young's modulus versus clay content, for plasticized wheat starch with MMT-Na (microbiocomposite) and OMMT-CS (exfoliated nanobiocomposite), stabilized at 57%RH at room temperature. *Source*: Chivrac et al. (2008).

MMT-Na (Chivrac et al., 2008) (with corn and wheat starch) or OMMT-CS (Zhang et al., 2007) (with wheat starch). The mechanical improvement depends on the nanobiocomposites structure. The modulus increase is highest in the case of exfoliation with modified MMT. For instance, Figure 12.7 shows the modulus difference determined by uniaxial tensile test between an exfoliated structure based on CS and a nonexfoliated one, based on MMT-Na. Similar behavior is obtained on the elongation at break, which slightly increases with the clay content in the case of exfoliation and decrease in the case of a microcomposites structure. In the same way, the energy at break increases in the case of exfoliation and decrease in the case of a microcomposite structure.

From thermomechanical measurements based on DMTA characterization, we can determine the influence of the nanofillers on the local mobility of the chains and thus on their relaxation temperatures which could in turn be associated with the glass transitions. In the case of plasticized starch/MMT-Na nanobiocomposites, the temperatures of two main relaxation peaks shifted toward higher temperatures indicating that the layered clays strongly restricted the starch chain mobility. This tendency (Park et al., 2002) was attributed to the MMT-Na higher affinity with the starch chains. The same trends were observed by DSC (Huang et al., 2006), meaning that starch/clay hybrids were strongly affected by the clay surface polarity and the clay/matrix interactions.

Some authors studied in detail the thermal stability of nanobiocomposites by TGA. Park et al. (2003) showed that the potato starch/MMT-Na and OMMT-OH hybrids have a higher degradation temperature than the neat matrix. The MMT-Na thermal stability was higher than the OMMT-OH nanobiocomposites one. Such a result highlighted a relationship between the clay dispersion state and the thermal stability. Such behavior is observed in most nanocomposite systems and is linked to the clay aspect ratio and the dispersion state. The exfoliation of the MMT nanoplatelets into the starch matrix increases the tortuosity of the combustion gas diffusion pathway and the formation of a char at the surface.

Nanoclays also impact the water vapor permeability of the corresponding nanocomposite materials. Park et al. (2002) examined the potato starch nanobiocomposite water vapor permeabilities with different type of clays. According to these authors, all the clay-based films showed lower water vapor permeability compared to the neat matrix. Best results are obtained with MMT-Na nanobiocomposites, which present the higher dispersion state according to these authors. The same trends are observed in different papers (Park et al., 2003; Huang and Yu, 2006). Unfortunately, a recent paper shows that with OMMT-CS, results obtained with high glycerol content are rather poor (Chivrac et al., 2010a, 2010b). The relatively high plasticizer content (23 wt% glycerol) induces a phase separation, with plasticizer-rich and carbohydrate-rich phases, resulting in the nanoclay being preferentially located in the carbohydrate-rich domains. As a consequence, a preferential way for water transfer was more likely created in the very hydrophilic glycerol-rich domains where the nanoclay platelets were almost totally absent. Thus, even if exfoliated morphology is achieved, the heterogeneous clay distribution and phase separation phenomena explain the lack of improvement and even the decline in the moisture barrier properties for these glycerol-plasticized starch nanobiocomposites. For the microbiocomposites based on OMMT-Bz, OMMT-Alk2, or OMMT-OH, the barrier properties enhancement was linked to a decrease in the water solubility due to the surfactant hydrophobic nature. This behavior is induced by these two distinct phenomena, that is, (i) the silicate layers dispersion and (ii) the solubility of the gas into the material (Alexandre and Dubois, 2000).

12.5 SUMMARY

In conclusion, we have examined the effects of sustainable materials on the macro- or nanostructure and subsequent processing, thermomechanical properties, and performance properties of plasticized starch polymers. This examination included a detailed review of the complexity of starch polymers, recent advances in novel starch modifications and compounds, and a detailed examination of the effects of plasticized starch macrobiocomposites and nanobiocomposites. Specific structures and subsequent properties are controlled by many specific factors such as filler shape, size, and surface chemistry; processing conditions; and environmental aging. In the case of nanobiocomposites, it is evident that nanomaterials–polymer matrix interfacial interactions are extremely important to the final nanostructures and performance of these materials.

The different structures we can obtain can fulfill the requirements of different applications, such as packaging or other short-lived applications (catering, agriculture, leisure, hygiene) where long-lasting polymers are not entirely adequate. In some special cases, these materials can also find biomedical applications linked with their intrinsic properties.

REFERENCES

Alexandre, M., and P. Dubois, (2000). Polymer-layered silicate nanocomposites: preparation, properties and uses of a new class of materials. *Mater. Sci. Eng. Rep.* **28**:1–63.

Angellier, H., S. Molina-Boisseau, P. Dole, et al. (2006). Thermoplastic starch-waxy maize starch nanocrystals nanocomposites. *Biomacromolecules* **7**:531–539.

Angles, M. N., and A. Dufresne (2000). Plasticized starch/tunicin whiskers nanocomposites. 1. Structural analysis. *Macromolecules* **33**:8344–8353.

Angles, M. N., and A. Dufresne (2001). Plasticized starch/tunicin whiskers nanocomposite materials. 2. Mechanical behavior. *Macromolecules* **34**:2921–2931.

Avella, M., J. J. De Vlieger, M. E. Errico, et al. (2005). Biodegradable starch/clay nanocomposite films for food packaging applications. *Food Chem.* **93**:467–474.

Averous, L., (2004). Biodegradable multiphase systems based on plasticized starch: a review. *J. Macromol. Sci. Polym. Rev.* **C44**:231–274.

Averous, L., (2007). Cellulose-based biocomposites: comparison of different multiphasic systems. *Compos. Interf.* **14**:787–805.

Averous, L., and N. Boquillon, (2004). Biocomposites based on plasticized starch: thermal and mechanical behaviours. *Carbohydr. Polym.* **56**:111–122.

Averous, L., N. Fauconnier, L. Moro, et al. (2000a). Blends of thermoplastic starch and polyesteramide: processing and properties. *J. Appl. Polym. Sci.* **76**:1117–1128.

Averous, L., and C. Fringant, (2001). Association between plasticized starch and polyesters: processing and performances of injected biodegradable systems. *Polym. Eng. Sci.* **41**:727–734.

Averous, L., C. Fringant, and L. Moro, (2001). Plasticized starch–cellulose interactions in polysaccharide composites. *Polymer* **42**:6565–6572.

Averous, L., and P. J. Halley, (2009). Biocomposites based on plasticized starch. *Biofuels Bioprod. Biorefining Biofpr.* **3**:329–343.

Averous, L., and F. Le Digabel, (2006). Properties of biocomposites based on lignocellulosic fillers. *Carbohydr. Polym.* **66**:480–493.

Averous, L., L. Moro, P. Dole et al. (2000b). Properties of thermoplastic blends: starch–polycaprolactone. *Polymer* **41**:4157–4167.

Azizi Samir, Ma.S., F. Alloin, and A. Dufresne, (2005). Review of recent research into cellulosic whiskers, their properties and their application in nanocomposite field. *Biomacromolecules* **6**:612–626.

Azizi Samir Ma.S F. Alloin, J.-Y. Sanchez, et al. (2004). Preparation of cellulose whiskers reinforced nanocomposites from an organic medium suspension. *Macromolecules* **37**:1386–1393.

Baumberger, S., C. Lapierre, and B. Monties, (1998a). Utilization of pine kraft lignin in starch composites: impact of structural heterogeneity. *J. Agric. Food Chem.* **46**:2234–2240.

Baumberger, S., C. Lapierre, B. Monties, et al. (1998b). Use of kraft lignin as filler for starch films. *Polym. Degrad. Stabil.* **59**:273–277.

Bordes, P., E. Pollet, and L. Avérous, (2009). Nano-biocomposites: biodegradable polyester/nanoclay systems. *Progr. Polym. Sci. (Oxf.)* **34**:125–155.

Chen, B. and J. R. G. Evans, (2005). Thermoplastic starch–clay nanocomposites and their characteristics. *Carbohydr. Polym.* **61**:455–463.

Chen, M., B. Chen, and J. R. G. Evans, (2005). Novel thermoplastic starch–clay nanocomposite foams. *Nanotechnology* **16**:2334–2337.

Chiou B.-S. E. Yee, G. M. Glenn, et al. (2005). Rheology of starch–clay nanocomposites. *Carbohydr. Polym.* **59**:467–475.

Chiou, B.-S., E. Yee, D. Wood, et al. (2006). Effects of processing conditions on nanoclay dispersion in starch–clay nanocomposites. *Cereal Chem.* **83**:300–305.

Chiou, B. S., D. Wood, E. Yee, et al. (2007). Extruded starch–nanoclay nanocomposites: effects of glycerol and nanoclay concentration. *Polym. Eng. Sci.* **47**:1898–1904.

Chivrac, F., H. Angellier-Coussy, V. Guillard, et al. (2010a). How does water diffuse in starch/montmorillonite nano-biocomposite materials? *Carbohydr. Polym.* **82**:128–135.

Chivrac, F., E. Pollet, P. Dole, et al. (2010b). Starch-based nano-biocomposites: plasticizer impact on the montmorillonite exfoliation process. *Carbohydr. Polym.* **79**:941–947.

Chivrac, F., E. Pollet, and L. Averous, (2008). New approach to elaborate exfoliated starch-based nanobiocomposites. *Biomacromolecules* **9**:896–900.

Chivrac, F., E. Pollet, and L. Averous, (2009). Progress in nano-biocomposites based on polysaccharides and nanoclays. *Mater. Sci. Eng. Rep.* **67**:1–17.

Curvelo, Aa.S., A. J. F. De Carvalho, and Ja.M. Agnelli, (2001). Thermoplastic starch-cellulosic fibers composites: preliminary results. *Carbohydr. Polym.* **45**:183–188.

Cyras, V. P., L. B. Manfredi, M. T. Ton-That, et al. (2008). Physical and mechanical properties of thermoplastic starch/montmorillonite nanocomposite films. *Carbohydr. Polym.* **73**:55–63.

Dai, J. C. and J. T. Huang, (1999). Surface modification of clays and clay-rubber composite. *Appl. Clay Sci.* **15**:51–65.

De Carvalho, A. J. F. Aa.S. Curvelo, and Ja.M. Agnelli, (2001). A first insight on composites of thermoplastic starch and kaolin. *Carbohydr. Polym.* **45**:189–194.

Dean, K., L. Yu, and D. Y. Wu, (2007). Preparation and characterization of melt-extruded thermoplastic starch/clay nanocomposites. *Compos. Sci. Technol.* **67**:413–421.

Dole, P., L. Avérous, C. Joly, et al. (2005). Evaluation of starch-PE multilayers: processing and properties. *Polym. Eng. Sci.* **45**:217–224.

Doungjai, T. and C. Sanguansri, (2007). Aging effects on sorbitol- and non-crystallizing sorbitol-plasticized tapioca starch films. *Starch/Staerke* **59**:493–497.

Dufresne, A., D. Dupeyre, and M. R. Vignon, (2000). Cellulose microfibrils from potato tuber cells: processing and characterization of starch-cellulose microfibril composites. *J. Appl. Polym. Sci.* **76**:2080–2092.

Dufresne, A. and M. R. Vignon, (1998). Improvement of starch film performances using cellulose microfibrils. *Macromolecules* **31**:2693–2696.

Fischer, H. R., L. H. Gielgens, and T. P. M. Koster, (1999). Nanocomposites from polymers and layered minerals. *Acta Polym.* **50**:122–126.

Fringant, C., J. Desbrieres, and M. Rinaudo, (1996). Physical properties of acetylated starch-based materials: relation with their molecular characteristics. *Polymer* **37**:2663–2673.

Fringant, C., M. Rinaudo, M. F. Foray, et al. (1998). Preparation of mixed esters of starch or use of an external plasticizer: two different ways to change the properties of starch acetate films. *Carbohydr. Polym.* **35**:97–106.

Funke, U., W. Bergthaller, and M. G. Lindhauer, (1998). Processing and characterization of biodegradable products based on starch. *Polym. Degrad. Stabil.* **59**:293–296.

Gaudin, S., D. Lourdin, D. Le Botlan, et al. (1999a). Effect of polymer–plasticizer interactions on the oxygen permeability of starch-sorbitol-water films. *Macromol. Symp.* **138**:245–248.

Gaudin, S., D. Lourdin, D. Le Botlan, et al. (1999b). Relationships between oxygen permeability and polymer–plasticizer interactions on starch-sorbitol-water films. *Biopolym. Sci. Food Non Food Appl.* 173–178.

Gaudin, S., D. Lourdin, D. Le Botlan, et al. (1999c). Plasticisation and mobility in starch-sorbitol films. *J. Cereal Sci.* **29**:273–284.

Godbillot, L., P. Dole, C. Joly, et al. (2006). Analysis of water binding in starch plasticized films. *Food Chem.* **96**:380–386.

Huang, M. and J. Yu, (2006). Structure and properties of thermoplastic corn starch/montmorillonite biodegradable composites. *J. Appl. Polym. Sci.* **99**:170–176.

Huang, M., J. Yu, and X. Ma, (2005a). Studies on properties of the thermoplastic starch/montmorillonite composites. *Acta Polym. Sin.* 862–867.

Huang, M. F., J. G. Yu, and X. F. Ma, (2005b). Preparation of the thermoplastic starch/montmorillonite nanocomposites by melt-intercalation. *Chin. Chem. Lett.* **16**:561–564.

Huang, M. F., J. G. Yu, X. F. Ma, et al. (2005c). High performance biodegradable thermoplastic starch—EMMT nanoplastics. *Polymer* **46**:3157–3162.

Huang, M., J. Yu, and X. Ma, (2006). High mechanical performance MMT-urea and formamide-plasticized thermoplastic cornstarch biodegradable nanocomposites. *Carbohydr. Polym.* **63**:393–399.

Huang, M. F., J. G. Yu, and X. F. Ma, (2004). Studies on the properties of montmorillonite-reinforced thermoplastic starch composites. *Polymer* **45**:7017–7023.

Hulleman, S. H. D., F. H. P. Janssen, and H. Feil, (1998). The role of water during plasticization of native starches. *Polymer* **39**:2043–2048.

Kaṁpeerapappun, P., D. Aht-Ong, D. Pentrakoon, et al. (2007). Preparation of cassava starch/montmorillonite composite film. *Carbohydr. Polym.* **67**:155–163.

Ke, Y., J. Lü, X. Yi, et al. (2000). The effects of promoter and curing process on exfoliation behavior of epoxy/clay nanocomposites. *J. Appl. Polym. Sci.* **78**:808–815.

Lagaly, G. (1986). Interaction of alkylamines with different types of layered compounds. *Solid State Ion* **22**:43–51.

Lagaly, G. (1999). Introduction: from clay mineral-polymer interactions to clay mineral-polymer nanocomposites. *Appl. Clay Sci.* **15**:1–9.

Lourdin, D., L. Coignard, H. Bizot, et al. (1997). Influence of equilibrium relative humidity and plasticizer concentration on the water content and glass transition of starch materials. *Polymer* **38**:5401–5406.

Lourdin, D., S. G. Ring, and P. Colonna, (1998). Study of plasticizer-oligomer and plasticizer-polymer interactions by dielectric analysis: maltose-glycerol and amylose-glycerol-water systems. *Carbohydr. Res.* **306**:551–558.

Ma, X. F., J. G. Yu, and F. Jin, (2004). Urea and formamide as a mixed plasticizer for thermoplastic starch. *Polym. Int.* **53**:1780–1785.

Ma, X. F., J. G. Yu, and Y. B. Ma, (2005). Urea and formamide as a mixed plasticizer for thermoplastic wheat flour. *Carbohydr. Polym.* **60**:111–116.

Ma, X. F., J. G. Yu, and J. J. Wan, (2006). Urea and ethanolamine as a mixed plasticizer for thermoplastic starch. *Carbohydr. Polym.* **64**:267–273.

Martin, O., L. Averous, and G. Della Valle, (2003). In-line determination of plasticized wheat starch viscoelastic behavior: impact of processing. *Carbohydr. Polym.* **53**:169–182.

Mathew, A. P., and A. Dufresne, (2002a). Morphological investigation of nanocomposites from sorbitol plasticized starch and tunicin whiskers. *Biomacromolecules* **3**:609–617.

Mathew, A. P., and A. Dufresne, (2002b). Plasticized waxy maize starch: effect of polyols and relative humidity on material properties. *Biomacromolecules* **3**:1101–1108.

Mathew, A. P., W. Thielemans, and A. Dufresne, (2008). Mechanical properties of nano-composites from sorbitol plasticized starch and tunicin whiskers. *J. Appl. Polym. Sci.* **109**:4065–4074.

Noel, T. R., S. G. Ring, and M. A. Whittam, (1992). The structure and gelatinisation of starch. *Food Sci. Technol. Today* **6**:159–162.

Pandey, J. K. and R. P. Singh, (2005). Green nanocomposites from renewable resources: effect of plasticizer on the structure and material properties of clay-filled starch. *Starch-Starke* **57**:8–15.

Park, H. M., W. K. Lee, C. Y. Park, et al. (2003). Environmentally friendly polymer hybrids. Part I: mechanical, thermal, and barrier properties of thermoplastic starch/clay nanocomposites. *J. Mater. Sci.* **38**:909–915.

Park, H. M., X. Li, C. Z. Jin, et al. (2002). Preparation and properties of biodegradable thermoplastic starch/clay hybrids. *Macromol. Mater. Eng.* **287**:553–558.

Raquez, J. M., Y. Nabar, R. Narayan, et al. (2007). New developments in biodegradable starch-based nanocomposites. *Int. Polym. Process.*. **22**:463–470.

Russell, P. L. (1987). Gelatinization of starches of different amylose amylopectin content—a study by differential scanning calorimetry. *J. Cereal Sci.* **6**:133–145.

Shah, P. B., S. Bandopadhyay, and J. R. Bellare, (1995). Environmentally degradable starch filled low-density polyethylene. *Polym. Degrad. Stabil.* **47**:165–173.

Shen, Z., G. P. Simon, and Y. B. Cheng, (2002). Comparison of solution intercalation and melt intercalation of polymer-clay nanocomposites. *Polymer* **43**:4251–4260.

Shogren, R. L. (1992). Effect of moisture-content on the melting and subsequent physical aging of cornstarch. *Carbohydr. Polym.* **19**:83–90.

Shogren, R. L., C. L. Swanson, and A. R. Thompson, (1992). Extrudates of cornstarch with urea and glycols—structure mechanical property relations. *Starch-Starke* **44**:335–338.

Sinha Ray, S. and M. Okamoto, (2003). Polymer/layered silicate nanocomposites: a review from preparation to processing. *Progr. Polym. Sci. (Oxf.)* **28**:1539–1641.

Smits, A. L. M., S. H. D. Hulleman, J. J. G. Van Soest, et al. (1999a). The influence of polyols on the molecular organization in starch-based plastics. *Polym. Adv. Technol.* **10**:570–573.

Smits, A. L. M., S. H. D. Hulleman, J. J. G. Van Soest, et al. (1999b). The influence of plasticisers on the molecular organisation in starch based products. *Biopolym. Sci. Food Non Food Appl.* 179–182.

Thirathumthavorn, D. and S. Charoenrein, (2007). Aging effects on sorbitol- and non-crystallizing sorbitol-plasticized tapioca starch films. *Starch-Starke* **59**:493–497.

Vaia, R. A. and E. P. Giannelis, (1997). Polymer melt intercalation in organically-modified layered silicates: model predictions and experiment. *Macromolecules* **30**:8000–8009.

Van Tuil, R., P. Fowler, M. Lawther, et al. (2000). Properties of biobased packaging materials, in *Biobased Packaging Materials for the Food Industry—Status and Perspectives*. Frederiksberg, Denmark: KVL.

Van Soest, J. J. G., S. H. D. Hulleman, D. Dewit, et al. (1996). Crystallinity in starch bioplastics. *Ind. Crops Prod.* **5**:11–22.

Van Soest, J. J. G. and N. Knooren, (1997). Influence of glycerol and water content on the structure and properties of extruded starch plastic sheets during aging. *J. Appl. Polym. Sci.* **64**:1411–1422.

Wilhelm, H.-M., M.-R. Sierakowski, G. P. Souza, et al. (2003). Starch films reinforced with mineral clay. *Carbohydr. Polym.* **52**:101–110.

Wollerdorfer, M. and H. Bader, (1998). Influence of natural fibres on the mechanical properties of biodegradable polymers. *Ind. Crops Prod.* **8**:105–112.

Zhang, Q. X., Z. Z. Yu, X. L. Xie, et al. (2007). Preparation and crystalline morphology of biodegradable starch/clay nanocomposites. *Polymer* **48**:7193–7200.

13

THE POTENTIAL OF XYLANS AS BIOMATERIAL RESOURCES

ANNA EBRINGEROVA

13.1 INTRODUCTION

Continuous awareness of the increasing environmental pollution by synthetic materials based on petroleum feedstock and limiting reserves of this resource has led to a shift on the use of biodegradable materials. Products based on silvicultural, annually renewable agricultural and secondary biomass sources from various industries form the basis for sustainable, eco-efficient products that can compete and capture markets currently dominated by products based exclusively on petroleum feedstock. This is substantiated by an estimated production of 10–50 billion ton dry matter of plant biomass each year. Really, the plant polysaccharides and their derivatives represent promising sources of biomaterials, and it has been documented by several recent reviews outlining opportunities of their uses in medicine and pharmacy (Belyaev, 2000; Miraftab et al., 2001; Hooper and Cassidy, 2006; Fowler, 2006) and in the field of food, cosmetics, and other nonfood technologies (Dumitriu, 2002; Miraftab et al., 2001).

Xylans belong to the hemicellulose cell wall components of all plants that are second to cellulose, the most abundant polysaccharide in nature. Xylans constitute about 20–30% of the biomass of dicotyl plants (hardwoods and herbaceous plants) and in some tissues of monocotyl plants (cereals and grasses) they occur up to 50%. Therefore, they can be regarded as an enormous renewable resource of biopolymers expected, nowadays, not only for food and energy but also for chemical materials through biomass refinery (Wyman et al., 2005; Thomsen et al., 2008).

Polysaccharide Building Blocks: A Sustainable Approach to the Development of Renewable Biomaterials, First Edition. Edited by Youssef Habibi and Lucian A. Lucia.
© 2012 John Wiley & Sons, Inc. Published 2012 by John Wiley & Sons, Inc.

The interest in research of xylans has been motivated by various imputes. At the beginning, it was the impact of woody hemicelluloses on the production and functional properties of pulp, paper, and rayon fibers (Ebringerova, 1992). Later, xylans from cereals attracted the attention as dietary fibers with a broad range of physiological effects, which are indispensable components of healthy nutrition (Kritchevsky et al., 1990). During the past few decades, an outstanding increase in research activities in the field of hemicelluloses, including xylans, can be noticed. The demands for healthy food, alternative medicines, and a more effective utilization of the plant biomass without or with a minimum impact on the environment have been the main driving forces. During the years, several reports on xylan-type hemicelluloses have been published in the form of review articles (Stscherbina and Phillip, 1991; Ebringerova, 1992; Whistler, 1993; Ebringerova and Hromadkova, 1999; Ebringerova and Heinze, 2000; Heinze et al., 2004; Sun et al., 2004a; Izydorczyk, 2005; Ebringerova et al., 2005; Hansen and Plackett, 2008) that focus on the isolation, structural, molecular, and functional properties of xylans isolated from a large variety of plant materials, and the chemical modification of xylans. Recently, emphasis has been placed on xylan-rich hemicelluloses of hitherto noninvestigated agricultural (cereals, vegetables, fruits, grasses) and woody (hardwoods, softwoods) plants as well as of by-products resulting from their techno- logical processing. Regardless of the broadened knowledge about xylans, they have not yet found industrial exploitation in the right sense of the word. The main reason is that synthetic polymers are currently much less expensive than plant and microbial polysaccharides. The second reason is the prerequisite of a perfect characterization of the chemical structure and physicochemical properties of xylan preparations in order to generate reproducible technological processes for them. However, this makes the use of individual xylans difficult due to their structural diversity.

The aim of this chapter is to focus attention on the most important functional properties of various xylans and their suggested and realized applications. This chapter updates and extends the above-mentioned reviews and focuses on the importance of xylan-type hemicelluloses as resources of biomaterials and functional polymers applied with and without targeted chemical and/or biochemical modifications.

13.2 STRUCTURAL DIVERSITY AND OCCURRENCE OF XYLANS IN PLANTS

Xylans represent a group of the noncellulosic plant cell wall components traditionally named as hemicelluloses, because they were, in contrast to cellulose, extractable from plants by dilute alkali solutions. At present, they are classified as xyloglycans because their main polymer chain backbone is built up by glycosidically linked β-D-xylopyranosyl (Xyl$p$) residues. The very early term "pentosan," derived from the xylose and arabinose components prevailing in hemicelluloses isolated from cereal and grasses, is still in use, particularly in reports from the areas of food and medicine.

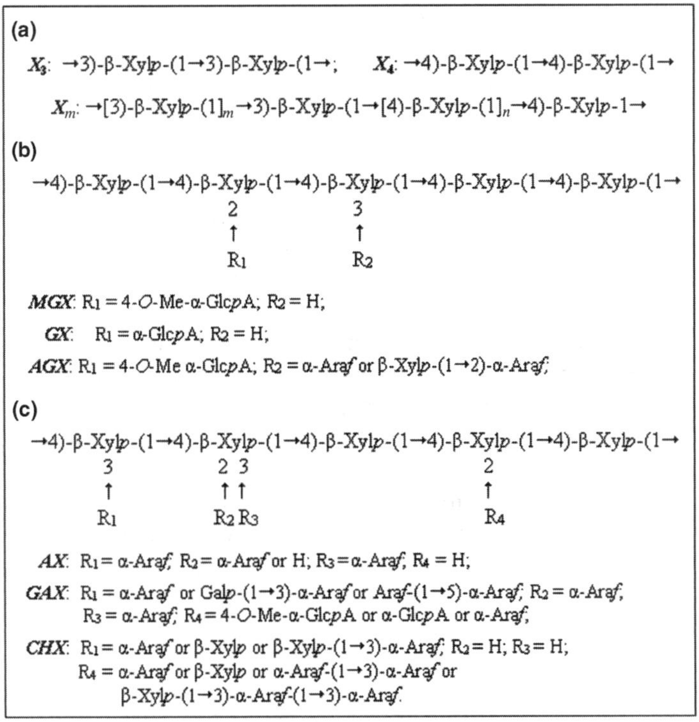

(a)

X_3: →3)-β-Xyl$p$-(1→3)-β-Xyl$p$-(1→; X_4: →4)-β-Xyl$p$-(1→4)-β-Xyl$p$-(1→

X_m: →[3)-β-Xyl$p$-(1]$_m$→3)-β-Xyl$p$-(1→[4)-β-Xyl$p$-(1]$_n$→4)-β-Xyl$p$-1→

(b)

→4)-β-Xyl$p$-(1→4)-β-Xyl$p$-(1→4)-β-Xyl$p$-(1→4)-β-Xyl$p$-(1→4)-β-Xyl$p$-(1→
        2   3
        ↑   ↑
        R₁   R₂

MGX: R₁ = 4-*O*-Me-α-Glc$p$A; R₂ = H;

GX: R₁ = α-Glc$p$A; R₂ = H;

AGX: R₁ = 4-*O*-Me α-Glc$p$A; R₂ = α-Ara$f$ or β-Xyl$p$-(1→2)-α-Ara$f$;

(c)

→4)-β-Xyl$p$-(1→4)-β-Xyl$p$-(1→4)-β-Xyl$p$-(1→4)-β-Xyl$p$-(1→4)-β-Xyl$p$-(1→
    3   2 3   2
    ↑   ↑ ↑   ↑
    R₁   R₂ R₃   R₄

AX: R₁= α-Ara$f$; R₂= α-Ara$f$ or H; R₃=α-Ara$f$; R₄ = H;

GAX: R₁ = α-Ara$f$ or Gal$p$-(1→3)-α-Ara$f$ or Ara$f$-(1→5)-α-Ara$f$; R₂ = α-Ara$f$;
R₃ = α-Ara$f$; R₄ = 4-*O*-Me-α-Glc$p$A or α-Glc$p$A or α-Ara$f$;

CHX: R₁= α-Ara$f$ or β-Xyl$p$ or β-Xyl$p$-(1→3)-α-Ara$f$; R₂ = H; R₃ = H;
R₄ = α-Ara$f$ or β-Xyl$p$ or α-Ara$f$-(1→3)-α-Ara$f$ or
β-Xyl$p$-(1→3)-α-Ara$f$-(1→3)-α-Ara$f$.

FIGURE 13.1 Structural features of (a) homoxylans, (b) heteroxylans MGX, GX, and AGX, and (c) heteroxylans AX, GAX, and the CHX from psyllium husks (Guo et al., 2008).

Xylans are a structurally very diverse group of polysaccharides differing in the type of glycosyl side chains, their proportions and locations in the Xyl$p$ monomers, and distribution along the main polymer chain (Ebringerova and Heinze, 2000). Xylans can be classified into several subgroups according to their structural variations (Figure 13.1):

- homoxylans consisting of 3-linked Xyl$p$ residues (X_3), mixed 3- and 4-linked Xyl$p$ residues (X_m), and 4-linked Xyl$p$ residues (X_4);
- glucuronoxylans (MGX, GX) having the X_4 backbone partially substituted at position 2 with a single 2-*O*-linked 4-*O*-methyl-α-D-glucopyranosyl uronic acid (MeGlc$p$A) residue or its nonmethylated form (Glc$p$A), respectively;
- (arabino)glucuronoxylan (AGX) having the backbone of MGX branched with single α-L-arabinofuranosyl (Ara$f$) side chains attached at position 3;
- arabinoxylans (AX) having the X_4 xylan backbone branched with single Ara$f$ residues at position 3 (monosubstitution) and both positions 2 and 3 (disubstitution);
- (glucurono)arabinoxylans (GAX) having the AX backbone branched at position 2 with MeGlc$p$A or Glc$p$A residues;

- complex heteroxylans (CHX) with the X_4 backbone heavily substituted by various mono- and oligomeric chains composed of Xyl$p$, Ara$f$, Gal$p$, and/or Glc$p$A residues (represented in Figure 13.1 by CHX from psyllium husks) (Guo et al., 2008).

In the native state, xylans are partially acetylated and phenolic acids (mainly ferulic and coumaric acids) are esterified to O-5 of the arabinofuranosyl (Ara$f$) constituents. Depending on the extraction conditions, xylan–lignin covalent linkages existing in the plant cell walls (Iiyama et al., 1994) may resist in xylan preparations, similarly as the bound phenolic acids.

As illustrated in a former review (Ebringerova and Heinze, 2000), the occurrence of xylans in the plant kingdom can be traced from the youngest plants (Spermatophyta)—angiosperms—up to the oldest plants (Sporophyta)—the algae—which was also established in a recent study (Carafa et al., 2005). Whereas MGX and GX occur in dicotyledonous plants (hardwoods, shrubs, and herbal plants), AGX are found in grasses, softwoods and their forerunner ginkgo, and ferns. Cereals and grasses contain several xylan types—AX, GAX, AGX, and CHX. In contrast to terrestrial plants with cellulose as the main cell wall component, homoxylans are typical skeletal constituents of some kinds of algae, where they were assumed to substitute the function of cellulose.

13.3 APPLICATION POTENTIAL OF POLYMERIC AND OLIGOMERIC XYLAN ISOLATES

Xylans are becoming more and more attractive as biopolymers as they can be utilized in their isolated form in various areas including food and nonfood applications. They are distributed within the plants and their tissues and organs, where they perform various biological functions (structural and/or storage, might be also defending) (Waldron et al., 2003). The structural diversity of xylan types even in different plant tissues within one plant indicated close structure–function relationships of these polysaccharides.

The functional properties of the xylans, similarly as other polysaccharides, result from their structural features and physicochemical and physical properties that are, to a various extent, preserved in the isolated preparations. However, the properties of the xylan products are in any case dependent on the isolation, separation, purification, and even drying conditions.

13.3.1 Food and Nonfood Applications

The most important functionalities of xylan isolates and their suggested application possibilities are summarized in Table 13.1, updated by data not mentioned earlier in the reviews (Stscherbina and Phillip, 1991; Ebringerova, 1992; Ebringerova and Hromadkova, 1999; Ebringerova and Heinze, 2000; Ebringerova et al., 2005). As shown, the functional properties of the xylan isolates affecting their potential uses

TABLE 13.1 Reviewed Data on Functional Properties and Potential Applications of Various Xylan Isolates

Xylan Type	Plant/Tissue	Functional Properties	Application Fields
X_m	Red seaweed	Dietary fiber nutritional characteristics	Functional food (Ebringerova et al., 2005)
MGX, Wis-MGX	Beech wood, beech pulp[a], aspen wood	Film forming, thickening, gelling, emulsifying, foaming, foam stabilizing	Papermaking, breadmaking, textile printing, filler for drug forms (Ebringerova and Heinze, 2000)
MGX-XOS	Birch wood	Antimicrobial activity	Antimicrobial additive (Ebringerova et al., 2005)
MGX	Medicinal herbs	Cytokinin-like activity Immunostimulatory, antitussive effect	Plant growth regulators (Katapodis et al., 2002) Immunoenhancing supplement (Ebringerova and Hromadkova, 1999)
GX	Quince seeds	Gelling, viscosity enhancing	Food additive (Ebringerova and Heinze, 2000)
AGX	Corncobs	Pseudoplasticity, plastic at higher concentrations	Papermaking, textile printing, filler for drug forms (Ebringerova, 1992); drug carrier (Ebringerova et al., 2005)
Ws-AGX	Corncobs	Film forming, viscoelasticity, surface activity, immunostimulatory activity	Pharmaceutical auxiliary materials; immunoenhancing supplement (Ebringerova and Hromadkova, 1999)
Wis-AGX	Corncobs	Aggregation behavior, resistance to digestion	Nanoparticle drug carrier (Ebringerova et al., 2005)
AX	Rice bran	Oxygen scavenging, immunostimulatory activity	Immunoenhancing supplement—BioBran[b] (Ebringerova and Hromadkova, 1999)
AX[c]	Flour (wheat sorghum, rye)	Film forming, gas retention, foamability	Breadmaking (Ebringerova and Hromadkova, 1999)
AX, AX-XOS[c]	Cereal grains	Inhibition of ice formation, antioxidant activity, bifidobacteria growth stimulation	Additives for ice cream, frozen foods (Ebringerova and Hromadkova, 1999); food antioxidant (Katapodis et al., 2003); prebiotics (Yuan et al., 2005a)

(Continued)

335

TABLE 13.1 *(Continued)*

Xylan Type	Plant/Tissue	Functional Properties	Application Fields
GAX	Corn husk	Immunostimulatory activity	Immunoenhancing supplement (Ogawa et al., 2005)
GAX	Corn fiber[d], cereal bran	Viscosity enhancing, film forming, gel forming	Thickener, food additive (Ebringerova et al., 2005)
GAX[c]	Corn bran	Hydrogel formation, biological activity	Wound management aid—Sterigel[e], branan ferulate/alginate fiber (Ebringerova et al., 2005)
GAX-XOS[c]	Wheat bran, corn bran	Antioxidative activity	Antioxidant food additive (Ohta et al., 1984; Yuan et al., 2005b)
CHX, GAX	Psyllium seed (*P. ovata*)	Laxative and cholesterol lowering activities, gelling	Dietary fiber supplement and plant-based pharmaceuticals (Sandhu et al., 1981)
CHX, AX	Psyllium husk (*P. major*)	Viscosity enhancing, anticomplementary activity	Plant-based pharmaceuticals (Ebringerova et al., 2005)
CHX	Psyllium seed	Gel forming	Alternative microbial culture media (Jain et al., 1997)
CHX	Gum exudate of tropical plants	Thickening, viscosity enhancing	Food gums (Maurer-Menestrina et al., 2003; Simas et al., 2004)

Ws, water-soluble; Wis, water-insoluble; XOS, xylooligosaccharides.

[a] Spent liquor from the viscose process.

[b] Producer: Daiwa Pharmaceutical Co., Ltd.

[c] Containing phenolic acids (ferulic, coumaric, etc.).

[d] Fiber from the wet milling process.

[e] Producer: Seton Scholl.

comprise solubility, rheological behavior of their solutions and dispersions, surface-active properties, and interactions with polysaccharides or other biopolymers. For xylans isolated from cereal grains, some edible grass seeds, and medicinal herbs or other plants used in traditional medicines, their physiological effects and/or various biological activities are also essential.

From the data in Table 13.1, four application fields can be recognized comprising food, pharmacy, medicine, and special technical applications related to the former fields. In general, the xylan research aimed toward (i) nutrition and health care and (ii) solution of environmental problems by substituting petrochemical feedstuffs with plant biomass has been on a rapid increase as will be documented in the following subsections.

13.3.1.1 Dietary and Functional Fibers and Food Additives   Xylans from cereals grains are components of dietary fibers contributing to their effects on some biochemical and beneficial physiological processes in humans and animals (Kritchevsky et al., 1990). The viscous character of water-insoluble xylans and high water absorption of the water-insoluble ones play a significant role in these processes. High-fiber diets are important in the prevention and management of obesity, chronic diseases, and cancer (Kendall et al., 2010). Cereal grain xylans represent ingredient of the so-called functional foods (Hasler, 2001). This term refers to the practice of fortifying foods with added ingredients that can confer health effects on the consumer (Charalampopoulos et al., 2002).

Nowadays, xylans from various plants have been recognized as source for the production of xylooligosaccharides (XOS) that had been classified as functional foods as well (Moure et al., 2006). Similarly as the known prebiotics oligofructose and inulin, they caused prebiotic effects derived from their ability to modulate the intestinal function by promoting the growth of bifidobacteria. Acidic XOS were prepared from corncobs (Yang et al., 2005; Vazquez et al., 2006), almond shells (Nabarlatz et al., 2007a), and rice hulls (Gullon et al., 2010), mainly aimed to apply them as prebiotic food ingredients. Prebiotic characteristics were reported also for feruloylated XOS isolated from wheat bran (Yuan et al., 2005a). The phenolic acids (mainly ferulic acid) that are effective scavengers of free radicals (Garcia-Conesa et al., 1997) are linked to cereal xylans. Persisting in the derived feruloylated XOS, they imparted them antioxidant activities (Ohta et al., 1984; Yuan et al., 2005b; Katapodis et al., 2003; Rao and Muralikrishna, 2006). These feruloylated XOS (feraxans) represent natural antioxidants applicable as food ingredients. Hence, the interest to produce biologically active XOS by optimal and environment-friendly technological processes shows a permanently increasing trend (Akpinar et al., 2009; Guilloux et al., 2009; Gullon et al., 2010; Rose and Inglett, 2010).

However, being small molecules, most antioxidants, including free ferulic acid and feruloyl oligosaccharides, are absorbed in the small intestine and do not enter enterohepatic circulation (Zhao et al., 2003). It can be assumed that phenolics closely associated with polysaccharides might overcome this problem. Recently, the phenolics containing MGX isolated from buckwheat hulls (Hromadkova et al., 2005) and almond shells (Ebringerova et al., 2008), similarly as the water-soluble AX from wheat bran (Patel et al., 2007), were reported to exhibit significant antioxidant activities. These preparations might be included into the group of antioxidant dietary fibers (AODFs), which were defined as a fiber containing significant amounts of natural antioxidants associated with the fiber matrix (Saura-Calixto, 1998).

Considerable attention has been paid to the role of cereal AX, beech wood MGX, and psyllium CHX in breadmaking (Ebringerova, 1992; Ebringerova et al., 2005; Izydorczyk, 2005). Interactions of the cereal AX with starch and proteins were suggested to be responsible for the highly positive effects of the xylans (present as flour components or additives) on the mechanical quality of dough as well as on the texture and other end product parameters of baked products. Gluten-free breads with a specific volume up to and beyond that of conventional wheat bread can be made with rice flour (not containing gluten) by addition of CHX from psyllium husks and hydroxypropylmethylcellulose (Haque et al., 1994).

A significant role was ascribed to phenolic acids containing xylans (Izydorczyk, 2005). By oxidative coupling of the ferulic acid substituents *in vivo* within and between the xylan chains through a dimerization reaction, hydrogel networks are formed (Ng et al., 1997) that were assumed to contribute to the above-mentioned effects in the breadmaking process. In a recent paper (Hromadkova et al., 2007), phenolics-containing MGX isolated from the seed hulls of buckwheat (a pseudocereal) were applied as additives in amounts between 0.3 and 0.5 wt% to wheat flours. Significant improvements were observed in processing of dough, and the sensory and other properties of the fresh bread and after its longer storage. Moreover, these effects were achieved also with medium-quality wheat flours.

Studies on the corn fiber gum, a by-product of the wet milling process, revealed galactose-containing highly branched GAX with small amounts of protein to be its essential component (Yadav et al., 2007a). Xylan fractions separated from the gum displayed significant emulsifying activity tested in a model oil-in-water emulsification system (Yadav et al., 2007b). The corn fiber gum was proposed as a gum arabic replacer for beverage flavor emulsification (Yadav et al., 2007c).

Attention has been paid to various highly branched CHX that present many features in common with gum arabic—a commercial food gum. The CHX from psyllium husk (*Plantago ovata*) has been used traditionally as a laxative agent and, currently, as a dietary fiber supplement. The strong gelling activities of the polysaccharide were related to their unusual linkage pattern and/or the high density of its branching resulting in resistance to digestion by the intestinal microflora (Fischer et al., 2004). In a recent study, Guo et al. (2008) isolated from the psyllium gum two acidic water-soluble arabinoxylan fractions and one insoluble neutral fraction, all of very high molecular weight (MW). The xylans differed in degree of branching and the type and length of branching units. The coexistence of these fractions in aqueous dispersion and mutual intermolecular interactions might contribute to the viscosity enhancement and gelling activity of psyllium gum. A mixture of wheat fiber and psyllium husk fiber plus three antioxidants was proposed to fortify bread and other baking products (Park et al., 1997).

CHX with thickening and viscosity enhancing properties were isolated also from the gum exudates of various tropical plants such as *Livistona chinensis*

(Maurer-Menestrina et al., 2003). The mucilages exuded from leaves of the flax species *Phormium tenax* and *P. cookianum* have many features in common with gum arabic (a food gum). They contain acidic CHX with a highly branched and partially *O*-acetylated backbone and their MW values ranged between 1280 and 1680 kDa (Sims and Newman, 2006). The acidic CHX isolated from the gum exudate of the Queen palm (*Scheelea romanzoffiana*) had a lower molar mass (MW = 140 kDa) and contained a large amount of fucosyl residues. It formed highly viscous aqueous solutions (Simas et al., 2004). All these polysaccharides represent potential gums for food and nonfood applications.

13.3.1.2 Biologically Active Substances   Medicinal plants are known to be a source of biologically active polysaccharides; however, most of them belong to the pectic polysaccharides (Paulsen, 2002). Various xylans possessing immunostimulatory activities were isolated from medicinal herbs used in traditional medicine (MGX, AGX, and highly branched AX) and also from nonmedicinal plants such as the ws-AGX from corncobs (Ebringerova et al., 2005). Phenolics-containing MGX from buckwheat hulls (Hromadkova et al., 2005) and AX from wheat bran (Patel et al., 2007) and the MGX-derived XOS from almond shells (Nabarlatz et al., 2007b) were shown to be immunologically active in various *in vitro* tests. Recently, acidic heteroxylans with antiulcer activity have been isolated from various medicinal plants (Cipriani et al., 2008). A purified MGX isolated from the chestnut (*Castanea sativa*) wood exhibited cytotoxic properties. It inhibited the proliferation of carcinoma cells, and their migration and invasion (Barbat et al., 2008).

Many of the immunologically active MGX from herbal plants exhibited antioxidant activities as well (Kardosova and Machova, 2006). At present, a close relation between the immunological and antioxidant activities has been assumed based on the fact that free radicals are involved in many disorders such as asthma, inflammation, arthritis, atherosclerosis, and so on (Vaya and Aviram, 2001). Accordingly, antioxidants may have the potential to prevent the early development of atherosclerosis and other health problems associated with radical-mediated damages.

The AX from rice bran (MW $\sim$ 3–5 kDa) that had been enzymatically modified with the extract from *Hyphomycetes mycelia* showed immunoenhancing effects and anti-HIV activity without any notable side effects (Ghoneum and Jewett, 2000). It found a commercial application as BioBran (MGN-3) that was claimed to be a nontoxic glyconutritional supplement. A novel CHX with anticomplementary activity was isolated from defatted rice bran (Wang et al., 2008). The partially hydrolyzed AX from corn husks (MW about 53 kDa) showed significant *in vivo* immunopotentiating effects (Ogawa et al., 2005).

XOS from various plant sources exhibit (alone or as active components of pharmaceutical preparations) a large variety of biological activities such as immunomodulatory and anti-infection properties, blood- and skin-related effects, and antiallergic, antimicrobial, and selective cytotoxic activities (Moure et al., 2006). Besides biological effects concerning human health, uronic acid-containing XOS have been employed as plant growth regulators (Katapodis et al., 2002).

MGX isolated from medicinal plants displayed antitussive activities (Ebringerova et al., 2005), which are common for rhamnogalacturonans isolated from the mucilage of *Althaea* species, a herbal expectorant (Nosalova et al., 2005). In comparison with codeine and other synthetic antitussive drugs, these xylans showed comparable cough-suppressing effects, but no unwanted side effects.

13.3.1.3 Additives in Pharmacy, Papermaking, and Other Technical Applications

The CHX isolated from the mucilage of tropical flax species exhibiting wound healing effects was recommended as additive in skin care products (Sims and Newman, 2006). MGX- and AGX-type xylans were due to their rheological properties suggested as pharmaceutical auxiliary materials for the production of various drug forms (Ebringerova, 1992). These xylans when applied in papermaking affected positively the mechanical properties and printability of paper due to their film-forming properties and interactions with cellulose fibers (Ebringerova and Hromadkova, 1999). The highly branched CHX from psyllium seeds form a thick mass when exposed to fluids. It was suggested for application as flocculant for sewage and tannery effluent treatment (Mishra et al., 2002) and textile wastewater treatment (Mishra and Bajpai, 2002, 2005).

Nowadays, polymer/clay biocomposites have become a popular research subject. Montmorillonite (NaMt) as a member of smectite group clay minerals has large adsorption capacities for polymer molecules due to its unique crystal structure and therefore is the most widely used layered silicate in polymer nanocomposites. The corncob AGX was used for the preparation of xylan/NaMt biocomposites by changing the xylan or NaMt concentrations (Unlu et al., 2009). It was found that lower amounts of xylan interacted with NaMt on the surface; however, at higher amounts, intercalation of NaMt occurred. These biocomposites showed better thermal and rheological behaviors with respect to the starting materials. They can find application especially in cosmetic formulations as both thickener and cleaner agents.

13.3.2 Biomaterials for Targeted Applications

Due to the negative environmental impact of synthetic polymers and future shortage of nonrenewable resources, polysaccharides started to be used as "new" biopolymeric materials ready for direct application or after modification of their functional properties into materials until now produced from synthetic polymers. Recent research in this field has been directed toward development of films, hydrogels, fibers, and biocomposites for food and nonfood applications (Lloyd et al., 1998; Belyaev, 2000; Tharanathan, 2003; Coviello et al., 2007; Lindblad et al., 2007). Results concerning the utilization of hemicelluloses as food packaging films and coatings and hydrogels and films for biomedical applications have been evaluated in a very recent review (Hansen and Plackett, 2008). The functional properties and application possibilities of various xylans applied in their isolated form (xylan isolates) are summarized in Table 13.2. Further information concerning non-included and later reports will be presented in the following subsections.

13.3.2.1 Films and Coatings for Packaging, Foodstuffs, and Other Purposes

Important characteristics of films for food packaging are the mechanical properties, oxygen permeability, and reduced water vapor permeability (WVP). Most of xylans gave no self-supporting films; they were crystalline and brittle. The data in Table 13.2 indicated that films (without plasticizers), applicable for food packaging, were obtained only with MGX from cotton stalks and AX from barley husks. Both xylans contained small amounts of residual lignin (Goksu et al., 2007; Hoije et al., 2005). In case of the other xylans, modification of the films by various plasticizers and/or in combination with additives, for example, various lipids or sucroesters, was necessary to achieve the functional properties of packaging films (Peroval et al., 2002; Phan The et al., 2002a, 2002b). With respect to oxygen permeability, the produced xylan films are comparable to other biopolymer films such as amylose and amylopectin. However, the water vapor permeabilities are several magnitudes higher than those of other polymers currently used for this purpose.

Edible films and coatings have been particularly considered in food preservation and technology. They have to form an actual barrier against oxygen, aroma, oil, or moisture.

Very useful for preparation of edible films is the corn hull AX that forms films of high strength and lubricity after addition of various plasticizers in amounts up to 20 wt% (Zhang and Whistler, 2004). These films were stable, strong, smooth, or transparent, and had mechanical properties, moisture content, and WVP controlled by the type and amount of the plasticizer. All plasticized arabinoxylan films produced in this study had lower WVP than those of unplasticized films. The film coating preservation experiment with grapes showed that arabinoxylan-based film coatings, arabinoxylan–sorbitol (AX/sorbitol) films in particular, had the best moisture barrier ability. However, it has to be stressed that the tested xylan is not of the AX type obtained from the cereal flours, but of the GAX type as indicated by the described structural features. The presence of uronic acid substituents may significantly affect the observed physical properties of the films.

Improved mechanical properties of polysaccharide films have been achieved by blending them with other natural polymers. Edible films were obtained by mixing wheat gluten with birch wood MGX or AGX from corncobs and grass in the presence of sorbitol as plasticizer (Kayserilioglu et al., 2003). AGX from corncobs gave films more stretchable and less stiff than the films from birch wood MGX and grass AGX. As suggested, the wheat gluten/xylan compositions were suitable for film production due to formation of stable protein–polysaccharide polymer networks in which the uronic acid constituents of the xylan are involved. Corncob AGX added films had a more uniform surface than the other films, which explains their high stretchability. Films from grass AGX showed little globular formation on the surface, whereas the use of birch wood MGX resulted in more heterogeneous films.

The corn fiber AX was applied in blends with the spruce wood hemicellulose galactoglucomannan (GGM) in various mass ratios using glycerol as plasticizer (Mikkonen et al., 2008). The blending did not bring the expected improvement of the mechanical properties of the GGM films. The homogeneous blend films were very sensitive to high relative humidity. Due to the highly branched structure, the

TABLE 13.2 Polymeric Xylan Isolates for Special Applications

Xylan Type	Plasticizing Additive	Properties	Application	References
MGX	None	Brittle films	None	Grondahl et al. (2004)
	Xylitol, sorbitol (20%, 35%, or 50%)	Semicrystalline film, strength and elongation depending on the percentage of plasticizer, low OP	Packaging for oxygen-sensitive products	
AX	Xylitol, sorbitol (20%, 35%, or 50%)	Transparent films, mechanical properties controlled by the plasticizer, low OP	Packaging for oxygen-sensitive products	Grondahl and Gatenholm (2007)
MGX[a]	None	Coherent film	Food packaging	Goksu et al. (2007)
	None, MGX[b]	Cracked film	None	Goksu et al. (2007)
MGX	None, mixing MGX with 1% lignin and glycerol (2%)	Film thickness and mechanical properties, slightly increased WVTR	Water-soluble films for packaging	Hoije et al. (2005)
AX[c]	None	Homogeneous film, highly hygroscopic, high strength	Food packaging films	Kayserilioglu et al. (2003)
MGX	Glycerol (2%), gluten + MGX (0–40%)	Tensile strength (8 MPa), elongation (50%), tensile strength (1.5–3 MPa)	Edible films	
AGX	Gluten + AGX	Elongation (600%)		
GAX	Various plasticizers (0–20%)	Stable films, good moisture barrier (WVP) and mechanical parameters	Coatings, edible films	Zhang and Whistler (2004)
GAX	Glycerol	Opaque films	None	Peroval et al. (2002)
	Glycerol, lipids, or sucroesters	Transparent films show WVP and WVTR	Edible films	Phan The et al. (2002a, 2002b)

342

MGX	None, mixing with chitosan	Chitosan 10%—self-supporting films; chitosan 30%—hydrogels	Biomedical applications	Gabrielii and Gatenholm (1998)
MGX	None	Brittle films	None	Gabrielii et al. (2000)
	None, mixing with chitosan	Chitosan <20%—hydrogels; chitosan >20% ws—films	Biomedical applications	
AGX	None	Film formation, resistance to digestion, 10-fold decrease in magnetite dissolution	Coating for magnetite microparticles	Silva et al. (2007)

ws, water-soluble; OP, oxygen permeability; WVP, water vapor permeability; WPTR, water vapor transport.

[a] Containing 1% residual lignin.

[b] Lignin-free.

[c] Containing 3% lignin.

xylan may absorb more water, which leads to plasticization and softening of the films.

MGX xylans in their native O-acetylated form were isolated from birch, beech, maple, and willow by a combined biodelignification and hot water extraction procedure (Stipanovic et al., 2006). To enhance the mechanical properties of these xylans as biomaterials, they were mixed with solutions of commercially available cellulose esters followed by casting into solid films. The acetylated xylan/cellulose triacetate "blends" showed mechanical properties comparable to those of the cellulose triacetate itself up to 25 wt% of xylan. Plasticizers were effective in increasing the strain to break for these materials but lowered the modulus at 1% strain. By mixing the native partially O-acetylated MGX isolated from delignified aspen wood (holocellulose) with bacterial cellulose, strong transparent composite films were obtained without a plasticizer (Dammström et al., 2005). A pronounced softening effect at about 85% relative humidity was observed for the pure xylan film. The composite film made from bacterial cellulose with aspen MGX as matrix also showed a prominent softening, but at a slightly lower surrounding relative humidity as compared to the pure xylan. No clear interactions could be observed between xylan and cellulose in the composite.

Biodegradable composite films with water transmission barrier properties were prepared employing oat spelt xylan (GAX), sulfonated cellulose whiskers (obtained by sulfuric acid hydrolysis of kraft pulp), and sorbitol as plasticizer (Saxena and Ragauskas, 2009). The mechanical properties of the films were evaluated using tensile testing under controlled temperature and humidity conditions. Addition of cellulose whiskers leads to a substantial improvement in strength properties. The results showed that xylan films reinforced by 10% whiskers exhibited a 74% reduction in specific water transmission properties with respect to xylan film and a 362% improvement with respect to xylan films reinforced with 10% softwood kraft fibers. Furthermore, films to which 7% cellulose whiskers were added showed that nanocellulose whiskers produced with sulfuric acid (sulfonated whiskers) were significantly better at increasing film strength than cellulose whiskers produced by hydrochloric acid (Saxena et al., 2009).

In a very recent paper (Viota et al., 2010), organically modified montmorillonite clay nanoparticles (Nanofil 8) were used as stabilizers of bioplastic composite xylan films. To improve the binding between xylan molecules and the nanoparticle, the later were coated with a nonionic surfactant (partially alkylated inulin). Suspensions of the coated nanoparticles were dried to give films. These were prepared from oat spelt GAX and birch MGX. The film formation was pH dependent. A thorough characterization of the wettability of the xylan films demonstrated that it is dominated by acid–base interactions and that incorporation of inulin-coated montmorillonite leads to a considerable reduction of the hydrophilic character of the films.

13.3.2.2 Hydrogels, Films, and Coatings for Biomedical Purposes   Hydrogels, that is, polymeric material that swells in water but does not dissolve, are materials that can be applied in various coating systems. As already mentioned earlier, cereal arabinoxylans containing ferulic acid substituents are able to form hydrogels by

cross-linking of the xylan chains under oxidative conditions. The use of enzymatic free radical generating agents such as laccase/O_2, peroxidase/H_2O_2, or chemical systems was described by Figueroa-Espinoza et al. (1998) and recently by Nino-Medina et al. (2010). The AX from corn bran (branan ferulate) was cross-linked by peroxidase forming a three-dimensional network (Ng et al., 1997). In the presence of water, the product gave a hydrogel suitable for use as a wound management aid, which was already commercialized as Sterigel®.

The impact of the feruloylation degree on the structure and properties of cereal flour AX gels has been later extensively studied by Carvajal-Millan et al. (2005). From a water-extractable cereal xylan (WEAX), a series of fractions (PF-WEAX) differing in ferulic acid (FA) content were prepared by chemical deesterification. The gelling ability of these materials after laccase treatment was suggested to be related to the FA content, whereas the gel structure and properties depended on the density of newly formed covalent cross-links. In addition to di-FA and tri-FA, higher ferulate cross-linking and physical entanglements were suggested to contribute to the final WEAX gel structure. In a further study (Carvajal-Millan et al., 2006), the rheological properties of gels prepared from the parent WEAX and derived PF-WEAX fractions were investigated. The differences in structural and rheological characteristics of the gels were reflected in their capacity to load and release proteins of MW ranging from 43 to 669 kDa. The possibility of modulating protein release from WEAX gels makes these gels potential candidates for the controlled delivery of proteins.

The molecular architecture and degree of substitution of arabinoxylans varies among different crops. The importance of the distribution of arabinose substituents along the backbone, which affects the conformation and the capacity of arabinoxylans to interact with one another, has been well documented (Izydorczyk, 2005). Sternemalm et al. (2008) investigated the behavior of AX and a series of fractions differing in the arabinose content prepared by controlled acid hydrolysis. These materials were tested by solution behavior and film properties. The removal of arabinose resulted in a gradual association of unsubstituted chains and aggregate formation was observed at Ara/Xyl ratios between 0.31 and 0.23, and precipitation at ratios below 0.1. The films prepared from debranched arabinoxylan were all translucent. The moisture content varied between 57% and 72%, which indicated the arabinoxylans to be very hygroscopic. This agreed with the observations of Hoije et al. (2005) who observed in barley husk AX a moisture content of 82% at 100% relative humidity. The material is plasticized by increased moisture content. A decrease in arabinose content resulted in the loss of a plasticizing effect, as determined by dynamic mechanical analysis correlating with the water binding capacity.

Arabinoxylan gels have received increasing attention as colon-specific protein delivery vehicles also due to their macroporous structure, aqueous environment, and dietary fiber nature (Carvajal-Millan et al., 2006; Berlanga-Reyes et al., 2009). Swellable hydrogels or water-soluble films were prepared by blending MGX from birch or aspen wood with chitosan (without a plasticizer), which depended on the xylan/chitosan compositions (Gabrielii and Gatenholm, 1998; Gabrielii et al., 2000). The authors suggested the complexation between glucuronic acid functionalities of xylan and amino groups of chitosan to be responsible for the network formation entailing the swelling behavior. The films are suitable for biomedical applications.

The corncob AGX applied as coating for supermagnetic particles is another example for biomedical application (Silva et al., 2007). The coating process was performed by coprecipitation of iron salts in alkaline medium and following emulsification/cross-linking reaction yielding magnetic polymeric particles. The product protected from gastric dissolution might be very promising for oral administration applied in gastric ulcer therapy.

13.4 APPLICATION POTENTIAL OF XYLAN DERIVATIVES

The chemical modification of polysaccharides is the most important route to modify the properties of the naturally occurring biopolymers and to use these renewable resources in the context of sustainable development (Tharanathan, 2003; Coviello et al., 2007). Recent research and development is focused on the improvement of known products and new synthesis paths as well as on new derivatives and alternative synthesis concepts. However, compared to cellulose, starch, and dextrans that are the most often modified commercial polysaccharides (Heinze et al., 2001; Heinze, 2004), the modification of xylans has been investigated to a much lower extent. A large variety of xylan derivatives are briefly summarized in the following sections, whereby emphasis is placed on xylan derivatives that are of interest for potential applications.

13.4.1 Biomaterials Based on Highly and Partially Functionalized Xylans

The first chemical modifications concerned the preparation of xylans with entirely substituted hydroxyl groups by methylation, benzylation, and acetylation. The derivatives served for analysis of the structural and molecular properties of the xylans. Later, the modification of xylans was aimed to improve and/or impart new functionalities that could be finally attractive for industrial applications. Most of all, chemical modification of xylans has been investigated as an alternative to existing aliphatic and aromatic ether and organic and inorganic ester derivatives of the commercial polysaccharides cellulose and starch. The variety of published modified xylans comprised sulfoalkyl, hydroxypropyl, and quaternized derivatives, acetates, various fatty acid esters, and sulfates. Corresponding reports from past research were summarized in several reviews (Ebringerova, 1992; Ebringerova and Hromad-kova, 1999; Sun et al., 2004a; Heinze et al., 2004; Ebringerova et al., 2005; Hansen and Plackett, 2008). The research activities were focused primarily on the modification process, various xylan sources, and elucidation of the physicochemical, physical, and other properties of the obtained derivatives. Only in some of the reports attention has been paid to investigate the derivatives in view of their future application possibilities. In fact, most of the useful xylan derivatives have already been included in the numerous patent applications and existing patents (not reviewed in this chapter). An overview on earlier reports concerning xylan derivatives, their functional properties, and suggested applications is given in Table 13.3. Non-included and later reports will be presented in the following subsections.

13.4.1.1 Additives in Various Technologies   Some of the known xylan derivatives shown in Table 13.3 have been reinvestigated to modify the reaction conditions and characterize the structural and physicochemical properties of the products. New reports appeared concerning carboxymethylation and cationization of xylans. The carboxymethylation reaction is generally used to improve or impart water solubility and other valuable properties to polysaccharides. This method was investigated in detail on xylans from various sources such as birch, beech, and eucalyptus wood and from oat husk, rye bran, and corncobs using different activation procedures (Petzold et al., 2006). Reactions were performed with sodium mono-chloroacetate in the presence of aqueous NaOH under completely heterogeneous conditions using 2-propanol as slurry medium or starting with xylan dissolved in aqueous alkali. One-step reactions lead to carboxymethyl xylans (CMX) with DS from 0.13 to 1.22, whereas the two-step synthesis (e.g., repeated reaction) yielded DS values up to 1.65. The CMX were water soluble at a minimum DS of 0.3 and decreased the surface tension of water from 7.27 to 45 mM/m. Interestingly, CMX obtained under completely heterogeneous conditions not only lead to higher DS but also showed highest surface tension depressing effects. The different substitution pattern of the CMX derivatives prepared from the various xylan types was suggested to be responsible for the observed differences. The CMX prepared under similar reaction conditions by Ren et al. (2008a) from the sugarcane bagasse hemicellulose (AGX type) had the maximum DS of 0.56, but significantly degraded polymer chains.

The xylan isolated from Ispaghula seed husks (CHX type) showed high swelling ability in water (Saghir et al., 2008). The carboxymethylation of the xylan carried out heterogeneously in the presence of alkali in the slurry medium yielded CMX with DS as high as 1.81 and, due to the incorporated ionic functions, water solubility starting at DS of 0.33. Future investigations are planned on the biological activity and rheological properties of these anionic polyelectrolytes for application as thickening agent in pharmacy and other industries.

The positive effects of cationized xylans in papermaking industry have been already demonstrated in earlier reports (Ebringerova, 1992; Heinze et al., 2004). Further research of cationized xylans was focused on improvement of the synthetic pathways. In a one-step reaction, water soluble 2-hydroxypropyltrimethylammonium (TMAHP) derivatives of birch wood MGX were prepared by reaction with 2,3-epoxypropyltrimethylammonium chloride in alkaline 1,2-dimethoxyethane slurry medium (Schwikal et al., 2006). The DS of the derivatives achieved values up to 1.64. From the derivatives, films with controlled pore size were formed by tailoring the concentration of casting solution. Due to the uronic acid side chains of MGX, the resulting TMAHP xylans have a polyionic structure with a permanent positively charged ammonium group and a pH-sensitive charged carboxyl group. Various reaction media were tested in the above-mentioned cationization reaction. Ren et al. (2006) used a two-step reaction in aqueous medium with sodium hydroxide as catalyst. The TMAHP derivatives prepared from sugarcane bagasse AGX showed

TABLE 13.3 Xylan Derivatives, Their Functional Properties, and/or Proposed Applications

Xylan Type	Derivative	Functional Properties and/or Application	References
X$_m$, AX	Carbamates, carbanilates	Thermoplasic materials; chiral stationary phases for HPLC; ligands for the coordination of Cu(II)	Ebringerova and Hromadkova (1999)
MGX, AGX	TMAHP derivatives	Beater additive in papermaking; flocculant; antimicrobial additive	Heinze et al. (2004)
AGX	Polycarboxylates	Detergency, biodegradability	Heinze et al. (2004)
AGX	Benzoates, caprates, myristates, laurate	Thermoplastic materials	Sun et al. (2004a)
AGX	Succinate, DS < 0.26	Thickener, metal ion binder	Heinze et al. (2004)
CHX	Laurylamine-grafted xylan (DS = 0.6–1.1)	Transparent homogeneous films, plastic at ambient temperature	Ebringerova et al. (2005)
MGX	Acetate, DS = 1.2	Thermoproccesable material	Grondahl et al. (2003)
CHX	Grafted copolymer with PAA	Capability of flocculating small solid particles; hydrogels; flocculants	Agarwal et al. (2002), Mishra et al. (2004), Kaith and Kumar (2007)
MGX	Furan-2-carboxylic acid esters	Self-supporting films; membranes for separation processes	Hansen and Plackett (2008)
MGX	Polysulfates	Heparinoid properties, antiulcer, antiangiogenic, anticholesteremic, anticancer, antiviral activities	Ebringerova and Hromadkova (1999), Sun et al. (2004a)

TMAHP, trimethylammonium-2-hydroxypropyl; PAA, polyacrylamide.

relatively low DS values (0.01–0.54) and the reaction was accompanied with significant degradation of the polymers. In a later report (Ren et al., 2008b), the authors showed that TMAHP-AGX derivatives with a low DS (<0.25) can be obtained without significant degradation by performing the synthesis in DMSO media containing NaOH.

Very recently, the effects in sulfate kraft pulp of cationic and carboxymethy-lated xylan derivatives prepared from the sugarcane bagasse AGX were studied in detail by Ren et al. (2009). It was found that cationic (DS = 0.37) and carbox-ymethyl xylans (DS = 0.35) could improve the physical properties of hand sheets, while a 1% dosage of the mixture of both derivatives (1:1, w/w) could enhance sharply their physical properties. The results implied that the both derivatives be utilized as wet-end additives in papermaking industry. Further derivatives for application as wet-end additives were prepared from wheat straw hemicelluloses (AGX type) by reaction with acrylamide under alkaline conditions in a one-step synthesis (Ren et al., 2008c). The novel bifunctional water-soluble xylan deriva-tives showed DS values between 0.12 and 0.58 and contained (under optimum reaction conditions) carboxymethyl and carbamoylethyl groups in ratio of 1:4.8. The products could also be used as flocculants in papermaking and sewage treatment plants.

13.4.1.2 Materials for Miscellaneous Applications

Methyl derivatives with DS up to 1.7 were prepared from wheat straw hemicellulose (AGX type) by the classical methylation method used in linkage analysis, for example, with methyl iodide using sodium hydride as a catalyst in anhydrous dimethyl sulfoxide (DMSO), by Fang et al. (2002). However, this method is not acceptable for large-scale production. Petzold et al. (2008) reported on new synthetic pathways. By methylation of birch wood MGX with methyl chloride under pressure in the presence of 40% aqueous NaOH, a DS of 0.94 was obtained. Using methyl iodide under homogeneous conditions in 25% aqueous NaOH or after addition of acetone leading to a heterogeneous slurry, the DS of the methyl xylans was about 0.5. The water-soluble derivatives showed remarkable surface tension depressing effects that might evoke further research focused on their use as anionic biosurfactants.

Amphoteric, pH-sensitive materials were obtained by reacting birch wood MGX with various 2-chloro-N,N-dialkylethylamine hydrochloride agents under heteroge-neous conditions (Schwikal and Heinze, 2007). The water solubility of the dialky-laminoethyl xylans depended on the pH value, DS, and type of the alkyl chains. The derivatives might be utilized in applications reported for similar starch and cellulose derivatives, for example, as chromatographic materials in protein separation or for immobilizing enzymes.

From the Ispaghula seed husk CHX, an ethyl derivative (EAX) was prepared heterogeneously with ethyl iodide in the presence of aqueous sodium hydroxide (Saghir et al., 2009). The DS values were as high as 0.61. The samples were soluble in DMSO at 80°C. The intrinsic viscosity (η) of the EAX samples depended on the reaction conditions applied. Further studies are planned to investigate the rheological characteristics, thermal stability, and biological activity of the derivatives.

Intense research activities have been focused on the hydrophobization of xylans by esterification (Table 13.3). The reaction of water-soluble beech wood MGX with lauroyl chloride in the DMF/pyridine system with or without dimethylaminopyridine as a catalyst yielded partially lauroylated MGX derivatives (Skalkova et al., 2006a). Due to the hydrophobization effect, the xylan esters lost solubility in water at DS $\sim$ 0.1, which coincided with the degree of branching of the xylan by the glucuronic acid side chains (MeGlcA:Xyl = 0.08) responsible for the hydrophilicity of the derivative. With increasing DS, the derivatives became soluble only in DMSO and at DS > 1.2 in organic solvents. The thermal transitions were DS dependent and the thermal stability increased at DS > 0.5. The lauroylated MGX derivatives might be incorporated into mixtures with other polysaccharides or synthetic polymers and as additives in copolymerization reactions, thus offering possibilities to obtain novel and, eventually, biodegradable composite materials.

Hydrophobization of xylans from various agricultural wastes (sugarcane bagasse, wheat straw, etc.) with acyl halogenides or acid anhydrides has been intensively studied to increase hydrophobicity, film-forming properties, and thermal stability (Heinze et al., 2004). Recently, lauroylated wheat straw AGX with DS ranging between 0.46 and 1.54 were synthesized with lauroyl chloride in various homogeneous media under mild reaction conditions (Peng et al., 2008). The thermal properties of the highest substituted xylan laurate obtained under optimal conditions (40°C, 35 min) were considered suitable for the production of plastic films. A very rapid acylation procedure aimed to prepare plastic films was developed using microwave irradiation as heating source (Xu et al., 2008; Ren et al., 2008d). Under these conditions, the xylan esters achieved the highest DS (1.34–1.63) at 78–85°C in a few minutes. Although the microwave irradiation has been effective in reaction time shortening, it should be noted that it resulted in higher degradation of the polymers than the conventional heating technique.

Acetylation of aspen MGX to various DS was performed by reaction with acetic anhydride in formamide/pyridine (Grondahl et al., 2003). The derivatives up to DS = 0.6 were partially soluble in hot water and with increasing DS became soluble in DMSO or chloroform. Acetylation prevented thermal degradation and at a DS of 1.2 resulted in a glass transition temperature, making it possible to thermoprocess the acetylated xylan. Acetylation of sugarcane bagasse AGX under mild reaction conditions yielded derivatives useful for the production of oil sorption-active materials (Sun et al., 2004b).

Esterification of wheat straw AGX performed with succinic anhydride in the N,N'-dimethylformamide/lithium chloride system yielded monoesters with DS between 0.4 and 1.5 (Sun et al., 2001). Due to the novel introduced carboxyl groups, the succinoylated AGX derivatives were soluble at higher DS. Increasing thermal stability was observed at DS $\geq$ 0.7. Succinoylation in aqueous systems (Sun et al., 2002) yielded water-soluble derivatives with DS up to 0.26, which were proposed for applications as thickening agents and metal ion binders. Water-soluble xylan succinates with DS up to 0.2 prepared from oat spelt AX exhibited remarkable gel-forming properties (Hettrich et al., 2006).

Novel hydrophobic xylan derivatives were prepared from the corn bran CHX by a two-step procedure (Fredon et al., 2002). The xylan was subjected to oxidation with sodium periodate followed by reductive amination with dodecylamine and sodium cyanoborohydride. Both steps were performed in water. The derivatives with DS values of 0.6–1.1 gave transparent and homogeneous plastic films at ambient temperature justified by the assessed glass transition temperature (T_g) values between −30 and 0°C. Reductive amination was recently demonstrated as a tool to access highly engineered xylan and other carbohydrates (Daus et al., 2010). In this paper, xylans of low molar mass were modified at the reducing end with mono- and bifunctional amines. The derivatives were capable of reacting with unmodified xylan and cellodextrins. These head–head linked polysaccharides showed increased molar mass that is essential for certain applications.

A graft copolymer of psyllium mucilage (containing CHX) and polyacrylamide has been synthesized in the presence of nitrogen using ceric ion–nitric acid redox initiator (Agarwal et al., 2002). The obtained Psy-g-PAM derivative showed a significantly higher flocculation efficiency in tannery and domestic wastewater than the pure psyllium mucilage and is, in contrast to synthetic flocculants, biodegradable and less expensive. This graft copolymer was reported to be effective also in flocculation and dye removal from textile wastewater (Mishra et al., 2004). In a further study (Kaith and Kumar, 2007), the psyllium CHX grafted with acrylamide using persulfate–hexamethylenetetramine as the initiator–cross-linker system formed hydrogels that are applicable for the selective absorption of water from different water–oil emulsions.

The research of xylan sulfates (called also pentosan polysulfates) has a long tradition in the field of medicine. They exhibit anticoagulant properties similar to those of naturally occurring sulfated polysaccharides as well as other biological activities such as antiviral and anti-HIV effects (Ebringerova and Hromadkova, 1999). Recently, xylan sulfates with DS between 0.93 and 1.95 were prepared from xylans (X_4 type) obtained by alkaline extraction from the red algae *Scinaia hatei* (Mandal et al., 2010). The derivatives displayed strong anti-HSV activity and a main inhibitory effect on viral entry. Hettrich et al. (2006) reported on xylan sulfates and mixed acetate/sulfate esters prepared from oat spelt AX by homogeneous and quasihomogeneous syntheses. Depending on their DS ranging between 0.70 and 1.97, applications as gels or plastics were suggested, which will be described in more detail in Section 13.4.2.

As mentioned in Section 13.3.1.1, polysaccharides with associated lignin or phenolic acids became attractive as polysaccharide antioxidants. Carbohydrate ester conjugates of various phenolic acids are known to exhibit antioxidant activities in addition to anticarcinogenic and anti-inflammatory effects (Kylli et al., 2008). The presence of the –CH=CH–COOR chain in hydroxycinnamic acids and their derivatives facilitates the radical stabilization, initially formed by dehydrogenative oxidation of phenolic hydroxyl group. The subject of a recent study (Wrigstedt et al., 2010) was the derivatization of oat spelt AX and birch wood MGX by antioxidative hydroxycinnamic acids. Feruloylated xylans with various DS (0.05–0.39) were prepared and their antioxidant activity evaluated using the emulsion lipid oxidation test. Determination of antioxidant activity by methods performed in aqueous media was not possible due to the insufficient solubility of the modified polymers. It was found that the MGX and AX derivatives inhibited lipid oxidation notably better than the native xylans (free of phenolic acids). The ferulic acid esters of MGX were more efficient antioxidants than those of AX, and the sinapic acid ester of AX (DS = 0.09) showed the strongest antioxidant activity.

13.4.2 Xylan Derivatives for Targeted Applications

The research activities on xylan modification for targeted applications have been rather sporadic and the number of scientific studies significantly increased only in the past few years. The research included mainly biocompatible xylan derivatives for applications as hydrogels, selective membranes, films, coatings, nanocomposites, and biosurfactants. Of interest have been also materials based on polymer cross-linking and graft copolymerization of natural polymers with synthetic monomers, which are alternatives of value in biodegradable packaging films (Tharanathan, 2003).

13.4.2.1 Films, Coatings, Hydrogels, and Nanoparticles   A detailed review summarizing the results of past research on sustainable films and coatings made from hemicelluloses and hemicellulose derivatives is given by Hansen and Plackett (2008). Some fundamental studies of films and other materials from modified hemicelluloses of this review are included in Table 13.3 and discussed earlier. The targeted applications (packaging and coating materials for foodstuff and biomedical applications) concerning xylans are briefly summarized in Table 13.4.

Further reports from ongoing research on modified xylans have been noticed in the field of hydrogels and nanoscaled materials. Hettrich and Fanter (2010) thoroughly characterized gels prepared from oat spelt xylan (AX type with less than 1.5% UA) and its derivatives reported in a former paper (Hettrich et al., 2006). Gels/pastes were formed in H_2O_2-containing alkaline medium, even though the xylan was free of phenolic acid cross-linking groups. The hydrophobic and hydrophilic nature of the xylan gels affecting their mechanical properties was varied by introducing different substituents. Hydrophobic gels were prepared from fatty acid esters (DS = 0.05–0.14), anionic gels from xylan sulfates (DS = 0.4–0.15), and carboxymethyl xylan (DS = 0.07) and cationic gels from TMAHP-xylan (DS = 0.11) and 2-hydroxy-3-dimethyllaurylammoniumpropylxylan (DS = 0.13). The gels with dry contents varying between 30% and 36% were characterized by porosity measurements, dynamic vapor sorption, scanning electron microscopy, and dispersion stability. The gels can be utilized in many areas including application in the cosmetic, pharmaceutical, and medical industries.

The possibility to prepare hydrogels by blending beech wood MGX with poly (vinyl alcohol) (PVA) was investigated by Tanodekaew et al. (2006). At first, maleic acid xylan esters (xylan-MA) were prepared. The novel introduced carboxylic groups provided reactive sites for further chemical reactions. The blends of xylan-MA with PVA generated swellable gels under thermal treatment. The contents of both polymer

TABLE 13.4 Xylan Derivatives Tailored for Special Applications

Xylan Type	Derivative/Modification	Properties	Application	References
AX	Surface fluorination	Hydrophobized film	Food packaging films	Grondahl et al. (2006)
AX	Grafted with SA, SM, or fatty acids and 20% glycerol[a]	High WVP using linseed oil	Edible water vapor barrier films	Peroval et al. (2003)
AGX	Acetate, propionate, and butyrate	DS = 1.6–2.3	Not proposed	Buchanan et al. (2003)
ws-AGX	Esters mixed with CA	Low thermal stability	Composite films	
AX	Hydroxypropyl xylan (HPX)	Water soluble at DS = 0.2–0.5	Hydrophobic film	Jain et al. (2001)
MGX, CHX	Acetoxypropyl xylan	Organosoluble	Polystyrene additive	
	Xylan laurates (DS = 1.3 and 1.2)	Elastic deformation up to 80°C	Hydrophobic films	Moine et al. (2004)

SA, stearyl acrylate; SM, stearyl methacrylate; CA; cellulose acetate (DS = 2.47); WVP, water vapor permeability.
[a] Additive.

components affected the cross-linking density and hydrophilicity of the gels, which are responsible for the swelling and strength behaviors of the gels. The response of L929 cells cultured on the xylan-MA/PVA gel suggested that the gel was noncytotoxic and hence has a potential for biomedical application.

The birch wood MGX was esterified with furan-2-carboxylic acid in DMSO containing N,N'-carbonyldiimidazole as activator in order to prepare photocrosslinkable polymers (Hesse et al., 2006). The xylan derivatives had DS of 0.09–0.86. From the most substituted sample, soluble in DMSO, films were cast and their surface roughness and pore sizes were determined by atomic force microscopy and scanning electron microscopy, respectively. Based on the results, the authors suggested that the furan-2-carboxylic acid esters of MGX are well suited to form self-supporting films, which can be applied as membrane materials for different separation processes.

Singh et al. (2008) synthesized a hydrogel from the psyllium husks consisting of CHX (Fischer et al., 2004) and acrylic acid by radiation cross-linking. The free radicals generated on both CHX and polyacrylic acid by gamma irradiation induced cross-linked polymerization and created three-dimensional networks. The polymers thus formed gave hydrogels that were pH sensitive and showed a good degree of swelling in the pH 7.4 buffer. Hence, these hydrogels have potential to deliver the drug in the colon in a controlled manner. Moreover, the hydrogels can act as double potential drug delivery devices because psyllium itself has therapeutic importance for its anticancer action.

Micro- and nanoparticles are currently employed in a wide range of applications such as packaging and drug delivery systems (Rhim and Ng, 2007). Nowadays, the structure formation of polysaccharides on the nanometer scale is a rapidly growing field of research. Similarly as defined functionalized derivatives of cellulose and dextran as well as of commercially produced cellulose esters, modified xylans were recently reported to self-assemble into nanoparticles (Kaya et al., 2009). Different techniques of nanoprecipitation have been elaborated by Heinze and Hornig (2009). The nanoparticles ranged in size from 60 to 600 nm. By a tailored functionalization of the polysaccharides, the nanoparticles are applicable for specific requests. In a further paper (Heinze et al., 2009), the authors presented several reaction pathways for the synthesis of self-assembled polysaccharides mentioned earlier, including xylan. Daus and Heinze (2010) studied spherical nanoparticles with a controlled balance between hydrophobicity and hydrophilicity. For this purpose, the partially hydrolyzed neutral xylan obtained by hydrothermal treatment of beech wood (average MW = 1.915 Da) was esterified homogeneously in DMSO with the anti-inflammatory drug ibuprofen via activation of the carboxylic acid with N,N'-carbonyldiimidazole. Subsequently, the xylan ibuprofen ester was sulfated in order to influence its hydrophobicity. The resulting xylan derivatives self-assemble into spherical nanoparticles with mean diameters ranging from 162 to 472 nm. Preliminary stability measurements indicated that hydrolytic stability decreases with increase in degree of substitution by sulfate groups. The results might be of importance, in general, for drug release from polysaccharide prodrugs.

Recently, birch wood MGX was reported to be useful as additive in the production of highly conducting polypyrrole/cellulose nanocomposite films (Sasso et al., 2010). Nanofibrillar cellulose (NFC) and carboxymethyl cellulose (CMC) were used as binders for the polypyrrole (PPy) particles, which were synthesized by oxidative polymerization in the presence of MGX or MGX mixed with NFC. Freestanding films were obtained by film casting from aqueous dispersions and characterized by morphology, particle size, and mechanical properties. Comparing the dissolved additives (CMC and MGX), smaller and more compact PPy particles were obtained with xylan. These properties were influenced by the additive charge availability and physical state of the polysaccharides.

13.4.2.2 Biosurfactants

Due to environmental regulations, a definite trend toward biomass-derived surfactants—the so-called green surfactant molecules—has been established. Biosurfactants are environment-friendly materials that show compatibility with biological systems (Burczyk, 2002; Piasecki, 2002). Surfactants are amphiphilic substances with structures consisting of a hydrophobic and a hydrophilic part. Next to the low molar mass compounds, water-soluble associative amphiphilic polysaccharides have been investigated as well.

Poly(carboxylic acid) obtained by oxidation of oat spelt xylan and containing blocks of nonoxidized Xyl$p$ residues was reported to behave as a biodegradable detergent (Matsumura et al., 1990). Further biosurfactants were prepared by partial hydrophobization of several commercial polysaccharides and the noncommercial one, the beech wood MGX. Long alkyl chains have been introduced onto the xylan and its sulfoethyl derivative by O-alkylation with lauryl bromide yielding La-GX and La-SEGX derivatives, respectively (Ebringerova et al., 1998). Both derivatives showed a moderate surface tension depressing effect; however, the emulsifying activity of La-GX was comparable with that of the commercial surfactant Tween 20 and higher in comparison to La-SEGX. Both derivatives stabilized protein foams against thermal disruption as effective as the food gum xanthan. Also, MGX containing p-carboxybenzyl substituents (DS $= 0.11$) displayed remarkable emulsifying and protein foam stabilizing activities (Ebringerova et al., 2000).

Very effective biosurfactants were obtained from MGX (differing in the uronic acid content) by transesterification of vinyl laurate under homogeneous and heterogeneous reaction conditions (Skalkova et al., 2006b). The water-soluble xylan laurates exhibited significant surface tension depressing effects and excellent emulsifying activity. They also showed high detergent performance properties tested by the washing power; it reached 76–89% of the values found for the control—sodium dodecyl sulfate. Some of the derivatives possessed acceptable antiredeposition efficiency as well. Due to the low esterification degrees (< 0.1), the derivatives should not suffer from loss of biodegradability. Based on the functional properties of the xylan-based biosurfactants, they were suggested for application as additives in detergent and other industries.

13.5 SUMMARY AND FUTURE PERSPECTIVES

A plenty of plants and plant wastes coming as by-products from industrial and nonindustrial processing are sustainable resources for the production of xylan polymers and oligosaccharides. A huge variety of functionalization reactions were carried out on xylans over the past five decades and a multitude of different applications of isolated xylans and xylan derivatives have already been investigated, published, and patented. During the past two decades, the research of xylans has spread out over the world. The reviewed applications for xylans include a large variety of possibilities such as their use as biomaterials (hydrogels, films, coatings, sorbents), specialty and fine chemicals (tensides, additives for food industry, pharmacy and cosmetics, fillers for drug forms, carrier material of drug delivery systems, immobilized enzymes and for protein transport), novel pharmaceuticals, nutraceuticals, antioxidant dietary fibers, and other health promoting products.

An essential prerequisite for the production and application of xylans and their derivatives on an industrial scale is the characterization of their chemical structure and physicochemical properties, the chemical purity, and the development of reproducible production processes. However, this is a problem when taking into account the diverse structural variations of xylans in different plants and in the tissues within the plant. Moreover, the isolated xylans usually contain some amounts of coextracted cell wall polymers and low molar mass substances that might affect the targeted functional properties of the products. This means that the isolation procedure must be simplified and special separation and purification steps, which are necessary in fundamental research, must be reduced to a minimum.

In view of the data presented, xylans and xylan derivatives will remain important biomacromolecules to be studied extensively in the future. The research areas will concern novel not yet investigated plant sources and plant wastes, polysaccharide isolation and purification technologies with the implementation of novel nonconventional techniques such as the use of ultrasound and microwave irradiation, enzymes, and/or their effective combinations, and also tailored modification processes. These topics are all intended in compliance with the intense trend toward a "green chemistry" and "healthy" products.

Even if some xylan polymers are shown to have environmental characteristics that are preferable to conventional synthetic polymers, it is indispensable to bring down the costs of biologically derived materials. All in all, much effort must be made in the future to compete and overcome the advantage of the low-cost synthetic materials.

ACKNOWLEDGMENTS

This work was financially supported by the Slovak Grant Agency VEGA 2/0062/09 and the Grant SAV-FM-EHP-2008-03-05.

REFERENCES

Agarwal, M., R. Srinivasan, and A. Mishra (2002). Synthesis of *Plantago psyllium* mucilage grafted polyacrylamide and its flocculation efficiency in tannery and domestic wastewater. *J. Polym. Res.* **9**:69–73.

Akpinar, O., K. Erdogan, and S. Bostanci (2009). Production of xylooligosaccharides by controlled acid hydrolysis of lignocellulosic materials. *Carbohydr. Res.* **344**:660–666.

Barbat, A., V. Gloaguen, C. Moine, O. Sainte-Catherine, M. Kraemer, H. Rogniaux, D. Ropartz, and P. Krausz (2008). Structural characterization and cytotoxic properties of a 4-*O*-methylglucuronoxylan from *Castanea sativa*. 2. Evidence of a structure–activity relationship. *J. Nat. Prod.* **71**:404–409.

Belyaev, E. Y. (2000). New medical materials based on modified polysaccharides. *Pharm. Chem. J.* **34**:36–41.

Berlanga-Reyes, C. M., E. Carvajal-Millan, J. Lizardi-Mendoza, A. Rascon-Chu, J. A. Marquez-Escalante, and A. L. Martinez-Lopez (2009). Maize arabinoxylan gels as protein delivery matrices. *Molecules* **14**:1475–1482.

Buchanan, C. M., N. L. Buchanan, J. S. Debenham, P. Gatenholm, M. Jacobsson, M. C. Shelton, T. L. Watterson, and M. D. Wood (2003). Preparation and characterization of arabinoxylan esters and arabinoxylan ester/cellulose ester polymer blends. *Carbohydr. Polym.* **52**:345–357.

Burczyk, B. (2002). Biodegradable and chemodegradable nonionic surfactants. In: *Encyclopedia of Surface and Colloid Science*, New York: Taylor & Francis, pp. 724–752.

Carafa, A., J. G. Duckett, J. P. Knox, and R. Ligrone (2005). Distribution of cell-wall xylans in bryophytes and tracheophytes: new insights into basal interrelationships of land plants. *New Phytologist* **168**:231–240.

Carvajal-Millan, E., V. Landillon, M.-H. Morel, X. Rouau, J.-L. Doublier, and V. Micard (2005). Arabinoxylan gels: impact of the feruloylation degree on their structure and properties. *Biomacromolecules* **6**:309–317.

Carvajal-Millan, E., S. Guilbert, J.-L. Doublier, and V. Micard (2006). Arabinoxylan/protein gels: structural, rheological and controlled release properties. *Food Hydrocolloids* **20**:53–61.

Charalampopoulos, D., R. Wang, S. S. Pandiella, and C. Webb (2002). Application of cereals and cereal components in functional foods: a review. *Int. J. Food Microbiol.* **79**:131–141.

Cipriani, T. R., C. G. Mellinger, L. M. de Souza, C. H. Baggio, C. S. Freitas, M. C. A. Marques, P. A. J. Gorin, G. L. Sassaki, and M. Iacomini (2008). Acidic heteroxylans from medicinal plants and their anti-ulcer activity. *Carbohydr. Polym.* **74**:274–278.

Coviello, T., P. Matricardi, C. Marianecci, and F. Alhaique (2007). Polysaccharide hydrogels for modified release formulations. *J. Control. Release* **119**:5–24.

Dammström, S., L. Salmen, and P. Gatenholm (2005). The effect of moisture on the dynamical mechanical properties of bacterial cellulose/glucuronoxylan nanocomposites. *Polymer* **46**:10364–10371.

Daus, S., and Th. Heinze (2010). Xylan-based nanoparticles: prodrugs for ibuprofen release. *Macromol. Biosci.* **10**:211–220.

Daus, S., Th. Elschner, and Th. Heinze (2010). Towards unnatural xylan based polysaccharides: reductive amination as a tool to access highly engineered carbohydrates. *Cellulose* **17**:825–833.

Dumitriu, S. (2002). Polysaccharides as biomaterials. In: S. Dumitriu,editor, *Polymeric Biomaterials*, New York: Marcel Dekker, pp. 1–61.

Ebringerova, A. (1992). Hemicellulosen als biopolymere Rohstoffe. *Das Papier* **46**:726–732.

Ebringerova, A., and Th. Heinze (2000). Xylan and xylan derivatives—biopolymers with valuable properties. 1. Naturally occurring xylans structures, isolation procedures and properties. *Macromol. Rapid Commun.* **21**:542–556.

Ebringerova, A., and Z. Hromadkova (1999). Xylans of industrial and biomedical importance. *Biotechnol. Genet. Eng. Rev.* **16**:325–346.

Ebringerova, A., I. Srokova, P. Talaba, M. Kacurakova, and H. Hromadkova (1998). Amphiphilic beechwood glucuronoxylan derivatives. *J. Appl. Polym. Sci.* **67**:1523–1530.

Ebringerova, A., J. Alfoldi, Z. Hromadkova, G. M. Pavlov, and S. E. Harding (2000). Water-soluble *p*-carboxybenzylated beechwood 4-*O*-methylglucuronoxylan: structural features and properties. *Carbohydr. Polym.* **42**:123–131.

Ebringerova, A., Z. Hromadkova, and Th. Heinze (2005). Hemicellulose. *Adv. Polym. Sci.* **168**:1–68.

Ebringerova, A., Z. Hromadkova, Z. Kostalova, and V. Sasinkova (2008). Chemical valorization of agricultural by-products: isolation and characterization of xylan-based antioxidants from almond shell biomass. *Bioresources* **3**:60–70.

Fang, J. M., P. Fowler, J. Tomkinson, and C. A. S. Hill (2002). Preparation and characterization of methylated hemicelluloses from wheat straw. *Carbohydr. Polym.* **47**:285–293.

Figueroa-Espinoza, M.-C., M.-H. Morel, and X. Rouau (1998). Effect of lysine, tyrosine, cysteine, and glutathione on the oxidative cross-linking of feruloylated arabinoxylans by a fungal laccase. *J. Agric. Food Chem.* **46**:2583–2589.

Fischer, M. H., N. Yu, G. R. Gray, J. Ralph, L. Anderson, and J. A. Marlett (2004). The gel-forming polysaccharide of psyllium husk (*Plantago ovata* Forsk). *Carbohydr. Res.* **339**:2009–2017.

Fowler, M. (2006). Plants, medicines and man. *J. Sci. Food Agric.* **86**:1797–1804.

Fredon, E., R. Granet, R. Zerrouki, P. Krausz, L. Saulnier, J. F. Thibault, J. Rosier, and C. Petit (2002). Hydrophobic films from maize bran hemicelluloses. *Carbohydr. Polym.* **49**:1–12.

Gabrielii, I., and P. Gatenholm (1998). Preparation and properties of hydrogels based on hemicellulose. *J. Appl. Polym. Sci.* **69**:1661–1667.

Gabrielii, I., P. Gatenholm, W. G. Glasser, R. K. Jain, and L. Kenne (2000). Separation, characterization and hydrogel-formation of hemicellulose from aspen wood. *Carbohydr. Polym.* **43**:367–374.

Garcia-Conesa, M. T., G. W. Plumb, K. W. Waldron, J. Ralph, and G. Williamson (1997). Ferulic acid dehydrodimers from wheat bran: isolation, purification and antioxidant properties of 8-*O*-4-diferulic acid. *Redox Rep.* **3**:319–323.

Ghoneum, M., and A. Jewett (2000). Production of tumor necrosis factor-alpha and interferon-gamma from human peripheral blood lymphocytes by MGN-3, a modified arabinoxylan from rice bran, and its synergy with interleukin-2 *in vitro*. *Cancer Detect. Prev.* **24**:314–324.

Goksu, E. I., M. Karamanlioglu, V. Bakir, L. Yilmaz, and V. Yilmazer (2007). Preparation and characterization of films from cotton stalk xylan. *J. Agric. Food Chem.* **55**:10685–10691.

Grondahl, M., and P. Gatenholm (2007). Oxygen barrier films based on xylans isolated from biomass. *ACS Symp. Ser.* **954**:137–152.

Grondahl, M., A. Teleman, and P. Gatenholm (2003). Effect of acetylation on the material properties of glucuronoxylan from aspen wood. *Carbohydr. Polym.* **52**:359–366.

Grondahl, M, L. Eriksson, and P. Gatenholm (2004). Material properties of plasticized hardwood xylans for potential application as oxygen barrier films. *Biomacromolecules* **5**:1528–1535.

Grondahl, M., A. Gustafsson, and P. Gatenholm (2006). Gas-phase surface fluorination of arabinoxylan films. *Macromolecules* **39**:2718–2721.

Guilloux, K., I. G. Gaillard, J. Courtois, B. Courtois, and E. Petit (2009). Production of arabinoxylan-oligosaccharides from flaxseed (*Linum usitatissimum*). *J. Agric. Food Chem.* **57**:11308–11313.

Gullon, P., M. J. Gonzalez-Munoz, M. P. van Gool, H. A. Schols, J. Hirsch, A. Ebringerova, and J. C. Parajo (2010). Production, refining, structural characterization and fermentability of rice husk xylooligosaccharides. *J. Agric. Food Chem.* **58**:3632–3641.

Guo, Q., S. W. Cui, Q. Wang, and J. C. Young (2008). Fractionation and physicochemical characterization of psyllium gum. *Carbohydr. Polym.* **73**:35–43.

Hansen, N. M. L., and D. Plackett (2008). Sustainable films and coatings from hemicelluloses: a review. *Biomacromolecules* **9**:1493–1505.

Haque, A., E. R. Morris, and R. K. Richardson (1994). Polysaccharide substitutes for gluten in non-wheat bread. *Carbohydr. Polym.* **25**:337–344.

Hasler, C. M. (2001). Functional foods. In: B. A. Bowman and R. M. Russell,editors, *Present Knowledge in Nutrition*, Washington, DC: ILSI Press, pp. 740–749.

Heinze Th. (2004). Chemical functionalization of cellulose. In: S. Dumitriu,editor, *Polysaccharide: Structural Diversity and Functional Versatility*, 2nd ed., New York: Marcel Dekker, pp. 551–590.

Heinze Th., and S. Hornig (2009). Versatile concept for the structure design of polysaccharide-based nanoparticles. *ACS Symp. Ser.* **1017**:169–183.

Heinze, U., V. Haack, and Th.Heinze (2001). New highly functionalized starch derivatives. In: E. Chiellini, H. Gil, G. Braunegg, J. Buchert, P. Gatenholm, and M. van der Zee, editors, *Biorelated Polymers: Sustainable Polymer Science and Technology*, Berlin: Springer, pp. 205–218.

Heinze, Th., A. Koschella, and A. Ebringerova (2004). Chemical functionalization of xylan: a short review. In: P. Gatenholm and M. Tenkanen,editors, *Hemicellulose: Science and Technology*, ACS Symposium Series, Vol. **864**, Washington, DC: American Chemical Society, pp. 312–325.

Heinze Th., S. Hornig, N. Michaelis, and K. Schwikal (2009). Polysaccharide derivatives for the modification of surfaces by self-assembly. *ACS Symp. Ser.* **1019**:195–221.

Hesse, S., T. Liebert, and Th. Heinze (2006). Studies on the film formation of polysaccharide based furan-2-carboxylic acid esters. *Macromol. Symp.* **232**:57–67.

Hettrich, K., and C. Fanter (2010). Novel xylan gels prepared from oat spelts. *Macromol. Symp.* **294**:141–150.

Hettrich, K., S. Fischer, N. Schroder, J. Engelhardt, U. Drechsler, and F. Loth (2006). Derivatization and characterization of xylan from oat spelts. *Macromol. Symp.* **232**:37–48.

Hoije, A., M. Grondahl, K. Tommeraas, and P. Gatenholm (2005). Isolation and characterization of physicochemical and material properties of arabinoxylans from barley husks. *Carbohydr. Polym.* **61**:266–275.

Hooper, L., and A. Cassidy (2006). A review of the health care potential of bioactive compounds. *J. Sci. Food Agric.* **86**:1805–1813.

Hromadkova, Z., A. Ebringerova, and J. Hirsch (2005). An immunomodulatory xylan–phenolic complex from the seed hulls of buckwheat (*Fagopyrum esculentum* Moench). *Chem. Pap.* **59**:223–224.

Hromadkova, Z., A. Stavova, A. Ebringerova, and J. Hirsch (2007). Effect of buckwheat hull hemicelluloses addition on the bread-making quality of wheat flour. *J. Food Nutr. Res.* **46**:158–166.

Iiyama, K., T. B.-T. Lam, and B. A. Stone (1994). Covalent cross-links in the cell wall. *Plant Physiol.* **104**:315–320.

Izydorczyk, M. (2005). *Food Carbohydrates: Chemistry, Physical Properties and Applications*, New York: Taylor & Francis.

Jain, N., S. Gupta, and S. B. Babbar (1997). Isabgol as an alternative gelling agent for microbial culture media. *J. Plant Biochem. Biotechnol.* **6**:129–131.

Jain, R. K., M. Sjostede, and W. G. Glasser (2001). Thermoplastic xylan derivatives with propylene oxide. *Cellulose* **7**:319–336.

Kaith, B. S., and K. Kumar (2007). In air synthesis of Psy-cl-poly(AAm) network and its application in water-absorption from oil–water emulsions. *eXPRESS Polym. Lett.* **1**:474–480.

Kardosova, A., and E. Machova (2006). Antioxidant activity of medicinal plant polysaccharides. *Fitoterapia* **77**:367–373.

Katapodis, P., A. Kavarnou, S. Kintzios, E. Pistola, D. Kekos, B. J. Macris, and P. Christakopoulos (2002). Production of acidic xylo-oligosaccharides by a family 10 endoxylanase from *Thermoascus aurantiacus* and use as plant growth regulators. *Biotechnol. Lett.* **24**:1413–1416.

Katapodis, P., M. Vardakou, E. Kalogeris, D. Kekos, B. J. Macris, and P. Christakopoulos (2003). Enzymic production of a feruloylated oligosaccharide with antioxidant activity from wheat flour arabinoxylan. *Eur. J. Nutr.* **42**:55–60.

Kaya, A., D. A. Drazenovich, W. G. Glasser, Th. Heinze, and A. R. Esker (2009). Hydroxypropyl xylan self-assembly at air/water and water/cellulose interfaces. *ACS Symp. Ser.* **1019**:173–191.

Kayserilioglu, B. S., U. Bakir, L. Yilmaz, and N. Akkas (2003). Use of xylan, an agricultural by-product, in wheat gluten based biodegradable films: mechanical, solubility and water vapor transfer rate properties. *Bioresour. Technol.* **87**:239–246.

Kendall, C. W. C., A. Esfahani, and D. J. A. Jenkins (2010). The link between dietary fibre and human health. *Food Hydrocolloids* **24**:42–48.

Kritchevsky, D., C. Bonfield, and J. W. Anderson (1990). *Dietary Fibre: Chemistry, Physiology and Health Effects*, New York: Plenum.

Kylli, P., P. Nousiainen, P. Biely, J. Sipila, M. M. Tenkanen, and M. Heinonen (2008). Antioxidant potential of hydroxycinnamic acid glycoside esters. *J. Agric. Food Chem.* **56**:4797–4805.

Lindblad, M. S., J. Sjoberg, A.-C. Albertsson, and J. Hartman (2007). Hydrogels from polysaccharides for biomedical applications. *ACS Symp. Ser.* **954**:153–167.

Lloyd, L. L., J. F. Kennedy, P. P. Methacanon, M. Paterson, and C. J. Knill (1998). Carbohydrate polymers as wound management aids. *Carbohydr. Polym.* **37**:315–322.

Mandal, P., C. A. Pujo, E. B. Damonte, T. Ghosh, and B. Ray (2010). Xylans from *Scinaia hatei*: structural features, sulfation and anti-HSV activity. *Int. J. Biol. Macromol.* **46**:173–178.

Matsumura, S., S. Maeda, and S. Yoshikawa (1990). Molecular design of biodegradable functional polymers. 2. Poly(carboxylic acid) containing xylopyranosediyl groups in the backbone. *Makromol. Chem.* **191**:1269–1275.

Maurer-Menestrina, J., G. L. Sassaki, F. F. Simas, P. A. J. Gorin, and M. Iacomini (2003). Structure of a highly substituted β-xylan of the gum exudate of the palm *Livistona chinensis* (Chinese fan). *Carbohydr. Res.* **338**:1843–1850.

Mikkonen, K. S., M. P. Yadav, P. Cooke, S. Willfor, K. B. Hicks, and M. Tenkanen (2008). Films from spruce galactoglucomannan blended with PVA, corn arabinoxylan, and konjac glucomannan. *Bioresources* **3**:178–191.

Miraftab, M., Q. Qiao, J. F. Kennedy, S. C. Anand, and G. J. Collyer (2001). Advanced materials for wound dressings. In: S. C. Anand,editor, *Biofunctional Mixed Carbohydrate Polymers*, Cambridge: Woodhead, pp. 164–172.

Mishra, A., and M. Bajpai (2002). Flocculation of textile wastewater by *Plantago psyllium* mucilage. *Macromol. Mater. Eng.* **287**:592–596.

Mishra, A., and M. Bajpai, (2005). Flocculation behaviour of model textile wastewater treated with a food grade polysaccharide. *J. Hazard. Mater.* **118**:213–217.

Mishra, A., A. Mishra, M. Agarwal, M. Bajpai, S. Rajani, and R. P. Mishra (2002). *Plantago psyllium* mucilage for sewage and tannery effluent treatment. *Iran. Polym. J.* **11**:381–386.

Mishra, A., R. Srinivasan, M. Bajpai, and R. Dubey (2004). Use of polyacrylamide-grafted *Plantago psyllium* mucilage as a flocculant for treatment of textile wastewater. *Colloid Polym. Sci.* **282**:722–727.

Moine, C., V. Gloaguen, J. M. Gloaguen, R. Granet, and P. Krausz (2004). Chemical valorization of forest and agricultural by-products. Obtention, chemical characteristics, and mechanical behavior of a novel family of hydrophobic films. *J. Environ. Sci. Health B* **39**:627–640.

Moure, A., P. Gullon, H. Dominguez, and J. C. Parajo (2006). Advances in the manufacture, purification and applications of xylo-oligosaccharides as food additives and nutraceuticals. *Process Biochem.* **41**:1913–1923.

Nabarlatz, D., A. Ebringerova, and D. Montane (2007a). Autohydrolysis of agricultural by-products for the production of xylo-oligosaccharides. *Carbohydr. Polym.* **69**:20–28.

Nabarlatz, D., D. Montane, A. Kardosova, S. Bekesova, V. Hribalova, and A. Ebringerova (2007b). Almond shell xylo-oligosaccharides exhibiting immunostimulatory activity. *Carbohydr. Res.* **342**:1122–1128.

Ng, A., R. N. Greenshields, and K. W. Waldron (1997). Oxidative cross-linking of corn bran hemicellulose: formation of ferulic acid dehydrodimers. *Carbohydr. Res.* **303**:459–462.

Nino-Medina, G., E. Carvajal-Millan, A. Rascon-Chu, J. A. Marquez-Escalante, V. Guerrero, and E. Salas-Munoz (2010). Feruloylated arabinoxylans and arabinoxylan gels: structure, sources and applications. *Phytochem. Rev.* **9**:111–120.

Nosalova, G., M. Sutovska, J. Mokry, A. Kardosova, P. Capek, and T. H. M. Khan (2005). Efficacy of herbal substances according to cough reflex. *Minerva Biotechnol.* **17**:141–152.

Ogawa, K., M. Takeuchi, and N. Nakamura (2005). Immunological effects of partially hydrolyzed arabinoxylan from corn husk in mice. *Biosci. Biotechnol. Biochem.* **69**:19–25.

Ohta, T., S. Yamasaki, Y. Egashira, and H. Sanada (1984). Antioxidative activity of corn bran hemicellulose fragments. *J. Agric. Food Chem.* **42**:653–660.

Park, H., P. A. Seib, and O. K. Chung (1997). Fortifying bread with a mixture of wheat fiber and psyllium husk fiber plus three antioxidants. *Cereal Chem.* **74**:207–211.

Patel, T. R., S. E. Harding, A. Ebringerova, M. Deszczynski, Z. Hromadkova, A. Togola, B. S. Paulsen, G. A. Morris, and A. J. Rowe (2007). Weak self-association in a carbohydrate system. *Biophys. J.* **93**:741–749.

Paulsen, B. S. (2002). Biologically active polysaccharides as possible lead compounds. *Phytochem. Rev.* **1**:379–387.

Peng, F., J.-L. Ren, B. Peng, F. Xu, R. C. Sun, and J.-X. Sun (2008). Rapid homogeneous lauroylation of wheat straw hemicelluloses under mild conditions. *Carbohydr. Res.* **343**:2956–2962.

Peroval, C., F. Debefaurt, D. Despre, and A. Voilley (2002). Edible arabinoxylan films. 1. Effects of lipid type on water vapor permeability, film structure, and other physical characteristics. *J. Agric. Food Chem.* **50**:3977–3983.

Peroval, C., F. Debeaufort, A. M. Seuvre, B. Chevet, D. Despre, and A. Voilley (2003). Modified arabinoxylan-based films. Part B. Grafting of omega-3 fatty acids by oxygen plasma and electrobeam irradiation. *J. Agric. Food Chem.* **51**:3120–3126.

Petzold, K., K. Schwikal, and Th. Heinze (2006). Carboxymethyl xylan: synthesis and detailed structure characterization. *Carbohydr. Polym.* **64**:292–298.

Petzold, K., W. Günther, M. Kötteritzsch, and Th. Heinze (2008). Synthesis and characterization of methyl xylan. *Carbohydr. Polym.* **74**:327–332.

Phan The, D., C. Peroval, F. Debeaufort, D. Despre, J. L. Courthaudon, and A. Voilley (2002a). Arabinoxylan-lipids-based edible films and coatings. 2. Influence of sucroester nature on the emulsion structure and film properties. *J. Agric. Food Chem.* **50**:266–272.

Phan The, D., C. Peroval, F. Debeaufort, D. Despre, J. L. Courthaudon, and A. Voilley (2002b). Arabinoxylan-lipids-based edible films and coatings. 3. Influence of drying temperature on film structure and functional properties. *J. Agric. Food Chem.* **50**:2423–2428.

Piasecki, A. (2002). Biodegradable and chemically degradable anionic surfactants. In: *Encyclopedia of Surface and Colloid Science*, New York: Taylor & Francis, pp. 701–723.

Rao, R. S. P., and G. Muralikrishna (2006). Water soluble feruloyl arabinoxylans from rice and ragi: changes upon malting and their consequence on antioxidant activity. *Phytochemistry* **67**:91–99.

Ren, J. L., R. C. Sun, C. F. Liu, Z. Y. Chao, and W. Luo (2006). Two-step preparation and thermal characterization of cationic 2-hydroxypropyltrimethylammonium chloride hemicellulose polymers from sugarcane bagasse. *Polym. Degrad. Stabil.* **91**:2579–2587.

Ren, J. L., R. C. Sun, and F. Peng (2008a). Carboxymethylation of hemicelluloses isolated from sugarcane bagasse. *Polym. Degrad. Stabil.* **93**:786–793.

Ren, J. L., F. Peng, R. C. Sun, C. F. Liu, Z. N. Cao, W. Luo, and J. N. Tang (2008b). Synthesis of cationic hemicellulosic derivatives with a low degree of substitution in dimethyl sulfoxide media. *J. Appl. Polym. Sci.* **109**:2711–2717.

Ren, J. L., F. Peng, and R. C. Sun (2008c). Preparation and characterization of hemicellulosic derivatives containing carbamoylethyl and carboxyethyl groups. *Carbohydr. Res.* **343**:2776–2782.

Ren, J. L., F. Xu, R. C. Sun, B. Peng, and J. X. Sun (2008d). Studies of the lauroylation of wheat straw hemicelluloses under microwave heating. *J. Agric. Food Chem.* **56**:1251–1258.

Ren, J. L., F. Peng, E. C. Sun, and J. F. Kennedy (2009). Influence of hemicellulosic derivatives on the sulfate kraft pulp strength. *Carbohydr. Polym.* **75**:338–342.

Rhim, J. W., and P. K. W. Ng (2007). Natural biopolymer-based nanocomposite films for packaging applications. *Crit. Rev. Food Sci.* **47**:411–433.

Rose, D. J., and G. E. Inglett (2010). Production of feruloylated arabinoxylo-oligosaccharides from maize (*Zea mays*) bran by microwave-assisted autohydrolysis. *J. Food Chem.* **119**:1613–1618.

Saghir, S., M. S. Iqbal, M. A. Hussain, A. Koschella, and Th. Heinze (2008). Structure characterization and carboxymethylation of arabinoxylan isolated from Ispaghula (*Plantago ovata*) seed husk. *Carbohydr. Polym.* **74**:309–317.

Saghir, S., M. S. Iqbal, A. Koschella, and Th. Heinze (2009). Ethylation of arabinoxylan from Ispaghula (*Plantago ovata*) seed husk. *Carbohydr. Polym.* **77**:125–130.

Sandhu, J., S., G. J. Hudson, and J. F. Kennedy (1981). The gel nature and structure of the carbohydrate of Ispaghula husk *ex Plantago ovata* FORSK. *Carbohydr. Res.* **93**:247–259.

Sasso, C., E. Zeno, M. Petit-Conil, D. Chaussy, M. N. Belgacem, S. Tapin-Lingua, and D. Beneventi (2010). Highly conducting polypyrrole/cellulose nanocomposite films with enhanced mechanical properties. *Macromol. Mater. Eng.* doi: .

Saura-Calixto, F. (1998). Antioxidant dietary fiber product: a new concept and a potential food ingredient. *J. Agric. Food Chem.* **46**:4303–4306.

Saxena, A., and A. J. Ragauskas (2009). Water transmission barrier properties of biodegradable films based on cellulosic whiskers and xylan. *Carbohydr. Polym.* **78**:357–360.

Saxena, A., T. J. Elder, S. Pan, and A. J. Ragauskas (2009). Novel nanocellulosic xylan composite film. *Compos. Part B* **40**:727–730.

Schwikal, K., and Th. Heinze (2007). Dialkylaminoethyl xylans: polysaccharide ethers with pH-sensitive solubility. *Polym. Bull.* **59**:161–167.

Schwikal, K., Th. Heinze, A. Ebringerova, and K. Petzold (2006). Cationic xylan derivatives with high degree of functionalization. *Macromol. Symp.* **232**:49–56.

Silva, A. K. A., E. L. da Silva, E. E. Oliveira, T. Nagashima Jr., L. A. L. Soares, A. C. Medeiros, J. H. Araujo, I. B. Araujo, A. S. Carric, and E. S. T. Egito (2007). Synthesis and characterization of xylan-coated magnetite microparticles. *Int. J. Pharm.* **334**:42–47.

Simas, F. F., P. A. J. Gorin, M. Guerrini, A. Naggi, G. L. Sassaki, and C. L. Delgobo (2004). Structure of heteroxylan of gum exudate of the palm *Scheelea phalerata* (uricuri). *Phytochemistry* **65**:2347–2355.

Sims, I. A., and R. H. Newman (2006). Structural studies of acidic xylans exuded from leaves of the monocotyledonous plants *Phormium tenax* and *Phormium cookianum*. *Carbohydr. Polym.* **63**:379–384.

Singh, B., N. Chauhan, and S. Kumar (2008). Radiation crosslinked psyllium and polyacrylic acid based hydrogels for use in colon specific drug delivery. *Carbohydr. Polym.* **73**:446–455.

Skalkova, P., I. Srokova, A. Ebringerova, K. Csomorva, and I. Janigova (2006a). Thermostable polymers prepared from glucuronoxylan. *Cellulose Chem. Technol.* **40**:525–530.

Skalkova, P., I. Srokova, V. Sasinkova, and A. Ebringerova (2006b). Polymeric surfactants from beechwood glucuronoxylan. *Tenside Surf. Deterg.* **43**:137–141.

Sternemalm, E., A. Hoije, and P. Gatenholm (2008). Effect of arabinose substitution on the material properties of arabinoxylan films. *Carbohydr. Res.* **343**:753–757.

Stipanovic, A. J., T. E. Amidon, G. M. Scott, V. Barber, and M. K. Blowers (2006). Hemicellulose from biodelignified wood: a feedstock for renewable materials and chemicals. *ACS Symp. Ser.* **921**:210–221.

Stscherbina, D., and B. Phillip (1991). Neuere Ergebnisse zur Isolierung, Modifizierung und Anwendung von Xylanen (Literaturbericht). *Acta Polym.* **42**:345–351.

Sun, S. C., X. F. Sun, and F. Y. Zhang (2001). Succinoylation of wheat straw hemicelluloses in *N,N'*-dimethylformamide/lithium chloride systems. *Polym. Int.* **50**:803–811.

Sun, R. C., X. F. Sun, and X. Bing (2002). Succinoylation of wheat straw hemicelluloses with a low degree of substitution in aqueous systems. *J. Appl. Polym. Sci.* **83**:757–766.

Sun, R. C., X. F. Sun, and J. Tomkinson (2004a). Hemicelluloses and their derivatives. *ACS Symp. Ser.* **864**:2–22.

Sun, X. F., R. C. Sun, and J. X. Sun (2004b). Acetylation of sugarcane bagasse using NBS as a catalyst under mild reaction conditions for the production of oil sorption-active materials. *Bioresour. Technol.* **95**:343–350.

Tanodekaew, S., S. Channasanon, and P. Uppanan (2006). Xylan/polyvinyl alcohol blend and its performance as hydrogel. *J. Appl. Polym. Sci.* **100**:1914–1918.

Tharanathan, R. N. (2003). Biodegradable films and composite coatings: past, present and future. *Trends Food Sci. Technol.* **14**:71–78.

Thomsen, M. H., A. Thygesen, and A. B. Thomsen (2008). Hydrothermal treatment of wheat straw at pilot plant scale using a three-step reactor system aiming at high hemicellulose recovery, high cellulose digestibility and low lignin hydrolysis. *Bioresour. Technol.* **99**:4221–4228.

Unlu, C. H., E. Gunister, and O. Atici (2009). Synthesis and characterization of NaMt biocomposites with corn cob xylan in aqueous media. *Carbohydr. Polym.* **76**:585–592.

Vaya, J., and M. Aviram (2001). Nutritional antioxidants: mechanisms of action, analyses of activities and medical applications. *Curr. Med. Chem. Immunol. Endocr. Metab. Agents* **1**:99–117.

Vazquez, M. J., J. L. Alonso, H. Dominguez, and J. C. Parajo (2006). Enhancing the potential of oligosaccharides from corncob autohydrolysis as prebiotic food ingredients. *Ind. Crops Prod.* **24**:152–159.

Viota, J. L., M. Lopez-Viota, B. Saake, K. Stana-Kleinschek, and A. V. Delgado (2010). Organoclay particles as reinforcing agents in polysaccharide films. *J. Colloid Interface Sci.* **347**:74–78.

Waldron, K. W., M. L. Parker, and A. C. Smith (2003). Plant cell walls and food quality. *Compr. Rev. Food Sci. F* **2**:128–146.

Wang, L., H. Zhang, X. Zhang, and Z. Chen (2008). Purification and identification of a novel heteropolysaccharide RBPS2a with anti-complementary activity from defatted rice bran. *Food Chem.* **110**:150–155.

Whistler, R. L. (1993. Hemicelluloses. In: R. L. Whistler and J. N. BeMiller, editors, *Industrial Gums, Polysaccharides and Their Derivatives*. Orlando, FL: Academic Press, pp. 295–308.

Wrigstedt, P., P. Kylli, L. Pitkanen, P. Nousiainen, M. Tenkanen, and J. Sipil (2010). Synthesis and antioxidant activity of hydroxycinnamic acid xylan esters. *J. Agric. Food Chem.* **58**:6937–6943.

Wyman, C. E., B. E. Dale, R. T. Elander, M. Holtzapple, M. R. Ladisch, and Y. Y. Lee (2005). Comparative sugar recovery data from laboratory scale application of leading pretreatment technologies to corn stover. *Bioresour. Technol.* **96**:2026–2032.

Xu, F., J. X. Jiang, R. C. Sun, D. She, B. Peng, J. X. Sun, and J. F. Kennedy (2008). Rapid esterification of wheat straw hemicelluloses induced by microwave irradiation. *Carbohydr. Polym.* **73**:612–630.

Yadav, M. P., D. B. Johnston, and K. B. Hicks (2007a). Structural characterization of corn fiber gums from coarse and fine fiber and a study of their emulsifying properties. *J. Agric. Food Chem.* **55**:6366–6371.

Yadav, M. P., M. L. Fishman, H. K. Chau, D. B. Johnston, and K. B. Hicks (2007b). Molecular characteristics of corn fiber gum and their influence on CPG emulsifying properties. *Cereal Chem.* **84**:175–180.

Yadav, M. P., D. B. Johnston, A. Y. Hotchkiss Jr., and K. B. Hicks (2007c). Corn fiber gum: a potential gum arabic replacer for beverage flavor emulsification. *Food Hydrocolloids* **24**:1022–1030.

Yang, S., Z. Xu, W. Wang, and W. Yang (2005). Aqueous extraction of corncob xylan and production of xylooligosaccharides. *LWT Food Sci. Technol.* **38**:677–682.

Yuan, X., J. Wang, and H. Yao (2005a). Feruloyl oligosaccharides stimulate the growth of *Bifidobacterium bifidum. Anaerobe* **11**:225–229.

Yuan, X., J. Wang, and H. Yao (2005b). Antioxidant activity of feruloylated oligosaccharides from wheat bran. *Food Chem.* **90**:759–764.

Zhang, P., and R. L. Whistler (2004). Mechanical properties and water vapor permeability of thin film from corn hull arabinoxylan. *J. Appl. Polym. Sci.* **93**:2896–2902.

Zhao, Z., Y. Egashira, and H. Sanada (2003). Digestion and absorption of ferulic acid sugar esters in rat gastrointestinal tract. *J. Agric. Food Chem.* **51**:5534–5539.

14

MICRO- AND NANOPARTICLES FROM HEMICELLULOSES

EMMERICH HAIMER, FALK LIEBNER, ANTJE POTTHAST, AND THOMAS ROSENAU

14.1 INTRODUCTION

Hemicelluloses represent an immense natural and renewable resource of biopolymers, but still there are rather few processes that utilize the potential of these molecules. Apart from using hemicelluloses as a basis for chemicals, their utilization for the production of novel designed materials is challenging. The building blocks of hemicelluloses (plant polyoses) are glucose, xylose, mannose, galactose, arabinose, and rhamnose which are specifically combined according to the different plant species (e.g., xyloglucans for hardwood and straw, glucomannans for soft wood).

Industrial or technical supply of pure hemicellulose fractions is still a very demanding and laborious process. Current wood pulping, such as sulfate pulping with its prehydrolyzates or from TMP/groundwood, generate fractions of hemicellulose that are still "contaminated" with cellulose and/or lignin residues. Therefore, many applications are limited to a so-called low-tech sector, such as energy regeneration, production of feed yeast, adhesives for chipboards, or fermentation to alcohol (Schliephake et al., 1986).

The accessibility of hemicelluloses is dependent on the quality of the pretreatment process (Taherzadeh and Karimi, 2008). Available potential sources for almost pure hemicellulose fractions are (Kamm et al., 2006) as follows:

Polysaccharide Building Blocks: A Sustainable Approach to the Development of Renewable Biomaterials, First Edition. Edited by Youssef Habibi and Lucian A. Lucia.
© 2012 John Wiley & Sons, Inc. Published 2012 by John Wiley & Sons, Inc.

- prehydrolysis (acid hydrolysis) (Brennan et al., 1986; Hinman et al., 1992);
- precipitation component/solution component (Organosolv process) (Pan et al., 2007; Rodríguez and Jiménez, 2008);
- steam-/pressure-processes in water (Bousaid et al., 1999; Leppänen et al., 2010)
- aqueous extraction at higher temperature (Kim and Mazza, 2006; Zhao et al., 2009);
- solvent extraction (at higher temperature) (Ebringerová and Hromádková, 2010; Zhang et al., 2010).

In this chapter, an overview on possible applications for polymeric hemicelluloses is given. Especially particulate hemicellulose materials are of increasing interest, as such particles can be used for many different applications. We first introduce the concept of micro- and nanoparticles and the principles of obtaining them. After that the preparation routes for hemicellulosic micro- and nanoparticles are described and examples for their application are given.

14.2 BACKGROUND

14.2.1 Why Small Scale Particles and How to Get Them?

Materials in micro- and nanometer size often exhibit outstanding properties which cannot be found in the bulk form of the respective substance. Especially for particles smaller than 100 nm, the physical properties may vary extremely from the bulk material. Due to this change in properties, nanomaterials offer a broad range of potential applications from electronics, optics, material sciences to application in biological systems. Research and development in the field of inorganic nanotechnology has become a big topic in science during the past years as these materials are extremely interesting for electronics and material sciences. Another field of research which is highly interested in micro- and nanoparticles is the field of pharmaceutics; especially drug delivery systems based on small particles gain more and more interest (Kohane, 2007).

The major effect of small particle sizes on particulate drug delivery systems is the disproportionality of the surface area-to-volume ratio to the particle radius. By decreasing the particle radius it is thus possible to decrease drug payload as the amount of drug in contact with the surrounding release medium is increased. Diffusive mass transfer and in principle all mass transfer-dependent release kinetics are accelerated in small particles. If the particle is optimized to bind to specific receptors, small particles will have a better binding behavior. Thus, it can often be beneficial to decrease the particle radius down to micro- or nanoscale. The major challenge is how such small particle radii can be obtained.

In principle, there exist two strategies towards micro- and nanoparticles. The first one, the top-down strategy, can be realized by milling or attrition, repeated quenching or lithography (Guozhong, 2004). Most of these processes have significant disadvantages, such as broad particle size distributions, difficult process design and control, or impurities, for example, from the milling medium. Thus, the second strategy, the bottom-up approach, has become far more popular for the synthesis of very small particles. For hemicelluloses the solution properties of the respective polymer vary strongly. Thus, different routes of solving and supersaturation in order to obtain small-scale particles in bottom-up approach have to be applied for different hemicellulose molecules. Different routes are given later in this chapter. In the following let us consider some theoretical aspects of this bottom-up approach.

According to Guozhong, bottom-up approaches can be grouped into two categories, namely thermodynamic equilibrium systems and kinetic systems. The kinetic systems are characterized by a limiting factor, such as a defined amount of precursors available, while the thermodynamic equilibrium systems are characterized by the well-known three process-steps: (1) generation of supersaturation, (2) nucleation, and (3) subsequent particle growth. Figure 14.1 illustrates the three process-steps of the thermodynamic equilibrium approach.

During the initial phase of the process, the solute concentration increases and even at concentrations higher than the equilibrium concentration of the respective substance no nucleation occurs. After reaching the concentration which is necessary for nucleation, the so-called labile zone, particle nuclei develop which immediately start to grow. By forming a solid phase, the solute concentration decreases below the nucleation level resulting in further growth of the existing nuclei as long as the solute concentration is beyond the solute equilibrium concentration within the so-called stable zone.

Both the rate constants for nucleation and the rate constant for particle growth correlate exponentially with solute supersaturation as is indicated in Figure 14.2. Within the stable zone the nucleation rate is almost zero while the growth rate is

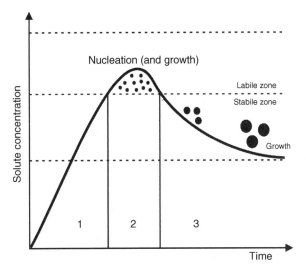

FIGURE 14.1 Three process-steps of the thermodynamic equilibrium approach (Haruta and Delmon, 1986).

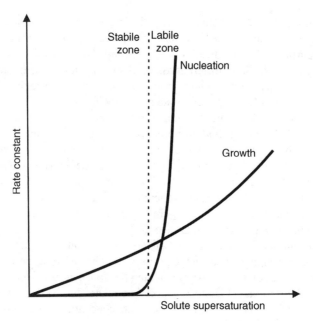

FIGURE 14.2 Dependence of rate constants for nucleation and particle growth in dependence of solute-supersaturation (Grassmann, 1967).

significantly larger than zero. As already mentioned, no particle nuclei develop within this zone, but existing nuclei tend to grow until supersaturation reaches zero. Supersaturation within the stabile zone can be used for kinetic particle generation systems. For thermodynamic equilibrium systems, supersaturation must reach the labile zone.

Immediately after reaching the labile zone the nucleation rate exceeds the growth rate strongly so that a huge amount of substance is consumed by the formation of nuclei at higher supersaturation values. Precipitation at high supersaturation values thus leads to formation of very small particles as due to the high nucleation rate the vast amount of solute is consumed for particle nucleation. Preparation of micro- and nanoparticles should be performed at such high supersaturation values.

Additionally to nucleation and particle growth, agglomeration or particle breakup and fragmentation may affect the morphology and particle size distribution of the precipitate. These effects strongly depend on the nature of the solute and may be either significant or insignificant.

The question now is how this supersaturation can be reached? Apart from traditional crystallization techniques such as cooling crystallization, evaporation crystallization, and evaporative-cooling crystallization more recent techniques can be applied for the preparation of micro- and nanoparticles. These techniques comprise salting-out or antisolvent crystallization, reactive crystallization, and crystallization by the means of supercritical fluids (Kirk and Othmer, 2008). Additionally lithographic methods have gained significant interest over the past years (Rabanel et al., 2009).

14.2.2 Why Hemicelluloses?

Within the field of drug delivery, polysaccharides as incorporation matrices have a long tradition (Sunamoto and Iwamoto, 1986; Schacht et al., 1990). Especially cyclodextrins which are used to entrap fragrances and cosmetic actives and starch-based systems are widely applied not only in delivery systems technology (Rosen, 2006). Also hemicelluloses have been considered to be potential polymeric carriers in the pharmaceutical sector for many years (Juslin and Paronen, 1984; Iwamoto et al., 1991; Baveja et al., 1991).

Hemicelluloses are abundant as they are reported to be the second most abundant biopolymer in the plant kingdom. The availability in pure form is, however, still challenging as mentioned in the introduction. Due to the diversity of occurrence and the resulting variety of the chemical structure of these polymers the application potential of hemicelluloses is quite huge. The hemicellulose polysaccharides are usually distinguished according to their sugar composition and are divided into four groups of structurally different types, namely xylans, mannans, xyloglucans, and mixed-linkage β-glucans (Ebringerová, 2006).

Xylans are typical hemicelluloses of hardwoods and herbal plants. The major fraction of xylans is only little branched and thus insoluble in water, but soluble in alkaline aqueous solutions and polar organic solvents, such as dimethyl sulfoxide (DMSO). Mannans, especially glucomannans, are the major hemicelluloses of softwood. Depending on branching and the degree of acetylation most of these polysaccharides are soluble in alkaline solutions. Some mannans, especially of the group of galactomannans, are highly branched and water soluble. These polysaccharides are frequently termed gums (guar, carob) and are highly utilized in the food industry as gelling or viscosity-enhancing additives (Englyst et al., 2007; Gáspár et al., 2005; Lunn and Buttriss, 2007). The groups of xyloglucans and mixed-linkage β-glucans are less ubiquitous and, depending on the chemical structure, fractions of each group are water soluble.

Apart from the advantage of general availability of hemicelluloses, the biological activity of these polymers is highly interesting for applications within the pharmaceutical sector. The beneficial effects of hemicelluloses reach from inhibitory action on mutagenicity and antiphlogistic effects to mitogenic and comitogenic activity (Ebringerová and Heinze, 2000). For drug delivery purposes a range of hemicelluloses exhibit the outstanding characteristic that they remain intact under physiological conditions in the stomach environment and even in the small intestine (Sinha and Kumria, 2001; Rubinstein, 1995). The polymers can only be degraded by colon microflora and are thus well suited for colon-specific drug delivery applications (Oliveira et al., 2010).

14.3 PREPARATION STRATEGIES FOR HEMICELLULOSE PARTICLES

The preparation strategies for hemicellulose particles strongly depend on the isolation procedure of the polymers. For the vast majority of hemicelluloses,

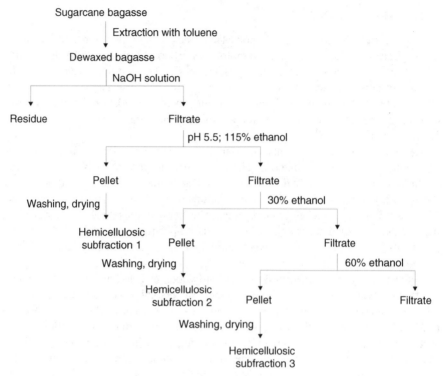

FIGURE 14.3 Example of a hemicellulose purification strategy based on alkaline extraction and precipitation with ethanol (Peng et al., 2009).

especially glucuronoxylans, extraction at alkaline conditions is the standard procedure (Sun et al., 2000; Fang et al., 2000). Precipitation and fractionation of the polysaccharides is usually done by neutralization and subsequent addition of ethanol (see the example in Figure 14.3).

Mainly due to the glucuronic acid residues the glucuronoxylan polymers are soluble under alkaline conditions. Upon neutralization and subsequent addition of alcohols (methanol or ethanol) the solubility decreases drastically resulting in supersaturation and precipitation of the hemicellulose polymer. This relatively simple method can be used to prepare hemicellulosic micro- and nanoparticles (Garcia et al., 2001). Parameters influencing the particle size are the hemicellulose concentration, the concentration of the neutralizing acid, the pH-value reached and the presence of additives, e.g., lipophilic phase builders (Nagashima et al., 2008) or lignin monomers and polymerization enzymes (Barakat et al., 2007).

The size of the particles obtained by neutralization of an alkaline solution varies strongly with the polysaccharide concentration, as was shown by Garcia et al. (2001). Barakat et al. (2007) were able to reach smaller particles by polymerization of lignin monomers in a hemicellulose solution. Nagashima et al. (2008) investigated the effect of different external lipophilic phases on the particle size. The results are given in Table 14.1.

TABLE 14.1   Comparison of Hemicellulose Particle Size Obtained by Neutralization of Alkaline Hemicellulose Solutions with and without Additives

Xylan Concentration (mg/cm^3)	pH	Additives	Particle Size (nm)	References
2.85	0.5	–	1790 ± 264	Garcia et al. (2001)
2	7.0	–	371 ± 106	Garcia et al. (2001)
4	7.0	–	985 ± 260	Garcia et al. (2001)
16	7.0	–	677 ± 260	Garcia et al. (2001)
1	5.0	Coniferyl/sinapyl alcohol, peroxidase	26 ± 5	Nagashima et al. (2008)
n.a.	4.5	Chloroform/ cyclohexane	21.7 ± 1.8	Barakat et al. (2007)
n.a.	2.8	Medium chain triglycerides	71.7 ± 2.9	Barakat et al. (2007)
n.a.	4.1	Soybean oil	13.3 ± 2.1	Barakat et al. (2007)

Another solvent suitable for hemicellulose extraction is subcritical water (Buranov and Mazza, 2007, 2010). At temperatures above 160°C liquid water can efficiently be used to extract hemicelluloses (Song et al., 2008). Solved polysaccharides can be precipitated by cooling the solution. Depending on the extraction conditions fragmented lignin can be observed within the precipitates (Leschinsky et al., 2007).

Aqueous extraction of hemicelluloses can additionally be performed using microwave support (Lundqvist et al., 2003). In this case, very short dissolution times can be reached. Even if extraction with liquid hot water in principle is a relatively simple method, the extracts have up to now not been used to directly prepare and characterize particles from such solutions.

In recent work, new hemicellulose-particle preparation routes have been developed. Starting from xyloglucans dissolved in polar aprotic solvents such as DMSO, which is a suitable solvent for water-insoluble hemicelluloses (see Figure 14.4), N,N-dimethylacetamide (DMAc) or acetone, either supercritical CO_2 antisolvent precipitation (Haimer et al., 2008, 2010) or dialysis (Heinze et al., 2008) can be used to prepare micro- and nanoparticles.

The utilization of supercritical carbon dioxide ($scCO_2$) for the preparation of small particles has been constantly developed over the past two decades (Jung and Perrut, 2001). Several processes based on $scCO_2$ have been found to be applicable for the preparation of pharmaceutical powders (Byrappa et al., 2008; Mishima, 2008). Depending on the physicochemical properties of the used materials the substances can directly be dissolved in and precipitated from $scCO_2$. Alternatively, methods such as particles from gas saturated solutions (PGSSs) or supercritical antisolvent-precipitation (SAS) have to be applied.

Micro- and nanoparticles can be relatively easily prepared from supercritical solutions via a process called rapid expansion of supercritical solution (RESS). In this

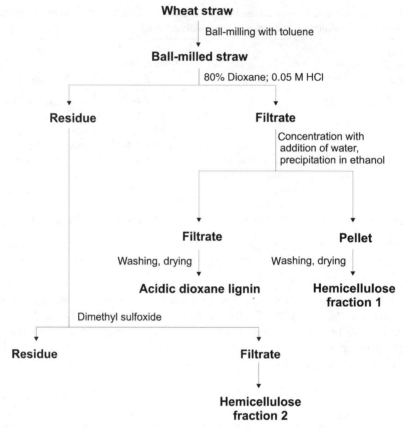

FIGURE 14.4 Example of a hemicellulose purification strategy based on DMSO as extraction solvent (Sun et al., 2005).

case, a supercritical solution is depressurized via a nozzle, leading to almost instantaneous supersaturation and thus high nucleation rates (Sun et al., 2005).

Substances exhibiting low solubility in supercritical carbon dioxide but a significant solubility of carbon dioxide within the substance often change their physicochemical properties at higher amounts of dissolved CO_2 (Aionicesei et al., 2008). Solid polyethylene, for instance, can be melted by dissolving CO_2 in the solid phase. This gas saturated solution can then be depressurized over a nozzle leading to immediate solidification and formation of small particles (PGSS) (Rodrigues et al., 2004).

For molecules which interact very weakly with carbon dioxide neither RESS nor PGSS can be applied. The method of choice is then supercritical antisolvent-precipitation. The mechanism of this particle formation technique is in principle the dilution of the solvent with carbon dioxide in order to lower its capacity for the dissolved substance. Therefore, the only important prerequisite for applying this method is the miscibility of the solvent with carbon dioxide (Kim et al., 2008). For hemicelluloses which are usually not dissolvable in $scCO_2$ the SAS-process can

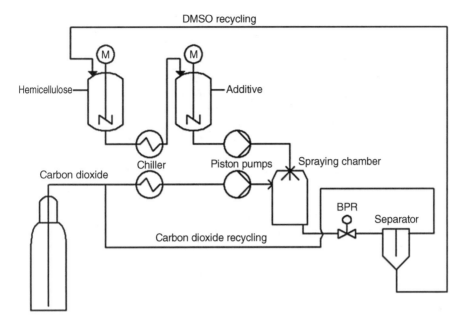

FIGURE 14.5 Scheme of the semicontinuous antisolvent precipitation with supercritical carbon dioxide.

effectively be used for preparing micro- and even nanoparticles. DMSO is, as well as several other aprotic polar solvents, well miscible with CO_2 at elevated pressures. By increasing the CO_2 content within the solvent the solubility of the hemicelluloses can be lowered which leads to precipitation of the polymer. The process can also be used to co-precipitate $scCO_2$ insoluble substances together with hemicelluloses in order to coat the former with the latter. Process variables such as solute concentration, pressure, temperature and, for the semicontinuous process, nozzle geometry can be varied in order to influence particle size and morphology. The semicontinuous supercritical antisolvent precipitation process is given in Figure 14.5.

In this case, the spraying chamber is pressurized with carbon dioxide at a constant temperature beyond 31°C and the hemicellulose (with or without additive) solution is sprayed into the pressurized vessel. Immediately after entering the CO_2-atmosphere CO_2 dissolves within the solvent leading to a rapid decrease of hemicellulose solubility. The emerging precipitate is collected by a filter at the bottom of the spraying chamber. The solvent as well as the major amount of CO_2 can be recycled during this process. Xylan particles are precipitated by such a method in Figure 14.6. Using supercritical carbon dioxide as an antisolvent, particle sizes above or below 1 μm can be obtained, depending on the process conditions.

A rapid solute supersaturation is usually the best route to obtain very small particles. However, slow supersaturation, by using dialysis, can also lead to small particles with a narrow particle size distribution (Heinze et al., 2008). Chemical modification, by esterification, can be effectively applied to avoid agglomeration and hydrogen bonding during the slow supersaturation process.

FIGURE 14.6 Xylan-particles precipitated by semicontinuous supercritical antisolvent precipitation at 40°C and 150 bar.

Another possibility to prepare micro- and nanoparticles from carbohydrate polysaccharides is the emulsion technique. Particles can be obtained by cross-linking (Kotsaeng et al., 2010; Gomaa et al., 2010) or by precipitation (Shah and Ravula, 2000). In the latter case, aqueous solutions of hemicelluloses are mixed at high mechanical stirring rates with an oil phase containing a surfactant. By adding salt, this emulsion subsequently is broken and the resulting particles can be collected and stored.

14.4 EXAMPLES FOR HEMICELLULOSE PARTICLE APPLICATION

Hemicellulose particles and gels have found several application purposes, four of them which are pertinent to the particles discussed above are presented in the following chapter. We distinguish between adsorption, surface modification, and coating; drug targeting and controlled release; and an application field of evolving importance, namely protein–hemicellulose interactions.

14.4.1 Adsorption

Activated carbon is one of the most used adsorbents for industrial purposes and can easily be produced from agricultural by-products and polysaccharide particles (Demirbas, 2009). Agricultural by-products can furthermore be used to prepare anion exchanger adsorbents (Orlando et al., 2002; Gao et al., 2009).

Fibers and particles from agricultural by-products, especially the group of dietary fibers, exhibit significant binding tendencies for a multiplicity of potentially harmful substances. Pesticides, i.e., can be effectively adsorbed on hemicellulose (Ta et al., 1999).

Dietary fiber particles are able to efficiently bind to various substances such as bile acids (Eastwood and Hamilton, 1968) and lipid-soluble vitamins (Omaye et al., 1983). Animal tests have shown that these molecules decrease the toxicity of food carcinogens (Sjödin et al., 1992). According to Smith-Barbaro et al. (1981), a mechanism by which fiber may exert a protective effect in the digestive tract is adsorbing xenobiotics and thus lowering the effective concentration available for absorption, which finally favors their excretion in the feces. Morita et al. (1995) have shown that the amount of bound polychlorinated biphenyls (PCB) *in vitro* correlates with fecal PCB output in the presence of various fibers. While several *in vitro* investigations have reported the binding capacity of different particles to mutagens (Ferguson et al., 1992; Roberton et al., 1991) more information on the interactions between matrix components and pesticides is still needed.

14.4.2 Surface Modification and Coating

Application of hemicelluloses in the field of coating and packing materials has been explored for several years, for example arabinoxylans from barley husk (Höije et al., 2005) or glucuronoxylan from aspen wood (Gröndahl et al., 2004). Preparing moisture barriers on the basis of native hemicelluloses is problematic due to the hygroscopic nature of these polymers. Additionally, films from many of these polymers are usually semicrystalline and therefore, plasticizers have to be used. Nevertheless, chemically modified hemicelluloses can effectively be film-casted (Hansen and Plackett, 2008).

In natural lignocellulosic tissues, hemicelluloses are known to be the linker between the cellulose network and lignin. Thus during the pulping process, the purification of cellulose, hemicelluloses are largely separated from the cellulose polymers. This can be done by different processes usually at elevated temperatures.

It has been found, that hemicelluloses, especially xylans, tend to readsorb on the cellulose fibers at later stages of this cooking (Dahlman et al., 2003; Akerholm and Salmén, 2002). In terms of the kinetic precipitation approach, hemicellulose particles can thus be precipitated in the presence of cellulose. Cellulose fibers act as a precursor and can be modified by adsorption and particle formation of hemicelluloses on their surface (Henriksson and Gatenholm, 2001, 2002; Bodin et al., 2007).

Hemicelluloses can be used to encapsulate probiotic bacteria (Ding and Shah, 2009) in order to increase their acid tolerance. Especially mannans such as xanthan gum and carrageenan gum appear to be effective in protecting probiotic cells from harsh environmental conditions. Emulsion techniques can also be applied to obtain polysaccharide-coated oil droplets. These droplets can be used as carriers for lipophilic drugs (Iwamoto et al., 1991). The effectiveness of such carriers systems can be increased by modification of the carbohydrate-structure by the means of derivatization (Sato et al., 1986; Sunamoto and Iwamoto, 1986). Another field of application of the emulsification/cross-linking technique for hemicelluloses is the coating of magnetic resonance contrast agents and magnetic markers in order to protect them from gastric solution (Silva et al., 2007; Liu et al., 2009).

14.4.3 Drug Targeting and Controlled Release

An important field of application of micro- and nanoparticles is, as already mentioned, drug delivery and controlled release. Due to their physicochemical properties hemicelluloses are proper matrices for gastrointestinal delivery systems.

A drug which is frequently used as a reference substance for describing new processes is ibuprofen which was successfully connected to xylan by esterification (Daus and Heinze, 2010). Additionally, it was possible to modify the hydrophobicity of the hemicellulose esters in order to control their agglomeration tendency and to enable targeting to certain surfaces. Another possibility to modify the properties of xylan-based particles in order to achieve gastrointestinal release is the coating with Eudragit (Mi et al., 2005). In their work, Mi et al. were able to show, that the enteric-coated xyloglucan beads are suitable as a carrier for oral drug delivery of irritant drugs into the stomach as the release is strongly dependent on the systems pH.

For the release of large molecules the microscopic structure of the hemicellulose particles plays an important role. Frequently crosslinker- or spacer-agents are necessary to reach a good release performance. One example is the presence of lignin residues, if covalently bound to the hemicellulose polymer, which can be used to crosslink the hemicellulose chains and thus, build up gels or particles (Claudia et al., 2009). Such networks can be used as release matrices even for large molecules, such as proteins.

Didanosine is a widely used anti-HIV drug. However, its oral therapy is associated with poor gastrointestinal (GI) tolerability. In the study of Kaur et al. (2008) an attempt has been made to formulate sustained and targeted release nanoparticles of didanosine using gelatin as polymer and mannan coating to further enhance its macrophage uptake and its distribution in organs that act as major reservoirs of HIV. The coating was achieved simply by adsorption of mannan at room temperature, which was dissolved before at elevated temperatures.

Enalaprilate (Enal), another active pharmaceutical component being used as an ACE-inhibitor, was intercalated into a layered double hydroxide (Mg/Al-LDH) by an ion exchange reaction and afterwards this particles were coated with xyloglucan (Ribeiro et al., 2009) as the use of a layered double hydroxide (LDH) to release active drugs is limited by the low pH of the stomach. The xyloglucan (XG) was extracted from *Hymenaea courbaril* (jatobá) seeds, Brazilian species. The coating of the particles was simply achieved by adsorption from aqueous solutions.

Takahashi et al. explored the potential of an alternative *in situ* gelling material of natural origin, xyloglucan, for the topical delivery of nonsteroidal anti-inflammatory drugs (NSAIDs) such as ibuprofen and ketoprofen. Xyloglucan was derived from tamarind seeds, and partially degraded by β-galactosidase in order to achieve thermally reversible gelation in dilute aqueous solution, as the sol/gel transition temperature varies with the degree of galactose elimination (Takahashi et al., 2002). Miyazaki et al. (1998) successfully investigated the similar system for the release of indomethacin, another nonsteroidal anti-inflammatory drug commonly used to reduce fever, pain, stiffness, and swelling and diltiazem, a calcium blocker. Both were able to demonstrate the advantages of xyloglucan compared to usual release matrices, the most important ones being nontoxicity and biocompatibility (Burgalassi et al., 2000).

Grafting of xyloglucan was used to obtain conjugates which can be targeted to specific organs. Cao et al. grafted xyloglucan with doxorubicin, a drug used in cancer chemotherapy, and galactosamine, a terminal moiety that can be used to target polymeric conjugates to hepatocytes. The content of doxorubicin was over 5% (wt) in the conjugate. The conjugate was applied as nanoparticulate drug delivery systems (nanoDDSs) with an average size of 142 nm in diameter. In an *in vitro* cytotoxicity experiment, the nanoDDS has similar cytotoxicity as free doxorubicin against HepG2 cells. It was found, that these nanoDDS generated higher therapeutic effect than nontargeted doxorubicin nanoparticles or free doxorubicin in a human tumor xenograft nude mouse model (Cao et al., 2010).

Oil-in-water (o/w) emulsions are widely used in clinics for parenteral nutrition as they can be produced on a large industrial scale and are relatively stable. Such emulsions can solubilize a considerable amount of lipophilic drugs in the hydrophobic domain of the oil droplet and thus, they have potential for medical application as a carrier for lipophilic drugs. When oil emulsions are intravenously injected, these vehicles tend to accumulate largely in the reticuloendothelial system (e.g., liver and spleen). This is an undesirable characteristic for application of an oil-in-water emulsion as a therapeutic system. The tissue distribution can be successfully altered by modification of their surface structure. In their work, Iwamoto et al. (1991) coated the surface of oil droplets in an o/w emulsion with cholesterol-bearing polysaccharide derivatives.

14.4.4 Protein–Hemicellulose Interactions

The biological roles of carbohydrates are particularly important in the assembly of complex multicellular organs and organisms, which requires interactions between cells and the surrounding matrix. All cells and numerous macromolecules in nature carry an array of covalently attached monosaccharides or oligosaccharides. The chemistry and metabolism of carbohydrates were prominent matters of interest in the first part of the twentieth century. Although these topics engendered much scientific attention, carbohydrates were primarily considered as a source of energy or as structural materials and were believed to lack other biological activities. The development of glycobiology in the past 20 years induced the merging of the traditional disciplines of carbohydrate chemistry and biochemistry with a modern understanding of the cell and molecular biology of glycans and, in particular, their conjugates with proteins and lipids ("glycomics") (Varki, 2009). One important aspect of carbohydrates in biology is cellular recognition. Although known for much longer, the importance of oligosaccharides and polysaccharides for cellular recognition and intercellular adhesion has been recognized only during the past few decades (Taylor, 2006).

Pereira et al. (2008) investigated the influence of temperature, pH, and the presence of Ca^{2+} ions favor on the formation of xyloglucan–alginate layers onto Si/SiO$_2$ wafers. They were able to prepare stable films of mixtures of xyloglucan and alginate that were then analyzed considering their suitability as matrices for the adsorption of two glucose/mannose-binding seed (*Canavalia ensiformis* and *Dioclea altissima*) lectins. These covered surfaces were then used as supports for the immobilization of dengue virus particles. The key feature of the utilization of these polysaccharides in the test is that the diagnosis of dengue fever should be rapid, facile, and low-cost.

The lectins adsorbed irreversibly onto xyloglucan–alginate surfaces forming homogeneous monolayers. The adsorption was mainly driven by electrostatic interaction between the alginate carboxyl-groups on the surface and lectin positively charged residues. Xyloglucan–alginate negatively charged surfaces were suitable for the immobilization of the majority of the virus particles.

Protein–polysaccharide interactions can also be used for cell anchorage. Cell adhesion is a strict requirement for survival of most cell types, and it orchestrates critical roles in many cellular functions including migration, proliferation, differentiation, and apoptosis (Price, 1997). In their study, Seo et al. (2005) used xyloglucan as an extracellular matrix for hepatocyte adhesion based on Ca–alginate capsules. In their work, they focused on the ability of xyloglucan to specifically interact with hepatocytes adhering onto the coated surface, and on the enhanced liver-specific functions induced by hepatocyte spheroids in alginate/xyloglucan capsules. They were able to show, that formation of multicellular hepatocyte spheroids maintained high liver-specific functions after adhesion onto xyloglucan based on alginate capsules. The results suggest that the multicellular spheroid formation of the hepatocytes in the presence of xyloglucan as a new synthetic extracellular matrix can enhance the liver-specific functions in three-dimensional geometry.

Enzyme immobilization is another aspect of protein–polysaccharide interactions. The essence of enzyme immobilization is to attach the enzyme to a support material which will stabilize the enzyme and maintain its activity. There are three main methods to immobilize the enzymes, including (a) chemical immobilization of enzyme in which the enzyme is linked to the matrix by a covalent bond, (b) physical absorption on the surface of a carrier, and (c) entrapping the enzyme in a semipermeable supportive material.

Wang et al. (2008) used konjac glucomannan to immobilize L-asparaginase. Nanoparticles were prepared by dropping a carboxymethyl konjac glucomannans (CKGM) solution with L-asparaginase into a chitosan (CS) solution through a needle, while sonicating. The preparation of the nanocapsules was completely conducted in water and the immobilized L-asparaginase was found to maintain the original activity of the free enzyme. The encapsulation efficiency reached 68.0% when both the concentrations of CKGM and CS were 0.01% and the particle size was in a range 100–300 nm. Compared with the free L-asparaginase, the immobilized enzyme system showed significantly higher thermostability and had preferable resistance to acid and alkaline environments. These nanocapsules exhibited semipermeability and could thus be used to immobilize thermal and pH-sensitive enzymes.

14.5 FUTURE PERSPECTIVES

The utilization of by-products of the conversion of cellulosic and lignocellulosic biomass, such as fats, lignin, or hemicelluloses, is of increasing importance as these substances significantly affect the feasibility of biorefinery concepts.

If it succeeds to increasingly utilize the hemicelluloses fraction of lignocellulosic materials, the effectivity and competitiveness of biorefinery processes would largely benefit. "When processes are implemented that allow for the entire plant (virtually any plant) to be chemically converted—economically—to chemicals or energy, the bio-based economy will truly have arrived," as was stated by Davenport (2008). Having this in mind, the importance of searching for new separation and formulation technologies for biomass-based products and by-products becomes obvious.

The utilization of hemicelluloses as polymers is considered to have high potential, but still only few processes and products are fully developed at the moment. In this chapter, we tried to give an overview over the different routes for purification and formulation of water-soluble and -insoluble hemicellulose in order to show how different hemicellulose sources can be used as parent materials. The purified polymers can additionally be modified in order to improve their performance for the respective application.

Green polymers, such as cellulose and hemicellulose which are ubiquitous and cheap, offer a great chance for lowering the dependency on fossil resources. Research and development within the field of engineering of these polymers has been increasing steadily, still a lot of work is necessary and many new applications have to be discovered. In this chapter, it was our intention to demonstrate that the field of potential applications for hemicellulose particles is already quite broad and covers as well "high tech" application areas, although the future is yet to bring full and wide applicability of these hitherto somewhat neglected resources.

REFERENCES

Aionicesei, E., M. Škerget, and Ž. Knez (2008). Measurement and modeling of the CO_2 solubility in polyethylene glycol of different molecular weights. *J. Chem. Eng. Data* **53**(1):185–188.

Varki A., (2009). *Essentials of Glycobiology*, 2nd ed. Cold Spring Harbor, NY: Cold Spring Harbor Laboratory Press.

Akerholm, M., and L. Salmén, (2002). Dynamic FTIR spectroscopy for carbohydrate analysis of wood pulps. *J. Pulp Paper Sci.* **28**(7):245–249.

Barakat, A., J. L. Putaux, L. Saulnier, B. Chabbert, and B. Cathala, (2007). Characterization of arabinoxylan-dehydrogenation polymer (synthetic lignin polymer) nanoparticles. *Biomacromolecules* **8**(4):1236–1245.

Baveja, S. K., K. V. Ranga Rao, J. Arora, N. K. Mathur, and V. K. Vinayah, (1991). Chemical investigations of some galactomannan gums as matrix tablets for sustained drug delivery. *Ind. J. Chem. B* **30**(2):133–137.

Bodin, A., L. Ahrenstedt, H. Fink, H. Brumer, B. Risberg, and P. Gatenholm, (2007). Modification of nanocellulose with a xyloglucan-RGD conjugate enhances adhesion and proliferation of endothelial cells: implications for tissue engineering. *Biomacromolecules* **8**(12):3697–3704.

Bousaid, A., J Robinson, Y. J. Cai D. J. Gregg, and J. N. Saddler, (1999). Fermentability of the hemicellulose derived sugars from steam-exploded softwood (Douglas fir). *Biotechnol. Bioenerg.* **64**:284–289

Brennan, A. H., W. Hoagland, and D. J. Schell, (1986). High temperature acid hydrolysis of biomass using engineering—scale plug flow reactor: result of low solid testing. *Biotechnol. Bioenerg. Symp.* **17**:53–70

Buranov, A. U., and G. Mazza, (2010). Extraction and characterization of hemicelluloses from flax shives by different methods. *Carbohydr. Polym.* **79**(1):17–25.

Buranov, A. U., and G. Mazza, (2007). Fractionation of flax shives by water and aqueous ammonia treatment in a pressurized low-polarity water extractor. *J. Agric. Food Chem.* **55**(21):8548–8555.

Burgalassi, S., P. Chetoni, L. Panichi, E. Boldrini, and M. F. Saettone, (2000). Xyloglucan as a novel vehicle for timolol: pharmacokinetics and pressure lowering activity in rabbits. *J. Ocular Pharmacol. Therap.* **16**(6):497–509.

Byrappa, K., S. Ohara, and T. Adschiri, (2008). Nanoparticles synthesis using supercritical fluid technology—towards biomedical applications. *Adv. Drug Deliv. Rev.* **60**(3):299–327.

Cao, Y., Y. Gu, H. Ma, J. Bai, L. Liu, P. Zhao, and H. He, (2010). Self-assembled nanoparticle drug delivery systems from galactosylated polysaccharide-doxorubicin conjugate loaded doxorubicin. *Int. J. Biol. Macromol.* **46**(2):245–249.

Claudia, M. B. R., C. M. Elizabeth, L. M. Jaime, R. C. Agustin, A. M. E. Jorge, and L. M. L. Ana, (2009). Maize arabinoxylan gels as protein delivery matrices. *Molecules* **14**(4):1475–1482.

Dahlman, O., A. Jacobs, and J. Sjöberg, (2003). Molecular properties of hemicelluloses located in the surface and inner layers of hardwood and softwood pulps. *Cellulose* **10**(4):325–334.

Daus, S., T. Heinze, (2010). Xylan-based nanoparticles: prodrugs for ibuprofen release. *Macromol. Biosci.* **10**(2):211–220.

Davenport, R., (2008). Chemicals & polymers from biomass. *Ind. Biotechnol.* **4**(1):59–63.

Demirbas, A., (2009). Agricultural based activated carbons for the removal of dyes from aqueous solutions: a review. *J. Hazard. Mater.* **167**(1–3):1–9.

Ding, W. K., and N. P. Shah, (2009). Effect of various encapsulating materials on the stability of probiotic bacteria. *J. Food Sci.* **74**(2):M100–M107.

Eastwood, M. A., and D. Hamilton, (1968). Studies on the adsorption of bile salts to non-absorbed components of diet. *Biochim. Biophys. Acta* **152**:165–173.

Ebringerová, A., (2006). Structural diversity and application potential of hemicelluloses. *Macromol. Symp.* **232**:1–12.

Ebringerová, A., and T. Heinze, (2000). Xylan and xylan derivatives—biopolymers with valuable properties. 1: Naturally occurring xylans structures, isolation procedures and properties. *Macromol. Rapid Commun.* **21**(9):542–556.

Ebringerová, A., and Z. Hromádková, (2010). An overview on the application of ultrasound in extraction, separation and purification of plant polysaccharides. *Cent. Eur. J. Chem.* **8**(2):243–257.

Englyst, K. N., S. Liu, and H. N. Englyst, (2007). Nutritional characterization and measurement of dietary carbohydrates. *Eur. J. Clin. Nutr.* **61** (Suppl. 1): S19–S39.

Fang, J. M., R. C. Sun, and J. Tomkinson, (2000). Isolation and characterization of hemicelluloses and cellulose from rye straw by alkaline peroxide extraction. *Cellulose* **7**(1):87–107.

Ferguson, L. R., A. M. Roberton, R. J. McKenzie, M. Watson, and P. J. Harris, (1992). Adsorption of a hydrophobic mutagen to dietary fibre from taro (*Colocasia esculenta*), an imported food plant of the South Pacific. *Nutr. Cancer* **17**:85–95.

Gao, B. Y., X. Xu, Y. Wang, Q. Y. Yue, and X. M. Xu, (2009). Preparation and characteristics of quaternary amino anion exchanger from wheat residue. *J. Hazard. Mater.* **165**(1–3):461–468.

Garcia, R. B., T. Nagashima Jr. A. K. C. Praxedes, F. N. Raffin, T. F. A. L. Moura, and E. S. T. Do Egito, (2001). Preparation of micro and nanoparticles from corn cobs xylan. *Polym. Bull.* **46**(5):371–379.

Gáspár, M., T. Juhász, Zs. Szengyel, and K. Réczey, (2005). Fractionation and utilisation of corn fibre carbohydrates. *Proc. Biochem.* **40**(3–4):1183–1188.

Gomaa, Y. A., L. K. El-Khordagui, N. A. Boraei, and I. A. Darwish, (2010). Chitosan microparticles incorporating a hydrophilic sunscreen agent. *Carbohydr. Polym.* **81**(2):234–242.

Grassmann, P., (1967). *Einführung in die thermische Verfahrenstechnik*, Berlin: Walter de Gruyter & Co.

Gröndahl, M., L. Eriksson, and P. Gatenholm, (2004). Material properties of plasticized hardwood xylans for potential application as oxygen barrier films. *Biomacromolecules* **5**(4):1528–1535.

Guozhong C., (2004). *Nanostructures & Nanomaterials: Synthesis, Properties & Applications*, 1st edn., Imperial College Press.

Haimer, E., F. Liebner, M. Wendland, A. Potthast, and T. Rosenau, (2008). Precipitation of hemicelluloses from DMSO/water mixtures using carbon dioxide as an antisolvent. *J. Nanomater.* **2008**(1):art. no. 826974.

Haimer, E., M. Wendland, A. Potthast, U. Henniges, T. Rosenau, and F. Liebner, (2010). Controlled precipitation and purification of hemicellulose from DMSO and DMSO/water mixtures by carbon dioxide as anti-solvent. *J. Supercrit. Fluids.*

Hansen, N. M. L., and Plackett, D., (2008). Sustainable films and coatings from hemicelluloses: a review. *Biomacromolecules* **9**(6):1493–1505.

Haruta, M., and B. Delmon, (1986). Preparation of homodisperse solids. *J. Chim. Phys.* **83**:859–868.

Heinze, T., K. Petzold, and S. Hornig, (2008). Novel nanoparticles based on xylan. *Cellulose Chem. Technol.* **41**(1):13–18.

Henriksson, A., and P. Gatenholm, (2001). Controlled assembly of glucuronoxylans onto cellulose fibres. *Holzforschung.* **55**(5):494–502.

Henriksson, A., and P. Gatenholm, (2002). Surface properties of CTMP fibers modified with xylans. *Cellulose* **9**(1):55–64.

Hinman, N. D., D. J. Schell, C. J. Riley, P. W. Bergeron, and P. J. Walter, (1992). Preliminary estimate of the cost of ethanol production for SSF technology. *Appl. Biochem. Biotechnol.* **34–35**:639–649

Höije, A., M. Gröndahl, K. Tømmeraas, and P. Gatenholm, (2005). Isolation and characterization of physicochemical and material properties of arabinoxylans from barley husks. *Carbohydr. Polym.* **61**(3):266–275.

Iwamoto, K., T. Kato, M. Kawahara, N. Koyama, S. Watanabe, Y. Miyake, and J. Sunamoto, (1991). Polysaccharide-coated oil droplets in oil-in-water emulsions as targetable carriers for lipophilic drugs. *J. Pharmaceut. Sci.* **80**(3):219–224.

Jung, J., and M. Perrut, (2001). Particle design using supercritical fluids: literature and patent survey. *J. Supercrit. Fluids* **20**(3):179–219

Juslin, M., and P. Paronen, (1984). Xylan—a possible filler and disintegrant for tablets. *J. Pharm. Pharmacol.* **36**(4):256–257.

Kamm, B., P. R. Gruber, and M. Kamm, (2006). *Biorefineries—Industrial Processes and Products. Status Quo and Future Directions*, Weinheim: Wiley-VCH Verlag GmbH & Co. KGaA.

Kaur, A., S. Jain, and A. K. Tiwary, (2008). Mannan-coated gelatin nanoparticles for sustained and targeted delivery of didanosine: *in vitro* and *in vivo* evaluation. *Acta Pharmaceut.* **58**(1):61–74.

Kim, J. W., and G. Mazza, (2006). Optimization of extraction of phenolic compounds from flax shives by pressurized low-polarity water. *J. Agric. Food Chem.* **54**(20):7575–7584.

Kim, M. S., S. J. Jin, J. S. Kim, H. J. Park, H. S. Song, R. H. H. Neubert, and S. J. Hwang, (2008). Preparation, characterization and *in vivo* evaluation of amorphous atorvastatin calcium nanoparticles using supercritical antisolvent (SAS) process. *Eur. J. Pharmaceut. Biopharmaceut.* **69**(2):454–465.

Kirk, and Othmer, (2008). *Separation Technology.* 2nd edn., Wiley-Interscience.

Kohane, D. S., (2007). Microparticles and nanoparticles for drug delivery. *Biotechnol. Bioeng..* **96**(2):203–209.

Kotsaeng, N., J. Karnchanajindanun, and Y. Baimark, (2010). Chitosan microparticles prepared by the simple emulsification-diffusion method. *Partic. Sci. Technol.* **28**(4):369–378.

Leppänen, K., P. Spetz, A. Pranovich, K. Hartonen, V. Kitunen, and H. Ilvesniemi, (2010). Pressurized hot water extraction of Norway spruce hemicelluloses using a flow-through system. *Wood Sci. Technol.* 1–14.

Leschinsky, M., R. Patt, and H. Sixta, (2007). Water prehydrolysis of *E. globulus* with the main emphasis on the formation of insoluble components. In: Proceeding of the Pulp Paper Conf., Fibre Modifications and Brightening, Finnish Paper Engineers Association, Helsinki, Finland. pp. 7–14.

Liu, J., Y. Zhang, T. Yang, Y. Ge, S. Zhang, Z. Chen, and N. Gu, (2009). Synthesis, characterization, and application of composite alginate microspheres with magnetic and fluorescent functionalities. *J. Appl. Polym. Sci.* **113**(6):4042–4051.

Lundqvist, J., A. Jacobs, M. Palm, G. Zacchi, O. Dahlman, and H. Stålbrand, (2003). Characterization of galactoglucomannan extracted from spruce (*Picea abies*) by heat-fractionation at different conditions. *Carbohydr. Polym.* **51**(2):203–211.

Lunn, J., and J. L. Buttriss, (2007). Carbohydrates and dietary fibre. *Nutr. Bull.* **32**(1):21–64.

Taylor, M. E., (2006). *Introduction to Glycobiology*, 2nd edn. New York: Oxford University Press.

Mi, K. Y., K. C. Hoo, H. K. Tae, J. C. Yun, T. Akaike, M. Shirakawa, and S. C. Chong, (2005). Drug release from xyloglucan beads coated with Eudragit for oral drug delivery. *Arch. Pharm. Res.* **28**(6):736–742.

Mishima, K., (2008). Biodegradable particle formation for drug and gene delivery using supercritical fluid and dense gas. *Adv. Drug Deliv. Rev.* **60**(3):411–432.

Miyazaki, S., F. Suisha, N. Kawasaki, M. Shirakawa, K. Yamatoya, and D. Attwood, (1998). Thermally reversible xyloglucan gels as vehicles for rectal drug delivery. *J. Control. Release* **56**(1–3):75–83.

Morita, K., K. Hamamura, and T. Iida, (1995). Binding of PCB by several types of dietary fibre *in vivo* and *in vitro*. *Fukuoka Igaku Zasshi* **86**:212–217.

Nagashima Jr., T., E. E. Oliveira, A. E. da Silva, H. R. Marcelino, M. C. S. Gomes, L. M. Aguiar, I. B. de Araújo, L. A. L. Soares, A. G. de Oliveira, and E. S. T. Egito, (2008). Influence of the lipophilic external phase composition on the preparation and characterization of xylan microcapsules—a technical note. *AAPS Pharm. Sci. Tech.* **9**(3):814–817.

Oliveira, E. E., A. E. Silva, T. N. Júnior, M. C. S. Gomes, L. M. Aguiar, H. R. Marcelino, I. B. Araújo, M. P. Bayer, N. M. P. S. Ricardo, A. G. Oliveira, and E. S. T. Egito, (2010). Xylan from corn cobs, a promising polymer for drug delivery: production and characterization. *Bioresource Technol.* **101**(14):5402–5406.

Omaye, S. T., F. I. Chow, and A. A. Betschart, (1983). *In vitro* interactions between dietary fibre and 14C-vitamin D or 14C-vitamin E. *J. Food Sci.* **48**:260–261.

Orlando, U. S., A. U. Baes, W. Nishijima, and M. Okada, (2002). Preparation of agricultural residue anion exchangers and its nitrate maximum adsorption capacity. *Chemosphere* **48**(10):1041–1046.

Pan, X., D Xie, R. W. Yu, D Lam, and J. N. Saddler, (2007). Pretreatment of lodgepole pine killed by mountain pine beetle using ethanol organosolv process: fractionation and process optimization. *Ind. Eng. Chem. Res.* **46**:2609–2617.

Peng, F., J. L. Ren, F. Xu, J. Bian, P. Peng, and R. C. Sun, (2009). Comparative study of hemicelluloses obtained by graded ethanol precipitation from sugarcane bagasse. *J. Agric. Food Chem.* **57**(14):6305–6317.

Pereira, E. M. A., M. R. Sierakowski, T. A. Jó, R. A. Moreira, A. C. O. Monteiro-Moreira, R. F. O. França, B. A. L. Fonseca, and D. F. S. Petri, (2008). Lectins and/or xyloglucans/alginate layers as supports for immobilization of dengue virus particles. *Colloid. Surfaces B Biointerf.* **66**(1):45–52.

Price, L. S., (1997). Morphological control of cell growth and viability. *Bioessays* **19**:941–943.

Rabanel, J. M., X. Banquy, H. Zouaoui, M. Mokhtar, and P. Hildgen, (2009). Progress technology in microencapsulation methods for cell therapy. *Biotechnol. Progr.* **25**(4):946–963.

Ribeiro, C., G. G. C. Arizaga, F. Wypych, and M. R. Sierakowski, (2009). Nanocomposites coated with xyloglucan for drug delivery: *in vitro* studies. *Int. J. Pharmaceut.* **367**(1–2):204–210.

Roberton, A. M., L. R. Ferguson, H. J. Hollands, and P. J. Harris, (1991). Adsorption of a hydrophobic mutagen to five contrasting dietary fibre preparations. *Mutat. Res.* **262**:195–202.

Rodrigues, M., N. Peiriço, H. Matos, E. Gomes De Azevedo, M. R. Lobato, and A. J. Almeida, (2004). Microcomposites theophylline/hydrogenated palm oil from a PGSS process for controlled drug delivery systems. *J. Supercrit. Fluids* **29**(1–2):175–184.

Rodríguez, A., and L. Jiménez, (2008). Pulping with organic solvents other than alcohols. *Afinidad* **65**(535):188–196.

Rosen M., (2006). *Delivery System Handbook for Personal Care and Cosmetic Products: Technology. Applications and Formulations*, Norwich, NY: William Andrew.

Rubinstein, A., (1995). Approaches and opportunities in colon-specific drug delivery. *Crit. Rev. Therap. Drug Carrier Syst.* **12**(2–3):101–149.

Sato, T., K. Kojima, T. A. Ihda, J. Sunamoto, and R. M. Ottenbrite, (1986). Macrophage activation by poly(maleic acid-alt-2-cyclohexyl-1, 3-dioxap-5-ene) encapsulated in poly-saccharide-coated liposomes. *J. Bioactive Compat. Polym.* **1**(4):448–460.

Schacht, E., S. Vansteenkiste, J. Loccufier, and D. Permentier, (1990). *Use of dextran as drug carrier*, American Chemical Society, Polymer Preprints, Division of Polymer Chemistry **31** (2): pp. 717–718.

Schliephake, D. et al.; (1986). Nachwachsende Rohstoffe, Teil I Holz und Stroh [Landwirtschaft-Industrie e.V. (ed.), Verlag J. Kordt, Bochum] (a) I3–I226, (b) I–160, (c) I–166.

Seo, S. J., T. Akaike, Y. J. Choi, M. Shirakawa, I. K. Kang, and C. S. Cho, (2005). Alginate microcapsules prepared with xyloglucan as a synthetic extracellular matrix for hepatocyte attachment. *Biomaterials* **26**(17):3607–3615.

Shah, N. P., and R. R. Ravula, (2000). Microencapsulation of probiotic bacteria and their survival in frozen fermented dairy desserts. *Aust. J. Dairy Technol.* **55**(3):139–144.

Silva, A. K. A., E. L. da Silva, E. E. Oliveira, T. Nagashima Jr., L. A. L. Soares, A. C. Medeiros, J. H. Araújo, I. B. Araújo, A. S. Carriço, and E. S. T. Egito, (2007). Synthesis and characterization of xylan-coated magnetite microparticles. *Int. J. Pharmaceut.* **334**(1–2):42–47.

Sinha, V. R., and R. Kumria, (2001). Polysaccharides in colon-specific drug delivery. *Int. J. Pharmaceut.* **224**(1–2):19–38.

Sjödin, P., M. Nyman, L. L. Nielsen, H. Wallin, and M. Jagerstad, (1992). Effect of dietary fibre on the disposition of a carcinogen (2-14C-labeled MeIQx) in rats. *Nutr. Cancer* **17**:139–151.

Smith-Barbaro, P., D. Hanson, and B. S. Reddy, (1981). Carcinogen binding to various types of dietary fibre. *J. Natl. Cancer Inst.* **67**:495–497.

Song, T., A. Pranovich, I. Sumerskiy, and B. Holmbom, (2008). Extraction of galactoglucomannan from spruce wood with pressurised hot water. *Holzforschung* **62**(6):659–666.

Sun, R., J. M. Fang, and J. Tomkinson, (2000). Characterization and esterification of hemicelluloses from rye straw. *J. Agric. Food Chem.* **48**(4):1247–1252.

Sun, X. F., R. Sun, P. Fowler, and M. S. Baird, (2005). Extraction and characterization of original lignin and hemicelluloses from wheat straw. *J. Agric. Food Chem.* **53**(4):860–870.

Sun, Y. P., M. J. Meziani, P. Pathak, and L. Qu, (2005). Polymeric nanoparticles from rapid expansion of supercritical fluid solution. *Chem. Eur. J.* **11**(5):1366–1373.

Sunamoto, J., and K. Iwamoto, (1986). Protein-coated and polysaccharide-coated liposomes as drug carriers. *Crit. Rev. Therap. Drug Carrier Syst.* **2**(2):117–136.

Ta, C. A., J. A. Zee, T. Desrosiers, J. Marin, P. Levallois, P. Ayotte, and G. Poirier, (1999). Binding capacity of various fibre to pesticide residues under simulated gastrointestinal conditions. *Food Chem. Toxicol.* **37**(12):1147–1151.

Taherzadeh, M. J., K. Karimi, (2008). Pretreatment of lignocellulosic wastes to improve ethanol and biogas production: a review. *Int. J. Molec. Sci.* **9**(9):1621–1651.

Takahashi, A., S. Suzuki, N. Kawasaki, W. Kubo, S. Miyazaki, R. Loebenberg, J. Bachynsky, and D. Attwood, (2002). Percutaneous absorption of non-steroidal anti-inflammatory drugs from *in situ* gelling xyloglucan formulations in rats. *Int. J. Pharmaceut.* **246**(1–2):179–186.

Wang, R., B. Xia, B. J. Li, S. L. Peng, L. S. Ding, and S. Zhang, (2008). Semi-permeable nanocapsules of konjac glucomannan-chitosan for enzyme immobilization. *Int. J. Pharmaceut.* **364**(1):102–107.

Zhang, A. P., C. F. Liu, and R. C. Sun, (2010). Fractional isolation and characterization of lignin and hemicelluloses from triploid of *Populus tomentosa* Carr. *Ind. Crops Prod.* **31**(2):357–362.

Zhao, Y., W. J. Lu, H. T. Wang, and J. L. Yang, (2009). Fermentable hexose production from corn stalks and wheat straw with combined supercritical and subcritical hydrothermal technology. *Bioresource Technol.* **100**(23):5884–5889.

15

NONXYLAN HEMICELLULOSES AS A SOURCE OF RENEWABLE MATERIALS

DAVID PLACKETT AND NATANYA HANSEN

15.1 INTRODUCTION

The need for new and sustainable sources of the energy and materials which are so vital in today's world is now widely recognized, especially in view of the likely future depletion of fossil fuel reserves. As a result, and also because of their associated environmental benefits, there is much ongoing interest in polymers from renewable bioresources (Belgacem and Gandini, 2008). For example, since most bio-derived polymers are also biodegradable, increased development and use of these polymers can at least partly address the increasing problem of plastic wastes in the land and marine environments. This chapter reviews hemicelluloses and particularly nonxylan hemicelluloses in this context and also in respect to eventual high-value applications for these carbohydrates (e.g., in medicine).

Hemicelluloses are one of the most abundant types of polymer occurring in nature and typically constitute 20–30% of the mass of annual and perennial plants. These compounds are in fact not one single biopolymeric structure, but a diverse group of polysaccharides. Hemicelluloses in the plant cell wall are bound to cellulose and lignin (Figure 15.1) and complex isolation procedures are normally required in order to achieve separation from the plant raw material. Hemicellulose composition depends on the plant source and on other factors such as the plant subspecies and the growing conditions. This diversity and their hydrophilic character present significant challenges in respect to industrial applications and the practical use of

Polysaccharide Building Blocks: A Sustainable Approach to the Development of Renewable Biomaterials,
First Edition. Edited by Youssef Habibi and Lucian A. Lucia.

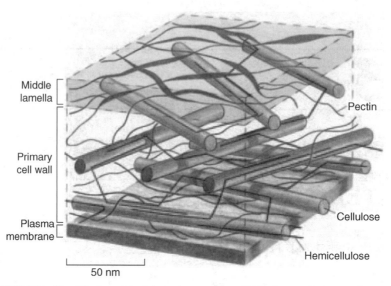

FIGURE 15.1 Simplified model of the primary cell wall (lignin not shown). Adapted from http://www.learner.org/courses/envsci/visual/visual.php?shortname=primary_cell_wall, US Department of Energy Genome Programs/genomics.energy.gov.

many hemicelluloses has thus far been focused largely on use as a feedstock and the conversion to products with lower molecular weights such as ethanol, xylitol, and furfuryl alcohol. The potential of hemicelluloses for applications in materials and higher-value uses has received some attention but not, for example, to the same extent as cellulose and lignin. In addressing this topic in respect to nonxylan hemicelluloses, it is instructive to cite research on xylans for comparative purposes and that is the approach adopted in this chapter.

15.2 BACKGROUND

Plant biomass is composed of cellulose, lignin, and hemicelluloses, along with starch and lesser amounts of pectins, lipids, proteins, and inorganics, depending upon species. When first discovered, hemicelluloses were erroneously believed to be a cellulose precursor; however, unlike the hemicelluloses, cellulose is a linear polysaccharide with a unique structure consisting of $\beta(1 \rightarrow 4)$-linked D-glucopyranoses. Cellulose from wood pulp has typical chain lengths between 300 and 1700 units, whereas cellulose from cotton and other plant fibers as well as bacterial celluloses have chain lengths ranging from 800 to 10,000 units (Klemm et al., 2005). In contrast, hemicelluloses are mostly branched polysaccharides of much lower molecular weight with a degree of polymerization (DP) in the range of 80–200 (Sun et al., 2004). Cellulose is known to provide the structural reinforcement in plant cell walls while hemicelluloses were thought to provide a more passive role by acting as a filler between cellulose fibrils and coupling to lignin; however, as suggested by

Atalla et al. (1993), it now also seems likely that hemicelluloses play an active role in regulating the structure of plant cell walls by influencing the patterns of aggregation of cellulose during biogenesis. The involvement of hemicelluloses in strengthening plant cell walls through interaction with cellulose and, in some cases, with lignin has been discussed in a recent review which also points out that, in addition to their function in the cell wall, hemicelluloses can also occur as seed storage carbohydrates (Scheller and Ulvskov, 2010). Traditionally, hemicellulose is defined as the alkali-soluble material which is left after the removal of pectic substances from plant cell walls (Selvendran and O'Neill, 1987; Sun et al., 2004) and the term is generally used to describe a number of noncrystalline hexose and pentose sugars. Four main groups of hemicelluloses may be defined according to their primary structure: xylans, man-noglycans (mannans), β-glucans, and xyloglucans (Ebringerova et al., 2005). Some hemicelluloses are also referred to as gums, especially in industrial applications (e.g., guar gum, a galactomannan, and maize bran gum, a xylan). In addition to their presence in plants, hemicelluloses can be found in other sources such as fungi and bacteria; however, the focus of this chapter will be on plant-derived hemicelluloses.

15.2.1 Hemicellulose Types and Structures

Hemicellulose type and structure can vary significantly between different feedstocks as well as between particular sources, depending on factors such as origin and plant growth stage (Sun et al., 2004). The source dependency of hemicellulose compo-sition has, for example, been illustrated in a study of four rice varieties, in which sugar compositional analysis showed significant variations in the main components including arabinose (5–23%), xylose (17–40%), and glucose (36–55%) in different cultivars (Lai et al., 2007). The impact of growth environment was likewise demonstrated in a recent study of hemicelluloses in the endosperm cell walls of barley, in which the ratio between the different hemicelluloses as well as the molecular weights varied significantly depending on the eco-zone where the barley was grown (Lazaridou et al., 2008). There may also be variations in hemicellulose structure as a function of location in the plant (i.e., stem, grain, etc.). A number of the most common hemicellulose building blocks are depicted in Figure 15.2. In some hemicelluloses there are various substitutions on the pyranose and furanose rings, of which *O*-acetylation is the most common.

15.2.1.1 Xylans   Although a quantitatively minor component of the primary walls of dicots and nongraminaceous monocots, xylans are the principal hemicelluloses found in annual plants and are the predominant type found in the secondary walls of woody plants, especially hardwoods (Darvill et al., 1980; Ebringerova and Heinze, 2000). These polysaccharides can also be found in the cell wall of certain species of red or green algae. Xylans consist of a D-xylopyranose backbone with side groups on the C-2 and/or C-3 positions, while the D-xylopyranose units are linked by β(1 → 3) or β(1 → 4) bonds (Figure 15.3). The backbone xylopyranose units may also be *O*-acetylated. Glucuronoxylans have a side chain of either α-D-glucuronic acid or its 4-*O*-methyl derivative, while arabinoxylans are substituted on C-2 and/or C-3 with α-L-arabinofuranosyl residues.

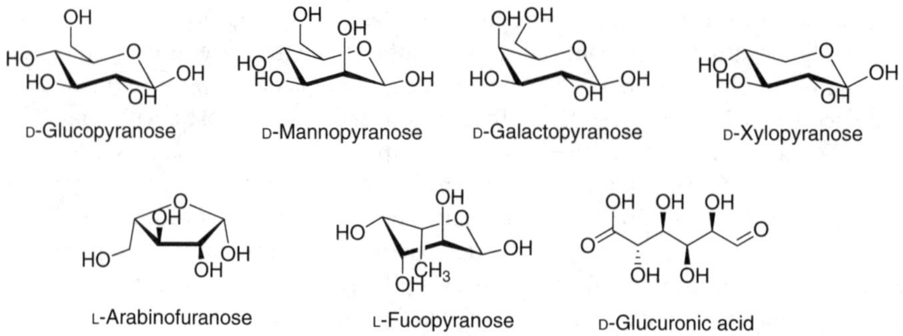

FIGURE 15.2 Key hemicellulose building blocks.

15.2.1.2 Mannans   Mannoglycans or mannans are divided into two groups: galactomannans and glucomannans. The former contain a $\beta(1 \rightarrow 4)$-linked D-mannopyranose backbone and the latter are comprised of D-mannopyranose and D-glucopyranose with $\beta(1 \rightarrow 4)$ linkages. Both types of mannoglycans may have varying degrees of branching with D-galactopyranose residues in the C-6 position of the mannose backbone, although glucomannans are mainly straight chain polymers. The term galactoglucomannan is generally applied to mannans consisting of randomly distributed $(1 \rightarrow 4)$-linked mannose and glucose units with $(1 \rightarrow 6)$-linked galactose units attached to mannose moieties and with more than 15% substitution at the C-6 position (Ebringerova, 2005; Willför et al., 2008). Figure 15.4 shows the chemical structure of galactoglucomannan extracted from spruce wood. Most galactomannans are similarly substituted with D-galactopyranose residues α-linked in the 6-position of the D-mannopyranose backbone with 30–96% substitution and are found in the storage tissues of seeds (Ebringerova et al., 2005). These hemicelluloses are water soluble and galactomannans with a high degree of branching are used industrially as gums (e.g., guar and locust bean gum).

Mannose-containing polysaccharides are found in the cell walls of many plants (Stephen, 1982). In hardwoods, mannans constitute less than 7% of total hemicellulose content and have not been studied as extensively as the mannans from softwoods, which contain galactoglucomannans as the main hemicellulosic component (60–70%) and glucomannans which are present to a much lesser extent (1–5%)

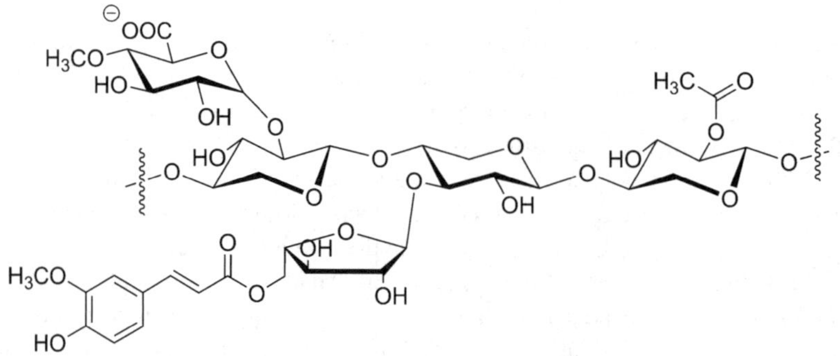

FIGURE 15.3 Schematic structure of a xylan.

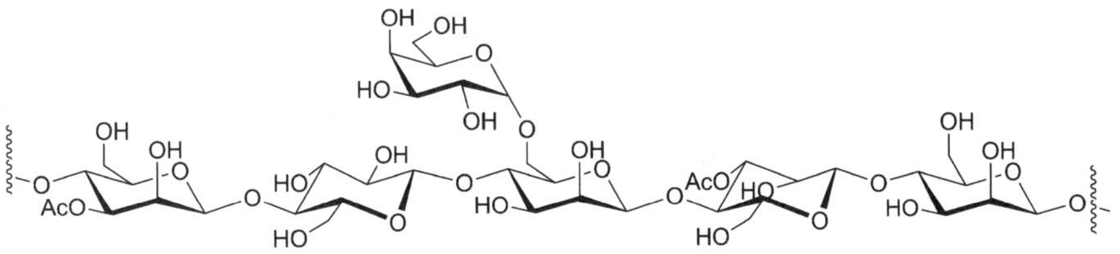

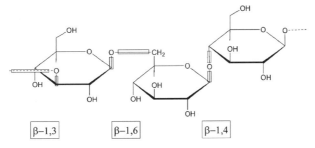

FIGURE 15.4 Chemical structure of spruce galactoglucomannan. *Source*: Kirsi Mikkonen, Department of Food and Environmental Sciences, University of Helsinki, Finland.

(Spiridon and Popa, 2008). Partially acetylated glucomannans are one of the active ingredients in aloe vera, where they are found in the skin tissue (Ebringerova et al., 2005). Some studies have also shown the presence of galactomannans in aloe vera, which may be due to variations in plant origin and growing conditions.

15.2.1.3 Mixed Linkage β-Glucans

Mixed linkage β-glucans occur in the cell walls of gramineae and are comprised of a D-glucopyranose backbone with mixed β(1 → 3) and β(1 → 4) linkages in different ratios (Carpita, 1996). β-Glucan content in cereals is reported as follows: rye (1–2%), oats (3–7%), wheat (<1%), and barley (3–11%) (Skendi et al., 2003). As noted by Westerlund et al. (1993), mixed linkage β-glucans consist of cellotriosyl and cellotetraosyl segments which are distributed differently for the various cereals. The ratio of cellotriosyl to cellotetraosyl segments is reported to be between 1.5 and 4.5 (Burton et al., 2010) and no branching is found in this type of hemicellulose. The β(1 → 3) linkages introduce irregularities in the otherwise linear chains rendering these polymers water soluble. Research has shown that insoluble (1,3/1,6) β-glucans (Figure 15.5) have greater biological activity than the (1,3/1,4) β-glucan counterparts (Ooi and Liu, 2000).

15.2.1.4 Xyloglucans

Xyloglucans are present as a major component in primary cell walls of all higher plants (Ebringerova et al., 2005) and have a backbone of β(1 → 4) linked D-glycopyranose residues with a distribution of D-xylopyranose in the C-6 position. The distribution of the side chains divides the xyloglucans in two categories: one with two xylopyranose-substituted units followed by two glucopyranose units (termed XXGG) and one with three xylopyranose-substituted units followed by a single glucopyranose unit (termed XXXG). In addition, a number of side chains occur on the C-2 position of the D-glycopyranosyl backbone as well as

FIGURE 15.5 Diagram showing orientation and location of different β-glucan linkages.

on the xylopyranosyl subunits. The most common substitutions include L-fucopyr-anose, L-arabinofuranose, D-galactopyranose, and D-glucopyranose as well as combinations thereof (Zhou et al., 2007), making characterization of this group of hemicelluloses especially difficult. As reviewed by Vincken et al. (1997), some xyloglucans are O-substituted. In certain seeds, such as those of tamarind, xyloglucans are present as a storage polysaccharide in place of starch. The wider abbreviated nomenclature for xyloglucans is presented in Figure 15.6.

15.2.2 Isolation Procedures

The exact nature of the interactions between cellulose and the different hemicelluloses in the plant cell wall is yet to be fully understood (Hill et al., 2009); however, it is well established that the ease of extraction and the preservation of the

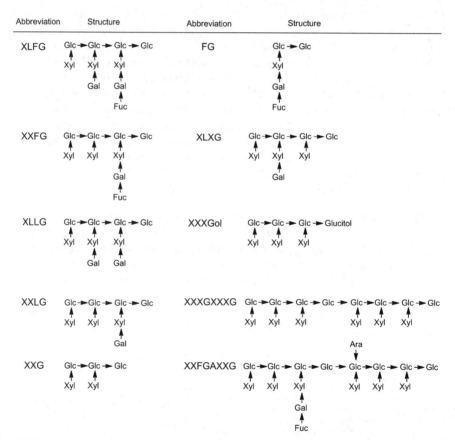

FIGURE 15.6 Abbreviated nomenclature for some commonly occurring xyloglucans (sugar residues and linkages shown are Ara = α-L-arabinofuranosyl (1 → 2), Fuc = α-L-fucopyranosyl (1 → 2), Gal = β-D-galactopyranosyl (1 → 2), Glc = β-D-glucopyranosyl (1 → 4) (or reducing D-glucose), Xyl = α-D-xylopyranosyl (1 → 6), Glucitol = reduced glucose).

polysaccharide structure during extraction can vary according to the plant material and the type of hemicellulose.

Hemicelluloses are thought to interact through both noncovalent and covalent bonding with cellulose, lignin, and proteins in both the primary and secondary plant cell walls. Detailed isolation procedures are therefore required in order to separate these constituents from one another and from the other plant components. The first step in hemicellulose isolation can involve alkaline hydrolysis of the ester bonds in the lignocellulosic matrix followed by extraction into aqueous media, which is an easier process for annual plants than for wood (Sun et al., 2004). After the initial separation step, the hemicellulose fraction may be subjected to a number of processes including sequential precipitation and refinement by chromatographic methods (e.g., ion-exchange, gel-permeation).

Glasser et al. (2000) utilized various procedures, including pretreatments, to extract noncellulosic carbohydrates from barley husks and yellow poplar chips. Extraction with aqueous alkali was preferred and initial steps such as prehydrolysis and delignification gave xylan-rich products with reduced polydispersity and adequate yield. However, steam explosion was not favored because severe hydrolytic depolymerization of hemicelluloses occurred in this case. Therefore, in addition to the natural variation in hemicelluloses, the isolation procedure also has an influence on the composition of the product obtained since unwanted side reactions such as deacetylation, degradation and contamination with lignin can sometimes occur.

In summary, a number of methods have been used to obtain hemicelluloses from plant sources including chemical extraction (e.g., using alkali, dimethyl sulfoxide, methanol/water) as well as steam or microwave treatment. The choice of extraction method depends on the source and on the interaction of the targeted hemicelluloses with the other components present in the plant.

15.2.2.1 Isolation of Mannans Research on isolation of mannans has mainly focused on softwoods because these have a relatively high content of these hemicelluloses, with *O*-acetylgalactoglucomannans as the principal species. Mannans are thought to be strongly associated with cellulose microfibrils in the cell wall and this interaction is dependent on the number and type of hemicellulose side groups. A higher affinity for cellulose occurs when there are fewer side groups and lower degrees of acetylation (Melton et al., 2009). Furthermore, mannans form complexes with lignin through covalent bonds, making the isolation of pure products very difficult (Willför et al., 2008). In order to isolate mannans from wood the source is first mechanically fragmented (i.e., chipping, grinding, etc.) followed by an extraction step. Some *O*-acetylgalatoglucomannans are water soluble and may be extracted at pH 5.5–7 (alkaline solutions cause deacetylation). The use of dimethyl sulfoxide is also an effective and nondestructive alternative (Willför et al., 2008). Other extraction methods have been proposed such as steam (Palm and Zacchi, 2003) or microwave treatment (Lundqvist et al., 2002, 2003). However, extensive degradation of mannans can occur when these approaches are used. The commercially important konjac glucomannan has been isolated from the flour of tubers by sulfur dioxide bleaching followed by numerous extractions (Chua et al., 2010).

15.2.2.2 Isolation of β-Glucans The isolation of β-glucans from cereals is complicated as separation from a number of components including starch, lipids, and proteins must take place and therefore a series of treatments is normally employed with extraction being the key step. The choice of isolation method influences the yield, composition, and molecular weight of β-glucans. The most frequently employed technique involves various alkaline extractions. It can be a challenge to isolate high molecular weight β-glucans; however, this was, for example, achieved by Immerstrand et al. (2009) using step-wise enzymatic treatment of oat bran with α-amylase and protease (pancreatin) to obtain β-glucans with molecular weights of up to 9.08×10^5 g/mol. Utilization of a similar enzymatic treatment was recently shown to be more efficient in β-glucan purification than alkaline or acidic extraction, resulting in fewer impurities and the highest hemicellulose recovery ($\sim$87%) from oat flour (Ahmad et al., 2010).

15.2.2.3 Isolation of Xyloglucans Isolation of xyloglucans is particularly challenging due to their strong interaction with cellulose in the plant cell wall. However, the reverse is true for xyloglucans in seeds, which may be extracted through hot water treatment (Ebringerova, 2005). Xyloglucans are traditionally extracted with strong alkaline solution (KOH or NaOH); however, alternative methods such as ultrasound-assisted extraction have also proven successful (Caili et al., 2006).

15.3 HEMICELLULOSES AS A SOURCE OF RENEWABLE MATERIALS

Given the great abundance of hemicelluloses in biomass it is no surprise that considerable research effort has been directed toward their use as a source of renewable materials. This research, aspects of which are outlined here, has taken one of several forms, including the production of films from hemicelluloses which has often necessitated chemical or physical modification to enhance their film-producing potential. In connection with this objective, reports in the literature have included studies on hydrophobizing hemicelluloses as well as attempts to introduce thermoplasticity.

Hartman et al. (2006a) produced films from acetylated galactoglucomannan with plasticizer addition or through blending with either sodium alginate or carboxy-methylcellulose. The blend films displayed better behavior than the plasticized films, exhibiting good mechanical properties at high humidity and relatively high storage moduli. For reasons to do with potential applications in packaging, oxygen and water vapor permeability is frequently of interest in such investigations and, in this case, the lowest oxygen permeability values were found for the blend films containing alginate or carboxymethylcellulose. These values were slightly higher than those recorded for amylose and amylopectin or for plasticized glucuronoxylan films (Rindlav-Westling et al., 1998; Gröndahl et al., 2004). *O*-Acetylgalactoglucomannan was also studied by Hartman et al. (2006b) as an oxygen barrier. In this work, *o*-acetylgalactoglucomannan was obtained from thermomechanical pulp process water which was subjected to sequential ultrafiltration and diafiltration to obtain a preconcentrate free of low molecular weight compounds. The result was a water-soluble product with M_w ~10,000 g/mol and a polydispersity (PDI) of 1.3. This preconcentrate was subsequently lyophilized to obtain a solid which was ~90% pure.

Nonxylan hemicelluloses have also been examined as edible coatings. For example, β-glucans from barley and oats have been plasticized with glycerol and tested as water vapor barriers (Tejinder, 2003). The resulting films were opaque with tensile strength values varying with solution concentration and β-glucan source. Films from oat β-glucans were shown to be water soluble while those prepared from barley were less soluble (e.g., films from hull-less barley remained intact after 24 h of water immersion). β-Glucans from different types of oats have also been evaluated as films (Skendi et al., 2003). In addition, the use of β-glucans from Shiitake mushrooms has explored as a coating on poly(D,L-lactic-*co*-glycolic acid) film in order to enhance the cell affinity of the surface. A plasma surface treatment prior to coating resulted in the best proliferation of human dermal fibroplast cells, rendering the films suitable as substrates for skin cell regeneration (Lee et al., 2007).

15.3.1 Chemical Modification of Hemicelluloses

The chemical modification of hemicelluloses to enhance key properties and expand their range of potential uses has been discussed in recent review articles (Hansen and Plackett, 2008; Cunha and Gandini, 2010a). As in the case of cellulose, the availability of hydroxyl groups in hemicelluloses provides a suitable target for a variety of classical chemical modification techniques. Although the preparation of hemicellulose acetate films was reported decades ago (Smart and Whistler, 1949), to the authors' knowledge the first study on hydrophobic plastic films from hemi-celluloses was reported in 2002, in which laurylamine chains were bound to maize bran heteroxylan using periodic oxidation followed by reductive amination (Fredon et al., 2002). The degree of substitution varied from 0.5 to as high as 1.1 and was calculated according to Equation 15.1 in which M_m = average molar mass of a sugar unit, %C = carbon percentage, M_c = average carbon mass of a sugar unit, $M_{c(amine)}$ = carbon mass of the amine chain, and $M_{(amine)}$ = molar mass of the amine:
$$DS = \frac{M_m(\%C) - M_c}{M_{c(amine)} - (M_{(amine)} - k \times M_{OH})(\%C)}.$$
Results from thermal and thermoanalytical studies on the amine-modified xylans showed a glass transition (T_g) of around $-30°C$, indicating plastic behavior at ambient temperature.

15.3.1.1 Esterification   Various research groups have investigated esterification of hemicelluloses using acyl chlorides under homogeneous conditions (Chaa et al., 2008; Fang et al., 1999; Moine et al., 2004; Peng et al., 2008a; Sun et al., 1999, 2000a, 2001). In one study, Sun et al. extracted hemicelluloses from rye straw and carried out esterification using acyl chlorides in an *N,N*-dimethylformamide (DMF)/LiCl solvent system and in the presence of triethylamine and 4-(dimethylamino)pyridine. Only minor degradation of the hemicellulose chains was detected (Sun et al., 2000b). Some research on hemicellulose esterification has specifically involved forest product- or forest by-product-derived hemicelluloses (Moine et al., 2004; Sun et al., 1999, 2001). Lauroylation of xylans in the presence of microwave irradiation was carried out with degrees of substitution in the range of 1.2–1.3 to give hydrophobic films that compared well with cellulose films in terms of tensile strength properties (Moine et al., 2004). In this case, xylan from beech wood was included in the investigation. Sun et al. (1999) used a similar homogeneous reaction system applied to delignified poplar chips and obtained minimal hemicellulose degradation. Subsequent research by Sun et al. (2001) extended the esterification studies to include formation of propionyl, hexanoyl, lauroyl, and palmitoyl derivatives.

Ren et al. (2007a) demonstrated the use of ionic liquids as a medium in which to acetylate wheat straw hemicelluloses, achieving yields between 70 and 90% and degree of substitution between 0.49 and 1.53. Ionic liquids are of interest in this respect as environmentally benign reaction media with strong capability to dissolve complex macromolecules. Reaction yields were calculated using the assumption that all the hemicelluloses were converted to diacetylated form and the acetylated hemicelluloses were found to have improved thermal stability. Peng et al. (2008b) carried out rapid phthalylation and succinylation of wheat straw hemicelluloses using N-bromosuccinimide as catalyst at 50°C for 5 min in the presence of microwave radiation. The structure of the reaction products is shown in Figure 15.7. In this work, the degree of substitution was between 0.43 and 1.47 and ^{13}C NMR studies showed that esterification occurred at both the C-3 and C-2 positions of the β-xylose units, but preferentially at the C-3 position. However, these researchers also discovered that microwave radiation had a negative effect in that some degradation of the hemicelluloses took place, leading to lower thermal stability in the derivatives as compared to those produced by a conventional heating technique.

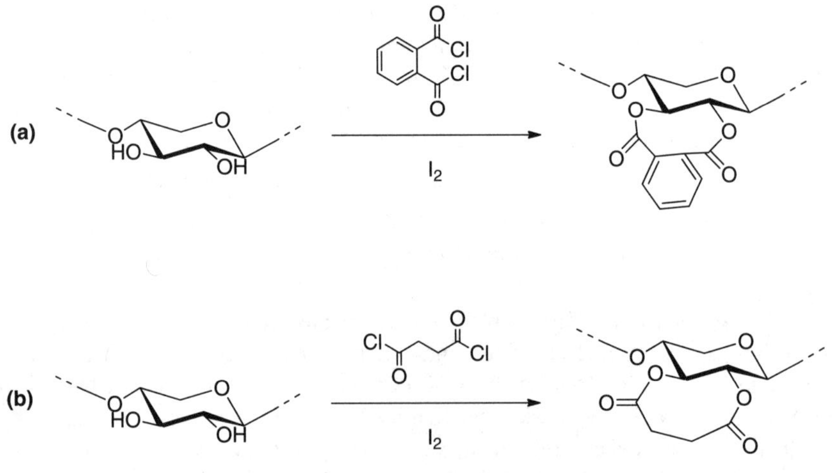

FIGURE 15.7 Scheme for rapid phthalylation (a) and succinylation (b) of hemicelluloses. *Source*: Peng et al. (2008b). Reactions were conducted in DMF.

Xylan esters such as acetate, propionate, butyrate, caprate, laurate, myristate, and palmitate were prepared by a classical method involving reaction with the corresponding acid anhydrides in the presence of a mildly basic catalyst such as pyridine (Carson and Maclay, 1948a). Direct reaction with succinic anhydride has also been used to produce esterified xylan from oat spelts (Hettrich et al., 2006). The same authors conducted sulfation of oat-derived xylan, either by reaction with amidosulfuric acid in DMSO solution or with chlorosulfuric acid in DMF solution with relatively high degrees of substitution in each case. It was suggested that the sulfated hemicelluloses might play a role in the biomedical field as anticoagulants.

15.3.1.2 Etherification
Chemically modified wheat straw and sugarcane bagasse hemicelluloses have been the focus of various etherification studies. For example, sugarcane bagasse hemicellulose consisting of D-xylose, L-arabinose, D-glucose, D-galactose, D-mannose, D-glucuronic acid, 4-O-methyl-D-glucuronic acid, D-galacturonic acid and, to a lesser extent, L-rhamnose, L-fucose, and various O-methylated neutral sugars, was etherified using 2,3-epoxypropyltrimethyl ammonium chloride in an aqueous medium and in the presence of sodium hydroxide as a catalyst (Ren et al., 2007b). Yields from such etherification processes involving sugarcane bagasse hemicelluloses were reported to be in the range 35–42% depending upon reaction conditions. The product, a cationic hemicellulose, has been proposed as an additive in papermaking. Another etherification procedure involves reaction with acrylamide in alkaline solution and this has been investigated using hemicelluloses extracted from wheat straw. Under alkaline conditions the result is a modified hemicellulose bearing carbamoylethyl substituents as well as a fraction of carboxyethyl groups (Ren et al., 2008a).

15.3.1.3 Methylation
Hemicelluloses can be methylated using standard methods such as reaction with dimethyl sulfate in NaOH or methyl iodide/silver oxide treatment (Dutton and Murata, 1961; Dutton and McKelvey, 1961). Isogai et al. (1985) subsequently developed a new method for holocellulose methylation using powdered sodium hydroxide and methyl iodide in a sulfur dioxide-diethylamine-methyl sulfoxide solvent system. In this research, the substrates were spruce and beech wood-meals as well as cell wall polysaccharides extracted from cultivated tobacco (*Nicotiana tabacum*).

15.3.1.4 Other Chemical Modification Procedures
Thermoplastic hemicellulose derivatives have been reported as a result of reacting propylene oxide (PO) with xylan (Jain et al., 2000). The method used by these researchers involved exposure of alkaline xylan solutions to PO. The resulting hydroxypropyl xylan (HPX) was a low molecular weight, water-soluble material exhibiting low intrinsic viscosity, and thermoplasticity. A water-insoluble product (APX), which was also thermoplastic, was obtained through peracetylation of HPX in formamide solution. Films cast from HPX and APX had higher tensile strength and lower toughness than corresponding cellulose derivative films. The authors of this work suggested that the xylan

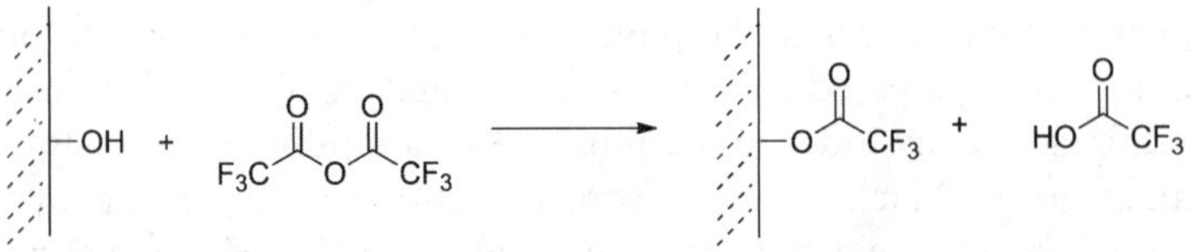

FIGURE 15.8 Gas-phase fluorination of arabinoxylan. *Source*: Gröndahl et al. (2006).

derivatives could be used as additives in melt processing of plastics and blends with polystyrene showed shear-thinning effects in the melt and a plasticization effect in the solid state.

Gröndahl et al. looked at the oxygen barrier properties of glucuronoxylan from aspen wood with plasticizer addition in the range of 20–50 wt% and also examined surface fluorination as a method to produce hydrophobic films (Gröndahl et al., 2006). The success of the surface fluorination method, which involved reaction with gaseous trifluoroacetic anhydride (TFAA), was confirmed using FTIR spectroscopy and surface contact angle measurements (Figure 15.8). However, as pointed out by Cunha et al. (2007) and also mentioned in the context of cellulose fiber esterification with perfluorinated carboxylic moieties (Cunha and Gandini, 2010b), trifluoroacetates are sensitive to water and this would limit the range of possible applications.

In a study by Hartman et al. (2006a, 2006b), O-acetylgalactoglucomannan hemicellulose isolate obtained from thermomechanical pulping (TMP) process water was modified using the Williamson benzylation as well as two methods of surface grafting and lamination of unmodified hemicellulose film with hydrophobic benzylated galactoglucomannan film. The benzylated films had oxygen-barrier properties which were less sensitive to moisture than those based on acetylated galactoglucomannan. Of the various strategies tested to obtain a combination of moisture tolerance and high-barrier properties, promising performance was obtained from a laminate of modified and unmodified hemicelluloses which was generated by soaking a blend film containing 70% galactoglucomannan and 30% alginate in a saturated solution of benzylated galactoglucomannan in DMF. The authors expressed the view that the laminate might be produced on conventional extrusion equipment (Hartman et al., 2006b).

The procedure typically used to carboxymethylate cellulose by reaction with sodium monochloroacetate in NaOH can also be applied to ethanolic suspensions of hemicelluloses. DS values of 0.1–0.56 have been reported in one such study; however, hemicellulose degradation was also reported (Ren et al., 2008b).

Hemicelluloses can be tosylated and then subjected to treatment with an alkyl thiocyanate to yield thiocyano-substituted chains with DS = 0.44–0.52. Carson and Maclay (1948b) carried out such reactions with corn cob and lima bean pod hemicelluloses as well as guar mannogalactan, cellulose and potato starch, with the objective of learning more about the structure and reactivity of these carbohydrates.

Hemicelluloses have been suggested as starting materials for production of novel biodegradable hydrogels (Lindblad et al., 2001, 2004). Low molecular weight

(< 3000) hemicelluloses extracted from spruce wood by steam explosion have been methacrylolated and then transformed into soft hydrogels by radical polymerization in the presence of redox initiators. The main structure of the hemicelluloses extracted in this case was O-acetylgalactoglucomannan, characterized by a high degree of acetylation with good solubility in both water and organic solvents. The methacryloylation procedure has also been described by Palm and Zacchi (2003). The resulting hydrogels were generally soft, elastic, and easily swellable in water. The research showed that it was possible to produce hemicellulose hydrogels with properties similar to those of poly(2-hydroxyethyl methacrylate) hydrogels. In a further development, alkenyl derivatives of hemicelluloses were formed, leading to a library of hemicellulose-based hydrogels (Voepel et al., 2009a).

15.3.2 Physical Modification of Hemicelluloses to Form Nanocomposites

There are few reports so far on the concept of converting hemicelluloses to useful films through formation of nanocomposites. Since hemicelluloses are hydrophilic, it would be a logical step to investigate combination with hydrophilic nanoclays as has been pursued with carbohydrates such as starch and chitosan (e.g., Chung et al., 2010; Oguzlu and Tihminlioglu, 2010). There have been two recent studies in this direction, although notably these have not been exclusively focused on use of unmodified nanoclay fillers. Viota et al. (2010) prepared hemicellulose-nanoclay films by combining an organomodified montmorillonite (OMMT) clay or an inulin nonionic surfactant-coated OMMT with an oat spelt xylan. A considerable reduction in the hydrophilic character of the films was noted when inulin-coated OMMT was used as an additive. The efficiency of film formation was strongly influenced by pH and xylan type. Ünlü et al. (2009) investigated the properties of various combinations of corn cob xylan and unmodified montmorillonite (MMT). The resulting biocomposites showed better thermal and rheological properties than nonreinforced xylan and the authors suggested that these materials might find use in cosmetics as thickeners and cleaning agents.

There have been recent reports on the use of nanocellulose in the form of acid-hydrolyzed cellulose whiskers (CWs) to produce reinforced xylan films. Saxena et al. (2009) combined CW generated by sulfuric acid extraction from a finely ground softwood pulp in aqueous suspension with oat spelt xylan and obtained cast films which had much higher tensile strength and tensile energy absorption characteristics than films prepared from unreinforced xylan. It was also concluded that the addition of sulfuric acid-extracted CW provided much better xylan film properties than when hydrochloric acid-extracted CW was utilized. The water vapor transmission of CW-reinforced xylan films was shown to exhibit a remarkable 74% reduction when compared with unreinforced xylan films and an even more impressive 362% decrease relative to films from xylan reinforced with 10% softwood kraft fibers. The authors reached the conclusion that the barrier properties when CW was used could be attributed to the high crystallinity of CW and the formation of a dense hydrogen-bonded composite structure within the cast films. By contrast, the kraft fibers were present as aggregates and this produced a more open structure with

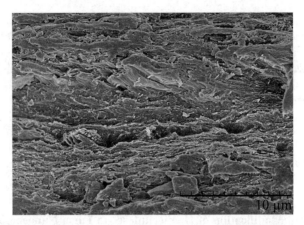

FIGURE 15.9 Field emission-scanning electron microscopy (FE-SEM) image showing the cross-section of a cast oat spelt xylan film containing 20% microfibrillated cellulose (MFC) and which had been subjected to tensile testing. Good dispersion of the MFC and some fiber pull-outs can be noted. The image was obtained by Thomas Blomfeldt at the Fiber and Polymer Technology Department of the Royal Institute of Technology (KTH) in Stockholm, Sweden using a Hitachi S4800 FE-SEM to examine specimens which were sputter coated with gold to a depth of 6 nm. The 10 μm scale bar on the image is divided to indicate 1 μm intervals.

reduced film barrier properties (Saxena and Ragauskas, 2009). The illustration in Figure 15.9 shows an SEM image of a cast nanocellulose-reinforced hemicellulose (xylan) film prepared at DTU using microfibrillated cellulose (MFC) produced at Innventia AB in Stockholm. In this case, the addition of MFC allowed the preparation of films which could be easily handled and measured for various properties (e.g., mechanical, permeability) which was not possible when using xylan alone.

15.4 APPLICATIONS OF HEMICELLULOSES

A few of the existing and possible future applications for hemicelluloses have been discussed earlier in this chapter but, compared to cellulose and lignin, there are still relatively few large-scale commercial uses for hemicelluloses in new materials. This is still the case even though, as reported, there have been extensive studies on hemicellulose modification with new potential applications in mind. Production of films which may ultimately be suitable for a niche in the commodity packaging market has been in focus, and as well as investigations on xylans, konjac gluco-mannan or galactoglucomannan from spruce wood (Figure 15.10) have been studied in this context (Zhang and Whistler, 2004; Hartman et al., 2006b; Mikkonen et al., 2008). From a commercial perspective, to the authors' knowledge the only company so far which has successfully transferred hemicellulose technology for packaging out of the laboratory and into a commercial setting is the Swedish company Xylophane AB (http://www.xylophane.com). This company has developed a new renewable and

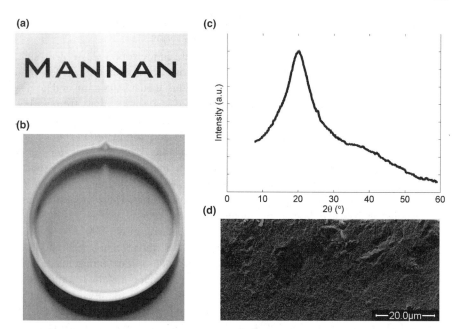

FIGURE 15.10 Illustration of spruce galactoglucomannan (GGM) films produced in the laboratory of Professor Maija Tenkanen, Department of Food and Environmental Sciences, University of Helsinki: (a) GGM film on printing paper, (b) GGM film on a plate, (c) X-ray diffraction pattern of GGM film plasticized with 40% glycerol (w/w of GGM) (XRD measurements involved use of Cu $K\alpha_1$ radiation and the figure was obtained from Kari Pirkkalainen, University of Helsinki), (d) scanning electron micrograph of GGM film. The scale bar is 20 μm.

biodegradable hemicellulose-based material for packaging with good barrier properties against oxygen, grease, and aromas. Potential uses for products based on the Xylophane technology are for packaging of oxygen-sensitive dairy products, greasy snacks, or pet foods, as well as aromatic products such as spices and coffee. A pilot plant for producing xylan as the basis for the barrier material was opened in February 2010 and Xylophane AB is now undertaking product development in cooperation with the packaging industry.

Higher-value uses for hemicelluloses such as in wound care and for other medical purposes have also received some attention (Lloyd et al., 1998; Ebringerova and Heinze, 2000; Haishi and Takeshita, 1997a, 1997b; Voepel et al., 2009b). In addition to use in wound management, gels based on xyloglucans have been explored as vehicles for controlled drug delivery (Miyazaki et al., 1998; Roos et al., 2008). Research has shown that alkenyl-functionalized galactogluocomannans could offer sustained delivery of incorporated substances mediated by diffusion and enzymatic hydrolysis. In one example of the potential use of nonxylan hemicelluloses in drug delivery, *O*-acetyl galactoglucomannan from spruce wood was functionalized and microspheres were then formed in an oil–water emulsion during *in situ* covalent cross-linking into a hydrogel. Sustained delivery of model drugs from such microspheres was demonstrated (Edlund and Albertsson, 2008).

There is a significant history of hemicellulose use in the food industry as stable, nontoxic, gel-forming, and water-soluble additives. Commercial thickeners and emulsifying agents include mannans such as guar gum, locust bean gum, and konjac glucomannan (Zhang et al., 2001). Beta-glucans, isolated, for example, from barley, rice, and oats, have attracted considerable interest because of their reported cholesterol-regulating effects (Kahlon and Chow, 1997) and commercial products are available (Morgan et al., 1999). Konjac glucomannan derived from the tuber *Amorphophallus konjac* has been widely used in the food industry for many years, but is the subject of continuing research because of its gel- and film-forming properties as well as its biocompatibility and biodegradability. These properties have led to studies on the use of konjac glucomannan in applications such as drug delivery, cellular therapy, encapsulation, and emulsification (Zhang et al., 2005).

β-Glucans can form highly viscous solutions and thermoreversible gels and have found use in the food industry as thickening and texture modification agents (e.g., in low-fat products). Originally, β-glucans were studied on account of the complications they gave rise to during brewing processes. However, today the positive effect of β-glucans in dietary fibers is well known and oat bran β-glucan is accepted by the US Food and Drug Administration (FDA) as a functional food ingredient. Among the positive physiological effects of β-glucans are the lowering of cholesterol levels, and the control of postprandial blood glucose and insulin levels, as well as reported antitumor properties (Gunness and Gidley, 2010; Cheung et al., 2002).

Xyloglucans have been used as thickening and stabilizing agents in foodstuffs and, due to their lack of toxicity and excellent gelling properties, recent research on applications of xyloglucans includes their use in drug delivery (Coviello et al., 2007; Brumer et al., 2004; Miyazaki et al., 1998).

In papermaking there have been reports that modified hemicellulose additives can improve paper quality in hand sheets (Ren et al., 1999). The use of 1% cationic or anionic hemicellulose addition in recycled fiber-based packaging paper and old corrugated container (OCC) pulp has been studied. In particular, the addition of the cationic hemicellulose with a DS of 0.5 provided significant improvement in the breaking length, burst index, tear index, and ring crush index of sheets prepared from OCC pulp. Hemicelluloses were, therefore, considered suitable as a source of new wet-end additives for the paper industry (Ren et al., 2009).

15.5 SUMMARY AND FUTURE PERSPECTIVES

Hemicelluloses are a class of branched polysaccharides which are widely dispersed in plant materials and represent a potential source for a vast range of new bio-based materials. The fact that, unlike say cellulose and lignin, there has been little commercial development of hemicellulose products to date, can probably be attributed to their variation in composition and availability according to plant type, location in the plant, time of harvest, growing conditions, and a number of other factors. However, with increasing attention to nonfossil fuel, "green" bio-materials, and a steady growth in research activity, hemicelluloses may come more into focus. In addition, the much-discussed biorefinery concept (Plackett and Ragauskas, 2010; Plackett, 2011) presents the possibility of extracting value from noncellulosic components such as hemicelluloses and lignin as a means of off-setting the cost of bioethanol production from lignocellulosics.

In discussing uses for hemicelluloses in modified or unmodified forms, the research literature tends not to distinguish between different types of hemicelluloses in terms of the application of a particular chemical modification pathway (e.g., acetylation). However, given the abundance of hydroxyl groups in the structure of hemicelluloses, it seems likely that particular methods of modification would be more or less equally applicable regardless of hemicellulose type.

Since much of the work on hemicelluloses for practical applications has been undertaken with xylans, a cross-section of that work has been presented here in addition to a discussion of the status of research on materials from nonxylan hemicelluloses such as the β-glucans, mannans, and xyloglucans.

Given the vast amounts of hemicelluloses from biomass which are potentially available from wood and agro-resources internationally and their proven adaptation as nontoxic, biocompatible, and sustainable bio-materials, it seems certain that research in the field of hemicellulose-based materials will continue and a variety of practical applications will be adopted by industry. These applications could include films and coatings in the packaging industry as well as films, coatings and hydrogels with potential uses in medicine (e.g., drug delivery). Past research has shown that mannans and β-glucans as well as xylans are suitable starting points for new biomaterials. In addition to these developments, it seems likely that there will be much continuing interest in the use of hemicelluloses as chemical feedstocks or as a component of new copolymers (Persson et al., 2009).

REFERENCES

Ahmad, A. F., M. Anjum, T. Zahoor, H. Nawaz, and Z. Ahmed (2010). Extraction and characterization of β-D-glucan from oat for industrial utilization. *Int. J. Biol. Macromol.* **46**:304–309.

Atalla, R. H., J. M. Hackney, I. Uhlin, and N. S. Thompson. (1993). Hemicelluloses as structure regulators in the aggregation of native cellulose. *Int. J. Biol. Macromol.* **15**:109–112.

Belgacem, M. N. and A. Gandini. (2008). *Monomers, Polymers and Composites from Renewable Resources*. Amsterdam: Elsevier.

Brumer, H. III, Q. Zhou, M. J. Baumann, K. Carlsson, and T. T. Teeri. (2004). Activation of crystalline cellulose surfaces through the chemoenzymatic modification of xyloglucan. *J. Am. Chem. Soc.* **126**:5715–5721.

Burton, R. A., M. J. Gidley, and G. B. Fincher. (2010). Heterogeneity in the chemistry, structure and function of plant cell walls. *Nat. Chem. Biol.* **6**:724–732.

Caili, F., T. Haijun, L. Quanhong, C. Tongyi, and D. Wenjuan. (2006). Ultrasound-assisted extraction of xyloglucan from apple pomace. *Ultrasonic. Sonochem.* **13**:511–516.

Carpita, N. C. (1996). Structure and biogenesis of the cell walls of grasses. *Ann. Rev. Plant Physiol. Plant Mol. Biol.* **47**:445–476.

Carson, J. F., and W. D. Maclay. (1948a). Esters of lima bean pod and corn cob hemicelluloses. *J. Am. Chem Soc.* **70**:293–295.

Carson, J. F., and W. D. Maclay. (1948b). The thiocyanation of polysaccharide tosyl esters. *J. Am. Chem. Soc.* **70**:2220–2223.

Chaa, L., N. Joly, V. Lequart, C. Faugeron, J-C. Mollet, P. Martin, and H. Morvan. (2008). Isolation, characterization and valorization of hemicelluloses from *Aristidia pungens* leaves as biomaterial. *Carbohydr. Polym.* **74**:597–602.

Cheung, N. K. V., S. Modak, A. Vickers, and B. Knuckles. (2002). Orally administered beta-glucans enhance anti-tumor effects of monoclonal antibodies. *Cancer Immunol. Immnuother.* **51**:557–564.

Chua, M., T. C. Baldwin, T. J. Hocking, and K. Cha. (2010). Traditional uses and potential health benefits of *Amorphophallus konjac* K. Koch ex N. E. *Br. J. Ethnopharmacol.* **128**:268–278.

Chung, Y-L., S. Ansari, L. Estevez, S. Hayrapetyan, E. P. Giannelis, and H-M. Lai. (2010). Preparation and properties of biodegradable starch–clay nanocomposites. *Carbohydr. Polym.* **79**:391–396.

Coviello, T., P. Matricardi, C. Marianecci, and F. Alhaique. (2007). Polysaccharide hydrogels for modified release formulations. *J. Control. Release* **119**:5–24.

Cunha, A. G., C. S. R. Freire, A. J. D. Silvestre, C. P. Neto, A. Gandini, E. Orblin, and P. Fardim. (2007). Characterization and evaluation of the hydrolytic stability of trifluoroacetylated cellulose fibers. *J. Colloid Interf. Sci.* **316**:360–366.

Cunha, A. G., and A. Gandini. (2010a). Turning polysaccharides into hydrophobic materials: a critical review. Part 2. Hemicelluloses, chitin/chitosan, starch, pectin and alginates. *Cellulose* **17**:1045–1065.

Cunha, A. G., and A. Gandini. (2010b). Turning polysaccharides into hydrophobic materials: a critical review. Part 1. Cellulose. *Cellulose* **17**:875–889.

Darvill, J. E., M. McNeil, A. G. Darvill, and P. Albersheim. (1980). Structure of plant cell walls. XI. Glucuronoarabinoxylan, a second hemicellulose in the primary cell walls of suspension-cultured sycamore cells. *Plant Physiol.* **66**:1135–1139.

Dutton, G. G. S., and S. A. McKelvey. (1961). The constitution of a hemicellulose of cherry wood (*Prunus avium* L. Var. Bing). *Can. J. Chem.* **39**:2582–2589.

Dutton, G. G. S., and T. G. Murata. (1961). The constitution of the hemicellulose of apple wood (*Malus pumila* L. Var. Golden Transparent). *Can. J. Chem.* **39**:1995–2000.

Ebringerova, A. (2005). Structural diversity and application potential of hemicelluloses. *Macromol. Symp.* **232**:1–12.

Ebringerova, A., and T. Heinze. (2000). Xylan and xylan derivatives—biopolymers with valuable properties 1. Naturally occurring xylans structures, isolation procedures and properties. *Macromol. Rapid Commun.* **21**:542–556.

Ebringerova, A., Z. Hromadkova, and T. Heinze. (2005). Hemicellulose. *Adv. Polym. Sci.* **186**:1–67.

Edlund, U., and A. C. Albertsson. (2008). A microspheric system: hemicellulose-based hydrogels. *J. Bioact. Compat. Polym.* **23**:171–186.

Fang, J. M., R. C. Sun, P. Fowler, J. Tomkinson, and C. A. S. Hill. (1999). Esterification of wheat straw hemicelluloses in the *N*,*N*-dimethylformamide/lithium chloride homogeneous system. *J. Appl. Polym. Sci.* **74**:2301–2311.

Fredon, E., R. Granet, R. Zerrouki, P. Krausz, L. Saulnier, J. F. Thibault, J. Rosier, and C. Petit. (2002). Hydrophobic films from maize bran hemicelluloses. *Carbohydr. Polym.* **49**:1–12.

Glasser, W. G., W. E. Kaar, R. K. Jain, and J. E. Sealey. (2000). Isolation options for non-cellulosic polysaccharides (HetPS). *Cellulose* **7**:299–317.

Gröndahl, M., L. Eriksson, and P. Gatenholm. (2004). Material properties of plasticized hardwood xylans for potential application as oxygen barrier films. *Biomacromolecules* **5**:1528–1535.

Gröndahl, M., A. Gustafsson, and P. Gatenholm. (2006). Gas-phase surface fluorination of arabinoxylan films. *Macromolecules* **39**:2718–2721.

Gunness, P., and M. J. Gidley. (2010). Mechanisms underlying the cholesterol-lowering properties of soluble dietary fibre polysaccharides. *Food Funct.* **1**:149–155.

Haishi, M., and T. Takeshita. (1997a). Studies on anti-tumor activity of wood hemicelluloses. 1. Anti-tumor effect of 4-*O*-methylglucuronoxylan on solid tumor in mice. *Agric. Biol. Chem.* **43**:951–959.

Haishi, M., and T. Takeshita. (1997b). Studies on anti-tumor activity of wood hemicelluloses. 2. The host-mediated anti-tumor effect of 4-*O*-methylglucuronoxylan. *Agric. Biol. Chem.* **43**:961–967.

Hansen, N. M. L., and D. Plackett. (2008). Sustainable films and coatings from hemicelluloses: a review. *Biomacromolecules* **9**:1493–1505.

Hartman, J., A.-C. Albertsson, and J. Sjöberg. (2006a). Surface- and bulk-modified galacto-glucomannan hemicellulose films and film laminates for versatile oxygen barr. *Biomacro-molecules* **7**:1983–1989.

Hartman, J., A.-C. Albertsson, M. S. Lindblad et al. (2006b). Oxygen barrier materials from renewable sources: Material properties of softwood hemicellulose-based films. *J. Appl. Polym. Sci.* **100**:2985–2991.

Hettrich, K., S. Fischer, N. Schröder, J. Engelhardt, U. Drechsler, and F. Loth. (2006). Derivatization and characterization of xylan from oat spelts. *Macromol. Symp.* **232**:37–48.

Hill, S. J., R. A. Franich, P. T. Callaghan, and R. H. Newman. (2009). Nature's nanocomposites: a new look at molecular architecture in wood cell walls. *N. Z. J. Forestry Sci.* **39**:251–257.

Immerstrand, T., B. Bergenståhl, C. Trägårdh, M. Nyman, S. Cui, and R. Öste. (2009). Extraction of β-glucan from oat bran in laboratory scale. *Cereal Chem.* **86**:601–608.

Isogai, A., A. Ishizu, J. Nakano, S. Eda, and K. Kato. (1985). A new facile methylation method for cell-wall polysaccharides. *Carbohydr. Res.* **138**:99–108.

Jain, R. K., M. Sjöstedt, and W. G. Glasser. (2000). Thermoplastic xylan derivatives with propylene oxide. *Cellulose* **7**:319–336.

Kahlon, T. S., and F. I. Chow. (1997). Hypocholesterolemic effects of oat, rice, and barley dietary fibers and fractions. *Cereal Foods World* **42**:86–92.

Klemm, D., B. Heublein, H.-P. Fink, and A. Bohn. (2005). Cellulose: fascinating biopolymer and sustainable raw material. *Angew. Chem. Int. Ed.* **44**:3358–3393.

Lai, V. M. F., S. Lu, W. H. He, and H. H. Chen. (2007). Non-starch polysaccharide compositions of rice grains with respect to rice variety and degree of milling. *Food Chem.* **101**:1205–1210.

Lazaridou, A., T. Chornick, C. G. Biliaderis, and M. S. Izydorczyk. (2008). Sequential solvent extraction and structural characterization of polysaccharides from the endosperm cell walls of barley grown in different environments. *Carbohydr. Polym.* **73**:621–639.

Lee, S. G., E. Y. An, J. B. Lee, J. C. Park, J. W. Shin, J. K. Kim, F. Z. Cui, G. Brauer, K.-M. Wang, and J. G. Han (2007). Enhanced cell affinity of poly(D,L-lactic-*co*-glycolic acid) (50/50) by plasma treatment with β-(1 → 3) (1 → 6)-glucan. *Surf. Coat Technol.* **201**:5128–5131.

Lindblad, M. S., E. Ranucci, and A-C. Albertsson. (2001). Biodegradable polymers from renewable sources. New hemicellulose-based hydrogels. *Macromol. Rapid Commun.* **22**:962–967.

Lindblad, M. S., E. Ranucci, and A. C. Albertsson. (2004). New hemicellulose-based hydrogels. In *Hemicelluloses: Science and Technology*, Eds. P. Gatenholm and M. Tenkanen, 347–359. Washington, DC: American Chemical Society.

Lloyd, L. L., J. F. Kennedy, P. Methacanon, M. Paterson, and C. J. Knill. (1998). Carbohydrate polymers as wound management aids. *Carbohydr. Polym.* **37**:315–322.

Lundqvist, J., A. Jacobs, M. Palm, G. Zacchi, O. Dahlman, and H. Stålbrand. (2003). Characterization of galactoglucomannan extracted from spruce (*Picea abies*) by heat-fractionation at different conditions. *Carbohydr. Polym.* **51**:203–211.

Lundqvist, J., A. Teleman, L. Junel, G. Zacchi, O. Dahlman, F. Tjerneld, and H. Stålbrand. (2002). Isolation and characterization of galactoglucomannan from spruce (*Picea abies*). *Carbohydr. Polym.* **48**:29–39.

Melton, L. D., B. G. Smith, R. Ibrahim, and R. Schröder. (2009). Mannans in primary and secondary plant cell walls. *N. Z. J. Forestry Sci.* **39**:153–160.

Mikkonen, K. S., M. P. Yadav, P. Cooke, S. Willför, K. B. Hicks, and M. Tenkanen. (2008). Films from spruce galactoglucomannan blended with poly(vinyl alcohol), corn arabinoxylan, and konjac glucomannan. *Bioresources* **3**:178–191.

Miyazaki, S., F. Suisha, N. Kawasaki, M. Shirakawa, K. Yamatoya, and D. Attwood. (1998). Thermally reversible xyloglucan gels as vehicle for rectal drug delivery. *J. Control. Release* **56**:75–83.

Moine, C., V. Gloaguen, J.-M. Gloaguen, R. Granet, and P. Krausz. (2004). Chemical valorization of forest and agricultural by-products. Obtention, chemical characteristics, and mechanical behavior of a novel family of hydrophobic films. *J. Environ. Sci. Health B* **39**:627–640.

Morgan, K. R., C. J. Roberts, S. J. B. Tendler, M. C. Davies, and P. M. Williams. (1999). A ^{13}C CP/MAS NMR spectroscopy and AFM study of the structure of Glucagel™, a gelling β-glucan from barley. *Carbohydr. Res.* **315**:169–179.

Oguzlu, H., and F. Tihminlioglu. (2010). Preparation and barrier properties of chitosan-layered silicate nanocomposite films. *Macromol. Symp.* **298**:91–98.

Ooi, V. E. C., and J. F. Liu. (2000). Immunomodulation and anti-cancer activity of polysaccharide-protein complexes. *Curr. Med. Chem.* **7**:715–729.

Palm, M., and G. Zacchi. (2003). Extraction of hemicellulosic oligosaccharides from spruce using microwave oven or steam treatment. *Biomacromolecules* **4**:617–623.

Peng, F., J-L. Ren, B. Peng, F. Xu, R-C. Sun, and J-X. Sun. (2008a). Rapid homogeneous lauroylation of wheat straw hemicelluloses under mild conditions. *Carbohydr. Res.* **343**:2956–2962.

Peng, F., J-L. Ren, X. F. Sun, F. Xu, R-C. Sun, B. Peng, and J-X. Sun. (2008b). Rapid phthaloylation and succinylation of hemicelluloses by microwave radiation. *e-Polymers* 108.

Persson, J, O. Dahlman, and A. C. Albertsson. (2009). Hemicellulose based polylactide copolymers, synthesis and characterization. EPNOE 2009 Advanced Polysaccharide Materials, September 21–24, 2009, Turku, Finland.

Plackett, D. (2011). Biorefinery (Wood). In *McGraw-Hill Yearbook of Science & Technology 2011*. 31–34. New York: McGraw-Hill.

Plackett, D., and A. Ragauskas. (2010). The biorefinery concept. In *Sustainable Development in the Forest Products Industry*, Eds. R. M. Rowell, F. C. Jorge and J. K. Rowell, 45–64. Porto, Portugal: Fernando Pessoa University Press.

Ren, J-L., F. Peng, R-C. Sun, and J. F. Kennedy. (1999). Influence of hemicellulosic derivatives on the sulphate kraft pulp strength. *Carbohydr. Polym.* **75**:338–342.

Ren, J-L., R-C. Sun, C. F. Liu, Z. N. Cao, and W. Luo. (2007a). Acetylation of wheat straw hemicelluloses in ionic liquid using iodine as a catalyst. *Carbohydr. Polym.* **70**:406–414.

Ren, J-L., R-C. Sun, and C. F. Liu. (2007b). Etherification of hemicelluloses from sugarcane bagasse. *J. Appl. Polym. Sci.* **105**:3301–3308.

Ren, J-L., F. Peng, and R-C. Sun. (2008a). Preparation and characterization of hemicellulosic derivatives containing carbamoylethyl and carboxyethyl groups. *Carbohydr. Res.* **343**:2776–2782.

Ren, J-L., R-C. Sun, and F. Peng. (2008b). Carboxymethylation of hemicelluloses isolated from sugarcane bagasse. *Polym. Degr. Stab.* **93**:786–793.

Ren, J-L., F. Peng, and R-C. Sun. (2009). The effect of hemicellulosic derivatives on the strength properties of old corrugated container pulp fibres. *J. Bio. Mater. Bioen.* **3**:62–68.

Rindlav-Westling, A., M. Stading, A-M. Hermansson, and P. Gatenholm. (1998). Structural, barrier and mechanical properties of amylose and amylopectin films. *Carbohydr. Polym.* **36**:217–224.

Roos, A. A., U. Edlund, J. Sjöberg, A.-C. Albertsson, and H. Stålbrand. (2008). Protein release from galactoglucomannan hydrogels: Influence of substitutions and enzymatic hydrolysis by β-mannanase. *Biomacromolecules* **9**:2104–2110.

Saxena, A., T. J. Elder, S. Pan, and A. J. Ragauskas. (2009). Novel nanocellulosic xylan composite film. *Compos. Part B Eng.* **40**:727–730.

Saxena, A., and A. J. Ragauskas. (2009). Water transmission barrier properties of biodegradable films based on cellulosic whiskers and xylan. *Carbohydr. Polym.* **78**:357–360.

Scheller, H. V., and P. Ulvskov. (2010). Hemicelluloses. *Annu. Rev. Plant Biol.* **61**:263–289.

Selvendran, R. R., and O'Neill, M. A. (1987). Isolation and analysis of cell walls from plant material. In *Methods of Biochemical Analysis*. Vol. 32, Ed. D. Glick, 25–123. New York: John Wiley & Sons.

Skendi, A., C. G. Biliaderis, A., Lazaridou, and M. S. Izydorczyk. (2003). Structure and rheological properties of water soluble β-glucans from oat cultivars of *Avena sativa* and *Avena bysantina*. *J. Cereal Sci.* **38**:15–31.

Smart, C. L., and R. L. Whistler. (1949). Films from hemicellulose acetates. *Science* **110**:713–714.

Spiridon, I. and Popa, V. I. (2008). Hemicelluloses: major sources, properties and applications, In *Monomers, Polymers and Composites from Renewable Resources*, Eds. M. N. Belgacem and A. Gandini, 289–304. Amsterdam: Elsevier.

Stephen, A. M. (1982). Other plant polysaccharides. In *The Polysaccharides*. Vol. 2, Ed. G. O. Aspinall, 97–123. New York: Academic Press.

Sun, R-C., J. M. Fang, J. Tomkinson, and C. A. S. Hill. (1999). Esterification of hemicelluloses from poplar chips in homogeneous solution of *N,N*-dimethylformamide/lithium chloride. *J. Wood Chem. Technol.* **19**:287–306.

Sun, R-C., J. M. Fang, and J. Tomkinson. (2000a). Stearoylation of hemicelluloses from wheat straw. *Polym. Degrad. Stab.* **67**:345–353.

Sun, R-C., J. M. Fang, and J. Tomkinson. (2000b). Characterization and esterification of hemicelluloses from rye straw. *J. Agric. Food Chem.* **48**:1247–1252.

Sun, R-C., J. M. Fang, J. Tomkinson, Z. C. Geng, and J. C. Liu. (2001). Fractional isolation, physico-chemical characterization and homogeneous esterification of hemicelluloses from fast-growing poplar wood. *Carbohydr. Polym.* **44**:29–39.

Sun, R-C., X. F. Sun, and J. Tomkinson. (2004). Hemicelluloses and their derivatives. In *Hemicellulose: Science and Technology*, Eds. P. Gatenholm and M. Tenkanen, 2–22. Washington, DC: American Chemical Society.

Tejinder. S. (2003). Preparation and characterization of films using barley and oat β-glucan extracts. *Cereal Chem.* **80**:728–731.

Ünlü, C. H., E. Günister, and O. Atici. (2009). Synthesis and characterization of NaMt biocomposites with corn cob xylan in aqueous media. *Carbohydr. Polym.* **76**:585–592.

Vincken, J. P., W. S. York, G. Beldman, and A. G. Voragen, (1997). Two general branching patterns of xyloglucan, XXXG and XXGG. *Plant Physiol.* **114**:9–13.

Viota, J. L., M. Lopez-Viota, B. Saake, K. Stana-Kleinschek, and A. V. Delgado. (2010). Organoclay particles as reinforcing agents in polysaccharide films. *J. Colloid Interf. Sci.* **347**:74–78.

Voepel, J., U. Edlund, and A.-C. Albertsson. (2009a). Alkenyl-functionalized precursors for renewable hydrogels design. *J. Polym. Sci. Part A Polym. Chem.* **47**:3595–3606.

Voepel, J., J. Sjöberg, M. Reif, A.-C. Albertsson, U.-K. Hultin, and U. Gasslander. (2009b). Drug diffusion in neutral and ionic hydrogels assembled from acetylated galactogluco-mannan. *J. Appl. Polym. Sci.* **112**:2401–2412.

Westerlund, E., R. Anderson, and P. Åman. (1993). Isolation and chemical characterization of water-soluble mixed-linked β-glucans and arabinoxylans in oat milling fractions. *Carbohydr. Polym.* **20**:115–123.

Willför, S., K. Sundberg, M. Tenkanen, and B. Holmbom. (2008). Spruce-derived mannans— A potential raw material for hydrocolloids and novel advanced natural materials. *Carbohydr. Polym.* **72**:197–210.

Zhang, H., M. Toshimura, K, Nishinari, M. A. K. Williams, T. J. Foster, and I. T. Norton. (2001). Gelation behaviour of konjac glucomannan with different molecular weights. *Biopolymers* **59**:38–50.

Zhang, P., and R. L. Whistler. (2004). Mechanical properties and water vapor permeability of thin film from corn hull arabinoxlan. *J. Appl. Polym. Sci.* **93**:2896–2902.

Zhang, Y-Q., B-J. Xie, and X. Gan. (2005). Advance in the applications of konjac glucomannan and its derivatives. *Carbohydr. Polym.* **60**:27–31.

Zhou, Q., M. W. Rutland, T. T. Teeri, and H. Brumer. (2007). Xyloglucan in cellulose modification. *Cellulose* **14**:625–641.

INDEX

Polysaccharide Building Blocks: A Sustainable Approach to the Development of Renewable Biomaterials,
First Edition. Edited by Youssef Habibi and Lucian A. Lucia.
© 2012 John Wiley & Sons, Inc. Published 2012 by John Wiley & Sons, Inc.